Faunal and Floral Migrations and Evolution
in SE Asia-Australasia

FAUNAL AND FLORAL MIGRATIONS AND EVOLUTION IN SE ASIA-AUSTRALASIA

IAN METCALFE, JEREMY M.B. SMITH, MIKE MORWOOD
AND IAIN DAVIDSON

A.A. BALKEMA PUBLISHERS / LISSE / ABINGDON / EXTON (PA) / TOKYO

Library of Congress Cataloging-in-Publication Data

Faunal and floral migrations and evolution in SE Asia-Australasia / Ian Metcalfe ... [et al.]
      p. cm.
    Includes bibliographical references.
    ISBN 9058093492
      1. Animal migration--Asia, Southeastern. 2. Plants--Migration--Asia, Southeastern. 3.
  Evolution (Biology)--Asia, Southeastern. 4. Animal migration--Australasia. 5.
  Plants--Migration--Australasia. 6. Evolution (Biology)--Australasia. I. Metcalfe, Ian,
  1949-

  QL754 .F38 2001
  591.56'8'0959--dc21

                                                                        2001031157

ISBN 90 5809 349 2

Contents

Preface
Penny van Oosterzee 9

Introduction
I. Metcalfe, J.M.B. Smith, M. Morwood, & I. Davidson 11

Section 1. Palaeogeographic Background 13

Palaeozoic and Mesozoic tectonic evolution and biogeography of SE Asia-Australasia
Ian Metcalfe 15

Cenozoic reconstructions of SE Asia and the SW Pacific: changing patterns of land and sea
Robert Hall 35

Section 2. Palaeozoic and Mesozoic geology and biogeography 57

Cambrian to Permian conodont biogeography in East Asia-Australasia
Robert S. Nicoll and Ian Metcalfe 59

Wallace Lines in eastern Gondwana: Palaeobiogeography of Australasian Permian Brachiopoda
N.W. Archbold 73

A review of the Early Permian flora from Papua (West New Guinea)
J. F. Rigby 85

A biogeographic comparison of the dinosaurs and associated vertebrate faunas from the Mesozoic of Australia and Southeast Asia
John A. Long and Eric Buffetaut 97

Early Middle Jurassic (Aalenian) radiolarian fauna from the Xialu chert in the Yarlung Zangbo Suture Zone, southern Tibet
Atsushi Matsuoka, Kenta Kobayashi, Toru Nagahashi, Qun Yang, Yujing Wang and Qinggao Zeng 105

Section 3. Wallace's Line

111

Why Wallace drew the line: A re-analysis of Wallace's bird collections in
the Malay Archipelago and the origins of biogeography
Danielle Clode and Rory O'Brien

113

The linear approach to biogeography: Should we erase Wallace's Line?
Walter R. Erdelen

123

Faunal exchange between Asia and Australia in the Tertiary as evidenced
by recent butterflies
Rienk de Jong

133

Why does the distribution of the Honeyeaters (Meliphagidae) conform so
well to Wallace's Line?
Hugh A. Ford

147

Human influences on vertebrate zoogeography: animal translocation and
biological invasions across and to the east of Wallace's Line
Tom Heinsohn

153

Wallace's line and marine organisms: the distribution of staghorn corals
(*Acropora*) in Indonesia
Carden C. Wallace

171

Section 4. Plant biogeography and evolution

183

Why are there so many Primitive Angiosperms in the Rain Forests of
Asia-Australasia?
R.J. Morley

185

Australian Paleogene vegetation and environments: evidence for palaeo-
Gondwanan elements in the fossil records of Lauraceae and Proteaceae
Anthony J. Vadala and David R. Greenwood

201

Vegetation and climate in lowland Southeast Asia at the Last Glacial
Maximum
A. Peter Kershaw, Dan Penny, Sander van der Kaars, Gusti Anshari and
Asha Thamotherampillai

227

The restiads invade the north: the diaspora of the Restionaceae
Barbara G. Briggs

237

Evolutionary history of *Alectryon* in Australia
Karen J. Edwards and Paul A. Gadek

243

Section 5. Non Primates 253

Australasian distributions in Trichoptera (Insecta) - a frequent pattern or
a rare case?
Wolfram Mey 255

Butterflies and Wallace's Line: faunistic patterns and explanatory hypotheses
within the south-east Asian butterflies
R. L. Kitching, R. Eastwood & K. Hurley 269

The vertebrate fauna of the Wallacean Island Interchange Zone:
the basis of inbalance and impoverishment
Allen Keast 287

Dispersal versus vicariance, artifice rather than contest
B. Michaux 311

The Australian rodent fauna, flotilla's, flotsam or just fleet footed?
H. Godthelp 319

Corroboration of the Garden of Eden Hypothesis
Thomas H. Rich, Timothy F. Flannery, Peter Trusler & Patricia Vickers-Rich 323

Mammals in Sulawesi: where did they come from and when, and what
happened to them when they got there?
Colin Groves 333

Section 6. Primates 343

Radiation and Evolution of Three Macaque Species, *Macaca fascicularis,
M. radiata* and *M. sinica*, as Related to the Geographic Changes in the
Pleistocene of Southeast Asia
R.-L. Pan and C.E. Oxnard 345

Borneo as a biogeographic barrier to Asian-Australasian migration
Douglas Brandon-Jones 365

Modelling Divergence, Inter-breeding and Migration: Species Evolution
in a Changing World
Charles Oxnard and Ken Wessen 373

Early hominid occupation of Flores, East Indonesia, and its wider
significance
Mike Morwood 387

The requirements for human colonisation of Australia
Iain Davidson
399

Did early hominids cross sea gaps on natural rafts?
J. M. B. Smith
409

Preface

Penny van Oosterzee
EcOz Australia, GPO Box 381, Darwin NT 0801, Australia

It all started with a letter written in 1858. A perplexed Alfred Russell Wallace in eastern Indonesia wrote to Henry Walter Bates in London that in Indonesia "there are two distinct faunas rigidly circumscribed, which differ as much as those of South America and Africa and more than those of Europe and North America. Yet there is nothing on the map or on the face of the islands to mark their limits. The boundary line often passes between islands closer than others in the same group". Later, in 1863, Wallace read a paper to the Royal Geographical Society in London on the geography of Indonesia. He drew a red line on the map passing down the Makassar Strait. To the west he wrote 'Indo-Malayan region' and to the east he wrote 'Australo-Malayan region'. This became known as the Wallace Line.

Wallace had an explanation to account for the extraordinary juxtaposition of two fauna which were apparently poles apart. He believed the western part to be a separated portion of continental Asia, the eastern the fragmentary prolongation of a former Pacific continent. This idea marks the birth of biogeography. Wallace had tried to explain where species come from, and why they occur where they do. He was the first to realise that to really understand the distribution of species over the face of the earth (and in particular Indonesia), you had to appreciate, not only the species' evolutionary history, but also the geological history of the region where they occurred. His Line merely underscored this important point.

Today, where it is relevant, the Wallace Line is used mainly as an heuristic tool to shed light on the evolution of elements of flora and fauna. In this work the Wallace Line is used as a springboard for interpreting the distribution and evolution of birds, mammals, marine organisms invertebrates, and one of the region's most recent arrivals, humans.

Wallace reasoned - incorrectly as it happens - that it would be impossible to tell the origins of the various fragments of land in this archipelagic jig-saw puzzle. The animals and plants, however, would reveal clues of their former history. The extent to which species distributions are limited by underlying geology, in this spectacular region of collision, is a question that is as relevant today as it was in Wallace's time.

This book provides the most up to date information on the geological evolution of the region and provides detailed insights into its palaeobiogeography and changing climate. The extent to which the present-day biogeography is permeated and altered by the activities of humans across and the east of the Wallace Line provides a fine counterpoint to this truly multidisciplinary volume.

Introduction

Ian Metcalfe, Jeremy Smith, Mike Morwood, Iain Davidson
(Editors)
University of New England, Armidale NSW 2351, Australia

Conception of this book began with a chance meeting between Ian Metcalfe and Iain Davidson in an Armidale supermarket one Saturday morning in 1998. Ian Metcalfe, the senior editor, had been developing the idea of a University of New England Asia Centre multidisciplinary conference on biogeography in SE Asia for some time, and Iain Davidson embraced the idea immediately over a supermarket trolly and plans were laid to obtain some seed funding for the venture. The conference crystallised when the Asia Centre agreed to provide funds to bring the keynote speaker, Prof. Robert Hall, to Armidale for the conference.

The international conference, *Where Worlds Collide:Faunal and floral migrations and evolution in SE Asia-Australasia* was held under the auspices of the University of New England Asia Centre (UNEAC) in Armidale, New South Wales, Australia from 29 November - 2 December, 1999, and this book contains thirty-one selected papers out of the forty-five oral and six poster papers presented at the conference. The multi-disciplinary conference brought together more than 100 scientists from 12 countries including geologists, palaeontologists, zoologists, botanists, entomologists, evolutionary biologists and archaeologists. Australian participants were from all over Australia and in particular, palaeontological groups from Macquarie and Deakin Universities were well represented. The conference was a formal contribution to UNESCO International Geological Correalation Program Projects 411 *Geodynamics of Gondwanaland-derived Terranes in E & S Asia*, and 421 *North Gondwana Mid-Palaeozoic biodynamics* and 13 papers and posters were presented by members of these two projects at the meeting.

The multi-disciplinary scientists at the conference examined the geological and biological history of the SE Asian-Australasian region over the last 540 million years, to relate its geological past to its present biological peculiarities. The "colliding worlds" which gave the conference its title are the various Gondwanaland-derived continental terranes of the region which assembled to form Asia during Carboniferous to Cretaceous times, and present-day Australasia, originally attached to Antarctica, and Eurasia into which it has crashed after 40 million years of steady northward drift.

The convergence of the Australian continent, and Eurasia has brought into close contact two contrasting faunas and floras. Although some animals and plants have succeeded in crossing the remaining narrowing seas, many others remain to one or other side of the shrinking intercontinental gap which is marked by the Wallace Line. To one side are wallabies, possums and cockatoos; to the other tigers, deer and woodpeckers. The line was first recognised, and is named for, the Victorian travelling naturalist and collector Alfred Russel Wallace, also well known as being co-discoverer with Charles Darwin of the theory of evolution by natural selection.

Forty-five oral presentations and six posters were delivered during the three days of the conference. In addition, Penny van Oosterzee, the Eureka Prize winning author, also delivered an evening public lecture entitled *Where Worlds Collide:Wallace in Wonderland.*

The book is structured into six sections which essentially reflect the sessions of the conference. Section 1 *Palaeogeographic Background* includes two papers by Ian Metcalfe and Robert Hall which provide overviews of the geological and tectonic evolution of SE Asia-Australasia, and changing patterns of land and sea for the last 540 million years. Section 2 *Palaeozoic and Mesozoic geology and biogeography* contains five papers discussing Palaeozoic and Mesozoic biogeography of conodonts, brachiopods, plants, dinosaurs and radiolarians and the recognition of ancient biogeographic boundaries or "Wallace Lines" in the region. Section 3 *Wallace's Line* contains six papers that specifically relate to biogeographic boundary established by Wallace, including the history of its establishment, its significance to biogeography in general and its applicability in the context of modern biogeography. Section 4 *Plant biogeography and evolution* includes papers on primitive angiosperms, the diaspora of the 'southern rushes', and environmental, climatic, and evolutionary implications of plants and palynomorphs in the region. The biogeography and migration of insects, butterflies, birds, rodents and other non-primate mammals is discussed in the seven papers of Section 5 *Non Primates*. The final Section 6 *Primates* focusses on the biogeographic radiation, migration and evolution of primates and includes papers on the occurrence and migration of early hominids and the requirements for human colonisation of Australia.

All papers in this book have been subject to rigorous international peer review by at least two referees in addition to review by the editors. We would like to thank the following for their kind assistance with this review process: Jonathan Aitchison, Neil Archbold, Mike Archer, Anthony J. Barber, Henry Barlow, Peter Bellwood, Bill Boyd, Douglas Brandon-Jones, Barbara Briggs, Jeremy Bruhl, David Bullbeck, Clive Burrett, Stephen Carey, Russell L. Ciochon, Joel Cracraft, Rienk de Jong, Roger Farrow, Hugh Ford, Robert Gargett, Henk Godthelp, Colin Groves, Neville Haile, Tom Heinsohn, Jeremy Holloway, Geoff Hope, Peter Jarman, Zerina Johanson, Peter Kershaw, Roger Kitching, Peter Linder, Virginie Millien, Bernard Michaux, Robert Morley, Ralph Molnar, N. Prakash, Tom Rich, John Rigby, Alex Ritchie John Roberts, Lesley Rogers, Brian Rosen, Andrew Simpson, Peter Stauffer, John Talent, Steve Trewick, Hubert Turner, Steve Van Dyck, Nigel Wace, Koji Wakita, Moyra Wilson, Yoram Yom-Tov, and Erwin Zodrow.

Assoc. Prof. Ian Metcalfe
Assoc. Prof. Jeremy Smith
Assoc. Prof. Mike Morwood
Prof. Iain Davidson
(Editors)

University of New England
Armidale NSW 2351
Australia
October, 2000.

Section 1

Palaeogeographic Background

Palaeozoic and Mesozoic tectonic evolution and biogeography of SE Asia-Australasia

I. Metcalfe
Asia Centre, University of New England, Armidale NSW 2351, Australia

ABSTRACT: Southeast Asia comprises a complex assembly of continental fragments (terranes), bounded by suture zones that represent former ocean basins. In the Early Palaeozoic (545-410 Ma), all the principal Southeast Asian continental terranes were located on the Himalayan-Australian Gondwanaland margin, where they formed part of a "Greater Gondwanaland". Multidisciplinary studies suggest that the Asian terranes rifted and separated from NW Australian Gondwanaland as three continental slivers around 350, 270, and 200-140 Ma. During the separation of these continental slivers, three ocean basins, the Palaeo-Tethys, Meso-Tethys and Ceno-Tethys, opened between each successive sliver and Gondwanaland. The first sliver, interpreted to have separated at 350 Ma, included North China, South China, Tarim, and Indochina. By 325 Ma, these terranes had no Gondwanaland faunas and floras, and palaeomagnetic data indicate that they were separated from the parent craton. Clockwise rotation of Gondwanaland around 320-300 Ma sent Australia into high southern latitudes, and glaciation affected much of Gondwanaland from 320 to 270 Ma. During this time, glacial ice reached the marine environment and glacial-marine sediments were deposited on the Sibumasu, Qiangtang, and Lhasa terranes now located in South and Southeast Asia. Gondwanaland cold-climate faunas and floras and cold water oxygen isotopic signatures are also found in Lower Permian rocks of these terranes. Rifting, starting at about 300 Ma, led to separation of the Sibumasu and Qiangtang terranes from Gondwanaland at approximately 270 Ma. These Cimmerian terranes drifted rapidly northwards between 270 and 250 Ma and significant changes in both brachiopod and fusulinid biogeographic provinces on these between 300 and 250 Ma are interpreted as being primarily due to the separation and northwards drift of the Cimmerian continental terranes, coupled with climatic amelioration following retreat of the Gondwanaland glaciation. Rifting and separation of a third continental sliver from Gondwanaland began 230 million years ago with the separation of the Lhasa block and progressed eastwards with separation of West Burma and possibly the small Sikuleh terrane of SW Sumatra and other small microcontinental fragments now located in Borneo and West Sulawesi around 160 Ma. Amalgamation and accretion of Gondwanaland-derived Asian terranes occurred progressively between the Late Devonian and Cretaceous.

1 INTRODUCTION

Faunal and floral links between SE Asia and Australasia go back more than 500 million years and the waxing and waning of biogeographic affinities between the two regions directly reflect the geological and tectonic histories of and interactions between these regions. Palaeozoic and Mesozoic biogeographic studies of the SE Asian region must be undertaken within a framework of the kinematic history of the allochthonous continental terranes of the region. Present-day distributions of animals and plants in SE Asia-Australasia are the result of both relatively recent dispersals, and more ancient vicariance events that can be related to the geological evolution of the region.

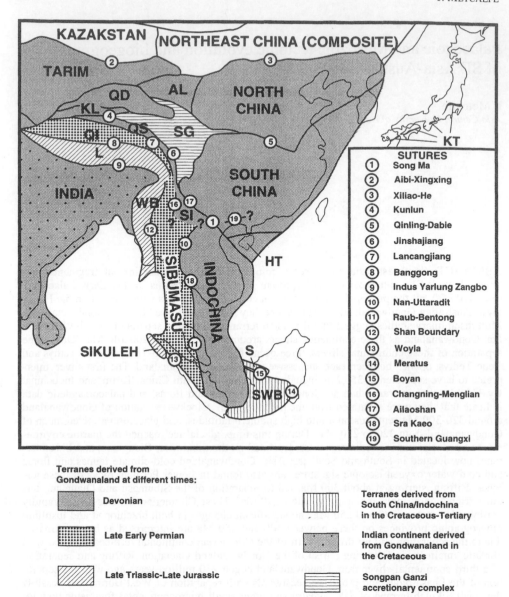

Figure 1. Distribution of principal continental terranes and sutures of East and Southeast Asia. WB = West Burma, SWB = South West Borneo, S = Semitau Terrane, HT = Hainan Island terranes, L = Lhasa Terrane, QI = Qiangtang Terrane, QS = Qamdo-Simao Terrane, SI= Simao Terrane, SG = Songpan Ganzi accretionary complex, KL = Kunlun Terrane, QD = Qaidam Terrane, AL = Ala Shan Terrane, KT = Kurosegawa Terrane.

East and Southeast Asia comprise a complex assembly of continental fragments (terranes), which are bounded by suture zones that represent the remnants or sites of former ocean basins (Figures 1 and 2). Biogeographical and other data suggest that in the Early Palaeozoic (545-410 Ma) all of the principal East and Southeast Asian continental terranes were located on the India-Australian margin of the ancient supercontinent Gondwanaland, where they formed part of a "Greater Gondwanaland" (Metcalfe, 1988, 1990, 1993, 1996, 1998, 1999; Burrett *et al.* 1990).

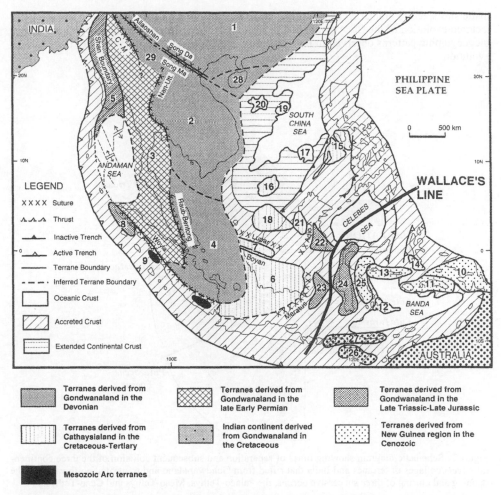

Figure 2. Distribution of continental blocks and fragments (terranes) and principle sutures of Southeast Asia in relation to Wallace's Line (m odified after Metcalfe, 1990). 1. South China 2. North Indochina 3. Sibumasu 4. Hainan Island terranes 5. West Burma 6. S.W. Borneo 7. Semitau 8. Sikuleh 9. Natal 10. West Irian Jaya 11. Buru-Seram 12. Buton 13. Bangai-Sula 14. Obi-Bacan 15. North Palawan 16. Spratley Islands-Dangerous Ground 17. Reed Bank 18. Luconia 19. Macclesfield Bank 20. Paracel Islands 21. Kelabit-Longbowan 22. Mangkalihat 23. Paternoster 24. West Sulawesi 25. East Sulawesi 26. Sumba 27. Qamdo-Simao terrane.

Multidisciplinary data (e.g., biogeographic, biostratigraphic, palaeomagnetic, structural/tectonic) suggest that the Asian terranes rifted and separated from NW Australian Gondwanaland as three continental slivers around 350, 270, and 200-140 million years ago, in the Devonian, Early Permian, and Late Triassic-Late Jurassic respectively. During the separation of these continental slivers, three ocean basins, the Palaeo-Tethys, Meso-Tethys, and Ceno-Tethys, are interpreted to have opened between each successive sliver and Australian Gondwanaland. Closure of these ocean basins by subduction processes led to the juxtaposition of once widely separated continental fragments, and the remnants of the ocean basins are now preserved in the suture zones of the region (Figure 3).

During this complex evolution, reorganisation of continent-ocean configurations, the formation of the supercontinent Pangea, and its subsequent breakup, coupled with major global climatic shifts, led to dramatic changes in the biogeography of organisms. This paper discusses the tectonic evolution of the SE Asian region in relation to Australasia, and the changing SE Asian biogeographic patterns observed in the Palaeozoic and Mesozoic that are related to this tectonic evolution.

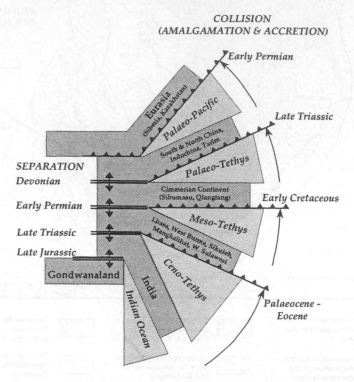

Figure 3. Schematic diagram showing times of separation and subsequent collision of the three continental slivers/collages of terranes and India that rifted from Gondwanaland and translated northwards by the opening and closing of three successive oceans, the Palaeo-Tethys, Meso-Tethys and Ceno-Tethys.

2 PALAEOZOIC AND MESOZOIC TECTONIC EVOLUTION AND BIOGEOGRAPHY

The East and Southeast Asian terranes represent three collages that successively rifted and separated from the India-Australian margin of Gondwanaland as three elongate continental slivers. As these three slivers separated from Gondwanaland in the Devonian, late Early Permian and Late Triassic-Late Jurassic, three ocean basins, the Palaeo-Tethys, Meso-Tethys and Ceno-Tethys were opened and subsequently closed (Metcalfe, 1990, 1996, 1998, 1999). Australia remained part of eastern Gondwanaland (attached to Antarctica and India) during the Palaeozoic and early Mesozoic and finally separated from Antarctica at about 45 Ma, after which it drifted northwards eventually to begin its currently ongoing collision with SE Asia. The principal tectonic events and the changing biogeography of organisms in the SE Asia-Australasian region during the Palaeozoic and Mesozoic are outlined below:

2.1 Cambrian-Ordovician-Silurian (545 - 410 Ma)

Close faunal affinities between the East and Southeast Asian terranes and Australia are seen in the Cambrian, Ordovician, and Silurian (Burrett, 1973; Burrett and Stait, 1985; Metcalfe, 1988, Burrett et al., 1990), and distinctive Asian-Australian faunal provinces can be recognised during

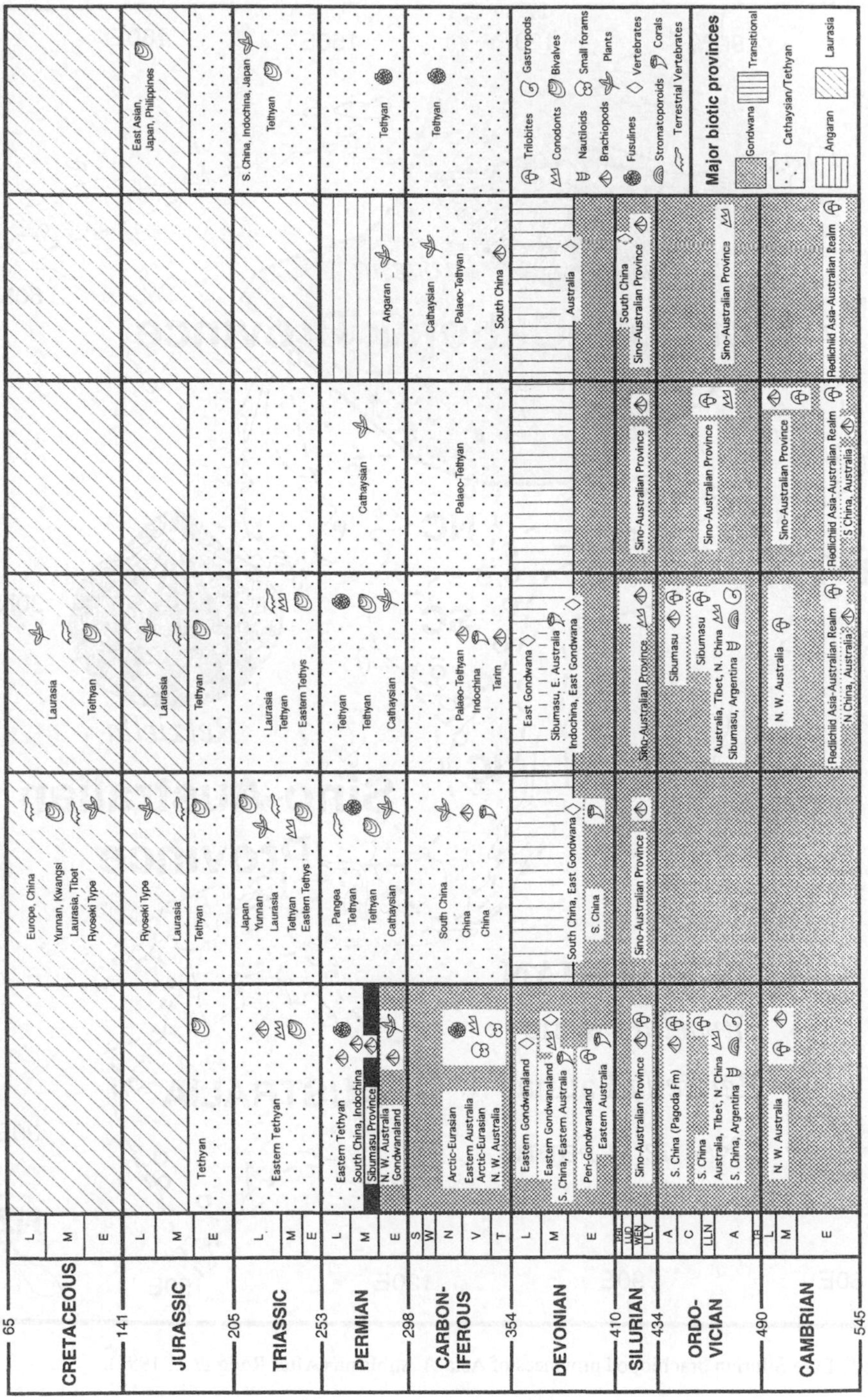

Figure 4. Palaeozoic and Mesozoic faunal and floral provinces and faunal/floral affinities vs. time for the principal East Asian continental terranes. For explanation see text.

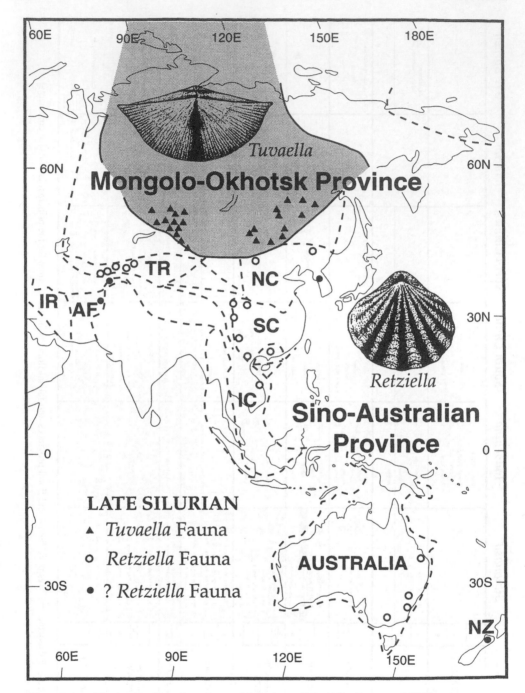

Figure 5. Late Silurian brachiopod provinces of Asia-Australasia (After Rong *et al.* 1995).

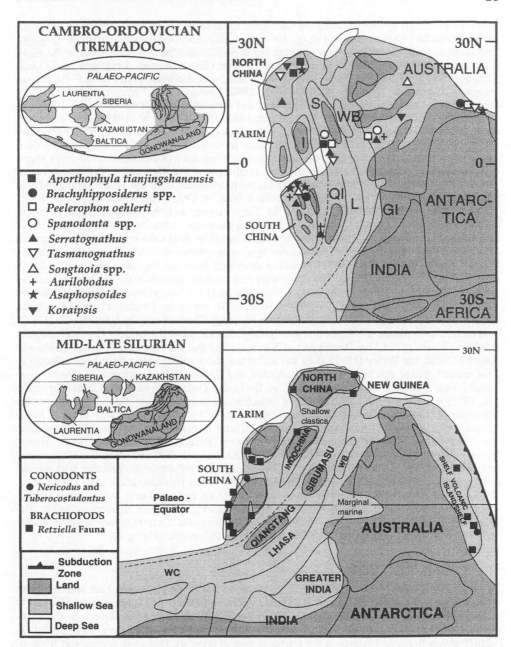

Figure 6. Reconstructions of eastern Gondwanaland for the Cambro-Ordovician (Tremadoc) and Mid-Late Silurian showing the postulated positions of the East and Southeast Asian terranes, distribution of land and sea, and shallow-marine fossils that illustrate Asia-Australia connections at these times. I = Indochina QI = Qiangtang L = Lhasa S = Sibumasu WB = West Burma WC = Western Cimmerian Continent GI = Greater India. Present day outlines are for reference only. Distribution of land and sea for Chinese blocks principally from Wang (1985). Land and sea distribution for Pangea/Gondwanaland compiled from Golongka *et al.* (1994), Smith *et al.* (1994); and for Australia from Struckmeyer & Totterdell (1990).

this time interval (Figure 4). During the Cambrian, trilobite faunas of Asia-Australasia define an Asian-Australian Faunal Realm (Yang, 1994) and archaeocyaths and in particular lingulate brachiopods suggest close proximity of both North and South China to Australasia in the Early Cambrian (Brock *et al.*, 1999). Late Cambrian trilobites (Shergold *et al.*, 1988) and brachiopods (Yang, 1994) of Sibumasu, South and North China, and Australia define a Sino-Australian province. Ordovician trilobites (Burrett *et al.*, 1990; Fortey, 1997; Fortey and Cocks, 1998), brachiopods (Laurie and Burrett, 1992), corals and stromatoporoids (Webby *et al.*, 1985; Lin and Webby, 1989), nautiloids (Stait and Burrett, 1982, 1984; Stait *et al.*, 1987), gastropods (Jell *et al.*, 1984), and conodonts (Burrett *et al.*, 1990; Nicoll and Totterdell, 1990; Nicoll and Metcalfe, 1994, 1999) again define a Sino-Australian province. Early Ordovician trilobites from Dali, West Yunnan, on the Qamdo-Simao terrane are regarded as Gondwanan cold-water forms by Zhou *et al.* (1998a) with close relatives in south-central Europe and the Yangtze region of South China. Ordovician (Late Areng-early Llanvirn) trilobites from the Tarim terrane indicate that Tarim and South China were close to each other in Ordovician times, with 80% forms common between the Xinjiang region of the Tarim terrane and the Yangtze region of South China. The presence of typical Gondwana forms suggests these terranes were peri-Gondwana in the Ordovician (Zhou *et al.* 1998b). Upper Ordovician (Caradoc) trilobite faunas from southern Thailand on the Sibumasu terrane, dominated by *Ovalocephalus*, nileids and remopleuridids, are identical, even at the species level to faunas described from the Pagoda Limestone of South China (Fortey, 1997), indicating close proximity of these two terranes in the Late Ordovician. In addition, Upper Ordovician (Ashgill) to Lower Silurian (Llandovery) brachiopods and trilobites from south Thailand and the Northern Shan States of Burma on the Sibumasu terrane are identical to those reported from the Yangtze region of South China (Cocks and Fortey 1997). Furthermore, the sequence of faunas and stratigraphy from at least the Middle Ordovician to the Middle Silurian of the Thai-Burma part of Sibumasu and the Yichang area of South China was said by Cocks and Fortey (1997) to be so similar that these regions must have been adjacent to one another during this time period. In addition, the Gondwanaland acritarch *Dicrodiacroium ancoriforme* Burmann has been reported from the Lower Ordovician of South China (Servais *et al.*, 1996). Cambrian-Ordovician faunas of Indochina are poorly known and their biogeographic affinities are not clear. However, Silurian brachiopods of Indochina, along with those of South China, North China, Eastern Australia, and the Tarim terrane, belong to the Sino-Australian province (Figure 5) characterised by the *Retziella* fauna (Rong *et al.*, 1995). In addition, Silurian conodonts provide links between South China and Australia (Nicoll and Metcalfe, 1994; Nicoll and Metcalfe,1999). These biogeographic links in the Early Palaeozoic, together with similar stratigraphies (Metcalfe, 1996; Fortey and Cocks, 1998) and palaeomagnetic data (Zhao *et al.*, 1996; Klootwijk, C. 1996) suggests that the principal East and Southeast Asian terranes (Sibumasu, Indochina, South China, North China and Tarim) were positioned adjacent to the Himalayan-NW Australian margin of Gondwanaland forming part of a "Greater Gondwanaland" which was located in low equatorial latitudes during the early Palaeozoic (Figure 6).

2.2 *Devonian (410 - 354 Ma)*

The Devonian period was one of biogeographic transition for the principal East and Southeast Asian terranes, except for the Sibumasu terrane which continued to be biogeographically part of Gondwanaland until the early Sakmarian (Early Permian). Early Palaeozoic Gondwanaland faunas of the Indochina, South China, North China, and Tarim terranes give way to Cathaysian/Tethyan floras and faunas in the Carboniferous (Figure 4). Devonian faunas and floras of these terranes appear to have mixed biogeographical affinities. Lower Devonian (Emsian) trilobites from Thailand on the Sibumasu terrane can be closely compared to those of the same age in Turkey, Bohemia, Germany, and Morocco, and they belong to the peri-Gondwana "Hercynian magnafacies" (Fortey, 1989). Middle Devonian rugose corals of Sibumasu have close species affinities with those of South China and eastern Australia (Aung, 1995) and Middle Devonian polygnathid conodonts of Sibumasu also indicate connections to South China and eastern Gondwanaland (Burrett *et al.*, 1990). Of particular interest is Devonian vertebrate biogeography and especially that of shallow-marine and fresh-water fish. Much debate has raged in recent years regarding late Palaeozoic global reconstructions based on biogeography vs. palaeomagnetism, with some apparent conflicts between vertebrate biogeography and palaeomagnetism. An example of this is the "Amorica" problem as discussed by Young (1987) where Devonian vertebrate biogeography suggests that Gondwana and Euramerica were contiguous by Middle Devonian times, in contrast to palaeogeographic reconstructions based on palaeomagnetic data which imply a wide ocean separating these regions during the Devonian. Alternative explanations of the biogeography involve a continental block "Amorica" which traversed the ocean during the Devonian. Provincialism of Devonian vertebrates has long been known, and Young (1981) recognised five major vertebrate provinces for Early Devonian times (Figure 7A). At that time, the Devonian vertebrate faunal affinities of the North China, Indochina, Tarim, and Sibumasu terranes were largely unknown. Since then, much new information has come to light and was recently summarised by Young and Janvier (1999). It is now known that Middle to Late Devonian fish of Sibumasu (e.g., the thelodont *Turinia pagoda* Wang, Deng & Turner) have strong affinities with closely allied forms in the Officer Basin of Australia (Long, 1993) and also similarities with fish faunas of South China and east Gondwanaland (Long and Burrett 1989). It is also now known that galeaspids and sinolepids occur on North China, but interestingly, are somewhat younger than the South China forms. South China Devonian "endemic" fish are also now known to occur on the Tarim block and on Indochina (Tong-Dzuy and Cai 1995; Tong-Dzuy *et al.* 1996; Young and Janvier 1999). New discoveries of Devonian fish faunas linking the principal Asian terranes led Young and Janvier to propose that these terranes formed an "Asian Superterrane" in the Middle Palaeozoic and suggested that this superterrane was either located off NW Australia or more likely (in their view) outboard of eastern Australia (Figure 7B). The configuration of Young and Janvier's "Asian Superterrane" presents Tarim, North China, South China, and Indochina in their current relative positions. This is untenable in view of our extensive knowledge from the suture zones between these terranes indicating that the Qinling-Dabei suture zone segment of Palaeo-Tethys between North and South China did not close until Late Permian-Triassic times! Similarly, suturing between Indochina and South China probably occurred in the early Carboniferous, and the North China and Tarim terranes could not have had their current relative positions in the Middle Palaeozoic based on palaeomagnetic and other data. However, the positioning of the Tarim, North and South China, and Indochina terranes off NW Australia, but with rather different relative positions, is consistent with a range of multidisciplinary data (Metcalfe, 1996, 1998). Young and Janvier, however, considered that this position, "within the palaeo-Tethys" (sic) would not allow for sufficient isolation to develop the observed endemic continental fish faunas. A position off eastern Australia (position B in Figure 7B), was therefore favoured; this leans heavily on the controversial position of South China between Eastern Australia and Laurentia in a Late Proterozoic reconstruction of Rodinia (Li *et al.* 1995, 1996). Such a position in the Middle Palaeozoic would be precluded by our knowledge of the early Palaeozoic biogeography of the Asian terranes (see above) and the geology of eastern Australia which implies westwards directed subduction of the "Palaeo-Pacific" beneath eastern Australia. That would have resulted in the "Asian superterrne" being accreted to eastern Australia! I therefore prefer to position these terranes (though not in their current relative positions) off NW Australian Gondwana in the Devonian (Figure 8), and explain the waxing and waning biogeographic patterns observed on and between these terranes and Gondwanaland during the Middle to Late Palaeozoic by the process of rifting and separation of these terranes from Gondwanaland, their northwards drift, and the opening of the Palaeo-Tethys. Early Devonian endemic continental fish faunas could have developed because of shifting distributions of land and sea and the development of intracontinental marine barriers during these processes and are not necessarily the result of continental isolation in a true tectonic sense.

East Australia continued to reside in low southern latitudes during the Devonian but rotated counterclockwise. This counterclockwise rotation was concurrent with the clockwise rotation of the Chinese terranes as they separated from Gondwanaland as an elongate continental sliver. As this sliver separated from Gondwanaland, the Palaeo-Tethys ocean opened up (Figure 8). Late Devonian faunas on the Chinese terranes still have some Australian connections which are well illustrated by the occurrence of the antiarch (placoderm) group Sinolepidoidei (including *Grenfellaspis*) described exclusively from Grenfell in southeast Australia and from the North and South China terranes (Ritchie *et al.* 1992; Johanson and Ritchie 1999).

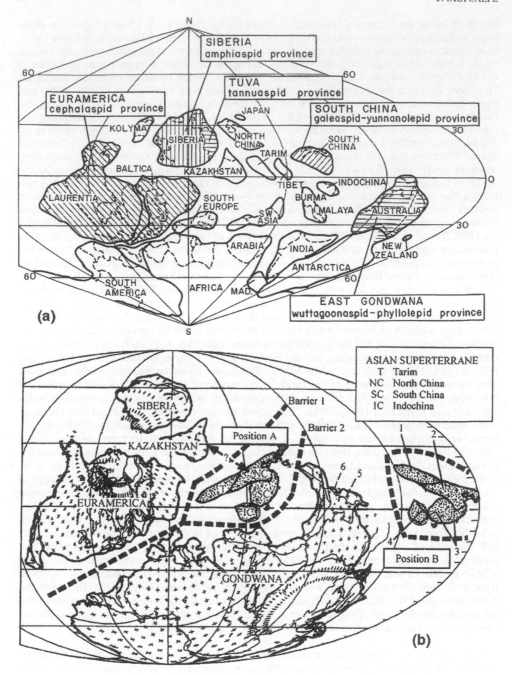

Figure 7. A. The five Devonian fish provinces of Young (1981). B. Alternative Late Palaeozoic positions proposed for the "Asian Superterrane" by Young and Janvier (1999).

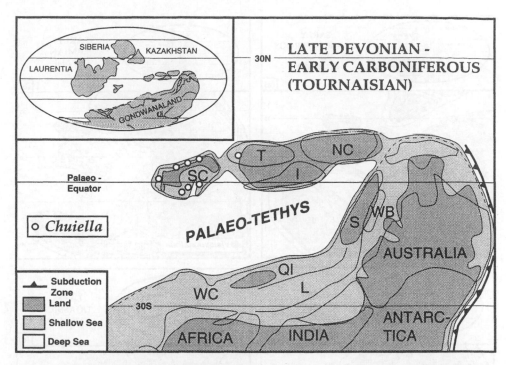

Figure 8. Reconstruction of eastern Gondwanaland for the Late Devonian to Lower Carboniferous (Tournaisian) showing the postulated positions of the East and Southeast Asian terranes, distribution of land and sea, and opening of the Palaeo-Tethys ocean at this time. Also shown is the distribution of the endemic Tournaisian brachiopod genus *Chuiella* (from Chen and Shi 1999). Present-day outlines are for reference only. Distribution of land and sea for Chinese blocks principally from Wang (1985). Land and sea distribution for Pangea/Gondwanaland compiled from Golongka *et al.* (1994), Smith *et al.* (1994); and for Australia from Struckmeyer & Totterdell (1990). Symbols as for figure 6.

2.3 *Carboniferous (354 - 298 Ma)*

During the Carboniferous, Gondwanaland rotated clockwise and collided with Laurentia in the west to form Pangea. Australia, still attached to northeast Gondwanaland, drifted from low southern latitudes (0° to 40°S) in Tournaisian - Visean times to high southern latitudes (50° to 75°S) in middle-late Namurian times. The major Gondwanan glaciation commenced in the Namurian and extended through into the Early Permian. There were major global shifts in both plate configurations and climate during this time and a change from warm to cold conditions in Australasia. This is reflected in the change from high- to low-diversity faunas in Australasia (Jones *et al.*, 2000) and especially eastern Australia where low diversity, endemic faunas developed in the Upper Carboniferous (Figure 9). Carboniferous faunas and floras in North and South China, Indochina, and Tarim are tropical/sub-tropical Cathaysian/Tethyan types and show no Gondwanaland affinities. These terranes had already separated from Gondwanaland and were located in low-latitude/equatorial positions during the Carboniferous (Figure 9). Proximity of Tarim and South China in the Tournaisian is indicated by common endemic species of the brachiopod genus *Chuiella* (Chen and Shi 1999; Figure 8). Indochina and South China collided within the Tethys during the early Carboniferous and amalgamated along the Song Ma suture zone, now located in Laos and Vietnam. Late Lower Carboniferous (uppermost Mississippian) floras on these terranes are similar and suggest continental connection at this time (Laveine *et al.* 1999). Ice sheets and glaciers extended across much of eastern Gondwanaland during the Late Carboniferous.

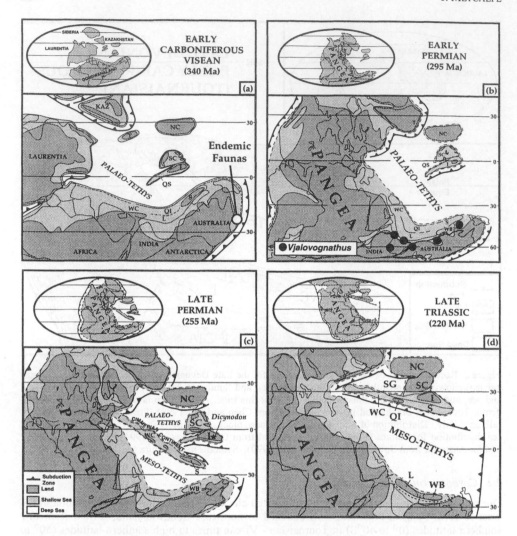

Figure 9. Palaeogeographic reconstructions of the Tethyan region for (a) Early Carboniferous (Visean), (b) Early Permian, (c) Late Permian, and (d) Late Triassic showing relative positions of the East and Southeast Asian terranes and distribution of land and sea. The distribution of the Lower Permian cold-water-tolerant conodont genus *Vjalovognathus* and the location of the Late Permian *Dicynodon* from Laos are also shown. Present day outlines are for reference only. Distribution of land and sea for Chinese blocks principally from Wang (1985). Land and sea distribution for Pangea/Gondwanaland compiled from Golongka *et al.* (1994), Smith *et al.* (1994); and for Australia from Struckmeyer & Totterdell (1990). Symbols as for figure 6.

2.4 *Early Permian (298 - 270 Ma)*

During the Permian, Australia remained in high southern latitudes. Glacial ice reached the marine environment of the northeast Gondwanaland margin and glacial-marine sediments were deposited on the Sibumasu, Qiangtang, and Lhasa terranes. Gondwanaland cold-climate faunas and floras characterised the Sibumasu, Qiangtang, and Lhasa terranes at this time. In addition, the distinctive cool-water tolerant conodont genus *Vjalovognathus* defines an eastern peri-Gondwanaland cold-water province (Nicoll and Metcalfe, 1998; Figure 9). Early Permian (Asselian-Early Sakmarian) brachiopod faunas of the North China, South China, Tarim, and Indochina terranes defined a Cathaysian province, whilst coeval brachiopod faunas of the Sibumasu,

Qiangtang, and Lhasa terranes formed part of the peri-Gondwana Indoralian Province (Shi and Archbold, 1998). The biogeographic contrast seen in the Lower Permian between terranes south west and northeast of the Palaeo-Tethyan Lancangjian, Changning-Menglian, Nan-Uttaradit, Sra Kaeo, and Bentong-Raub suture zones (Figure 10) is as striking as, if not more striking than that seen either side of Wallace's Line. This remarkable juxtaposition of Lower Permian cold-climate Gondwanaland fauanas and floras with warm-climate, equatorial Cathaysian faunas and floras is the result of the closure of the Palaeo-Tethys ocean which had separated them by thousands of kilometers. This is graphically illustrated by the distribution of Lower Permian plants in the region (Figure 11) in which three floral provinces are recognised, the southern hemisphere cool-climate Gondwana province (on India and the Lhasa, Qiangtang, and Sibumasu terranes), the palaeoequatorial warm, subtropical to tropical Cathaysian province (on North China, South China, and Indochina) and the northern hemisphere cool-climate Angaran province (on the Tarim terrane, Kazakstan, and Northeast China). Plotting of these floral provinces on a Lower Permian reconstruction of Pangea (Figure 11B) brings these floras into appropriate latitudinal positions and also demonstrates the relative isolation of the low latitude Cathaysian flora on the intra-Tethyan North China, South China, and Indochina terranes.

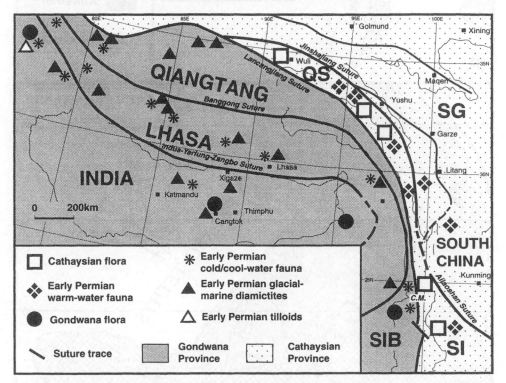

Figure 10. Distribution of Lower Permian Gondwana and Cathaysian province faunas and floras of the Tibet-Yunnan region, showing the remarkable juxtaposition of these highly contrasting cool- and warm-climate biotas either side of the main Palaeo-Tethyan divide represented by the Lancangjian and Changning-Menglian (C.M.) suture zones. QS = Qamdo-Simao terrane, SIB = Sibumasu terrane, SI = Simao terrane, SG = Songpan Ganzi accretionary complex.

2.5 Late Permian (270 - 252 Ma)

During the late Early Permian to Late Permian (Late Sakmarian-Midian), the Cimmerian continental sliver (Sengor *et al.*, 1988), including the Sibumasu and Qiangtang terranes, rifted away from the northeastern margin of Gondwanaland, and by the early Late Permian the Sibumasu and Qiangtang terranes had separated from Gondwanaland and the Meso-Tethys opened

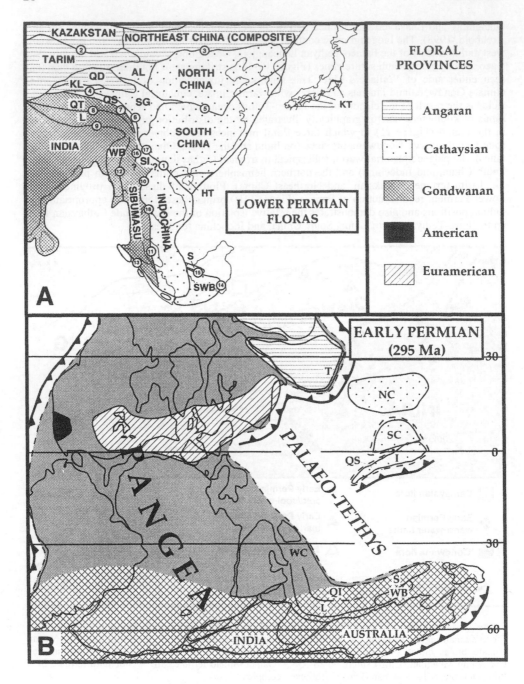

Figure 11. A. Distribution of Lower Permian floras on East and Southeast Asia terranes. B. Early Permian reconstruction of Pangea and Tethys showing positions of East and Southeast Asian terranes and floral provinces. WB = West Burma, SWB = South West Borneo, S = Semitau Terrane, HT = Hainan Island terranes, WC = Western Cimmerian Continent, L = Lhasa Terrane, QI = Qiangtang Terrane, QS = Qamdo-Simao Terrane, SI= Simao Terrane, SG = Songpan Ganzi accretionary complex, KL = Kunlun Terrane, QD = Qaidam Terrane, AL = Ala Shan Terrane, KT = Kurosegawa Terrane, T = Tarim Terrane, NC = North China, SC = South China, I = Indochina. Suture numbers as for Fig.1.

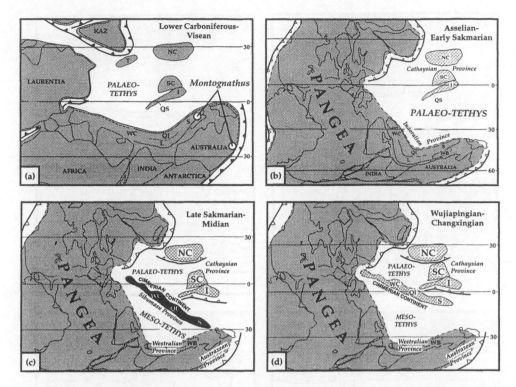

Figure 12. Palaeogeographic reconstructions for the (a) Lower Carboniferous (Visean) showing the occurrence of the eastern Gondwanaland endemic conodont genus *Montognathus*; and (b) Asselian-Early Sakmarian, (c) Late Sakmarian-Midian and (d) Wujiapingian-Changxingian, illustrating a tectonic vicariant model interpreting the change in marine provinciality of the Sibumasu and other elements of the Cimmerian continent during the Permian. Note that as Sibumasu separates from Gondwanaland and drifts northwards it loses its Indoralian (Gondwanaland) Province faunas, then develops endemic faunas representing an independent Sibumasu province, and finally becomes assimilated into the intra-Tethyan Cathaysian Province. After Shi and Archbold (1998).

between this continental sliver and Gondwanaland (Figure 9). The Palaeo-Tethys continued to be destroyed by northwards subduction beneath Laurasia, North China, and the amalgamated South China and Indochina terranes. During their northwards drift, after separation from Australian Gondwana, the Sibumasu and Qiangtang terranes developed initially a Sibumasu Province fauna and then were absorbed into the palaeoequatorial Cathaysian Province (Figure 12). This is particularly seen in the brachiopod faunas, but also in the fusulinid faunas, which are initially low-diversity cooler climate faunas characterised by *Monodiexodina* and *Eopolydiexodina* but are then replaced by higher diversity typical Tethyan fusulinid assemblages (Ueno, 1999). Brachiopod faunas of Australasia during Middle to Upper Permian times represent two provinces, a western Westralian Province and an eastern Austrazean Province (Archbold, 1996; Shi and Archbold, 1998). North and South China begin to collide during the Late Permian, and a connection between mainland Pangea and Indochina, via South and North China or via the western part of the Cimmerian continent, is indicated by the occurrence of the terrestrial vertebrate genus *Dicynodon* in the Upper Permian of Indochina (Figure 9).

2.6 Triassic (252 - 205 Ma)

Australia was in low to moderate southern latitudes during the Triassic. The Sibumasu and Qiangtang terranes collided and sutured to the Indochina/South China amalgamated terrane along the Lancangjian, Changning-Menglian, Nan-Uttaradit, and Bentong-Raub suture zones (Liu *et al.*, 1991, 1996; Metcalfe, 2000; Singharajwarapan and Berry, 2000).

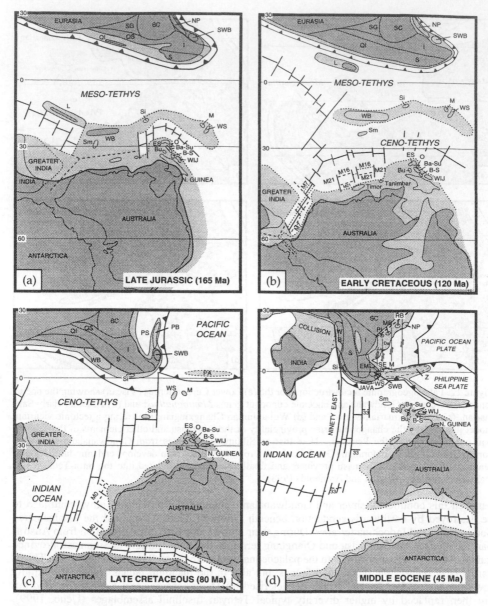

Figure 13. Palaeogeographic reconstructions for the East Asia-Australia region in (a) Late Jurassic, (b) Early Cretaceous, (c) Late Cretaceous, and (d) Middle Eocene times showing distribution of land and sea. SG = Songpan Ganzi accretionary complex SWB = South West Borneo (includes Semitau) NP = North Palawan and other small continental fragments now forming part of the Philippines basement Si = Siku-leh M = Mangkalihat WS = West Sulawesi ES = East Sulawesi O = Obi-Bacan Ba-Su = Bangai-Sula Bu = Buton B-S = Buru-Seram WIJ = West Irian Jaya Sm = Sumba PA = Insipient Philippine Arc. M numbers represent Indian Ocean magnetic anomalies. Other terrane symbols as in Figures 6 and 9. Modified from Metcalfe (1990, 1996, 1998) and partly after Smith *et al.* (1981), Audley-Charles (1988) and Audley-Charles *et al.* (1988). Present day outlines are for reference only. Distribution of land and sea for Chinese blocks principally from Wang, (1985). Land and sea distribution for Pangea/Gondwanaland compiled from Golongka *et al.* (1994), Smith *et al.* (1994); and for Australia from Struckmeyer & Totter-dell (1990).

The collision of North and South China was nearly complete, with exhumation of ultra-high-pressure metamorphics along the Qinling-Dabie suture zone. Sediment derived from the North-South China collisional orogen poured into the Songpan Ganzi accretionary basin producing huge thicknesses of flysch turbidites (Nie *et al.*, 1995).

2.7 Jurassic (205 - 141 Ma)

Australia remained in low to moderate southern latitudes in the Jurassic. Rifting and separation of the Lhasa, West Burma, Sikuleh, Mangkalihat, and West Sulawesi terranes from NW Australia occurred progressively during the Late Triassic to Late Jurassic. The Ceno-Tethys ocean basin opened behind these terranes as they separated from Gondwanaland (Figure 13). Final welding of North China to Eurasia also took place in the Jurassic and the initial pre-breakup rifting of the main Gondwanaland supercontinent began at this time.

2.8 Cretaceous (141 - 65 Ma)

The Lhasa terrane collided and sutured to Eurasia in earliest Cretaceous times (Figure 13). Gondwanaland broke up and India drifted north, making initial contact with Eurasia at the end of the Cretaceous. This early contact between India and Eurasia is indicated by palaeomagnetic data from the ninetyeast Ridge (Klootwijk *et al.*, 1992) and also by Late Cretaceous Eurasian frogs occurring in India. The small West Burma and Sikuleh terranes accreted to Sibumasu during the Cretaceous. Australia began to separate from Antarctica at about 85 Ma (early Late Cretaceous) and drift northwards, but a connection with Antarctica via Tasmania still remained until about 45 Ma.

ACKNOWLEDGEMENTS

The Australian Research Council is gratefully acknowledged for financial support under its Large Grants Scheme. The Asia Centre, University of New England is also thanked for facilities provided. I would like to thank Dr P.H. Stauffer and Prof. N.W. Archbold for their helpful reviews of this paper.

REFERENCES

Archbold, N.W. 1996. Paleobiogeography of Australian Permian brachiopod faunas. In P. Copper & J. Jin (eds), *Brachiopods*. (Proceedings of the Third International Brachiopod Congress): 19-23. Rotterdam: A.A. Balkema.

Audley-Charles, M.G. 1988. Evolution of the southern margin of Tethys (North Australian region) from Early Permian to Late Cretaceous. In M.G. Audley-Charles & A. Hallam (eds), *Gondwana and Tethys*, Geological Society Special Publication No. 37: 79-100. Oxford: Oxford University Press.

Audley-Charles, M.G., Ballantine, P.D. & Hall, R. 1988. Mesozoic-Cenozoic rift-drift sequence of Asian fragments from Gondwanaland. *Tectonophysics* 155: 317-330.

Aung, A.K. 1995. New Middle Devonian (Eifelian) rugose corals from Myanmar. *Journal of Southeast Asian Earth Sciences* 11: 23-32.

Brock, G.A., Engelbretsen, M. & Cockle, P. 1999. The significance of changing biogeogeographic signals for the East Gondwana-China connection during the Cambrian. In I. Metcalfe (ed), *Where Worlds Collide: Faunal and floral migrations and evolution in SE Asia-Australasia, Abstracts:* 12. Asia Centre, University of New England.

Burrett, C. 1973. Ordovician biogeography and continental drift. *Palaeogeography, Palaeoclimatology, Palaeoecology* 13: 161-201.

Burrett, C. & Stait, B. 1985. South-East Asia as part of an Ordovician Gondwanaland - a palaeobiogeographic test of a tectonic hypothesis. *Earth and Planetary Science Letters* 75: 184-190.

Burrett, C., Long, J. & Stait, B. 1990. Early-Middle Palaeozoic biogeography of Asian terranes derived from Gondwana. In W.S. McKerrow & C.R. Scotese (eds), Palaeozoic Palaeogeography and Biogeography. *Geological Society Memoir* 12: 163-174.

Chen, Z.Q & Shi, G.R. 1999. *Chuiella* gen. nov. (Brachiopoda) and palaeoecology from the Lower Carboniferous of the Kunlun Mountains, NW China. *Alcheringa* 23: 259-275.

Cocks, L.R.M. & Fortey, R.A. 1997. A new *Hirnantia* fauna from Thailand and the biogeography of the latest Ordovician of South-East Asia. *GEOBIOS* 20: 117-126.

Fortey, R.A. 1989. An Early Devonian trilobite fauna from Thailand. *Alcheringa* 13: 257-267.

Fortey, R.A. 1997. Late Ordovician trilobites from southern Thailand. *Palaeontology* 40: 397-449.

Fortey, R.A. & Cocks, L.R.M. 1998. Biogeography and palaeogeography of the Sibumasu terrane in the Ordovician: a review. In R. Hall, & J.D. Holloway (eds), *Biogeography and Geological Evolution of SE Asia:* 43-56. , Amsterdam, The Netherlands: Backhuys Publishers.

Golonka, J., Ross, M.I. & Scotese, C.R. 1994. Phanerozoic paleogeographic and paleoclimatic modeling maps. In A.F. Embry, B. Beauchamp & D.J. Glass (eds), *Pangea: Global Environments and Resources*. Canadian Society of Petroleum Geologists, Memoir 17: 1-47.

Jell, P.A., Burrett, C.F., Stait, B. & Yochelson, E.L. 1984. The Early Ordovician bellerophontoid *Peelerophon oehlerti* (Bergeron) from Argentina, Australia and Thailand. *Alcheringa* 8: 169-176.

Johanson, Z. & Ritchie, A. 1999. Sarcopterygian fishes from the Hunter Siltstone (Late Famennian) near Grenfell, NSW, Australia. In I. Metcalfe (ed), *Where Worlds Collide: Faunal and floral migrations and evolution in SE Asia-Australasia, Abstracts:* 37. Asia Centre, University of New England.

Jones, P.J., Metcalfe, I., Engel, B. A., Playford, G., Rigby, J., Roberts, J., Turner, S. & Webb, G.E. 2000. Carboniferous Palaeobiogeography of Australasia. In T. Wright, G. Young, & J. Talent (eds) *Palaeobiogeography of Australasian Faunas and Floras,* Australasian Association of Palaeontologists Memoir 23: 259-286.

Klootwijk, C. 1996. Phanerozoic configurations of Greater Australia: Evolution of the North West Shelf. Part Two: Palaeomagnetic and geologic constraints on reconstructions. *AGSO Record* 1996/52: 85p.

Klootwijk, C.T., Gee, J.S., Peirce, J.W., Smith, G.M. & McFadden, P.L. 1992. An early India-Asia contact: Paleomagnetic constraints from Ninetyeast Ridge, ODP Leg 121. *Geology* 20: 395-398.

Laurie, J.R. & Burrett, C. 1992. Biogeographic significance of Ordovician brachiopods from Thailand and Malaysia. *Journal of Paleontology* 66: 16-23.

Laveine, J.P., Ratanasthein, B. & Azhar Haji Hussin. 1999. The Carboniferous floras of Southeast Asia: Implications for the relationships and timing of accretion of some Southeast Asian Blocks. In I. Metcalfe (ed) *Gondwana dispersion and Asian accretion*, Final Results Volume for IGCP Project 321: 229-246. Rotterdam: A.A. Balkema.

Li, Z.X., Zhang, L. & Powell, C.McA. 1995. South China in Rodinia: Part of the missing link between Australia-East Antarctica and Laurentia? *Geology* 23: 407-410.

Li, Z.X., Zhang, L. & Powell, C. McA. 1996. Positions of the East Asian cratons in the Neoproterozoic supercontinent Rodinia. *Australian Journal of Earth Sciences* 43: 593-604.

Lin Baoyu & Webby, B.D. 1989. Biogeographic relationships of Australian and Chinese Ordovician corals and stromatoporoids. *Memoirs of the Association of Australasian Palaeontologists* 8: 207-217.

Liu Benpei, Feng Qinglai & Fang Nianqiao. 1991. Tectonic evolution of the Palaeo-Tethys in Changning-Menglian Belt and adjacent regions, western Yunnan. *Journal of China University of Geosciences* 2: 18-28.

Liu Benpei, Feng Qinglai, Fang Nianqiao, Jia Jinhua, He Fuxiang, Yang Weiping & Liu Diansheng. 1996. Tectono-paleogeographic framework and evolution of the Paleotethyan archipeligoes ocean in Changning-Menglian belt, Western Yunnan, China. In Fang Nianqiao & Feng Qinglai (eds) *Devonian to Triassic Tethys in Western Yunnan, China*: 1-12. China University of Geosciences Press.

Long, J.A. 1993. Palaeozoic vertebrate biogeography of south-east Asia and Japan. In J.A. Long (ed) *Palaeozoic Vertebrate Biostratigraphy and Biogeography*: 277-289. London, Belhaven Press.

Long, J.A. & Burrett, C.F. 1989. Fish from the Upper Devonian of the Shan-Thai terrane indicate proximity to east Gondwana and South China terranes. *Geology* 17: 811-813.

Metcalfe, I. 1988. Origin and assembly of Southeast Asian continental terranes. In: M.G. Audley-Charles & A. Hallam (eds), *Gondwana and Tethys*. Geological Society of London Special Publication No. 37: 101-118.

Metcalfe, I. 1990. Allochthonous terrane processes in Southeast Asia. *Philosophical Transactions of the Royal Society of London* A331: 625-640.

Metcalfe, I. 1993. Southeast Asian terranes: Gondwanaland origins and evolution. In. R.H. Findlay, R. Unrug, M.R. Banks, & J.J. Veevers (eds) *Gondwana 8 - Assembly, Evolution, and Dispersal* (Proceedings Eighth Gondwana Symposium, Hobart, 1991): 181-200. Rotterdam: A.A. Balkema.

Metcalfe, I. 1996. Pre-Cretaceous evolution of SE Asian terranes In R. Hall & D. Blundell, (eds) *Tectonic Evolution of Southeast Asia*. Geological Society Special Publication No. 106: 97-122.

Metcalfe, I. 1998. Palaeozoic and Mesozoic geological evolution of the SE Asian region: multidisciplinary constraints and implications for biogeography. In R. Hall, & J.D. Holloway (eds) *Biogeography and Geological Evolution of SE Asia*: 25-41. Amsterdam, The Netherlands: Backhuys Publishers.

Metcalfe, I. 1999. Gondwana dispersion and Asian accretion: an overview. In I. Metcalfe (ed) *Gondwana dispersion and Asian accretion*, Final Results Volume for IGCP Project 321: 9-28. Rotterdam: A.A. Balkema.

Metcalfe, I. 2000. The Bentong-Raub Suture Zone. *Journal of Asian Earth Sciences*, 18 (Part 6): 691-712.

Nicoll, R.S. & Metcalfe, I. 1994. Late Cambrian to Early Silurian conodont endemism of the Sinian-Australian margin of Gondwanaland. In Wang Zhi-hao & Xu Fang-ming (eds) *First Asian Conodont Symposium, Nanjing, China, Abstracts of Papers*: 7.

Nicoll, R.S. & Metcalfe, I. 1998. Early and Middle Permian conodonts from the Canning and Southern Carnarvon Basins, Western Australia: Their implications for regional biogeography and palaeoclimatology. *Proceedings Royal Society of Victoria* 110: 419-461.

Nicoll, R.S. & Metcalfe, I. 1999. Cambrian to Permian conodont biogeography in East Asia-Australasia. In I. Metcalfe (ed), *Where Worlds Collide: Faunal and floral migrations and evolution in SE Asia-Australasia, Abstracts*: 62-64. Asia Centre, University of New England.

Nicoll, R.S. & Totterdell, J.M. 1990. Conodonts and the distribution in time and space of Ordovician sediments in Australia and adjacent areas. *Tenth Australian Geological Convention, Geological Society of Australia, Abstracts* 25: 46.

Nie, Y.S., Yin, A., Rowley, D.B.& Jin, Y. 1995. Exhumation of the Dabie Shan ultra-high pressure rocks and accumulation of the Songpan-Ganzi flysch sequence, central China. *Geology* 22: 999-1002.

Ritchie, A., Wang Shitao, Young, G.C. & Zhang Guorui. 1992. The Sinolepidae, a family of Antiarchs (Placoderm Fishes) from the Devonian of South China and eastern Australia. *Records of the Australian Museum* 44: 319-370.

Rong Jia-Yu, Boucot, A.J., Su Yang-Zheng & Strusz, D.L. 1995. Biogeographical analysis of Late Silurian brachiopod faunas, chiefly from Asia and Australia. *Lethaia* 28: 39-60.

Sengör, A.M.C., Altiner, D., Cin, A., Ustaomer, T. & Hsu, K.J. 1988. Origin and assembly of the Tethyside orogenic collage at the expense of Gondwana Land. In M.G. Audley-Charles & A. Hallam (eds) *Gondwana and Tethys*. Geological Society Special Publication No. 37: 119-181. Oxford: Oxford University Press.

Servais, T., Brocke, R. & Fatka, O. 1996. Variability in the Ordovician acritarch *Dicrodiacrodium*. *Palaeontology* 39: 389-405.

Shergold, J., Burrett, C., Akerman, T. & Stait, B. 1988. Late Cambrian trilobites from Tarutao Island, Thailand. *New Mexico Bureau of Mines & Mineral Resources Memoir* 44: 303-320.

Shi, G.R. & Archbold, N.W. 1998. Permian marine biogeography of SE Asia. In R. Hall & J.D. Holloway (eds) Biogeography and Geological Evolution of SE Asia: 57-72. Amsterdam, The Netherlands: Backhuys Publishers.

Singharajwarapan, S. & Berry, R. 2000. Tectonic implications of the Nan Suture Zone and its Relationship to the Sukhothai Fold Belt. *Journal of Asian Earth Sciences* 18 (Part 6): 663-673.

Smith, A.G., Hurley, A.M. & Briden, J.C. 1981. *Phanerozoic palaeocontinental world maps*. Cambridge: Cambridge University Press.

Smith, A.G, Smith, D.G. & Funnell, B.M. 1994. *Atlas of Mesozoic and Cenozoic coastlines*. Cambridge, Cambridge University Press.

Stait, B. & Burrett, C.F. 1982. *Wutinoceras* (Nautiloidea) from the Setul Limestone (Ordovician) of Malaysia. *Alcheringa* 6: 193-196.

Stait, B. & Burrett, C.F. 1984. Ordovician nautiloid faunas of Central and Southern Thailand. *Geological Magazine* 121: 115-124.

Stait, B., Wyatt, D. & Burrett, C.F. 1987. Ordovician nautiloid faunas of Langkawi Islands, Malaysia and Tarutao Island, Thailand. *N. Jb. Geol. Palaont. Abh.* 174: 373-391.

Struckmeyer, H.I.M. & Totterdell, J.M. (Coordinators) and BMR Palaeogeographic Group. 1990. *Australia: Evolution of a continent*. Canberra: Bureau of Mineral Resources, Australia.

Tong-Dzuy Thanh & Cai Chongyang. 1995. Devonian flora of Vietnam. In Tran Van Tri (ed) Proceedings of the International Symposium Geology of Southeast Asia and adjacent areas. *Journal of Geology*, Series B (Geological Survey of Vietnam), No 5-6: 105-113.

Tong-Dzuy Thanh, Janvier, P. & Ta Hoa Phuong. 1996. Fish suggests continental connections between the Indochina and South China blocks in middle Devonian time. *Geology* 24: 571-574.

Ueno, K. 1999. Gondwana/Tethys divide in east Asia: solution from Late Paleozoic foraminiferal paleobiogeography. In Ratanasthein, B. & Rieb, S.L. (eds) *Proceedings of the International Symposium on Shallow Tethys (ST) 5, Chiang Mai, Thailand, February, 1999*: 45-54. Faculty of Science, Chiang Mai University.

Wang, H. 1985. *Atlas of the palaeogeography of China*. Beijing, Cartographic Publishing House.

Webby, B.D., Wyatt, D. & Burrett, C. 1985. Ordovician stromatoporoids from the Langkawi Islands, Malaysia. *Alcheringa* 9: 159-166.

Yang Jialu. 1994. Cambrian. In Yin Hongfu (ed). *The palaeobiogeography of China*. Oxford Biogeography Series No.8: 35-63. Oxford, Clarendon Press.

Young, G.C. 1981. Biogeography of Devonian vertebrates. *Alcheringa* 5: 225-243.

Young, G.C. 1987. Devonian palaeontological data and the Armorica problem. *Palaeogeography, Palaeoclimatology, Palaeoecology* 60: 283-304.

Young, G.C. & Janvier, P. 1999. Early-middle Palaeozoic vertebrate faunas in relation to Gondwana dispersion and Asian accretion. In I. Metcalfe (ed) *Gondwana dispersion and Asian accretion*, Final Results Volume for IGCP Project 321: 115-140. Rotterdam: A.A. Balkema.

Zhao, X., Coe, R.S., Gilder, S.A. & Frost, G.M. 1996. Palaeomagnetic constraints on the palaeogeography of China: implications for Gondwanaland. *Australian Journal of Earth Sciences* 43: 643-672.

Zhou, Z., Dean, W.T & Luo, H. 1998a. Early Ordovician trilobites from Dali, West Yunnan, China, and their palaeogeographical significance. *Palaeontology* 41: 429-460.

Zhou, Z., Dean, W.T, Yuan, W. & Zhou, T. 1998b. Ordovician trilobites from the Dawangou Formation, Kalpin, Xinjiang, North-West China. *Palaeontology* 41: 693-735.

Cenozoic reconstructions of SE Asia and the SW Pacific: changing patterns of land and sea

Robert Hall
SE Asia Research Group, Department of Geology, Royal Holloway University of London
Egham, Surrey TW20 0EX, U.K. Email: robert.hall@gl.rhul.ac.uk

ABSTRACT: The Cenozoic has seen the major tectonic events which have determined the present configuration of land and sea in SE Asia and the SW Pacific. Subduction throughout this period maintained volcanic arcs which formed discontinuously emergent island chains crossing the region. Early in the Cenozoic the major collision of India with SE Asia enlarged the area of land connected to Eurasia. Later, the continuing collision with Australia led to connections between Australia, Eurasia, and the Pacific. Despite long-term convergence of the major plates there have been important episodes of extension, forming ocean basins and causing subsidence within continental regions, which were probably driven by subduction. It is clear that very rapid changes in topography and distribution of land and sea have occurred. The geological and biogeographic interface of perhaps most interest is that between Sundaland and Australia, extending from Borneo to the Bird's Head of New Guinea. The biogeographic divides of Wallace and later workers have all been drawn through this region. Since the early Miocene the original deep water barrier between Australia and Sundaland has been eliminated but the process of convergence has never produced a simple route for the mixing of Australian and Asian floras and faunas. It is clear from the plate tectonic model summarised here that there have been multiple opportunities for dispersal and vicariance caused by regional tectonic processes. There are also more subtle geologically-related forces which may have modified biogeographic patterns, such as links between tectonics and sea level, the rise of mountains and global/local climate, and closure of seaways and oceanic circulation, which are suggested by the tectonic model. All these changes occurred within a framework of overall long-term cooling. Further, more extreme, changes in climate and sea level occurred during the Quaternary glacial and interglacial periods. A simple picture of convergence in which Australia and Sundaland collided, causing land to emerge, allowing colonisation by animals and plants from east and west is therefore probably too simple. Since the early Miocene Australia and Sundaland have moved closer together but as land emerged and mountains rose in some areas, new deep basins developed. The distribution of Australian and Asian plants and animals should therefore reflect this complexity, with further important modifications imposed by glacially-related sea level and climatic change in the Quaternary. In this picture, the zone of Wallace's Line is partly an ancient deep water barrier, partly a dynamic boundary marking a migration front, but also a relic of Neogene patterns which have been tectonically disrupted and modified by Quaternary climate change.

1 INTRODUCTION

Today, the waters of SE Asia contain the highest marine faunal diversity in the world, and the islands of the region contain some of the most diverse collections of plant and animal species found on Earth. The division between Asian and Australian floras and faunas in Indonesia, first recognised by Wallace in the nineteenth century, is now recognised as a biogeographic region of transition, named Wallacea (Figure 1), situated between areas with Asiatic and Australian floras and faunas,

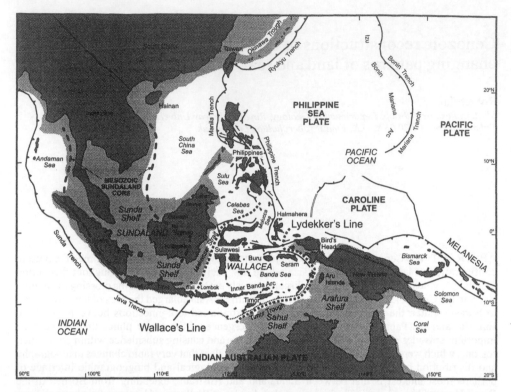

Figure 1. The region of Wallacea. The light shaded areas represent the shallow continental shelves of Sunda-land and northern Australia, drawn at the 200 m isobaths. The western boundary of Wallacea is Wallace's Line of 1863. The eastern boundary is Lydekker's Line which is the western limit of strictly Australian faunas.

with elements of both but where organisms show a high degree of endemism. This region extends from east of the Makassar Strait to west of the Bird's Head of New Guinea. It is worth noting that the boundaries of Wallacea are essentially the present edges of the shallow marine shelves of Sundaland and Australia.

Implicit in the concept of Wallacea is the idea that there were originally two principal biogeographic regions, those of Australasia and Asia, which were physically separated and which subsequently became connected. The region has changed as a result of the rapid plate movements during the Cenozoic and geological changes have driven changes in the distribution of land and sea. Wallace (1869) understood that geological processes were important in the development of present biogeographic patterns. However, the geological changes have not been unidirectional, and they have also influenced other factors which are likely to have influenced biogeographic patterns, such as ocean currents and local climatic patterns. In addition, the animals and plants themselves have changed as a result of evolutionary processes. Thus, a geological understanding of the region is likely to be of value to understanding its biogeography, but should be seen more as the essential background to a complex geological, physiographic, climatic and biotic evolution rather than as the explanation of the patterns observed at the present day.

2 GONDWANA TO WALLACEA

The SE Asian region owes its origin to the pre-Cenozoic break-up of Gondwana, the subsequent movement of Gondwana fragments northwards, and their eventual collision with Asia (e.g. Metcalfe 1998). Many fragments separated from Gondwana and amalgamated in SE Asia over a considerable period of time. The process of rifting led to formation of new oceans, and the northward motion of

Gondwana fragments required subduction of older oceanic crust at the edges of the growing Asian continent. By the late Mesozoic, fragments derived from Gondwana formed a composite Sundaland core surrounded by subduction zones. Further south and east, the northern Australian margin was a passive continental margin for most of the Mesozoic and it was from this region that many of the Gondwana fragments now found in SE Asia were derived. Between these two regions were wide oceans.

India and Australia separated from Gondwana in the Cretaceous and moved northwards as parts of different plates. India initially collided with the Asian continent about 50 million years ago, but continued to move north accompanied by complex internal deformation within Indochina and mainland SE Asia which continues to the present day. This enlarged the area of land connected to Eurasia. Throughout the entire region subduction processes maintained volcanic arcs which formed discontinuously emergent island chains. During the last 25 million years the collision of Australia with the Sundaland margin led to connections between Australia, Eurasia, and the Pacific. However, despite long-term convergence of the major plates there have been important episodes of extension, forming ocean basins and causing subsidence within continental regions, which were probably driven by subduction. In eastern Indonesia the northward movement of Australia during the Cenozoic has been marked by arc-continent collision, major strike-slip motion within the north Australian margin in northern New Guinea, and accretion of continental fragments derived from Australia. Fragments of arcs have been dispersed in New Guinea, east Indonesia and the Philippines by the movement of the Pacific plate. It is clear that very rapid changes in topography and distribution of land and sea have occurred.

3 SOME GEOLOGICAL FUNDAMENTALS

It is now agreed by Earth scientists that the exterior of the earth is formed of lithospheric plates which are more than 100 km thick. The continents are moving on these plates and the size of the globe has not changed in the past 100s of millions of years. The plate tectonic model (e.g. Kearey & Vine 1990) is so strongly supported by a huge range of geological evidence that it really cannot be examined as just another hypothesis. This is in contrast to some ideas that have been current at different times such as earth expansion which really do not pass the tests based on observational data. Explanations that have been put forward in the past for distributions of land masses and links between land masses which rely on such hypotheses as earth expansion or land bridges across the world's major oceans are not realistic.

The plates on the globe have moved in the past and it is possible to tell how they have moved because they have left behind them a pattern of lineations on the ocean floor. The polarity of the earth's magnetic field has changed irregularly through time and as igneous rocks have formed at the plate boundaries at mid-ocean ridges and frozen from melts they have left a trail on the ocean floor of the movement of the major plates in the form of magnetic lineations of alternating reversed and normal polarities. This is important for a number of reasons. It means that the movements of the major plates on the globe can be reconstructed for up to about 150 million years. The oldest crust in the oceans is in the western Pacific and is about 160 Ma old and for areas where there are magnetic lineations, and hence the age of the ocean is known, the history of plate motions can be reconstructed very precisely. In principle, mapping the ocean floors in detail provides the means to work out the motion paths of the plates which can be described in terms of simple mathematical parameters. It is possible to calculate rotation poles and rates of motion and thus build a global model of the history of plate motions.

Ocean crust is also important because it has a history that we know and understand very well. The crust is formed at mid-ocean ridges by the rise of hot magma which is extruded at the surface or frozen at depth. The sea bed is initially at depths of about 2.5 km below sea level and as the crust gets older it follows a very simple pattern of increasing depth with age. It is possible to map the age of the ocean crust around the major oceans very accurately using this simple age-depth relationship which is a function merely of the cooling of the outer earth, the lithosphere, which becomes denser and sinks. Similarly, the rifted passive margins formed during the breakup of continents also have a very predictable history of subsidence which reflects subsidence caused by the rifting followed by a long-term thermal subsidence due to cooling. Thus, for both oceanic regions and passive continental margins, it is possible to infer with some confidence the depth of sea at a particular age based on the plate tectonic model.

During closure of oceans the oceanic lithosphere is subducted at active margins and is once again predictably associated with given water depths, typically between 7 and 9 km at the deep trenches. The ocean lithosphere sinks deep into the mantle and at distances of about 100 km from trenches water from the subducting slab causes the mantle to melt, forming magmas which rise to the surface and produce island arc volcanoes. The arc volcanoes formed in such intra-oceanic island arc settings or active continental margin settings are not constantly active but over reasonably long periods of time it is likely that they will emerge above sea level. The older the arc, the thicker the crust, and the higher the probability of emergence. Volcanic arcs are ephemeral features, geologically at any rate. Young intra-oceanic arcs, such as those of the Izu-Bonin-Mariana arc, or Melanesian arcs, may never become emergent or be only locally and intermittently emergent at sites of active volcanicity. Such areas of land will disappear quite quickly after the volcanoes cease activity. In contrast, older arcs, and commonly those underlain by older continental or arc crust such as the Japanese islands or the islands of the Sunda arc, may be almost permanently emergent.

As this process proceeds it is possible to build volcanic magmatic mountain belts which are more substantial than those island arcs within oceans. These mountain belts may also grow from time to time by the accretion of objects carried along on oceanic plates, such as a large volcanic islands or microcontinental fragments, often called terranes. The Ontong Java plateau is one such example, and there are numerous other elevated regions of thickened crust throughout the Western Pacific which represent the products of mantle plumes, hotspots, or old arc remnants, and which when accreted will be described geologically as terranes. The idea of terranes is a popular one, and sometimes these are interpreted as the potential carriers of land plants and animals ('arks'), but in many cases these terranes, although very large, have spent their entire history beneath the waves. During the final stages in the plate tectonic or Wilson cycle, arcs and continents, and ultimately continents and continents, collide with one another and the consequence of this stage in the process is huge areas of uplifted land, as seen today north of India in the Himalayas and the Tibetan Plateau, where a very extensive area has been uplifted as a consequence of Asia and India collision.

4 SOME TECTONIC QUALIFICATIONS

Plate tectonics is very good at explaining what happens in the oceans; oceans seem to behave in a relatively rigid and predictable way in which all the deformation is concentrated at the edges of the plates. But continental regions are very different and the deformation is distributed in very odd ways. In the case of India-Asia collision some of the deformation has been taken up along large strike-slip faults within Asia. There is still a great deal of argument about exactly when the Himalayas and the mountain ranges to the north of India rose, exactly how the deformation was distributed, for example, how much of it was in uplift of the continental region and how much of it was taken up in rocks moving aside by strike-slip faulting (e.g. Peltzer & Tapponnier 1988; Houseman & England 1993). It is now accepted that there has been progressive indentation of India into Asia and that as a consequence the Asian continent has been deformed. We do not understand in detail exactly how this has happened. It is also very difficult to incorporate in a purely rigid plate tectonic model. This should be borne in mind in considering the regional reconstructions. The models describe quite well how plates move when they are oceanic but they do not yet describe well what is happening in the continental regions.

The complexity of the present-day tectonics of SE Asia and the SW Pacific means that three major (Pacific, Australia and Eurasia) and numerous smaller plates need to be considered to understand the development of the region. Present plate motions, based for example on GPS measurements and seismicity, appear to have only slight relevance to understanding the long-term kinematic development of the region, and in many areas it is possible to demonstrate significant and young changes in local plate motions. However, the rates of plate motions indicate that vast areas of oceanic crust have been lost, that many major and minor oceans have opened and closed, and the configuration of the region has changed significantly during the Cenozoic. Because so much oceanic lithosphere has been subducted, and because many of the small marginal basins lack well-developed magnetic lineations, there are many difficulties in reconstructing the region. Continental and arc crust has deformed in a non-rigid manner, and there is evidence of significant vertical axis rotations. Furthermore, there are numerous different time-scales, events which may or may not have been synchronous are often vaguely correlated, and the isotopic dating record for the whole region is inadequate. Finally, geological observations in the region where collisions are in progress at the present day show us that important tectonic features can disappear within short periods leaving almost no trace. One example will suffice from an area that I know in particular detail, the Halmahera and the Sangihe Arcs, where the present day Molucca Sea is disappearing by subduction in two directions (Hall et al. 1995a; Hall 2000). The consequence of this collision, which is occurring at the present day, is that the Halmahera arc is being eliminated and without doubt in 2 or 3 million years time only one arc will be preserved. Thus, a plate tectonic model must be regarded as an incomplete approximation which, like any other model, depends on an interpretation of a wide range of geological information from land, and from the basins on and off-shore.

5 THE GEOLOGICAL RECORD

Moving from tectonic reconstruction maps to detailed palaeogeographical maps involves further complexities. It is important to recognise when mapping land and sea distributions that the geological record that we deal with is essentially a marine record. Most of Earth history is recorded in rocks deposited below sea-level. Dating of rocks is largely based on fossils, and marine organisms generally provide the fossils of greatest biostratigraphic value which usually also provide some insight into the environment of deposition. Therefore in former marine areas there generally are sedimentary deposits, they have fossils in them, they can be dated and we can often infer a great deal about where those rocks were at different times. Geologists are therefore usually able to reconstruct the history of marine areas quite well.

On the other hand, the geological record as far as land is concerned presents very considerable problems. Uplift, erosion and periods of emergence are mainly recorded by negative evidence, such as unconformities and stratigraphic incompleteness. Even when there is a rock record it will often be difficult to date because sediments deposited on land typically represent restricted types of environments, and usually contain few fossils which have limited biostratigraphic value. It is also much more difficult to interpret continental environments. As one example, Death Valley in the western United States is below sea level but exactly the same sort of stratigraphic sequences could be formed in a continental setting of similar type if that basin were one or two thousand metres or more above sea level. In many ways the features of the rocks would be similar and of course rocks deposited in those sorts of environments cause major problems in dating. This is a major problem throughout Sundaland where we still lack an adequate understanding of the tectonics of basin formation simply because of our inability to date the sequences in the basins. The continuing debate about the timing of the rise of the Himalayas and the Tibetan plateau reflect geologists' uncertainties in dating, deducing topography on land, and interpreting geological evidence.

Nonetheless, despite the reservations about applicability of plate tectonics and the deficiencies of the geological record, what the plate tectonic cycle means from the point of view of distributions of plants and animals is that even though there may not be a complete geological record or a perfect model it is possible to say with some confidence something about depths of water and distribution of land. Thus, the mapping of land and sea follows from the mapping of tectonic elements, and this follows from the geological model which is based on a wide range of data. Broadly speaking, the maps of areas of land and sea should be regarded as maps of probability; for example, it may not be possible to know for certain if a particular area was land, but the knowledge that shallow marine clastic sediments are found in the area indicates that material was eroded from nearby land even though the land area cannot be delineated with certainty. By such reasoning it is possible to complete the gaps in maps using geological judgements and therefore, for an area shown as deep marine, the probability of that area being shallow marine is low, and of it being land is very low to zero. For the reasons outlined above, in many areas below sea level, such as passive continental margins and ocean basins, the tectonic history of the region defines the inferred depths quite well. However, for areas close to sea level assignments of depths are less certain. For example, areas of long-lived island arcs develop thickened crust, implying relative shallow water areas and local emergence. When volcanoes are active, magma production, thermal expansion and crustal buoyancy

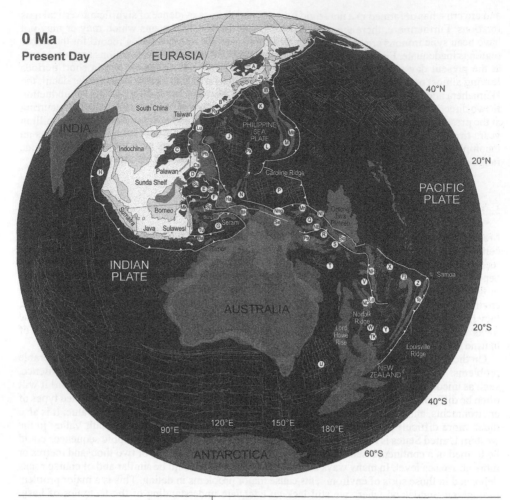

0 Ma
Present Day

Marginal Basins

A	Japan Sea	N	Ayu Trough
B	Okinawa Trough	P	Caroline Sea
C	South China Sea	Q	Bismarck Sea
D	Sulu Sea	R	Solomon Sea
E	Celebes Sea	S	Woodlark Basin
F	Molucca Sea	T	Coral Sea
G	Banda Sea	U	Tasman Sea
H	Andaman Sea	V	Loyalty Basin
J	West Philippine Basin	W	Norfolk Basin
K	Shikoku Basin	X	North Fiji Basin
L	Parece Vela Basin	Y	South Fiji Basin
M	Mariana Trough	Z	Lau Basin

Tectonic features

Ba	Banda Arc	Mk	Makassar Strait	Ry	Ryukyu Arc
BH	Bird's Head	Mn	Manus Island	Sa	Sangihe Arc
Ca	Cagayan Arc	NB	New Britain Arc	Se	Sepik Arc
Fj	Fiji	NC	New Caledonia	So	Solomons Arc
Ha	Halmahera Arc	NH	New Hebrides Arc	Sp	Sula Platform
IB	Izu-Bonin Arc	NI	New Ireland	Su	Sulu Arc
Ja	Japan Arc	Nng	North New Guinea	TK	Three Kings
Lo	Loyalty Islands		Terranes		Rise
Lu	Luzon Arc	Pa	Papuan Ophiolite	To	Tonga Arc
Ma	Mariana Arc	Pk	Palau-Kyushu Ridge	Tu	Tukang Besi
					Platform

Figure 2. Present-day tectonic features of SE Asia and the SW Pacific. Light straight lines are selected marine magnetic anomalies and active spreading centres. White lines are subduction zones and strike-slip faults. Labelled filled areas are mainly arc, ophiolitic, and accreted material formed at plate margins during the Cenozoic, and submarine arc regions, hot spot volcanic products, and oceanic plateaus. Pale grey areas represent submarine parts of the Eurasian continental margins. Dark grey areas represent submarine parts of the Australian continental margins.

can lead to emergence but individual volcanoes can be very short-lived on a geological time scale (typically less than one million years) even though an arc may have been a long-lived feature. Within volcanic arc sequences there may be indicators of age and regions of submergence and emergence from such features as weathered horizons, palaesols, and marine sediment intercalations but it is usually not possible to delineate precisely which areas were emergent, simply that there are likely to have been such areas.

6 THE PLATE TECTONIC MODEL

The plate tectonic model outlined here is essentially that described by Hall (1996, 1997, 1998) and the reader is referred to those papers for details and references. Previous reconstructions which cover all or parts of the region discussed here include those of Crook & Belbin (1978), Hamilton (1979), Briais et al. (1993), Burrett et al. (1991), Daly et al. (1991), Lee & Lawver (1995), Rangin et al. (1990), and Yan and Kroenke (1993). Animations and maps relevant to the tectonics and distribution of land and sea in the region are available via the World Wide Web from *http://www.gl.rhul.ac.uk/seasia/welcome.html*. Here I summarise only the key features of the regional tectonic model and various aspects which are relevant to the development of the region of Wallacea. The principal features of the region are shown on Figure 2 and a series of global reconstructions in Figures 3, 4 and 5.

6.1 *55-45 Ma*

Before 50 Ma (Figure 3) the continents of India and Australia were on separate plates. India collided with Asia in the early Tertiary but the exact age of collision and its consequences remain controversial (e.g. Rowley 1996). Many of the tectonic events in SE Asia are commonly attributed to the effects of Indian indentation into Asia and the subsequent extrusion of continental fragments eastwards along major strike-slip faults. However, this hypothesis (Tapponnier et al. 1982, 1990) and its predictions of major clockwise rotations, southeastward extrusion of fragments, and timing of events remain poorly supported by geological evidence in SE Asia.

Taiwan, Palawan and the extended South China Sea margins formed a passive margin, established during the late Cretaceous. Sundaland was separated from Eurasia by a wide proto-South China Sea probably floored by Mesozoic ocean crust. The Malay peninsula was closer to Indochina and the Malay-Sumatra margin was closer to NNW-SSE. East Borneo and West Sulawesi were part of Sundaland underlain by accreted arc and ophiolitic material as well as small Gondwana fragments which were accreted during the Cretaceous. Java and West Sulawesi were situated above a subduction zone where Indian plate lithosphere was subducting towards the north. The Java subduction system linked east into Pacific intra-oceanic subduction zones which included parts of the east Philippines and Halmahera. There was a north-dipping subduction zone at the southern edge of a Northern New Guinea plate. This area is difficult to reconstruct because so much of the West Pacific has been eliminated by subduction since 50 Ma but there is good evidence that this area resembled the present-day West Pacific in containing marginal basins, intra-oceanic arcs and subduction zones.

Australia was essentially surrounded by passive margins on all sides. To the west the passive margin was formed in the Late Jurassic and there was oceanic crust separating a Bird's Head microcontinent from Australia. Further east, Indian and Australian oceanic lithosphere had been subducting northwards beneath the Sepik-Papuan arc in the early Tertiary. During the Paleocene and early Eocene the New Guinea passive margin collided with this intra-oceanic arc causing emplacement of the Sepik and Papuan ophiolites (Davies 1971). After this event the New Guinea margin remained a passive margin for most of the Paleogene. The Tasman and Coral Seas were both fully open by the beginning of the Tertiary, and the Loyalty Rise and New Caledonia Rise were extended parts of the east Australasian margin.

After India-Asia collision, India moved more slowly northwards and India and Australia became part of a single plate. Northward subduction of Indian-Australian oceanic lithosphere continued beneath the Sunda-Java-Sulawesi arcs. Rift basins formed throughout Sundaland, but the timing of their initial extension is uncertain because they contain continental clastics which are poorly dated, and their cause is therefore also uncertain.

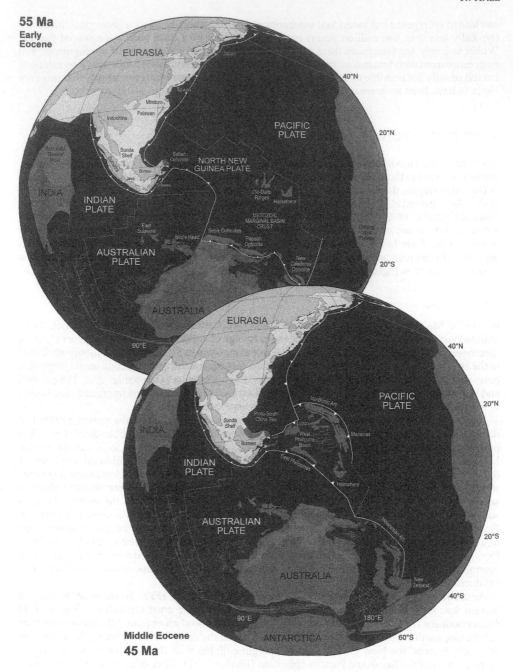

Figure 3. Reconstructions of the region at 55 and 45 Ma. The possible extent of Greater India and the Eurasian margin north of India are shown schematically. This was beginning of the period of collision between India and Asia, and between the north Australian continental margin and Pacific intra-oceanic island arcs which emplaced ophiolites on the north New Guinea margin, and later in New Caledonia. An oceanic spreading centre through the West Philippine basin, the Celebes Sea and the north Makassar Strait developed the deep water rift which became Wallace's Line.

The Java-Sulawesi subduction system continued into the West Pacific through the east Philippines and Halmahera arcs. During the Eocene the extended eastern Australasian passive margin had collided with an intra-oceanic arc resulting in emplacement of the New Caledonia ophiolite (Aitchison et al. 1995) followed by subduction polarity reversal. This led to the formation of a Melanesian arc system. Soon after 45 Ma south to southwest-directed subduction began beneath the eastern Australian margin, from Papua New Guinea to north of New Zealand, with major arc growth producing the older parts of the New Britain, Solomons and Tonga-Kermadec systems, leading to development of major marginal basins in the SW Pacific whose remnants probably survive only in the Solomon Sea.

Subduction of the Pacific-Northern New Guinea mid-ocean ridge led to massive outpouring of boninitic volcanic rocks (Stern & Bloomer 1992) which formed the Izu-Bonin-Mariana arc system, and the Philippine Sea plate became a recognisable entity. There was significant rotation of the Philippine Sea plate between 50 and 40 Ma and the motion history of this plate (Hall et al. 1995b) provides an important constraint on development of the eastern part of SE Asia. The West Philippine Basin, Celebes Sea, and Makassar Strait opened as single basin within the Philippine Sea plate. The opening of the West Philippine-Celebes Sea basin caused initiation of southward subduction of the proto-South China Sea beneath Luzon and the Sulu arc. It is this subduction which caused renewed extension along the South China margin, driven by slab-pull forces due to subduction between eastern Borneo and Luzon, and later led to sea-floor spreading in the South China Sea, rather than indentor-driven tectonics.

6.2 35-25 Ma

From 40-30 Ma (Figure 4) Indian ocean subduction continued at the Sunda-Java trenches, and also beneath the arc extending from Sulawesi through the east Philippines to Halmahera. Sea floor spreading continued in the West Philippine-Celebes Sea basin until about 34 Ma. By 30 Ma the Caroline Sea was widening above a subduction zone at which the newly-formed Solomon Sea was being destroyed as the Melanesian arc system migrated north. To the south of the Caroline Sea the South Caroline arc formed what later became the north New Guinea arc terranes. The backarc basins in the SW Pacific were probably very complex, as indicated by the anomalies in the South Fiji Basin, and will never be completely reconstructed because most of these basins have been subducted.

Within Sundaland deformation was complex and a plate tectonic model can only simplify the tectonics of the region by considering large and simple block movements and broadly predicting regional stress fields. In northern Indochina strike-slip motion was important (Wang and Burchfiel 1997) but deformation was not concentrated at the edge of rigid blocks. The Malay and Gulf of Thailand basins may have a significant component of strike-slip movement on faults controlling their development. However, they may have been initiated in a different tectonic setting, and in a region with an older structural fabric which influenced their development.

The period from 30-20 Ma saw the most important Cenozoic plate boundary reorganisation within SE Asia. At about 25 Ma, the New Guinea passive margin collided with the leading edge of the east Philippines-Halmahera-New Guinea arc system. The Australian margin, in the Bird's Head region, was also close to collision with the Eurasian margin in West Sulawesi and during this interval ophiolite was emplaced in SE Sulawesi. Soon afterwards the Ontong Java plateau collided with the Melanesian arc. These two major collisions caused a significant change in the character of plate boundaries in the region in the early Miocene. They linked the island arcs of Melanesia, the New Guinea terranes at the southern Caroline margin, and the Halmahera-Philippines arcs. This linkage seems to have coupled the Pacific to the marginal basins of the West Pacific, and the Caroline and Philippine Sea plates were subsequently driven by the Pacific.

Advance of the Melanesian arc system led to widening of the South Fiji basin and Solomon Sea basin (now mainly subducted). At the Three Kings Rise subduction seems to have been initiated soon after ocean crust was formed to the east, allowing the rise to advance east and spreading to propagate behind the rise into the Norfolk basin from a triple junction to the north.

The Caroline and Philippine Sea plates began to rotate, almost as a single plate, and the Izu-Bonin-Mariana trench system rolled back into the Pacific. Rifting of the Palau-Kyushu ridge began,

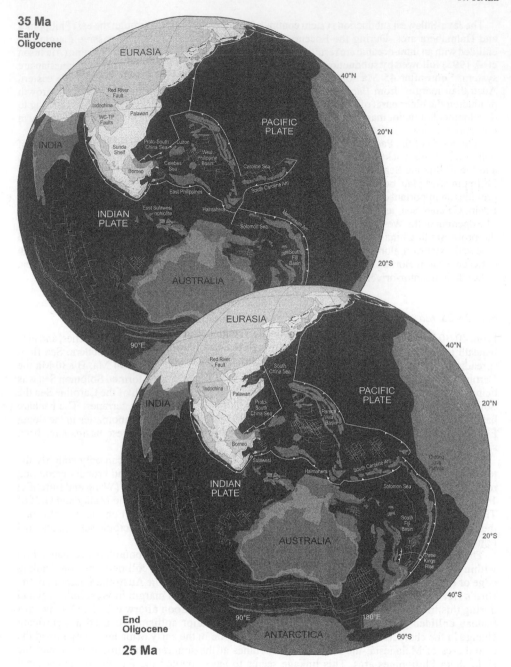

Figure 4. Reconstructions of the region at 35 and 25 Ma. India and Australia were parts of the same plate. Multiple arc systems extended from the Sundaland margin into the west Pacific, including the east Philippines-Halmahera arc, the Izu-Bonin-Mariana arc, and the South Caroline arc. Spreading also began after subduction flip in marginal basins around eastern Australasia producing the Solomon Sea and the island arcs of Melanesia. Slab pull due to southward subduction of the proto-South China Sea caused extension of the South China and Indochina continental margin and the present South China Sea began to open. By 25 Ma the east Philippines-Halmahera-South Caroline arc collided with the Australian margin and the Ontong Java plateau began to collide with the Melanesian arc. These two events caused major reorganisation of plate boundaries.

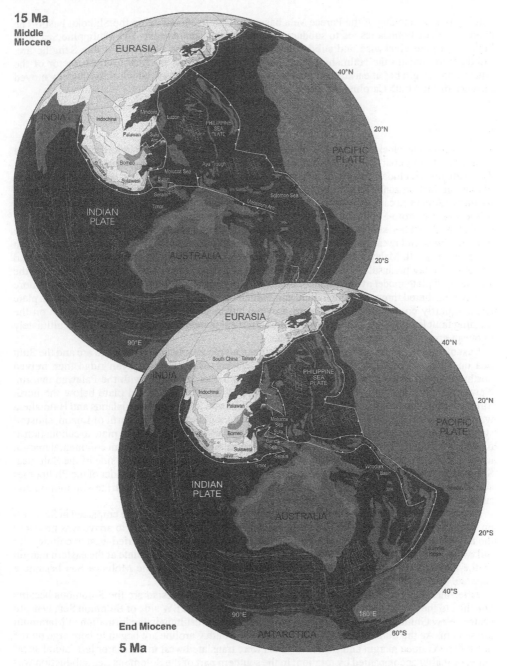

Figure 5. Reconstructions of the region at 15 and 5 Ma. The north Australian margin became a major left-lateral strike-slip system as the Philippine Sea-Caroline plate began to rotate clockwise with the Pacific. Movement on splays of the Sorong fault system led to the collision of Australian continental fragments in Sulawesi. This in turn led to counter-clockwise rotation of Borneo and parts of Sundaland, eliminating the proto-South China Sea. As the old oceanic lithosphere off NW Australia began to subduct extension in the overriding plate led to formation of deep water basins of the Banda Sea. The New Guinea terranes, formed in the South Caroline arc, docked in New Guinea but continued to move in a wide left-lateral strike-slip zone. The Solomon Sea was largely eliminated by subduction beneath eastern new Guinea and the New Hebrides arc but subduction there also led to development of new marginal basins within the last 10 Ma, including the Bismarck Sea, Woodlark basin, North Fiji basins, and Lau basin.

leading first to opening of the Parece Vela basin and later to spreading in the Shikoku basin. The change in plate boundaries led to subduction beneath the Asian margin. The Philippine Sea plate began to rotate clockwise and subduction began beneath north Sulawesi in the Sangihe arc. Subduction beneath the Halmahera-Philippines arc ceased and the New Guinea sector of the Australian margin became a strike-slip zone, the Sorong Fault system, which subsequently moved terranes of the South Caroline arc along the New Guinea margin.

6.3   15-5 Ma

After 20 Ma the clockwise rotation of the Philippine Sea plate necessitated changes in plate boundaries throughout SE Asia which resulted in the tectonic pattern recognisable today (Figure 5). These changes include the re-orientation of spreading in the South China Sea, and the development of new subduction zones at the eastern edge of Eurasia and in the SW Pacific. Continued northward motion of Australia caused the counter-clockwise rotation of Borneo. The remaining oceanic crust of the western proto-South China Sea, and thinned continental crust of the passive margin to the north, was thrust beneath Borneo. The rotation of Borneo was accompanied by counter-clockwise motion of west Sulawesi, and smaller counter-clockwise rotations of adjacent Sundaland blocks. In contrast, the north Malay peninsula rotated clockwise, but remained linked to both Indochina and the south Malay peninsula. This allowed widening of basins in the Gulf of Thailand although the simple rigid plate model overestimates extension in this region. This extension was probably more widely distributed throughout Sundaland and Indochina on many different faults. The Burma plate became partly coupled to the northward-moving Indian plate and began to move north on the Sagaing fault leading to stretching of the Sunda continental margin north of Sumatra, and ultimately to ocean crust formation in the Andaman Sea.

East of Borneo, the increased rate of subduction caused arc splitting in the Sulu arc and the Sulu Sea opened as a back-arc basin south of the Cagayan ridge. The Cagayan ridge then moved northwards, eliminating the eastern proto-South China Sea, to collide with the Palawan margin. New subduction had also begun at the west edge of the Philippine Sea plate below the north Sulawesi-Sangihe arc which extended north to south Luzon. The Philippine islands and Halmahera were carried with the Philippine Sea plate towards this subduction zone. North of Luzon, sinistral strike-slip movement linked the subducting west margin of the Philippine Sea plate to subduction at the Ryukyu trench. Collision of Luzon and the Cagayan ridge with the Eurasian continental margin in Mindoro and north Palawan resulted in a jump of subduction to the south side of the Sulu Sea. Southward subduction beneath the Sulu arc continued until 10 Ma. The remainder of the Philippines continued to move with the Philippine Sea plate, possibly with intra-plate strike-slip motion and subduction resulting in local volcanic activity.

As a result of changing plate boundaries fragments of continental crust were emplaced in Sulawesi on splays at the western end of the Sorong Fault system. The first of these to arrive was probably the SE Sulawesi fragment. Later, the Buton-Tukang Besi platform was carried west to collide with Sulawesi. Locking of splays of the Sorong fault caused subduction to initiate at the eastern margin of the Molucca Sea, producing the Neogene Halmahera arc. Thus the Molucca Sea became a separate plate as the double subduction system developed.

After the collision of the Ontong Java plateau with the Melanesian arc the Solomons became attached to the Pacific plate. Westward subduction began on the SW side of Solomon Sea, beneath eastern New Guinea, eliminating most of Solomon Sea and resulting in the formation of Maramuni arc system. As the Solomon Sea was eliminated the South Caroline arc began to converge on the north New Guinea margin and the arc terranes were translated west in the major left-lateral shear zone, probably accompanied by rotation. In the southern part of the Solomons Sea subduction was in the opposite direction (eastward) and created the New Hebrides arc system.

By 10 Ma SE Asia was largely recognisable in its present form. Rotation of Borneo was complete. This, with collision in the central Philippines and Mindoro, and continued northward movement of Australia, resulted in reorganisation of plate boundaries and intra-plate deformation in the Philippines. The Luzon arc came into collision with the Eurasian margin in Taiwan. Subduction continued at the Manila, Sangihe and Halmahera trenches, and new subduction began at the Negros and Philippine trenches. These subduction zones were linked by strike-slip systems active within the Philippines and this intra-plate deformation created many very small fragments which are difficult to describe using rigid plate tectonics. In west Sundaland, partitioning of convergence in Sumatra into orthogonal subduction and strike-slip motion effectively established one or more Sumatran forearc sliver plates. Extension on the strike-slip system linked to the spreading centre in the Andaman Sea (Curray et al. 1979).

The Molucca Sea continued to close by subduction on both sides. At present the Sangihe forearc has overridden the northern end of the Halmahera arc, and is beginning to over-thrust west Halmahera. In the Sorong fault zone, accretion of Tukang Besi to Sulawesi locked a strand of the fault and initiated a new splay south of the Sula platform. The Sula platform then collided with the east arm of Sulawesi, causing rotation of the east and north arms to their present position, leading to southward subduction of the Celebes Sea at the north Sulawesi trench.

The Eurasia-Philippine Sea plate-Australia triple junction was, and remains, a zone of microplates but within this contractional setting new extension began in the Banda Sea. The Bird's Head moved north relative to Australia along a strike-slip fault at the Aru basin edge. Mesozoic ocean crust north of Timor was eliminated at the eastern end of the Java trench by continued northern motion of Australia which brought the Australian margin into this trench as the volcanic inner Banda arc propagated east. Seram began to move east requiring subduction and strike-slip motion at the edges of this microplate. Since 5 Ma the southern Banda Sea has extended to its present dimensions and continental fragments are now found in the Banda Sea ridges within young volcanic crust.

North of the Bird's Head, and further east in New Guinea, transpressional movements were marked by deformation of arc and ophiolite slivers separated by sedimentary basins. Progressive westward motion of the South Caroline arc within the left-lateral transpressional zone led to docking of the north New Guinea terranes. This caused cessation of southward subduction of the Solomon Sea plate but resulted in its northward subduction beneath New Britain. The New Britain subduction led to rapid spreading in Woodlark basin as a consequence of slab-pull forces and rapid ripping open of continental crust beneath the Papuan peninsula. Elimination of most of the remaining Solomons marginal basin by eastward subduction led to formation of the New Hebrides arc and opening of the North Fiji basins.

7 LAND AND SEA IN WALLACEA

The geological and biogeographic interface of interest here is that between Sundaland and Australia, extending from Borneo to the Bird's Head of New Guinea (Figure 6). This area, separating Borneo and New Guinea, and including Sulawesi, the Banda Sea and the Moluccas, encapsulates many of the problems of the region. Figures 7 to 10 compile the general features of land and sea onto maps of the tectonic reconstructions for the region of Wallacea. The maps help to indicate the likely geographical connections and barriers and the periods when these were in existence. The period 30-0 Ma is of most interest to biogeographers since before then the separation between Asia and Australia was greater and for almost all land plants and animals it was probably not possible to cross this barrier. Essentially since 30 Ma (Figure 6) there has been a closure of the marine gap, and collision of the Sula Spur-Bird's Head microcontinental area with the eastern Sundaland margin. However, despite the continued convergence between the principal plates and the movement of fragments of continental crust into Sulawesi, at the same time there has been the opening of new deep ocean basins maintaining a difficult and indirect migration route between Australia and Asia.

In the west, Borneo formed part of Sundaland throughout the Cenozoic. Sundaland was mainly emergent, or intermittently transected by very shallow seas, and would presumably have been biogeographically linked to Asia for the whole of this period. Opening of the South China Sea, Celebes and Sulu Seas from the Eocene onwards had formed deep water barriers to the north and east of Borneo (Figure 6). Thus, from the Eocene, the Makassar Straits was the major barrier to the east because, although west Sulawesi was always close to Borneo it was largely submerged until at least the late Miocene. In the Middle Oligocene, about 30 Ma, there was still a deep oceanic gap between Sundaland and Australia (Figures 6 and 7). There must have been deep trenches along the eastern Sundaland margin extending into the west Pacific. There was certainly a lot of deep water. Virtually the only evidence for any land in the Sula Spur indicates a small emergent area on the island of Buru. In areas of volcanicity there is always the possibility there might have

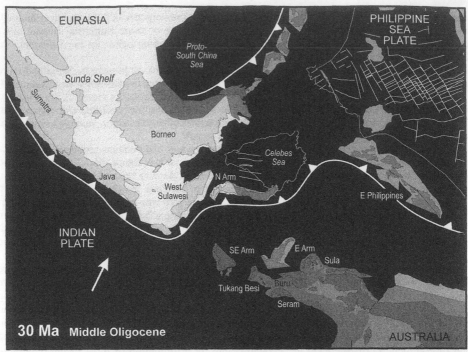

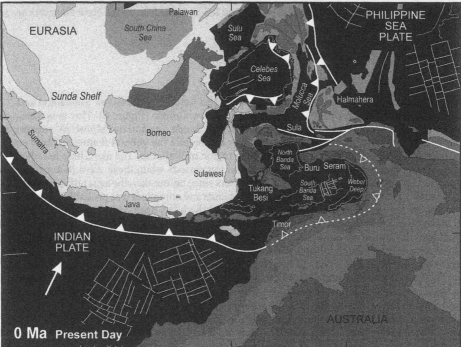

Figure 6. Reconstruction of Wallacea at 30 Ma and the present geological configuration of the area. The past 30 Ma has seen the elimination of oceanic lithosphere between Australia and Sundaland but the creation of the young deep basins of the north and south Banda Sea and the Weber deep. At the same time, Pacific terranes have moved by strike-slip movements along the north New Guinea margin into the Moluccas and Philippines, and deep marine parts of the Australian margin emerged from the sea to form high mountains in the islands of the outer Banda arc.

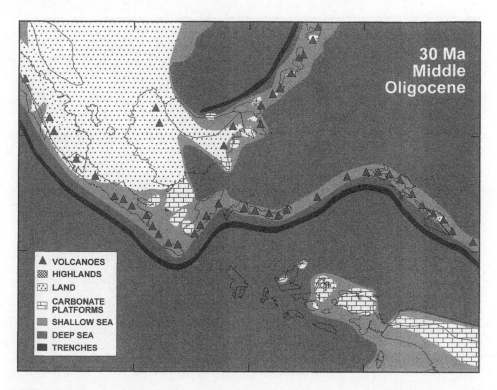

Figure 7. Postulated distribution of land and sea in the region of Wallacea at 30 Ma. Note that on these and subsequent maps modern coastal outlines are used for reference. Some coastal outlines only appear on some maps during the period 30-0 Ma, reflecting crustal growth, for example in the Sunda-Banda arcs. Volcanoes are shown schematically to indicate positions of arcs.

been ephemeral land that could have provided some connection between Sundaland and Australia but the possibility is very low; there is no significant volcanicity in the Sula Spur-Bird's Head area.

From the Early Miocene (Figure 8), mountains rose in Borneo, possibly as high as those now in New Guinea, expanding the area of land, and large deltas built out rapidly into the surrounding deep basins. However, the Makassar Straits remained wider than at present, with a very deep water central area and wide marine shelves, and was therefore the eastern limit of Asian floras and faunas. Recent work in west Sulawesi by the SE Asia Research Group indicates that emergence of land and uplift of mountains was quite recent (late Miocene or later) and rapid. There was no direct way of crossing between Borneo and west Sulawesi. However, the distribution of shallow marine carbonates, and the depths of water of the Sunda shelf, suggest there were always routes from Borneo via Java into Sulawesi, by way of other small islands, although west Sulawesi may itself have been little more than islands until the Pliocene. From the early Miocene there is good evidence for emergence in SE Sulawesi, but in western Sulawesi there is very little evidence of any land, in fact quite the contrary, there is good evidence of continuing marine deposition throughout much of west Sulawesi. So although the tectonic maps indicate that the straight line distances from Borneo to the Bird's Head were not much greater than at present there was probably very little land that might have provided a connection from Borneo to northern Australia. Even at 15 Ma ago the same situation applies. It is important to note that the evidence from the Miocene of Sulawesi particularly, and other parts of the Sula Spur-Bird's Head region, is often relatively poor, because later erosion has removed important parts of the stratigraphic record and because the younger clastic sequences are often difficult to date. However, it is also true that there is very little positive evidence for land throughout most of the area, in particular there is an absence of evidence for the extensive erosional products that would be expected had much of Sulawesi been mountains during

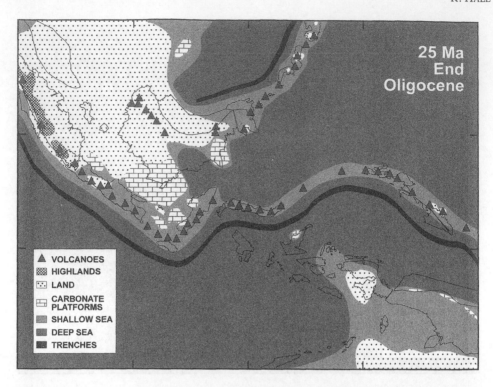

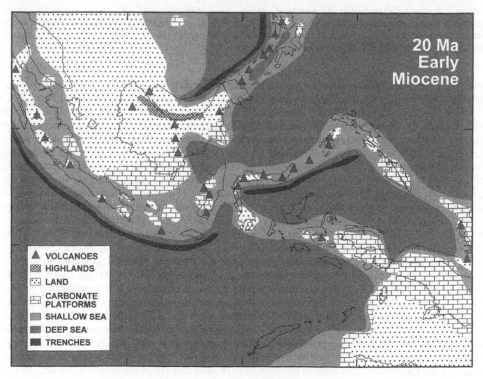

Figure 8. Postulated distribution of land and sea in the region of Wallacea at 25 Ma and 20 Ma.

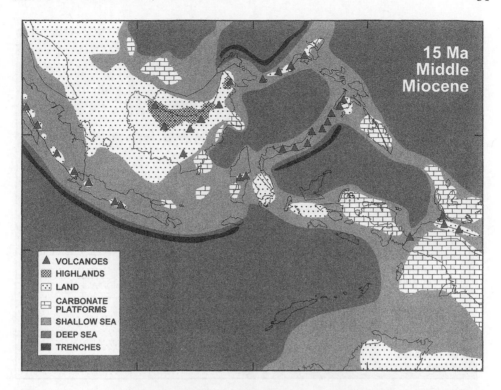

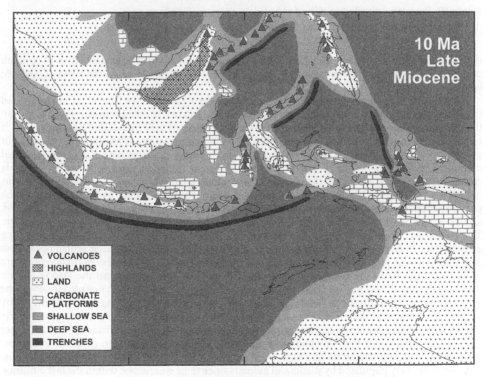

Figure 9. Postulated distribution of land and sea in the region of Wallacea at 15 Ma and 10 Ma.

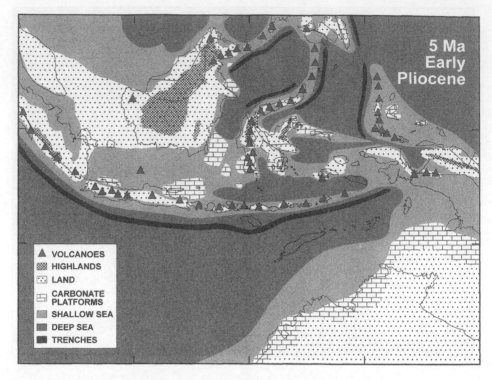

Figure 10. Postulated distribution of land and sea in the region of Wallacea at 5 Ma.

the early and middle Miocene, and there is considerable evidence for marine deposition over much of the region. I believe that the maps of Figures 7 to 10 are generous in assessing areas of possible land and shallow sea.

By about 10 Ma (Figure 9) the Australia-Sundaland gap seems to have been at its narrowest and the areas of possible land were relatively extensive. The Makassar Strait was still fairly wide but there is at that time the first good evidence for the emergence of land in much of Sulawesi. This is rather later than most previous workers suggest. This interpretation is partly a consequence of going carefully through the existing literature and also research in progress (S. J. Calvert, personal communication 1999) on the eastern side of the Makassar Straits in western Sulawesi which is dating sequences that are definitely continental and which are much younger than expected. It was not until 5 Ma (Figure 10) that there was substantial land in Sulawesi but by that time one of the pathways that may have existed previously which may have offered a Sundaland link into Sulawesi, started to be broken up because of the formation of the deep water basins in the Banda Sea region. The Banda basins probably opened in the last 10 million years by very rapid extension during convergence of Australia and Sundaland induced by roll-back of the subducting Indian ocean slab as the Java trench propagated east since the late Miocene. During the last few million years there have been significant movements of continental fragments into and around the Banda Sea on splays of the left-lateral Sorong fault system and local collisions and uplift as a result. However, the uplift has been accompanied by extension, partly driven by strike-slip faulting and partly driven by subduction forces and therefore deep water barriers have appeared as older ones disappeared. As these areas became deeper due to extension, mountains rose in Seram and Timor elevating former deep water deposits of the Australian margin. With the possible exception of small overthrust fragments of the Sundaland margin now found on Timor, the islands of the outer Banda arc must have been entirely populated by plants and animals since their emergence within the last 5 million years. During the same period the north Moluccan islands arrived from the east with Pacific island

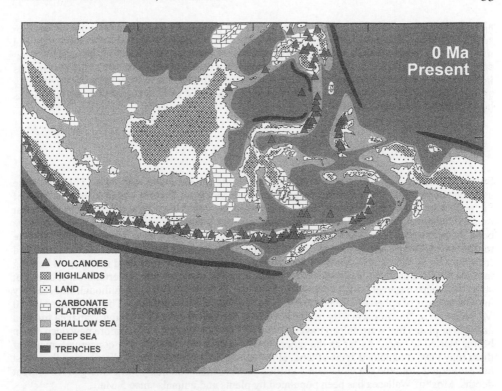

Figure 11. Simplified present day distribution of land and sea in the region of Wallacea for comparison with the palaeogeographic maps of earlier periods. Even a sea level fall of about 200m, the maximum probable fall during the Quaternary due to polar ice cap growth, would not provide a complete continuous link between the Australian and Sundaland continental margins.

arc fragments. They moved along the north New Guinea margin, remaining close to it at all times, and providing possible pathways for migration of Australian faunas and floras onto volcanic islands of the Halmahera arc. Thus by a strange irony, despite the convergence of Australia and Sundaland, geological processes have maintained the barriers to the mixing of Asian and Australian floras and faunas (Figure 11). There seems to have been no time when land plants and animals would have been able to avoid crossing water gaps in order to move between the two continents.

8 EFFECTS OF SEA LEVEL CHANGE

According to long term global sea level curves we currently live in a period of relatively low sea level but sea level has been much lower and higher during the very recent past due to melting and freezing of ice caps. Although there is now broad agreement on trends of Cenozoic changes in sea level there is still disagreement about the magnitude of eustatic (global sea level) changes (e.g. Haq et al. 1987; Kominz et al. 1998). It is currently not possible to be very precise about water depths on the maps presented here, and consequently assessing the effects of global sea level change is very difficult. Distinguishing the effects of global and tectonic contributions to sea level change is particularly problematical in tectonically active regions like SE Asia and the SW Pacific. However, as noted above, I believe that I have been generous in assessing the extents of land and shallow seas in these regions. I consider the boundary between shallow and deep water areas on the maps to approximate the 500 m isobath. If sea level did fall by 200 m, the maximum relative fall advocated for a single event in the Cenozoic with the exception of the Quaternary (Haq et al. 1987), it might have been possible to establish a short-lived land connection between Sundaland and Australia, although this seems unlikely. At other times before the Quaternary no such continuous connection seems possible. During the Quaternary, known sea level changes would at times have exposed most of the Sunda and Sahul-Arafura Shelves, as well as inducing significant biogeographic side-effects such as reduction in the areas of rain forest, but would never have allowed a continuous land link between Sundaland and Australia. I therefore consider this to have been the most likely situation during the earlier period of the Cenozoic, i.e. there were never continuous land links between Sundaland and Australia.

9 IMPLICATIONS

Sulawesi is an area of obvious interest because of its special fauna and flora. There is some evidence that at 20 Ma the SE parts of the island were emergent. There is a possibility that parts of the north arm of Sulawesi, where there were volcanoes, could have been emergent by then, but most of Sulawesi, certainly western Sulawesi, was not emergent at that time. There may have been the occasional island there but there could not have been much land. There is a good marine record during that period and it was not until much more recently that western Sulawesi emerged from the sea. By about 10–5 Ma there was probably a significant area of land in Sulawesi bearing in mind all the qualifications made earlier about negative evidence and uncertainties. It does not seem that there has ever been a continuous land link to northern Australia although it there may have been areas of ephemeral land which may have allowed island hopping. The period at about 10 Ma seems to be the time when there may have been the best chance of crossing Wallacea for land animals or plants that were able to traverse relatively narrow marine areas, before new deep ocean basins started to open. As far as the other smaller islands of Wallacea are concerned, many of them may have maintained areas of land, albeit ephemeral and changing in distribution, but most have emerged from the sea in the last 5 million years at most and many have emerged from great depths. Most of Wallacea has been populated by plants and animals since 5 Ma.

At the eastern end of this interface, northern New Guinea includes many fault-bounded terranes which accreted to the Australian margin during the late Cenozoic. Its mountains also emerged from the sea rapidly and very recently. New Guinea therefore provided wonderful opportunities for newly arrived plants and animals, in a climatic setting in which high diversity was encouraged. The rise of mountains probably provided a large range of new niches, at the same time forming new physical barriers, while their rapid rise may well have modified atmospheric circulation patterns contributing to drier climates in the Australian continent, and consequently forming climatic barriers to plant and animal movements. Geological processes contributed to biogeographic patterns by forming land and influencing climate but probably not by rafting unique biotas.

I summarised above some reservations about simplistic interpretations of geological data to explain biogeographic patterns. New Caledonia serves to remind us all of some fundamental geological and biogeographic problems in the region (Keast 1996). In New Caledonia there is apparently an ancient Gondwana flora, and other strange features of the flora and fauna seem to imply some land there since the late Cretaceous. On the other hand it is difficult to find any geological evidence that New Caledonia was above sea level until the late Eocene. How do we resolve this dilemma? I suggest that it indicates that we should remain cautious about apparently simple and definitive answers to our geological and biogeographic problems and we should all remain critical of our data and beliefs. The biogeographic patterns we observe today are the product of many factors, and geology, although fundamental, is only one of many important controls. It is clear from the geology of the region that the snapshot we see today is no less complicated than in the past. It is also clear that our geological data set is still not adequate to deal with the many questions we wish to answer. On the other hand, geology does have a historical record in the forms of fossils and rocks whereas many biogeographic patterns, and often those most enthusiastically interpreted, are nothing but present-day distributions which can be interpreted in numerous different ways. Molecular studies may in future provide a better historical record.

Since the early Miocene Australia and Sundaland have moved closer together but as land emerged and mountains rose in some areas, new deep marine basins developed. As these geologically-controlled changes occurred, oceanic and atmospheric circulation patterns changed, partly as the result of the closure of the Indo-Pacific seaway, and a host of new habitats were created. The distribution of Australian and Asian plants and animals should therefore reflect this complexity, with further important modifications imposed by glacially-related sea level and climatic change in the Quaternary. The zone of Wallacea is partly an ancient deep water barrier, partly a dynamic boundary marking a migration front, but also a relic of Neogene patterns which have been tectonically disrupted and modified by Quaternary climate change. Like the geology, the present biogeographic patterns need to be viewed as one image in a rapidly-changing scene which is still very far from achieving equilibrium.

ACKNOWLEDGEMENTS

Financial support for my work has at various times been provided by NERC, the Royal Society, the London University Central Research Fund, and the SE Asia Research Group. I thank the many geologists and biogeographers who have helped me in trying to understand this exciting region. I particularly thank Ian Metcalfe for his support which enabled me to attend the *Where Worlds Collide* conference.

REFERENCES

Aitchison, J.C., G.L. Clarke, S. Meffre & D. Cluzel 1995. Eocene arc-continent collision in New Caledonia and implications for regional Southwest Pacific tectonic evolution. *Geology* 23: 161-164.

Briais, A., P. Patriat & P. Tapponnier 1993. Updated interpretation of magnetic anomalies and sea floor spreading stages in the South China Sea: Implications for the Tertiary tectonics of Southeast Asia. *Journal of Geophysical Research* 98: 6299-6328.

Burrett, C., N. Duhig, R. Berry & R. Varne 1991. Asian and south-western Pacific continental terranes derived from Gondwana, and their biogeographic significance. In: Ladiges, P.Y., Humphries, C.J. & Martinelli, L.W. (Eds.), *Austral Biogeography. Australian Systematic Botany*. 4: 13-24.

Crook, K.A.W. & L. Belbin 1978. The Southwest Pacific area during the last 90 million years. *Journal of the Geological Society of Australia* 25: 23-40.

Curray, J.R., D.G. Moore, L.A. Lawver, F.J. Emmel, R.W. Raitt, M. Henry & R.M. Kieckhefer 1979. Tectonics of the Andaman Sea and Burma. In: Watkins, J., Montadert, L. & Dickenson, P.W. (Eds.), *Geological and Geophysical Investigations of Continental Margins, American Association of Petroleum Geologists Memoir* 29: 189-198.

Daly, M.C., M.A. Cooper, I. Wilson, D.G. Smith & B.G.D. Hooper 1991. Cenozoic plate tectonics and basin evolution in Indonesia. *Marine and Petroleum Geology* 8: 2-21.

Davies, H.L. 1971. Peridotite-gabbro-basalt complex in eastern Papua: an overthrust plate of oceanic mantle and crust. *Bureau of Mineral Resources, Geology and Geophysics Bulletin* 128: 48 pp.

Hall, R. 1996. Reconstructing Cenozoic SE Asia. In: Hall, R. & Blundell, D.J. (Eds.), *Tectonic Evolution of SE Asia. Geological Society of London Special Publication* 106: 153-184.

Hall, R. 1997. Cenozoic tectonics of SE Asia and Australasia. In: Howes, J.V.C. & Noble, R.A. (Eds.), *Proceedings of the International Conference on Petroleum Systems of SE Asia and Australia*. Indonesian Petroleum Association: 47-62.

Hall, R. 1998. The plate tectonics of Cenozoic SE Asia and the distribution of land and sea. In Hall, R. & Holloway, J. D. (Eds.), *Biogeography and Geological Evolution of SE Asia*. Backhuys Publishers, Leiden, The Netherlands: 99-131.

Hall, R. 2000. Neogene history of collision in the Halmahera region, Indonesia. *Proceedings Indonesian Petroleum Association, 27th Annual Convention*: 487-493.

Hall, R., J.R. Ali, C.D. Anderson & S.J. Baker 1995a. Origin and motion history of the Philippine Sea Plate. *Tectonophysics* 251: 229-250.

Hall, R., M. Fuller, J.R. Ali & C.D. Anderson 1995b. The Philippine Sea Plate: magnetism and reconstructions. In: Taylor, B. & Natland, J. (Eds.), *Active Margins and Marginal Basins of the Western Pacific American Geophysical Union, Geophysical Monograph* 88: 371-404.

Hamilton, W. 1979. Tectonics of the Indonesian region. *USGS Professional Paper* 1078: 345 pp.

Haq, B.U., J. Hardenbol & P.R. Vail 1987. Chronology of fluctuating sea levels since the Triassic. *Science* 235: 1156-1167.

Houseman, G. & P. England 1993. Crustal thickening versus lateral expulsion in the Indian-Asian continental collision. *Journal of Geophysical Research* 98: 12,233-12,249.

Kearey, P. & F.J. Vine 1990. *Global Tectonics*. Blackwell Scientific Publications: 302 pp.

Keast, A. 1996. Pacific biogeography: patterns and processes. In: Keast, A. & Miller, S. E. (Eds.) *The origin and evolution of Pacific Island Biotas: New Guinea to Eastern Polynesia: Patterns and Processes.* SPB Academic Publishing, Amsterdam, The Netherlands: 477-512.

Kominz, M.A., K.G. Miller & J.V. Brown 1998. Long-term and short-term Cenozoic sea-level estimates. *Geology* 26: 311-314.

Lee, T.Y. & L.A. Lawver 1995. Cenozoic plate reconstruction of southeast Asia. *Tectonophysics* 251: 85-139.

Metcalfe, I. 1998. Palaeozoic and Mesozoic geological evolution of the SE Asian region: multidisciplinary constraints and implications for biogeography. In: Hall, R. & Holloway, J. D. (Eds.), *Biogeography and Geological Evolution of SE Asia.* Backhuys Publishers, Leiden, The Netherlands: 25-41.

Peltzer, G. & P. Tapponnier 1988. Formation and evolution of strike-slip faults, rifts, and basins during the India-Asia collision: an experimental approach. *Journal of Geophysical Research* 93: 15085-15117.

Rangin, C., L. Jolivet & M. Pubellier 1990. A simple model for the tectonic evolution of southeast Asia and Indonesia region for the past 43 m.y. *Bulletin de la Société géologique de France* 8: 889-905.

Rowley, D.B. 1996. Age of initiation of collision between India and Asia: a review of stratigraphic data. *Earth and Planetary Science Letters* 145: 1-13.

Stern, R.J. & S.H. Bloomer 1992. Subduction zone infancy: Examples from the Eocene Izu-Bonin-Mariana and Jurassic California arcs. *Geological Society of America Bulletin* 104: 1621-1636.

Tapponnier, P., G. Peltzer & R. Armijo 1986. On the mechanics of the collision between India and Asia. In: Coward, M. P. & Ries, A. C. (Eds.) *Collision Tectonics. Geological Society of London Special Publication* 19: 115-157.

Tapponnier, P., G. Peltzer, A.Y. Le Dain, R. Armijo & P. Cobbold 1982. Propagating extrusion tectonics in Asia: new insights from simple experiments with plasticine. *Geology* 10: 611-616.

Wallace, A.R. 1869. *The Malay archipelago.* Dover, New York.

Wang, E. & B.C. Burchfiel 1997. Interpretation of Cenozoic tectonics in the right-lateral accommodation zone between the Ailao Shan shear zone and the eastern Himalayan syntaxis. *International Geology Review* 39: 191-219.

Yan, C.Y. & L.W. Kroenke 1993. A plate tectonic reconstruction of the SW Pacific 0-100 Ma. In: Berger, T., Kroenke, L.W., Mayer, L. et al., *Proceedings of the Ocean Drilling Program, Scientific Results* 130: 697-709.

Section 2

Palaeozoic and Mesozoic geology and biogeography

Cambrian to Permian conodont biogeography in East Asia-Australasia

Robert S. Nicoll
Department of Geology, Australian National University, Canberra, ACT 0200, Australia

Ian Metcalfe
Asia Centre, University of New England, Armidale, NSW 2351, Australia

ABSTRACT: Conodont faunas of the allochthonous East Asian terranes demonstrate bio-geographic affinities with Australasia during the Cambrian to Permian and suggest close proximity or attachment to Australian Gondwanaland in the Palaeozoic over a time frame of about 250 My, from about 500 Ma in the Late Cambrian to the end of the Permian at about 253 Ma. Lower Palaeozoic conodonts that define an East Asia-Australasia province include the Late Cambrian genus *Eodentatus* and Ordovician taxa such as *Serratognathus* spp., *Aurilobodus* spp., *Plectodina onychodonta* and *Tasmanognathus*. The distinctive Lower to Middle Silurian conodont genera *Tuberocostadontus* and *Nericodus* are also restricted to East Asia and Australasia. Devonian conodont faunas appear to be more cosmopolitan, which may be related to global oceanic current distribution patterns. At the same time endemic forms are common, but species restricted to East Asia – Australasia have not been widely recognised. Lower Carboniferous conodont faunas of the Asian terranes and Western Australia are generally cosmopolitan. However, the distribution of *Mestognathus beckmanni* appears restricted to Gondwanaland (including the Sibumasu terrane), and Laurentia (including the allochthonous terranes of N. America). Eastern Australian conodont faunas are highly endemic during the Visean and are dominated by the genus *Montognathus*, which has also recently been discovered on the Sibumasu terrane. Conodonts appear to be absent from the Permian of eastern Australia due to the cold conditions present there as a result of high latitudinal position and Carbo-Permian glaciation. Conodonts were also thought, until recently, to be absent from Western Australia due to cold climatic conditions. However, scarce low diversity conodont faunas are now known from Western Australia and these are characterized and dominated by the cold-water tolerant genus *Vjalovognathus* which invaded the shallow continental margin seas of Gondwanaland during periods of climatic amelioration. *Vjalovognathus* is also recorded from Timor and the Pamirs and defines a Tethyan-Gondwanaland conodont province during the Permian.

1 INTRODUCTION

Conodonts are an extinct group of small eel-shaped animals with chordate affinities (Aldridge et al., 1986; Pridmore et al., 1997). These animals are known almost exclusively from the small phosphatic "tooth-like" elements which are the only hard parts of the conodont animal. Normally, 15 different elements form an oral feeding apparatus in each individual animal and these conodont elements are utilized extensively for dating marine strata of Cambrian to Triassic ages. Preservation of the soft parts of the conodont animal is rare and was essentially unknown until the 1980s when unequivocal soft impressions of the conodont animal were reported from the Carboniferous Granton Shrimp Band in Scotland (Briggs et al., 1983; Aldridge et al., 1986). The conodont animal was restricted to holomarine environments and exhibited water depth-related biofacies. It was generally restricted to tropical and temperate warm marine waters but some cold water-tolerant forms are known. The group as a whole tends to be cosmopolitan, hence its extreme biostratigraphic value, but some shallow-marine conodont biofacies and temperature-restricted biotas display provinciality that can be useful for palaeobiogeographic input to global tectonic models (Bergström, 1990; Mei et al., 1999). We here present available information on Palaeozoic conodont biogeography for East Asia-Australasia and its implications for the tectonic evolution and palaeogeographic reconstructions of the region.

2 TECTONIC SETTING

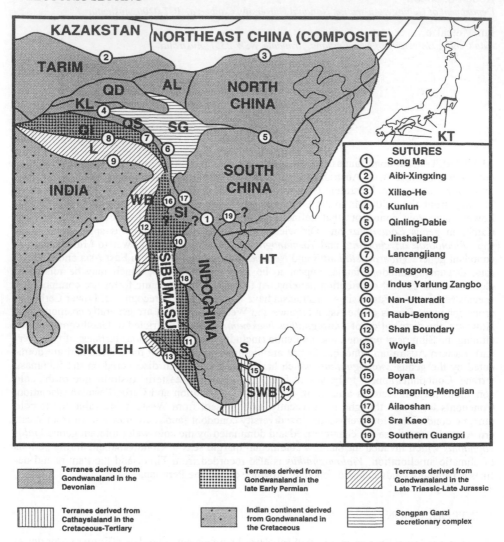

Figure 1. Distribution of principal continental terranes and sutures of East and Southeast Asia. AL = Ala Shan Terrane, HT = Hainan Island terranes, KL = Kunlun Terrane, KT = Kurosegawa Terrane, L = Lhasa Terrane, S = Semitau Terrane, SG = Songpan Ganzi accretionary complex, SI= Simao Terrane, SWB = South West Borneo, QD = Qaidam Terrane, QI = Qiangtang Terrane, QS = Qamdo-Simao Terrane, WB = West Burma.

Present day East Asia comprises an amalgamation of Gondwanaland-derived allochthonous continental lithospheric blocks (Figure 1) which were assembled between the Carboniferous and the early Cenozoic. Biogeographic, tectonostratigraphic and palaeomagnetic data suggest that these continental blocks (terranes) formed part of the India-Australian Gondwanaland margin in the Early Palaeozoic and that they rifted and separated from the margin of Gondwanaland as three continental slivers in the Devonian, late Early Permian and Late Triassic-Late Jurassic. As these three continental slivers separated from Gondwanaland, three successive ocean basins, Palaeo-Tethys, Meso-Tethys and Ceno-Tethys, were opened between the slivers and Gondwanaland. Each of these Tethyan oceanic basins were subsequently destroyed by northwards subduction processes and continental collisions to form the suture zones of East Asia (Metcalfe, 1998; 2001, this volume). Conodont biogeography provides valuable biogeographic constraints on the tectonic evolution of the region and discussions of the biogeographic aspects of conodont faunas of East Asia-Australasia are presented in terms of the above tectonic framework and to test palaeogeographic models.

3 CONODONT PALAEOBIOGEOGRAPHY

A preliminary assessment of conodont biogeographic distribution in East Asia – Australasia is presented below. Much taxonomic work remains to be done before a fully comprehensive evaluation of conodont biogeography in the region can be presented.

3.1 *Cambrian*

Conodonts are first recorded from the Upper Cambrian and while there seems to be some differentiation of biofacies relating to environmental factors, probably water temperature, there is little strong evidence for a distinct Gondwanan faunal province (Bergström, 1990). There are a few taxa that appear to be restricted to the northeast Gondwanan region. The Upper Cambrian conodont *Eodentatus bicuspatus* (Figure 6) was named based on specimens recovered from the Georgina Basin of central Australia (Nicoll & Shergold, 1991). This species has recently also been recovered from South China by Dong & Repetski (personal communication, 2000). Another species thus far reported only from North China (An et al., 1983) and the Georgina Basin (Nicoll & Shergold, 1991) regions is *Hirsutodontus nodus* (Zhang & Xiang, 1983).

3.2 *Ordovician*

Ordovician conodont biogeographic distribution in the region has been recently reviewed in some detail (Webby et al., 2000). Suggestions of the existence of an Australian Province were first made by Bergström (1971) and then expanded toward an Australasian or Gondwanan Province by Nicoll & Webby (1996). Nowlan et al. (1997) also proposed an Australasian Province based on the distribution of some Late Ordovician conodonts, principally *Taoqupognathus* and *Yaoxianognathus*. Recognition of Ordovician conodont provincialism is complicated by a strong environmental factor, probably water temperature, which controls two major biofacies; the North American Midcontinent Province (NAMCP) in warmer and shallower water and the North Atlantic Province (NATP) in colder and deeper water (Sweet & Bergström, 1974; Nicoll & Webby, 1996). Faunas representing both biofacies are found in the Gondwanan region (Nicoll & Webby, 1996).

Some Ordovician conodonts which define a probable East Asia-Australasia province include *Serratognathus* spp. and *Aurilobodus* spp. in the Lower Ordovician. Species of *Serratognathus* (Figure 6) are found throughout the northeastern Gondwanan region (Figure 2). The distribution of *Aurilobodus* appears to be more restricted. In the Middle Ordovician, species such as *Plectodina onychodonta* and *Tasmanognathus* (Figure 2) continue to indicate an East Asia and Australasia province. In the Upper Ordovician, conodonts are more limited in their distribution in Australia. The genera *Taoqupognathus* and *Yaoxianognathus* appear to be initially restricted to North China, South China and eastern Australia, but later have a limited distribution in Siberia and northwestern Canada (Nowlan et al., 1997).

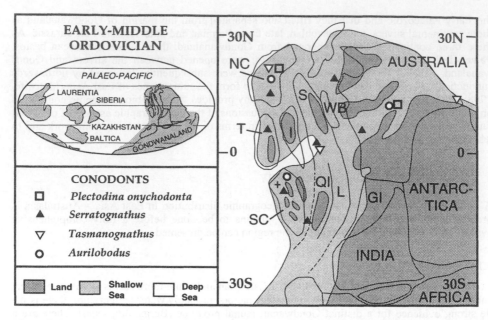

Figure 2. Reconstruction of eastern Gondwanaland for the Early-Middle Ordovician showing the postulated positions of the East and Southeast Asian terranes, distribution of land and sea, and the distribution
of the conodonts *Serratognathus*, *Tasmanognathus*, *Aurilobodus*, and *Plectodina onychodonta* that illustrate Asia-Australia connections at this time. GI = Greater India, I = Indochina, L = Lhasa, NC = North
China, S = Sibumasu, SC = South China, T = Tarim, QI = Qiangtang, WB = West Burma. Present day
outlines are for reference only. After Metcalfe (1998).

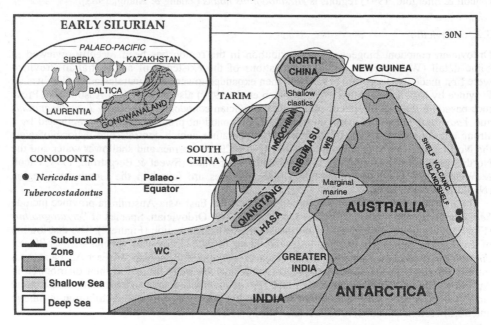

Figure 3. Reconstruction of eastern Gondwanaland for the mid-late Silurian showing the postulated positions of the East and Southeast Asian terranes, distribution of land and sea, and the distribution of the
conodonts *Nericodus* and *Tuberocostadontus* that appear to define an Australasian province at this time.
WB = West Burma, WC = Western Cimmerian Continent. After Metcalfe (1998).

3.3 Silurian

Lower to Middle Silurian conodont faunas (Figure 3) have not been extensively documented in East Asia-Australasia and where faunas have been studied they are dominantly composed of cosmopolitan species. However Zhou et al., (1981) and Bischoff (1986) described faunas which contained species that appear to be restricted to the region. Zhou et al., (1981) proposed two new genera, *Nericodina* and *Tuberocostadontus*, from South China (Guizhou Province). Two species of these genera, *N. cricostata* and *T. shiqianensis*, appear to be conspecific with *Pyrsognathus latus* and *P. obliquus*, respectively, established by Bischoff (1986) from eastern Australia. Other early to middle Llandovery faunas from eastern Australia may also contain endemic or Australasian Provincial faunas (Pickett et al., 2000), but more detailed studies are needed to confirm the extent of preliminary observations. Wenlock and younger Silurian faunas appear to be more cosmopolitan, but forms such as *Kockelella ranuliformis* appear to have much extended ranges in the Tasman Fold Belt of eastern Australia and this may portend some aspect of provinciality (Pickett et al.2000).

3.4 *Devonian*

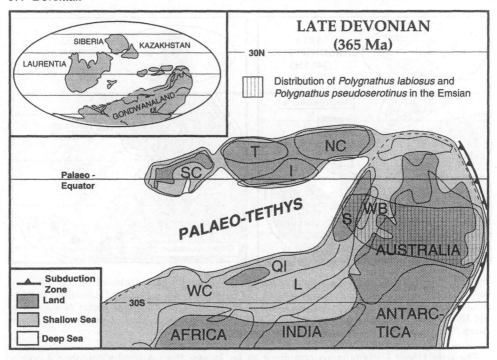

Figure 4. Reconstruction of eastern Gondwanaland for the Late Devonian showing the postulated positions of the East and Southeast Asian terranes, distribution of land and sea, and opening of the Palaeo-Tethys ocean at this time. Also shown is the distribution of *Polygnathus labiosus* and *Polygnathus pseudoserotinus* in the Emsian. Present day outlines are for reference only. See figures 2 or 3 for abbreviations. After Burrett et al. (1990) and Metcalfe (1998).

Devonian conodont biogeography has not been investigated in a manor which would easily assist the recognition of an East Asia – Australasian province. The principal studies of Klapper & Johnson (1980) and Klapper (1995) do not deal with localities in dispersed Gondwanan terranes and use a continental reconstruction by Heckel & Witzke (1979) that places many of these terranes already far removed from the Gondwanan margin. Klapper (1995) suggests that there may be a substantial number of Devonian endemic conodont species, but that the necessary detailed biogeographic analysis has yet to be studied. Peak Devonian conodont endemism is recorded in the Emsian (Klapper & Johnson, 1980). Burrett et al. (1990; Figure 4) suggest that the species

Polygnathus labiosus and *Polygnathus pseudoserotinus* are restricted in their biogeographic distribution to Australia and the Sibumasu block. Other East Asia – Australasian province species are certain to be established with further study.

3.5 *Carboniferous*

Australian Carboniferous conodont faunas are essentially restricted to the Tournaisian and Visean. These Lower Carboniferous conodont faunas of the Asian terranes and Western Australia are generally cosmopolitan. However, the distribution of *Mestognathus beckmanni* (Figures 5a, 7) appears restricted to Gondwanaland (including the Sibumasu terrane) and Laurentia (including the allochthonous terranes of N. America) but this species does not occur on the principal Chinese blocks (von Bitter et al., 1986) which separated from Gondwanaland in the Devonian. Eastern Australian conodont faunas are highly endemic during the Visean and are dominated by the genus *Montognathus*, which is also found in the Sibumasu terrane (Metcalfe, in press; Figures 5a, 7). This genus appears to be endemic to peri-East Gondwanaland and its occurrence on the Sibumasu block supports other evidence (Metcalfe, 1998) suggesting a marginal Gondwanaland position in the Carboniferous.

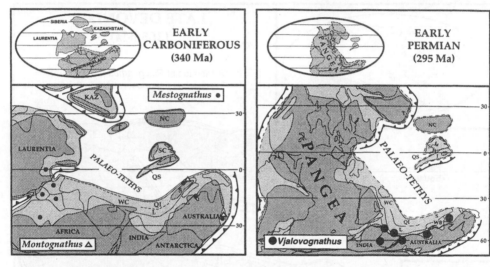

Figure 5. Palaeogeographic reconstructions of the Tethyan region for Early Carboniferous (a) and Early Permian (b) showing relative positions of the East and Southeast Asian terranes and Australasia and distribution of land and sea. The distribution of the Lower Carboniferous conodont genera *Mestognathus* and *Montognathus*, and the Lower Permian cold-water tolerant conodont genus *Vjalovognathus* are also shown. See figures 2 or 3 for abbreviations. After Metcalfe (1998).

The only Upper Carboniferous conodonts so far known from Australia comprise a poorly preserved, probably upper Namurian (lower Pennsylvanian) fauna from Murgon, southeast Queensland, which includes species of *Gnathodus*, *Declinognathodus* and possibly *Streptognathodus* or *Neognathodus* (Palmieri, 1969). The locality has been suggested to be part of an allochthonous block that constitutes part of the Gympie Terrane (Roberts, 1987). In addition, a small poorly preserved early Westphalian fauna, containing *Hindeodus minutus* (Ellison) and *Neognathodus* cf. *N. bassleri* (Harris & Hollingsworth) from float samples believed to have been derived from the Aimau Formation is located on the allochthonous Kemum Terrane of the Birds Head of Irian Jaya (Nicoll & Bladon, 1991). In New Zealand the only known Carboniferous fossils are Middle Pennsylvanian conodonts, including *Neognathodus bassleri* (Harris & Hollingsworth) and *Gondolella*, reported from the Kakahu Limestone of the South Island (Jenkins & Jenkins, 1971). The locality forms part of the Torlesse Terrane, the origin of which is unknown, and which could well have been situated some considerable distance from autochthonous Australasia in the Late Carboniferous, and in a different climatic zone. The allochthonous nature of the South Island New Zealand terranes is also suggested by the recent recovery of Early Permian conodonts from both the Torlesse and Caples Terranes (Ford et al., 1999).

3.6 *Permian*

Conodonts have so far not been reported from the Permian of eastern Australia. Recent sampling of Permian limestones of the Bowen Basin of Queensland by one of us (RSN, 1999) failed to recover conodonts, despite processing in excess of 10 kg. from each locality. Their absence is interpreted to be due to the cold conditions present there as a result of high latitudinal position and Carbo-Permian glaciation (Nicoll, 1976). Conodonts were also thought, until recently, to be absent from Western Australia due to cold climatic conditions. However, low diversity conodont faunas, of extremely low abundance, are now known from Western Australia and these are characterized and dominated by the cold-water-tolerant genus *Vjalovognathus*, and even less abundant *Hindeodus*, which invaded the shallow continental margin seas of Gondwanaland during periods of climatic amelioration (Metcalfe & Nicoll, 1993; Nicoll & Metcalfe, 1998). *Vjalovognathus* is also recorded from Timor, the Pamirs, Pakistan, and Tibet, and defines a Tethyan-Gondwanaland conodont province (Figure 5b) during the Permian (Nicoll & Metcalfe, 1998). Three species of *Vjalovognathus* are recognized in Western Australia, *Vjalovognathus australis* Nicoll & Metcalfe of Sakmarian-Artinskian age, *Vjalovognathus shindyensis* (Kozur) of early Kungurian age, and *Vjalovognathus* n. sp. A Nicoll & Metcalfe of late Kungurian-Roadian age (Nicoll & Metcalfe, 1998). A limited number of species of the genera *Hindeodus* and *Mesogondolella* also occur with the Gondwanaland endemic *Vjalovognathus*.

4 IMPLICATIONS FOR TECTONIC EVOLUTION AND PALAEOGEOGRAPHY

Lower Palaeozoic conodont faunas contain conodont genera and species that are restricted to the principal East Asian continental blocks and Australia and which define an Asian-Australian province of Gondwanaland at that time. This provinciality is consistent with that of other groups of shallow-marine organisms, especially trilobites (J.H. Shergold, pers. comm., 1994) which also define an Asian-Australian province in the early Palaeozoic. The biogeographic distribution of the Gondwanaland endemics is consistent with Sibumasu, Indochina, Tarim, Qiangtang, Lhasa, South China, North China and Hainan Island terranes being contiguous with the India-NW Australian Margin of Gondwanaland in the Early Palaeozoic (Figures 2 and 3).

Biogeographic, palaeomagnetic and other data suggest that North China, South China, Tarim, Indochina and Hainan terranes had separated from Gondwanaland by Carboniferous times (Metcalfe, 1998 and this volume). In particular, the Visean endemic genus *Montognathus* occurs on the Sibumasu terrane and in Australia but is absent from the terranes that had already separated from Gondwanaland. *Mestognathus beckmanni* also appears to be absent from the Visean of the intra-Tethyan continental terranes at this time (Figure 5a).

Biogeographically, the Tournaisian conodonts of both shallow- and deeper-water biofacies of western and eastern Australia are essentially cosmopolitan in nature and are closely linked with those of the northern continents. Upper Tournaisian deep-water conodonts representative of the *Scaliognathus anchoralis* zone (with re-worked shallow-water Carboniferous and older conodonts) are also known from the Peninsular Malaysia, Peninsular Thailand and North Thailand portions of the Sibumasu terrane (Baum et al., 1970; Igo, 1973; Metcalfe, 1983, in press), which was attached to NW Australia during the Carboniferous (Metcalfe, 1996, 1998; Figure 5a).

The general lack of conodonts in the post-Visean Carboniferous of Australia is interpreted to be the result of a combination of (a) the change from marine to largely non-marine conditions over much of the continent, (b) the southwards drift of eastern Gondwana (including Australia) into much higher palaeolatitudes during the overall clockwise rotation of Gondwana that resulted in collision with Laurasia to form Pangea, and (c) the cold climatic conditions due to glaciation on Gondwana which effectively precluded conodonts from the cold marine environment on, and marginal to, Australia. The abundance of conodonts in the Lower Carboniferous and their virtual absence from the Upper Carboniferous of Australia reflect major global shifts in both plate configurations and climate.

Figure 6. (Oposite) Biogeographically important Lower Palaeozoic conodonts from SE Asia-Australia. Repository of specimens: **CPC** – Commonwealth Palaeontological Collection, Australian Geological Survey Organisation, Canberra; **UTG** – University of Tasmania, Geology Department, Hobart; **A** – Department of Geology, University of Malaya, Kuala Lumpur. Number of microns represented by the scale bar is given in (brackets).

1, 2. *Eodentatus bicuspatus* Nicoll & Shergold (CPC 29046). 1. lateral view (50 µ); 2. aboral view (50µ). Late Cambrian (Payntonian), Ninmaroo Formation, Georgina Basin (from Nicoll & Shergold, 1991).

3. *Eodentatus bicuspatus* Nicoll & Shergold (CPC 29045). Lateral view (50 µ). Late Cambrian (Payntonian), Ninmaroo Formation, Georgina Basin (from Nicoll & Shergold, 1991).

4. *Eodentatus bicuspatus* Nicoll & Shergold (CPC 29048). Lateral view (50 µ). Late Cambrian (Payntonian), Ninmaroo Formation, Georgina Basin (from Nicoll & Shergold, 1991).

5 – 7. *Plectodina onychodonta* An & Xu. 5. Pa element (CPC 35827) lateral view. 6. Sa element (CPC 35828) posterior view (100 µ). 7. Sd element (CPC 35829) posterior view (100 µ). Middle Ordovician, Stokes Siltstone, Amadeus Basin.

8 – 10. *Nericodina cricostata* Zhou, Zhai & Xian = *Pyrsognathus latus* Bischoff. Pa element (CPC 35830). 8. Oblique oral-posterior view (100 µ), 9. Lateral view (100 µ), 10. Enlargement (20 µ) of lateral view showing ornamentation Early Silurian (Llandovery), Peppercorn Beds, Long Plain, NSW.

11. *Nericodina cricostata* Zhou, Zhai & Xian = *Pyrsognathus latus* Bischoff. Pa element (CPC 35831) oblique aboral-anterior view (100 µ). Early Silurian (Llandovery), Peppercorn Beds, Long Plain, NSW.

12 – 14. *Tuberocostatus shiqianensis* Zhou, Zhai & Xian = *Pyrsognathus obliquus* Bischoff. S element (CPC 35832), 12. Posterior view (100 µ), 13. Lateral view (100 µ), 14. Anterior view (100 µ). Early Silurian (Llandovery), Peppercorn Beds, Long Plain, NSW.

15, 16. *Tuberocostatus shiqianensis* Zhou, Zhai & Xian = *Pyrsognathus obliquus* Bischoff. S element (CPC 35833). 15. Aboral view (100 µ), 16. Lateral view (100 µ). Early Silurian (Llandovery), Peppercorn Beds, Long Plain, NSW.

17. *Tasmanognathus careyi* Burrett. Pa element (UTG 96850). Lateral view (100 µ). (From Burrett, 1979)

18. *Tasmanognathus careyi* Burrett. Pa element (UTG 96860). Lateral view (100 µ). (From Burrett, 1979)

19 – 22. *Serratognathus bilobatus* Lee. Pa element. 19. anterior view (A486; 100 µ); 20. lateral view (A488; 100 µ); 21. posterior view (A487; 100 µ); 22. detail of serrated ornament (A486; 20 µ). Lower Ordovician, Setul Limestone, Kaki Bukit, Peninsular Malaysia. From Metcalfe (1980).

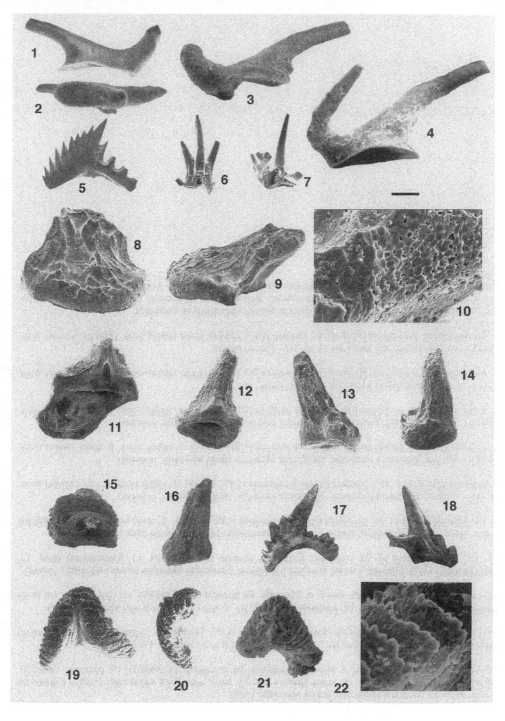

Figure 7. (oposite)
Biogeographically important Upper Palaeozoic conodonts from SE Asia-Australasia. Number of microns represented by the scale bar is given in (brackets). Repository of specimens: CPC – Commonwealth Palaeontological Collection, Australian Geological Survey Organisation, Canberra.

1. *Mestognathus beckmanni* Bischoff. Pa element (CPC 35834). Inner lateral view. (500 µ). Visean, Kanthan Limestone, Peninsular Malaysia (from Metcalfe, in press).

2. *Mestognathus beckmanni* Bischoff. Pa element (CPC 35835). Inner lateral view. (500 µ). Visean, Kanthan Limestone, Peninsular Malaysia (from Metcalfe, in press).

3, 4. *Mestognathus beckmanni* Bischoff. Pa element (CPC 35836). 3. upper view; 4. inner lateral view. (200 µ). Visean, Kanthan Limestone, Peninsular Malaysia (from Metcalfe, in press).

5, 6. *Mestognathus beckmanni* Bischoff. Pa element (CPC 35837). 5. upper view; 6. inner lateral view. (200 µ). Visean, Kanthan Limestone, Peninsular Malaysia (from Metcalfe, in press).

7, 8. *Montognathus* cf. *M. carinatus* Crane. Pa element (CPC 35838). 7. upper view; 8. inner lateral view. (500 µ). Visean, Kanthan Limestone, Peninsular Malaysia (from Metcalfe, in press).

9, 10. *Montognathus* cf. *M. carinatus* Crane. Pa element (CPC 35839). 9. inner lateral view; 10. oblique upper view. (500 µ). Visean, Kanthan Limestone, Peninsular Malaysia (from Metcalfe, in press).

11, 12. *Montognathus* cf. *M. carinatus* Crane. Pa element (CPC 35840). 11. inner lateral view; 12. oblique upper view. (500 µ). Visean, Kanthan Limestone, Peninsular Malaysia (from Metcalfe, in press).

13-16. *Vjalovognathus australis* Nicoll & Metcalfe. Pa element (CPC 34644). 13. upper view; 14. inner lateral view; 15. anterior view; 16. posterior view (200 µ). Timor (from Nicoll and Metcalfe, 1999).

17-18. *Vjalovognathus shindyensis* (Kozur). Pa element (CPC 34679). 17. upper view; 18. outer lateral view (200µ). Canning Basin, Western Australia (from Nicoll and Metcalfe, 1999).

19-24. *Vjalovognathus* sp. nov. A Nicoll & Metcalfe. Pa element (CPC 34693). 19. posterior view; 20. anterior views; 21. upper view; 22. outer lateral view; 23. inner lateral; 24. basal view (200µ). Carnarvon Basin, Western Australia (from Nicoll and Metcalfe, 1999).

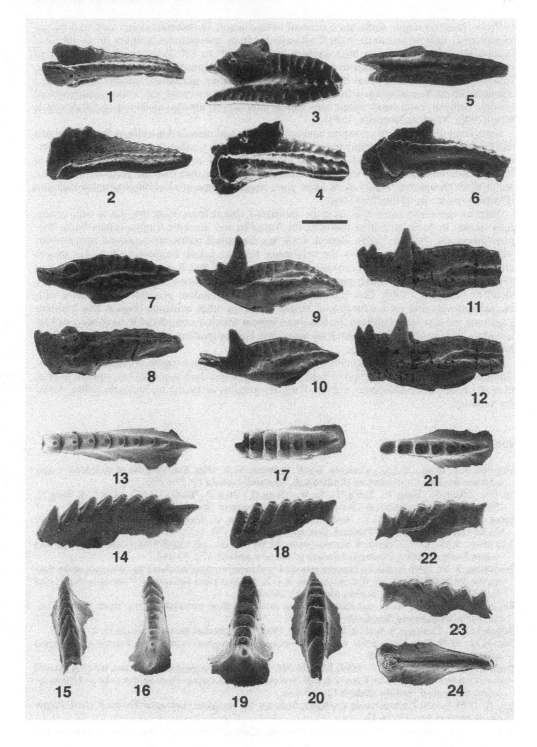

These observed major shifts are consistent with changes in palaeolatitudes indicated by palaeomagnetic data for Australia in the Carboniferous from low southern latitudes (0° to 40° S) in Tournaisian - Visean times to high southern latitudes (50° to 75° S) in middle-late Namurian times. Australia also continued to reside in high southern latitudes during the Early Permian. Cold conditions, resulting from high latitudes and persistent glaciation, effectively precluded conodonts from Australia, apart from some cold-water tolerant forms (eg. *Vjalovognathus*) that invaded Tethyan continental margin seas during intervals of climatic amelioration (Metcalfe & Nicoll 1995; Nicoll & Metcalfe, 1998).

Significant differences in conodont faunas of western and eastern Australia in the Tournaisian and Visean are interpreted mainly to reflect different biofacies rather than biogeographic differences (see Jones et al., in press). However, the high level of conodont endemism developed in the nearshore biofacies of the Visean of eastern Australia, together with the presence of endemism in other groups (eg. corals) at the same time, suggests some level of biogeographic isolation of eastern Australia during the Visean.

Permian conodonts were until recently considered absent from Australia due to cold conditions caused by high latitudinal positions for Australia and the Late Carboniferous-Early Permian glaciation (Nicoll, 1976). Recent work has identified restricted conodont faunas from Western Australia, dominated by the cold-water tolerant genus *Vjalovognathus* (see above) which, together with similar occurrences in the Pamirs, the Salt Range of Pakistan, Tibet and Timor, define a Tethyan-Gondwanan cold-water conodont province (Metcalfe & Nicoll, 1995; Nicoll & Metcalfe, 1998). This high-latitude cold-water conodont province is consistent with the palaeogeographic reconstruction of Pangea based on other evidence (Figure 5b). Permian conodonts are also recorded from the Sibumasu terrane which is considered to have been still attached to NW Australia until Sakmarian times, but *Vjalovognathus* has not been reported from the Permian of Sibumasu thus far.

Palaeozoic conodont biogeography in the SE Asia-Australasian region is consistent with and supports the Gondwanaland origin of East Asian terranes and their successive separation as continental slivers in the Devonian and late Early Permian as proposed by Metcalfe (1996, 1998, this volume).

REFERENCES

Aldridge, R.J., Briggs, D.E.G., Clarkson, E.N.K. & Smith, M.P. 1986. The affinities of conodonts – new evidence from the Carboniferous of Edinburgh, Scotland. *Lethaia* 19: 279-291.

An T.X., Zhang F., Xiang W., Zhang Y., Xu W., Zhang H., Jiang D., Yang C., Liu L., Cui Z. & Yang X. 1983. *The conodonts of North China and the adjacent regions.* Science Press of China, Beijing.

Baum, F., Braun, E. von, Hahn, L., Hess, A., Koch, K.E., Kruse, G., Quarch, H. & Siebenhuner, M. 1970. On the Geology of northern Thailand. *Geologishe Jahrbuch* 102: 1-23.

Bergström, S.M., 1971. Conodont biostratigraphy of the Middle and Upper Ordovician of Europe and eastern North America. *Geological Society of America Memoir* 127: 83-162.

Bergström, S. M. 1990. Relations between conodont provincialism and the changing paleogeography during the Early Palaeozoic. In W.S. McKerrow & C.R. Scotese (eds) *Palaeozoic Palaeogeography and Biogeography.* Geological Society Memoir 12: 105-121.

Bischoff, G.C.O. 1986. Early and Middle Silurian conodonts from midwestern New South Wales. *Courier Forschungsinstitut Senckenberg* 89: 1-337.

Briggs, D.E.G., Clarkson, E.N.K. & Aldridge, R.J. 1983. The conodont animal. *Lethaia* 16: 1-14.

Burrett, C., 1979. *Tasmanognathus*: a new Ordovician conodontophorid genus from Tasmania. *Geologica et Palaeontologica* 13: 31-38.

Burrett, C., Long, J. & Stait, B. 1990. Early-Middle Palaeozoic biogeography of Asian terranes derived from Gondwana. In W.S. McKerrow & C.R. Scotese (eds) *Palaeozoic Palaeogeography and Biogeography.* Geological Society Memoir 12: 163-174.

Igo, H. 1973. Lower Carboniferous conodonts from Ko Yo, Songkla, Peninsular Thailand. *Geol. Palaeont. Southeast Asia* 12: 29-42.

Jenkins, D. G. & Jenkins, T.B.H. 1971. First diagnostic Carboniferous fossils from New Zealand. *Nature* 233 (5315): 117-8.

Jones, P.J., Metcalfe, I. Engel, B. A., Playford, G., Rigby, J., Roberts, J., Turner, S. & Webb, G.E. 2000. Carboniferous Biogeography of Australasia In A.J. Wright, G.C. Young, J.A Talent & J.R. Laurie (eds), *Palaeogeography of Australasian Flora and Fauna*. Memoirs of the Australasian Association of Palaeontologists 23:

Klapper, G. & Johnson, J.G. 1980. Endemism and dispersal of Devonian conodonts. *Journal of Paleontology* 54: 400-455.

Mei, S., Henderson, C.M. & Jin, Y., 1999. Permian conodont provincialism, zonation and global correlation. *Permophiles* 35: 9-16.

Metcalfe, I. 1980. Ordovician conodonts from the Kaki Bukit area, Perlis, West Malaysia. *Warta Geologi* 6: 63-68.

Metcalfe, I., 1983. Southeast Asia. In C. Martinez Dias (ed.) *The Carboniferous of the world: 1. China, Japan and Southeast Asia.* International Union of Geological Sciences Publication 16: 213-143.

Metcalfe, I. 1988. Origin and assembly of Southeast Asian continental terranes. In M.G. Audley-Charles, & A. Hallam (eds), *Gondwana and Tethys*. Geological Society of London Special Publication 37: 101-118.

Metcalfe, I. 1990. Allochthonous terrane processes in Southeast Asia. *Philosophical Transactions of the Royal Society of London* A331: 625-640.

Metcalfe, I. 1993. Southeast Asian terranes: Gondwanaland origins and evolution. In R.H. Findlay, R. Unrug, M.R. Banks, & J.J. Veevers (eds) *Gondwana 8 - Assembly, Evolution, and Dispersal* (Proceedings Eighth Gondwana Symposium, Hobart, 1991): 181-200. Rotterdam: A.A. Balkema.

Metcalfe, I. 1996. Pre-Cretaceous evolution of SE Asian terranes In, R. Hall, & D. Blundell (eds) *Tectonic Evolution of Southeast Asia*. Geological Society Special Publication 106: 97-122.

Metcalfe, I. 1998. Palaeozoic and Mesozoic geological evolution of the SE Asian region: multidisciplinary constraints and implications for biogeography. In R. Hall, & J.D. Holloway (eds) *Biogeography and Geological Evolution of SE Asia*: 25-41. Amsterdam, The Netherlands: Backhuys Publishers.

Metcalfe, I. 2001. Palaeozoic and Mesozoic tectonic evolution and biogeography of SE Asia-Australasia. In I. Metcalfe, J.M.B. Smith, M. Morwood & I. Davidson (eds) *Faunal and floral migrations and evolution in SE Asia-Australasia*. Rotterdam, A.A. Balkema (this volume).

Metcalfe, I., in press. Mixed Devonian and Carboniferous conodonts from the Kanthan Limestone, Peninsular Malaysia and their stratigraphic and tectonic implications. Proceedings International Congress on Carboniferous-Permian, Calgary, 1999.

Metcalfe, I & Nicoll, R.S., 1995. Lower Permian conodonts from Western Australia, and their biogeographic and palaeoclimatological implications. In R. Mawson, & J. Talent, (eds.) Contributions to the first Australian Conodont Symposium (AUSCOS 1) held in Sydney, Australia, 18-21 July, 1995. *Courier Forschungsinstitut Senckenberg* 182: 559-560.

Nicoll, R.S., 1976. The effect of Late Carboniferous - Early Permian glaciation on the distribution of conodonts in Australia. In C.R. Barnes (ed). *Conodont Paleoecology*. Geological Association of Canada Special Paper 15: 273-278.

Nicoll, R.S. & Bladon, G.M. 1991. Silurian and Late Carboniferous conodonts from the Charles Louis Range and central Birds Head, Irian Jaya, Indonesia. *BMR Journal of Australian Geology & Geophysics*. 12: 279-286.

Nicoll, R.S. & Metcalfe, I. 1998. Early and Middle Permian conodonts from the Canning and Southern Carnarvon Basins, Western Australia: Their implications for regional biogeography and palaeoclimatology. *Royal Society of Victoria Proceedings* 110: 419-461.

Nicoll, R.S. & Shergold, J.H. 1991. Revised Late Cambrian (pre-Payntonian-Datsonian) conodont biostratigraphy at Black Mountain, Georgina Basin, western Queensland, Australia. *BMR Journal of Australian Geology & Geophysics* 12: 93-118.

Nicoll, R.S. & Webby, B.D. 1996. Ordovician. In G.C. Young & J.R. Laurie (eds) *AGSO Phanerozoic Timescale 1995*: 77-95 Melbourne: Oxford University Press.

Nowlan, G.S., McCracken, A.D. & McLeod, M.J., 1997. Tectonic and paleogeographic significance of Late Ordovician conodonts in the Canadian Appalachians. *Canadian Journal of Earth Sciences* 34: 1521-1537.

Palmieri, V. 1969. Upper Carboniferous conodonts from limestones near Murgon, south-east Queensland. *Geological Survey of Queensland Publication* 341, Palaeontological Paper 17: 1-13.

Pickett, J.W., Burrow, C.J., Holloway, D.J., Munson, T.J., Percival, I.G., Rickards, R.B., Sherwin, L., Simpson, A.J., Strusz, D.L. Turner S. & Wright A.J. 2000. Silurian Palaeobiogeography of Australia. In A.J. Wright, G.C. Young, J.A Talent & J.R. Laurie (eds), *Palaeogeography of Australasian Flora and Fauna*. Memoirs of the Australasian Association of Palaeontologists 23: 127-161.

Pridmore, P.A., Barwick, R.E. & Nicoll, R.S., 1997. Soft anatomy and the affinities of conodonts. *Lethaia* 29: 317-328.

Roberts, J., 1987. Carboniferous faunas: their role in the recognition of tectonostratigraphic terranes in the Tasman Belt, eastern Australia. In E.C. Leitch, & E. Scheibner (eds) *Terrane accretion and Orogenic Belts*. Geodynamics series 19: 93-102. American Geophysical Union, Washington D.C., Geological Society of America, Boulder, Colorado.

Sweet, W.C. & Bergström, S.M., 1974. Provincialism exhibited by Ordovician conodont faunas. In C.A. Ross (ed.), *Paleogeographic provinces and provinciality*. Society of Economic Paleontologists and Mineralogists Special Publication 21: 189-202.

von Bitter, P.H., Sandberg, C.A. and Orchard, M.J. 1986. Phylogeny, speciation, and palaeoecology of the Early carboniferous (Mississippian) conodont genus *Mestognathus*. Royal Ontario Museum Life Sciences Contributions 143: 115pp.

Webby, B.D., Percival, I.G., Edgecombe, G.D., Cooper, R.A., VandenBerg, A.H.M., Pickett, J.W., Pojeta, J., Playford, G., Winchester-Seeto, T., Young, G.C., Zhen, Y.-Y., Nicoll, R.S., Ross J.R.P. & Schallreuter, R. 2000. Ordovician biogeography of Australasia. In A.J. Wright, G.C. Young, J.A Talent & J.R. Laurie (eds), *Palaeogeography of Australasian Flora and Fauna*. Memoirs of the Australasian Association of Palaeontologists 23: 63-126.

Zhou X. Y., Zhai Z. Q. & Xian S. Y., 1981. On the Silurian conodont biostratigraphy, new genera and species in Guizhou Province. *Oil & Gas Geology* 2: 123-140 (in Chinese with English abstract)

Wallace lines in eastern Gondwana: Palaeobiogeography of Australasian Permian Brachiopoda

N. W. Archbold
School of Ecology and Environment, Deakin University, Rusden Campus, Clayton, Victoria 3168, Australia

ABSTRACT: The present Australian continent was a major component of the north eastern peninsula of Gondwana, itself the southern region of Pangaea, during the Permian period. Surrounding what is now Australia, were additional elements of north eastern Gondwana that are now incorporated into the tectonically complex regions of New Zealand, New Caledonia, the island of New Guinea, Timor, south east Asia, the Himalaya and southern Tibet. India was to the west and south west and Antarctica to the south. Marine water temperatures ranged from cold to temperate and tropical as Permian global climates ameliorated, global surface ocean circulation systems warmed, and due to rifting and northward drifting of some terranes.

Provincialism of global marine faunas was pronounced during the Permian and hence refined biostratigraphical correlations are often fraught with difficulty. The 'middle' Permian stratotypes approved by the International Subcommission on the Permian System have little direct relevance to correlations within the Gondwanan Region at the level of operational biostratigraphical zonal schemes. Brachiopoda are a dominant marine benthonic faunal element of Permian Gondwanan faunas and they provide refined correlations between marine basins within a specific faunal province. Modern faunal provinces are recognised by the distribution patterns of species and genera belonging to a single family or superfamily such as the Papilionoidea within the Insecta. This review provides an example from Permian Brachiopoda, using the distribution data of genera and subgenera of the superfamily Ingelarelloidea, in order to demonstrate the ability to define provinces and their 'Wallace lines' of demarcation between provinces in the geological past.

1 INTRODUCTION

During the Permian Period, the present Australian continent was a component of the eastern portion of Gondwana which itself was the southern region of Pangaea. Australia was surrounded by elements of New Zealand to the east and southeast, New Caledonia to the northeast, Irian Jaya to the north, Timor and the Cimmerian continental fragments to the northwest, southern Tibet, the Himalaya and Peninsular India to the west and southwest and Antarctica to the south. Marine and terrestrial faunal and terrestrial floral relationships existed between all these regions.

The cold to warm-temperate climate change during the Permian in the Australasian region and the combination of cold eastern marine surface currents and cool to warm western surface currents (Archbold, 1998a) collectively explain many of the palaeobiogeographical features of Australasian Permian faunas and floras. Marine faunas invariably demonstrate an increase in diversity from south (Tasmania) to north (Bowen Basin, Queensland) in eastern Australia. Endemic genera are developed within the Austrazean Province. The Westralian marine faunas demonstrate strong links with southern Tethyan, peripheral Gondwanan (Cimmerian) and Peninsular Indian faunas. Eastern Australian marine faunas, particularly those of Queensland, show strong links with the faunas of New Zealand (excluding the fusulinid bearing terranes) and with the limited faunas of New Caledonia. Terrestrial floras demonstrate links to the floras north and west of Australia but tend to be less well studied in terms of their palaeobiogeographical patterns.

The Permian geology of New Zealand is complex and has been interpreted as a series of displaced terranes (Adams et al. 1998). Several of these terranes were geographically close to the eastern margin of Gondwana during the Permian and hence share faunas with eastern Australia as part of the Austrazean Province (Campbell et al. 1998). A few Permian conodonts have been recorded from the Torlesse and Caples terranes which appears to indicate that these terranes may have been influenced by cool temperate marine currents (Ford et al. 1999). Tropical waters are considered to be the source of the allochthonous limestone blocks with fusulinids (Leven & Grant-Mackie, 1997; Leven & Campbell 1998) associated with two terranes and a palaeotethys - panthalassic original location has been proposed for these blocks. They are regarded as being allochthonous to the region of the Austrazean Province.

2 PROVINCIALISM, BIOSTRATIGRAPHY AND CORRELATIONS

In a review such as this, it is not possible to discuss in detail questions of preferred continental reconstructions or details of individual stratigraphical schemes for individual basins or tectonic blocks. References cited will provide the reader with sources for these details but it can be noted that Ziegler et al. (1998) provide a series of nine Permian time slice models of global continental reconstructions which include the Australasian region during the Permian. Archbold & Shi (1996) provide a summary of recent statistical investigations on the provincialism of the marine faunas of the Australasian region and the pertinent literature on statistical methodologies and history of the terminology of provincialism for the region. Recent reviews by Archbold (1996a), Shi (1996, 1998) and Grunt & Shi (1997), also summarise theoretical considerations underlying the determinations of provincial relationships.

The present review notes the debt owed to Teichert's (1951) classic paper on the biogeographical relationships of the Western Australian Permian faunas. Provincialism of Australian Permian marine faunas has been recognised since early this century and attention is drawn to the early review of Benson (1923). Individual palaeontological studies often emphasise generic faunal or floral links between adjacent provincial regions, obviously often for purposes of correlation, for example the recent studies on Permian echinoderms (Webster & Jell 1992, 1999), and brachiopods (Archbold, 1996b; Archbold & Shi, 1995, 1996), rather than discuss the total fauna of adjacent regions. These types of studies are standard procedure for the progress of palaeontological knowledge and particularly improved global correlations. Nevertheless, as indicated in the study by Archbold (1983) on the provincialism of chonetidine brachiopods for the Gondwanan region, it is important for provincialism studies to consider not only the genera present in a province but also the species diversity of the genus. The chonetid genus *Neochonetes* is now known to be represented by some nine named species in the Western Australian Permian where species are abundant and two rare unnamed species in eastern Australia and one named species in New Zealand. This type of generic 'similarity' between two faunal regions serves 'but to emphasise the differences that exist between east and west, due doubtless to the absence of a direct marine communication between the two regions' (David, 1950, p.373). Modern faunal provinces are invariably recognised on the basis of both species and generic distributions within individual families or superfamilies such as the Psittacidae within Aves or the Hesperioidea and Papilionoidea within Insecta.

Brachiopod faunas of the Australian Permian (Archbold, 1996b) are reasonably typical of most fossil groups in terms of the palaeobiogeographical relationships that they demonstrate. Some fossil groups are totally or primarily restricted to western basins (eg. conodonts, many ammonoids and the elasmobranch *Helicoprion*) with connections to the Tethyan and Uralian seas. Terrestrial faunal elements such as crustaceans, insects and tetrapods are unknown from the west of Australia. Some eastern Australian fossil groups developed endemic families and genera, eg. some small foraminiferans, crinoids, stellaroids and brachiopods, in adapting to cool to cold water conditions. Permian marine faunal assemblages from geographical areas that were adjacent to Western Australia demonstrate significant linkages to Westralian faunas eg. brachiopod faunas from Peninsular India (Archbold, et al., 1996), the eastern Himalaya (Singh & Archbold, 1993), Malaysia (Shi et al., 1997) western Yunnan (Shi et al. 1995, 1996), Irian

Jaya (Archbold, 1992), the Shan-Thai terrane (Archbold & Shi, 1995), the northwest shelf (Archbold, 1988) and Timor (Archbold & Barkham, 1989; Archbold & Bird, 1989).

Detailed time control for biostratigraphical and correlation purposes is required for meaningful discussions of provincialism (Archbold & Shi, 1996). However, it is the provincialism of Australasian Permian faunas and floras that has hampered the correlation of Australasian Permian sequences with Tethyan, Uralian and North American reference sections. Western Australian ammonoid occurrences have been accorded a primary role for international correlations (Archbold 1998b, 1999). Few ammonoids are known from eastern Australia. A few records of conodonts are now known from western Australia (Nicoll & Metcalfe 1998) but precise time ranges and the provincialism of these occurrences yield differing interpretations for correlation purposes (cf Archbold 1998a; Nicoll & Metcalfe 1998). The review by Archbold & Dickins (1996) provides some insight into the problems of correlation. Faunal stage schemes based on molluscan fossils, with some other faunal elements have been proposed for western and eastern Australia (eg. see Archbold & Dickins, 1996 for references) and more elaborate zonal schemes based on brachiopods have also been proposed for western Australia (Archbold, 1998a), eastern Australia (Briggs, 1998) and New Zealand (Waterhouse, 1982, 1998). Most brachiopod faunas have been documented for western Australia and New Zealand but numerous eastern Australian faunal elements require formal description. Zonation schemes using palynomorphs are widely applied (Mory & Backhouse, 1997, Price, 1997) and are progressively being integrated with marine faunal zonation schemes (Archbold 1999), but further work is required. Tables 1 and 2 summarise the current understanding of the ages of the critical western Australian biostratigraphical data (brachiopod zones, conodont and ammonoid data and palynological data) and the correlation of the western Australian data (using palynology) with the eastern Australian brachiopod zones. Table 3 provides the possible correlation of the eastern Australian brachiopod zones with the New Zealand brachiopod zones and the tropical water global scale proposed by the International Subcommission on the Permian System.

3 THE AUSTRALASIAN BRACHIOPOD FAUNAS

Brachiopods, at the generic level, are the numerically most abundant group in Australasian Permian marine faunas with well over 100 genera being recognised. Just over a dozen genera are shared between eastern and western faunas including *Streptorhynchus, Neochonetes, Echinalosia, Strophalosia, Taeniothaerus, Megasteges, Costatumulus, Lyonia, 'Cyrtella', Trigonotreta, Fusispirifer, Sulciplica, Cleiothyridina, Gilledia* and *Fletcherithyris*.

The palaeobiogeography of Australasian Permian brachiopods has recently been reviewed by Archbold (1996a), who also provided references current to the end of 1995 for Australasian brachiopod faunas. Since that review, the studies by Archbold (1996c, 1997) provide additional taxonomic data for western Australian faunas and those of Briggs (1998) and Waterhouse (1998a, 1998b) for eastern Australian and New Zealand faunas. The brachiopod faunas exhibit a striking contrast in generic and specific composition on either side of the continent, clearly reflecting warmer water currents affecting the western margin and cooler water currents affecting the eastern margin throughout most of the Permian. The Westralian and Austrazean Provinces were established initially (Archbold, 1983) on the basis of key brachiopod faunas of the suborder Chonetidina. Westralian faunas include a high diversity of Chonetoidea with the Anopliidae represented by *Tornquistia, Demonedys* and *Gatia* and the Rugosochonetidae by abundant *Neochonetes (Sommeriella), Svalbardia, Quinquenella*, rare *Waagenites* and a possible *Chonetinella*. Austrazean faunas possess a lower diversity with no anopliids and the Rugosochonetidae represented by *Svalbardia, Tivertonia*, probable *Capillonia* and two undescribed species of *Neochonetes*.

On a broader level, when considering the total brachiopod fauna, Westralian brachiopod faunas include mixtures of three generic groups. Genera characteristic of widely distributed Gondwanan faunas, including those of Peninsular India, the Himalaya and southern Tibet, are *Arctitreta, Costatumulus, Strophalosia, Taeniothaerus, Cyrtella, Tomiopsis* (subgenus *Ambikella*) and *Trigonotreta* and may indicate cooler water influences. A second group of genera are either endemic to the Westralian Province or demonstrate endemic lineages within the province. This group includes *Permorthotetes, Neochonetes (Sommeriella), Gatia, Notolosia*,

WEST AUSTRALIAN KEY BIOSTRATIGRAPHICAL DATA

STAGE Substage/Horizon	BRACHIOPOD ZONES	CONODONTS	AMMONOIDS	PALYNOLOGY
DORASHAMIAN – ? – = CHANG-HSINGIAN				
	W.(W.) imperfecta		• Cyclolobus parsulcatus	
DJHULFIAN	Liveringia magnifica			
MIDIAN				Dulhuntyispora parvithola
– ? –				
KAZANIAN	Sulciplica occidentalis			Didecitriletes ericianus
Sheshminsk				Dulhuntyispora granulata
UFIMIAN	F. coolkilyaensis		• D.goochi, Agathiceras applanatum	Microbaculispora villosa
Solikamsk	Neochonetes (S.) afanasyevae	▲ Vjalovognathus sp.	• Popanoceras sp. • B. australe • Daubichites goochi	
Iren	Svalbardia thomasi		Paragastrioceras • wandageense B.australe, D.goochi	
	Neochonetes (S.) nalbiaensis			
KUNGURIAN	Fusispirifer wandaceensis	▲ Vjalovognathus sp.	• Bamyaniceras australe	Praecolpatites sinuosus
Filippov	Fusispirifer cundlegoensis	▲ Mesogondolella idahoensis	• Bamyaniceras australe	
	Tornquistia magna	Vjalovognathus shindyensis	• Bamyaniceras australe	
Saranin	Fusispirifer byroensis	Hindeodus sp.3	• Bamyaniceras sp.	
Sargin	W. colemani		Bamyaniceras sp. • Aricoceras sp. Pseudoschistoceras simile	
	E. prideri	▲ V. shindyensis Hindeodus sp.2		
ARTINSKIAN	M. anomala		• Bamyaniceras sp. Aricoceras sp.	Microbaculispora trisina
	N.(S.)magnus		• Pseudoschistoceras simile	
Aktastinian	Strophalosia jimbaensis	▲ Vjalovognathus australis Hindeodus sp.1	Metalegoceras striatum &australe • Thalassoceras wadei	Didecitriletes byroensis
			Propopanoceras ruzhencevi	Striatopodocarpites fusus
Steriltamakian	Strophalosia irwinensis	▲ Vjalovognathus australis	• Metalegoceras sp. Uraloceras irwinense Mesolites sp.	Pseudoreticulatispora pseudoreticulata
SAKMARIAN	Trigonotreta occidentalis		• Uraloceras irwinense Juresanites jacksoni	
Tastubian	– ? – Lyonia lyoni		• Juresanites jacksoni	Pseudoreticulatispora confluens
Shikhan	?			Microbaculispora tentula
Uskalyk				
ASSELIAN				Protohaploxypinus spp.
Sjuren				

Table 1. Key biostratigraphical data for the Western Australian Permian (brachiopod and palynological zones and conodont and ammonoid data).

TRANS-AUSTRALIAN BRACHIOPOD CORRELATIONS

STAGE Substage/Horizon	BRACHIOPOD ZONES (W.A.)	EAST AUSTRALIAN PALYNOLOGY	BRACHIOPOD ZONES (E.A.)
DORASHAMIAN ? = CHANGHSINGIAN (TATARIAN)		APP6 — Triplexisporites playfordii	
DJHULFIAN	W. (W.) imperfecta	APP5006	
	Liveringia magnifica	Lycopodiumsporites 'crassus' ?	Echinalosia voiseyi
MIDIAN		APP5 — Dulhuntyispora parvithola	Echinalosia ovalis
			Echinalosia deari
			P. crassa
KAZANIAN			P. ingelerensis
		? APP43 Dulhuntyispora spp.	P. clarkei
		APP42	P. blakei
	Sulciplica occidentalis	Didecitriletes ericianus	Echinalosia wassi
			Echinalosia hanloni
Sheshminsk (UFIMIAN)		APP41 — Dulhuntyispora granulata	Echinalosia discinia
	F. coolkilyaensis	APP33	
Solikamsk	Neochonetes (S.) afanasyevae	Microbaculispora villosa	Echinalosia davidi
			Echinalosia maxwelli
Iren (KUNGURIAN)	Svalbardia thomasi	APP32	
	Neochonetes (S.) nalbiaensis		
	Fusispirifer wandageensis	Praecolpatites sinuosus	Echinalosia preovalis
Filippov	Fusispirifer cundlegoensis		
	Tornquistia magna		
Saranin (Baigendzhinian)	Fusispirifer byroensis	? APP31 ?	
Sargin	W. colemani	Praecolpatites spp.	Echinalosia warwicki
	E. prideri		Echinalosia curtosa
ARTINSKIAN	M. anomala	APP22	Tomiopsis strzeleckii
	N. (S.) magnus	Microbaculispora trisina	
Aktastinian	Strophalosia jimbaensis	APP21 — D. byroensis	Bandoproductus walkomi
Sterlitamakian	Strophalosia irwinensis	Pseudoreticulatispora pseudoreticulata	Strophalosia subcircularis
SAKMARIAN	Trigonotreta occidentalis	APP122	Strophalosia concentrica
Tastubian	?	Pseudoreticulatispora confluens	— ? — — ? —
	Lyonia lyoni		Lyonia bourkei
Shikhan	?		?
Uskalyk (ASSELIAN)		APP121 — Microbaculispora tentula	
Sjuren		APP11 — Protohaploxypinus spp.	

Table 2. Trans-Australian correlations based on brachiopod zones and key palynological zones.

EAST AUSTRALIAN - NEW ZEALAND CORRELATIONS

STAGE Substage/Horizon	BRACHIOPOD ZONES (E.A.)	BRACHIOPOD ZONES (N.Z.)	'GLOBAL STAGES'
DORASHAMIAN — ? — = CHANG-HSINGIAN		Wairakiella rostrata	CHANGHSINGIAN
		Marginalosia planata	
DJHULFIAN		Spinomartinia spinosa	WUCHIAPINGIAN
		Plekonella multicostata	
		Martiniopsis woodi	
	Echinalosia volseyi	Terrakea elongata	— ? —
MIDIAN	Echinalosia ovalis	Echinalosia ovalis	CAPITANIAN
	Echinalosia deeri		
	P. crassa		— ? —
KAZANIAN	P. ingelarensis		
	P. clarkei	Wyndhamia blakei	WORDIAN
	P. blakei		
	Echinalosia wassi		
	Echinalosia hanloni		— ? —
Sheshminsk UFIMIAN	Echinalosia discinia	Echinalosia maxwelli	ROADIAN
Solikamsk	Echinalosia davidi	Echinalosia discinia	— ? —
	Echinalosia maxwelli		
Iren		Spiriferella supplanta	
KUNGURIAN Filippov	Echinalosia preovalis	Terrakea concavum	KUNGURIAN
Saranin		Biconvexiella	
Sargin	Echinalosia warwicki	Echinalosia 'prideri'	ARTINSKIAN
	Echinalosia curtosa	Wyndhamia sp.	
ARTINSKIAN (Balgendzhinian)	Tomiopsis strzeleckii	Notostrophia homeri	
Aktastinian	Bandoproductus walkomi	Notostrophia zealandicus	
		Gondor fauna	
Sterlitamakian	Strophalosia subcircularis		SAKMARIAN
SAKMARIAN	Strophalosia concentrica		
Tastubian	– – ? – – – – – – – ? – – – –		
	Lyonia bourkei		
Shikhan	?		
Uskalyk ASSELIAN			ASSELIAN
Sjuren			

Table 3. Correlation of eastern Australian and New Zealand brachiopod zones and correlation with the 'global stages' of the Permian. Note that boundaries between the North American stages are precisely defined in their type regions but are not recognisable in the Gondwanan region.

Etherilosia, Coolkilella, Latispirifer, Crassispirifer, Fusispirifer, Imperiospira and *Tomiopsis (Geothomasia)*. The third group of genera consists of sporadic migrants into the Westralian Province from the warmer Cimmerian Province of the Cimmerian continental fragments (eg. see Archbold & Shi, 1995). Such genera include *Kiangsiella, Tornquistia, Demonedys, Waagenites, Stictozoster, Dyschrestia, Comuquia, Cimmeriella, Retimarginifera, Callytharrella, Costiferina, Waagenoconcha, Stenoscisma, Elivina, Spiriferella, Spirelytha, Gjelispinifera, Callispirina, Hustedia, Rhynchopora* and *Spirigerella*.

Eastern Australian Permian brachiopod faunas, with those of New Zealand, characterise the distinctive Austrazean Province. Increase in generic diversity from the Tasmanian faunas in the south to those of the Bowen Basin in the north, is a distinctive feature of the faunas. The Tasmanian faunas represent the most extreme example of cold-water Gondwanan faunas yet known. Core Gondwanan genera of the Austrazean Province include *Arctitreta, Echinalosia, Strophalosia, Terrakea, Trigonotreta, Sulciplica, Tomiopsis* (subgenera *Ingelarella* and *Johndearia*) and its allies, *Cyrtella, Fletcherithyris* and *Gilledia*. Tasmania and Queensland appear to have been two centres of evolution for the Ingelarelloidea with endemic genera and species of large size. Austrazean faunas also demonstrate high rates of evolution for the Productida (Briggs, 1998).

Australasian Permian Brachiopods have been analysed statistically as part of a series of studies on Western Pacific Permian provincialism, results of which are fully summarized in Archbold & Shi (1996). The Indoralian Province was found to encompass all faunas of Asselian-Sakmarian age from eastern and western Australia and peripheral Gondwanan regions of southeast Asia, then part of the southern Tethys. In the Late Sakmarian the Westralian, Austrazean and incipient Cimmerian Provinces became distinct. The Austrazean and Westralian Provinces existed until the end of the Permian whereas several terranes of the Cimmerian Province merged into the Cathaysian Province in the Late Permian (Shi, 1998).

In the Asselian - Aktastinian time slice, an Early Permian Himalayan Subprovince or Province has been proposed for Himalayan faunas and those from Afghanistan, the Karakorum, the Lhasa Terrane and possibly the Salt Range (Archbold & Shi 1996; Shi 1998). For the present review, the Himalayan Early Permian Subprovince is subsumed within the Indoralian Province which may be preferably defined as a Region in terms of hierarchical biogeography. This provincial area lacks an extensive development of mid-Permian marine faunas (except for the Salt Range and Afghanistan). However, the region becomes a readily recognised province again in the Late Permian (Lopingian), as discussed by Zhan & Lee (1977), Archbold & Thomas (1986) and Archbold (1987). During the Lopingian the remainder of the former Cimmerian Terranes were incorporated into the Cathaysian tropical province (Shi 1998).

4 WALLACE LINES AND THE INGELARELLOIDEA

The superfamily Ingelarelloidea, and its constituent genera, has been revised by Waterhouse (1998a) who raised the group to superfamilial status in a move forshadowed by Archbold (1995: 46). If the taxonomic revisions proposed by Waterhouse are accepted (as provisionally herein) it is evident that this particular superfamily of brachiopods evolved rapidly and developed many distinctive genera that occurred only in the cool marine waters of the Austrazean Province (Archbold 1983) of Permian Gondwana. Just as Alfred Russell Wallace based his famous line of demarcation between biogeographical regions primarily on a single group of vertebrate animals (birds, especially parrots, see van Oosterzee 1997: 36), the present author used a single group of brachiopods (the suborder Chonetidina) to define provinces for the Permian of Gondwana (Archbold, 1983). Modifications to the original proposals have been suggested after statistical analyses (Archbold & Shi, 1996), as briefly summarised above. Waterhouse's (1998a) taxonomic proposals provide an independent, if subjective, test of the current state of knowledge of Permian provincialism at the eastern margin of Gondwana during the Permian.

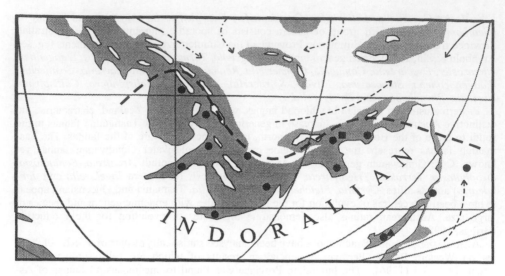

Figure 1. Eastern Gondwana, Asselian - Early Sterlitamakian time slice.
●⬤ = *Tomiopsis (Ambikella)*. ■ = *Fredericksia*.   ▲ = *Kelsovia*

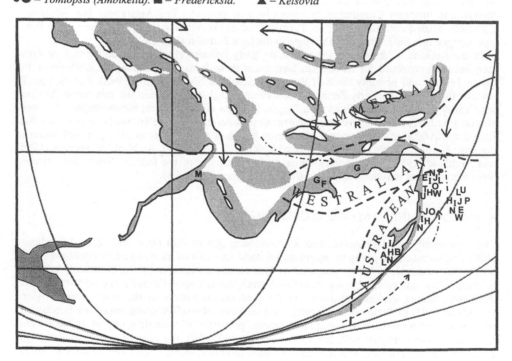

Figure 2. Eastern Gondwana, Kungurian - Kazanian time slice.
R = *Rorespirifer,*   **M** = *Martiniopsis,*   **F** = *Fredericksia,*
G = *Tomiopsis (Geothomasia),*   **I** = *Tomiopsis (Ingelarella),*   **J** = *Tomiopsis (Johndearia),*
E = *Mesopunctia,*   **O** = *Monklandia,*   **T** = *Tabellina,*
H = *Homevalaria,*   **A** = *Farmerella,*   **B** = *Birchsella,*
N = *Notospirifer,*   **L** = *Glendella,*   **W** = *Wairakispirifer,*
U = *Tigillumia,*   **P** = *Papulinella*

Remarkably, or perhaps not, the subjective data shown by plotting distributions of ingelarel-loid genera on three time slice reconstructions of eastern Gondwana (figures 1-3) conform with the statistical studies summarised by Archbold & Shi (1996) and Shi (1998). The Indoralian Province is delineated for the Asselian - Sakmarian (figure 1), the Cimmerian Province is a fea-ture (though poorly defined) for the Kungurian - Kazanian (figure 2) and the Westralian Prov-ince is delineated from the Kungurian until the end of the Permian. Once developed, the line of demarcation between the Austrazean and Westralian Provinces is an abrupt 'Wallace line' whereas the boundaries between the Westralian Province and the Cimmerian Province of the mid-Permian and Himalayan Province of the Late Permian (figure 3) are more transitional, but are more obvious when the total brachiopod faunas are considered.

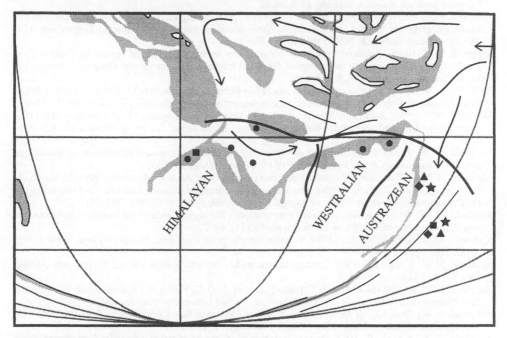

Figure 3. Eastern Gondwana, Tatarian time slice.
● = *Tomiopsis (Geothomasia),* ■ = *Martiniopsis,* ★ = *Mesopunctia,*
▲ = *Tomiopsis (Ingelarella),* ◆ = *Tomiopsis (Johndearia).*

Although *Martiniopsis* is recorded by Waterhouse (1998b) from New Zealand a separate subgenus is warranted for the New Zealand material based on details of thickening of the ven-tral adminicula and the subtle, narrow ventral sulcus.

ACKNOWLEDGEMENTS

Miss Heather Ansell word processed the manuscript and Monty Grover drafted the text figures and tables. The author's work on Late Palaeozoic faunas is supported by the Australian Research Council and Deakin University.

REFERENCES

Adams, C. J., Campbell, H.J., Graham, I.J. & Mortimer, N. 1998. Torlesse, Waipapa and Caples suspect terranes of New Zealand; Integrated studies of their geological history in relation to neighbouring ter-ranes. *Episodes* 21 (4): 235-240.

Archbold, N.W. 1983. Permian marine invertebrate provinces of the Gondwanan Realm. *Alcheringa* 7: 59-73.

Archbold, N.W. 1987. South Western Pacific Permian and Triassic marine faunas : Their distribution and implications for Terrane identification. *American Geophysical Union, Geodynamics Series* 19: 119-127.

Archbold, N.W. 1988. Permian Brachiopoda and Bivalvia from Sahul Shoals No. 1, Ashmore Block, Northwestern Australia. *Proceedings of the Royal Society of Victoria* 100: 33-38.

Archbold, N.W. 1992. Early Permian Brachiopoda from Irian Jaya. *BMR Journal of Australian Geology & Geophysics* 12: 287-296.

Archbold, N.W. 1995. Evolutionary and migration history of Australian Permian brachiopod faunas. *International Symposium 'Evolution of Permian marine biota', Abstracts*: 45-47. Moscow: Palaeontological Institute, Russian Academy of Sciences.

Archbold, N.W. 1996a. Gondwana and the Complex of Asia during the Permian: The Importance of Palaeobiogeographical Studies. In, *Gondwana Nine. Ninth International Gondwana Symposium:* 479-489. New Delhi: Oxford & IBH Publishing Co. Pvt. Ltd.

Archbold, N.W. 1996b. Palaeobiogeography of Australian Permian Brachiopod faunas. In, Copper, P. & Jin, Y. (eds), *Brachiopods*. Proceedings of the Third International Brachiopod Congress, Canada: 19-23. Rotterdam, Brookfield: A.A.Balkema.

Archbold, N.W. 1996c. Studies on Western Australian Permian Brachiopods 13. The fauna of the Artinskian Mingenew Formation, Perth Basin. *Proceedings of the Royal Society of Victoria* 108 (1): 17-42.

Archbold, N.W. 1997. Studies on Western Australian Permian brachiopods. 14. The fauna of the Artinskian High Cliff Sandstone, Perth Basin. *Proceedings of the Royal Society of Victoria* 109 (2): 199-231.

Archbold, N. W. 1998a. Correlations of the Western Australian Permian and Permian ocean circulation patterns. *Proceedings of the Royal Society of Victoria* 110 (1/2): 85-106.

Archbold, N.W. 1998b. Marine biostratigraphy and correlation of the West Australian Permian basins. In P. G., Purcell & R. R. Purcell (eds), *The Sedimentary Basins of Western Australia, 2, Proceedings of the Petroleum Exploration Society of Australia Symposium, Perth, WA, 1998*: 141-151.

Archbold, N. W. 1999. Permian Gondwanan correlations: The significance of the western Australian marine Permian. *Journal of African Earth Sciences* 28 (1): 62-75.

Archbold, N.W. & Barkham S.T. 1989. Permian Brachiopoda from near Bisnain Village, West Timor. *Alcheringa* 13: 125-140.

Archbold, N.W. &. Bird P.R 1989. Permian Brachiopoda from near Kasliu Village, West Timor. *Alcheringa* 13: 103-123.

Archbold, N.W. &. Dickins J.M 1996. Permian (Chart 6). In G.C. Young.& J.R. Laurie (eds), *An Australian Phanerozoic Timescale*: 127-135. Melbourne: Oxford University Press Australia.

Archbold, N.W., Shah S.C. & Dickins J.M. 1996. Early Permian Brachiopod faunas from Peninsular India: Their Gondwanan relationships. *Historical Biology* 11: 125-135.

Archbold, N.W. & Shi G.R. 1995. Permian brachiopod faunas of Western Australia: Gondwanan-Asian relationships and Permian climate. *Journal of Southeast Asian Earth Sciences* 11 (3): 207-215.

Archbold, N.W. & Shi G.R. 1996. Western Pacific Permian marine invertebrate palaeobiogeography. *Australian Journal of Earth Sciences* 43 (6): 635-641.

Archbold, N.W. & Thomas G.A. 1986. Permian Brachiopoda from Western Australia: A review in space and time. *Biostratigraphie du Paleozoique* 4: 431-438.

Benson, W.N. 1923. Palaeozoic and Mesozoic Seas in Australasia. *Transactions of the New Zealand Institute* 54: 1-62.

Briggs, D.J.C. 1998. Permian Productidina and Strophalosiidina from the Sydney - Bowen Basin and New England Orogen: systematics and biostratigraphic significance. *Memoir of the Association of Australasian Palaeontologists* 19: 1-258.

Campbell, H.J., Smale, D., Grapes, R., Hoke, I., Gibson G.N. &. Landis C.A 1998. Parapara Group: Permian - Triassic rocks in the Western Province, New Zealand. *New Zealand Journal of Geology & Geophysics* 41: 281-296.

David, T.W.E. (ed. Browne, W.R.), 1950. *The Geology of the Commonwealth of Australia*. London: Edward Arnold, 3 Volumes.

Ford, P.B., Lee D.E & Fischer P.T. 1999. Early Permian conodonts from the Torlesse and Caples Terranes, New Zealand. *New Zealand Journal of Geology & Geophysics* 42: 79-90.

Grunt, T.A. & Shi G.R. 1997. A hierarchical framework of Permian global marine biogeography. Proceedings of the 30th International Geological Congress, 12, *Palaeontology and Historical Geology*: 2-17. Utrecht: VSP.

Leven, E. Ja. & Campbell H.J. 1998. Middle Permian (Murgabian) fusuline faunas, Torlesse Terrane, New Zealand. *New Zealand Journal of Geology and Geophysics* 41: 149-156.

Leven, E. Ja., & Grant-Mackie J.A. 1997. Permian fusulinid Foraminifera from Wherowhero Point, Orua Bay, Northland, New Zealand. *New Zealand Journal of Geology and Geophysics* 40: 473-486.

Mory, A.J. & Backhouse J. 1997. Permian stratigraphy and palynology of the Carnarvon Basin, Western Australia. *Geological Survey of Western Australia, Report* 51: 1-41.

Nicoll, R.S. & Metcalfe I. 1998. Early and middle Permian conodonts from the Canning and Southern Carnarvon Basins, Western Australia: Their implications for regional biogeography and palaeoclimatology. *Proceedings of the Royal Society of Victoria* 110 (1/2): 419-461.

Price, P.L. 1997. Permian to Jurassic palynostratigraphic nomenclature. In P. M., Green, (ed.), *The Surat and Bowen Basins, South-east Queensland*: 137-178. Brisbane: Queensland Department of Mines and Energy.

Shi, G.R. 1996. A model of quantitative estimate of marine biogeographic provinciality. *Acta Geologica Sinica* 70 (4): 351-360.

Shi, G.R. 1998. Aspects of Permian marine biogeography: A review on nomenclature and evolutionary patterns, with particular reference to the Asian - Western Pacific region. *Palaeoworld* 9: 97-112.

Shi, G.R., Archbold N.W. & Fang Zongjie 1995. The biostratigraphical and palaeogeographical significance of an Early Permian brachiopod fauna from the Dingjiazhai Formation, Baoshan Block, Western Yunnan, China. *Journal of Geology, Geological Survey of Vietnam*, Series B 5-6: 63-74.

Shi, G.R., Fang Zongjie & Archbold N.W. 1996. An Early Permian brachiopod fauna of Gondwanan affinity from the Baoshan Block, western Yunnan, China. *Alcheringa* 20: 81-101.

Shi, G.R., Leman, Mohd Shafeea & Tan B.K. 1997. Early Permian brachiopods from the Singa Formation of Langkawi Island, northwestern Peninsular Malaysia: biostratigraphical and biogeographical implications. In P. Dheeradilok (ed.-in-chief), *Proceedings of the International Conference on Stratigraphy and Tectonic Evolution of Southeast Asia and the South Pacific*: 62-72. Bangkok: Department of Mineral Resources.

Singh, Trilochan & Archbold N.W. 1993. Brachiopoda from the early Permian of the Eastern Himalaya. *Alcheringa* 17: 55-75.

Teichert, C. 1951. The marine Permian faunas of Western Australia. (An Interim Review). *Palaontologische Zeitschrift* 24 (1-2): 76-90.

Van Oosterzee, P. 1997. *Where Worlds Collide, The Wallace Line*: 234pp. Kew: Reed Books Australia.

Waterhouse, J.B. 1982. New Zealand Permian Brachiopod Systematics, Zonation, and Paleoecology. *New Zealand Geological Survey Paleontological Bulletin* 48: 1-158, pls. 1-23.

Waterhouse, J.B. 1998a. Ingelarelloidea (Spiriferida: Brachiopoda) from Australia and New Zealand, and reclassification of Ingelarellidae and Notospiriferidae. *Earthwise* 1: 1-48.

Waterhouse, J.B. 1998b. Permian geology of Wairaki Downs, New Zealand, and the realignment of its biozones with the International Standards. *Proceedings of the Royal Society of Victoria* 110 (1/2): 235-245.

Webster, G.D. & Jell P.A. 1992. Permian echinoderms from Western Australia. *Memoirs of the Queensland Museum* 32 (1): 311-373.

Webster, G.D. & Jell P.A. 1999. New Permian crinoids from Australia. *Memoirs of the Queensland Museum* 43 (1): 279-339.

Ziegler, A.M., Gibbs M.T. &. Hulver M.L 1998. A mini-atlas of oceanic water masses in the Permian Period. *Proceedings of the Royal Society of Victoria* 110 (1/2): 323-343.

Zhan, Lipei & Lee, Li 1977. *The distribution of Permian brachiopods in China*: 15pp. Peking: Science Press.

A review of the Early Permian flora from Papua (West New Guinea)

J.F.Rigby

School of Natural Resource Sciences, Queensland University of Technology, Brisbane, Australia

ABSTRACT: A flora of approximately twenty species has been found in the Permian Aiduna Formation along the southern side of the main suture zone in Papua (West New Guinea). The flora has close affinities with both the Gondwanaland and the Cathaysian floras. All named species are endemic excepting *Trizygia speciosa* and the Gondwanan *Vertebraria indica*. Eleven species of *Glossopteris* from the Gondwanaland flora dominate the flora, however the Cathaysian-related *Gigantonoclea* and *Fascipteris* also occur. Pecopterid fronds do not emphasize any particular geographical relationship. The seed plants *Glossopteris* and *Gigantonoclea* needed a direct land connection with Gondwanaland and Cathaysia respectively during the time of growth or slightly earlier so they could spread to the area of modern New Guinea.

1 INTRODUCTION

Permian floras have been found in the Late Sakmarian to Early Artinskian Aiduna Formation on the southern side of the main suture zone in Papua, that is, on the northern edge of the Australasian Plate (Fig. 1). Shi *et al.* (1995) have dated the Aiduna Formation as lying within the range of Late Sakmarian to Early Artinskian based on associated Tethyan-related faunas. Previous investigations of the flora include those by Jongmans (1940), Hopping & Wagner (1962) and Rigby (1997a, 1999b). The present contribution discusses the sporophyte species found in the collections, and not previously reported on, and catalogues species published elsewhere. Possible palaeogeographical and palaeobotanical interpretations based on the complete floral assemblages, which are listed on Table 1, below, are considered.

2 SAMPLES CONSIDERED HERE AND THEIR PROVENANCE

Collections examined by Rigby (1983, 1997a, 1997b, 1999b, herein) were collected during field seasons in 1979 and 1980 by teams from the Irian Jaya Geological Mapping Project (Pigram & Panggabean, 1983), conducted jointly by the Geological Research and Development Centre (Bandung, Indonesia) and the Bureau of Mineral Resources, Australia (now the Australian Geological Survey Organisation, Canberra). Locality data available to me is limited to place names. Un-numbered localities plotted as plant-bearing on the Waghete map (Pigram & Panggabean, 1983) do not appear to agree with those on figs 2, 3 of Rigby (1983). I do not have access to the unpublished report by Lehner *et al.* (1955) whom I understand make reference to plant localities in the Aiduna Formation. My locality data, which is limited to numbered localities, is included in the legend to Table 1, so will not be repeated under each species.

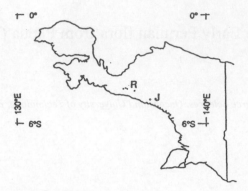

Figure 1. Papua (West New Guinea) showing areas where collections of plant fossils have been made within the Permian Aiduna Formation. J - Collections submitted to Jongmans (1940); R - Collections submitted to Rigby (1997a, 1997b, 1999b, herein). Localities for collections described by Hopping & Wagner (1962) are not known, but are believed to be in the same region as they also occurred within the Aiduna Formation. These are prefixed by W on Table 1. Map from Rigby (1997a).

3 PALAEOBOTANY

Originally (Rigby, 1997a) species were referred to as sp. A, sp. B etc. within the genera *Pecopteris* and *Glossopteris*. Since then some of the letter-designated species have been defined formally, and specimens identified as *Pecopteris* are now considered better referred to as pecopterid fronds. So as to maintain consistency, undefined species have continued to be listed using the original letter designations. The classification of genera follows Taylor & Taylor (1993) closely.

Division Sphenophyta
Order Sphenophyllales
Family Sphenophyllaceae
Genus *Trizygia* Royle 1833
Type species: *Trizygia speciosa* Royle 1833.

Trizygia speciosa Royle 1833
Previous records from New Guinea:
1940 *Sphenophyllum verticillatum* Schl., in Jongmans: 268; figs. 2, 3.
1962 *Sphenophyllum* cf. *speciosum* Royle, in Hopping & Wagner; figs 9 - 10a.
The specimens listed above are all identical with the many specimens figured from India (eg. Feistmantel, 1880) and to specimens from Queensland (Ball, 1912). The name *Trizygia speciosa*, or its synonym *Sphenophyllum speciosum*, has been applied to a number of sphenopsid remains from the Cathaysian region. I consider the species name to have been misapplied to the specimens from Cathaysia (as in Asama, 1970, and elsewhere).
Originally Royle (1833) figured and named specimens from India as *Trizygia speciosa*. M'Clelland (1850) described it. When Yabe (1922) described *Sphenophyllum sino-coreanum*, he compared it with *T. speciosa* pointing out the differences between the two species, herein quoted: "while the leaflets of *S. speciosum* 'replete with veins, which originate from two in the cuneiform base...' those of *S. sino-coreanum* are provided with four or five, sometimes even more, nerves at the base." Besides this difference, to me the Gondwanaland specimens have a bigger leaflet length ratio between the largest and smallest leaf of 3:1 whereas the ratio for leaflet length in *S. sino-coreanum* is nearer to 2:1. I consider the two to be separate species, disagreeing with Asama (1970) who included *S. sino-coreanum* as a synonym of *T. speciosa*. Specimens from Papua belong in my concept of the Gondwanaland *Trizygia speciosa*. Whether or not *Trizygia* is a junior synonym of *Sphenophyllum* is not resolvable at this stage. Asama

(1970) has presented a method for separating these genera, but for it to be accepted in Laurasia, a number of their Late Palaeozoic species now considered to belong in *Sphenophyllum* would have to be recombined into *Trizygia*. This is based on the variability of leaf size and shape within the bilaterally symmetrical leaf whorl in the latter, compared with uniformity of size and shape of the leaves within the radially symmetrical leaf whorl in *Sphenophyllum*. Gu & Shi (1974) separate these two species, keeping them in *Sphenophyllum*, but figuring (pl. 25, fig. 11; pl. 26, figs 1 - 3) the leaf shapes found in specimens of *S. sino-coreanum*.

Locality	J1	J2	W8	W10	W11	W12	W13	W14	R1	R2	R3	R4	R5	R6	R7	R8	R9	R10	R11	R12
Trizygia speciosa Royle	●				●											●	?			
pecopterid frond sp. A	●			●															●	●
pecopterid frond sp. B	●																			
fern frond gen. & sp. indet.									●	●										●
Ptychocarpus sp.	●																			
Cladophlebis sp.							●													
Fascipteris aidunae Rigby	●					●					●									
Glossopteris iriani Rigby											●		●		●					●
Glossopteris jongmansii Rigby	●										●									
Glossopteris skwarkoi Rigby	●										●									
Glossopteris wagneri Rigby										●	●								●	
Glossopteris sp. cf. *G. retifera* Feistmantel									●											●
Glossopteris sp. B-				●																
Glossopteris sp. C-									●		●		●							
Glossopteris sp. E-								●												
Glossopteris sp. F-													●							
Glossopteris sp. H													●							●
Glossopteris sp. indet													●				●			
Vertebraria indica Royle		●							●									●	●	●
Gigantonoclea iriani Rigby														●						
Koraua hartonoi Rigby														●						

Table 1. Distribution of Permian plants from the Aiduna Formation of Papua (West New Guinea). The localities are designated J from Jongmans (1940); W from Hopping & Wagner (1962); and R from Rigby (1997, 1998b, and herein). J1 - Oetakwa River, field locality P.2929. J2 - Setja River, field locality K.P.131. W8 - Aidoena River. W10 - Tipoema River. W11 - Kenataure River. W12 - Upper Aria River. W13 - Poeragi well, 2188-2194 m. W14 - Aifat River. R1 - 79 SS 7. R2 - 79 CP 201. R3 - 79 RY 188C. R4 - 79 RY 189A. R5 - 80 AG 64B. R6 - 80 BH 302D. R7 - 80 P 201A. R8 - 80 P 201B. R9 - 80 P 201C. R10 - 80 P 278A. R11 - 80 P 279A. R12 - 80 UH 200C. These number - letter combinations are the only locality data available to me for R1 to R12.

Pecopterid frond sp. A
Figures 2a - c
Previous records from New Guinea:
1940 *Pecopteris species* cf. *orientalis* Schenk, in Jongmans: 270; figs 6, 6a.
1940 *Pecopteris species* cf. *paucinervis* Jongmans, in Jongmans: 269; figs 5, 5a.
1962 *Pecopteris monyi* Zeiller, in Hopping & Wagner; figs 7, 7a.
1997a *Pecopteris* sp. A, in Rigby: 299, pl. 2, fig. 10; text fig. 2E.
1997a *Pecopteris* sp. C, in Rigby: 299.

These specimens are all sterile pecopterid frond fragments, bipinnate, pinnae sub-opposite to alternating, non-decurrent, pinnules with margins entire, apex obtuse to semicircular, secondary veins unbranched in smaller pinnules varying through dichotomizing only towards the base of the pinnule, to dichotomizing throughout in the largest pinnules. The progression is shown on Fig. 2d.

The gradational pinnule venation from unbranched in smaller pinnules to dichotomizing in larger pinnules explains why Jongmans (1940) with only a few specimens assigned the frond fragments to two species of *Pecopteris*, he did not observe transitional forms (Fig. 2c). I have examined all published references to *Pecopteris* in Gondwanaland literature, and none of the few known resemble my specimens. All Gondwanaland species are known from the then warmer, that is, more equatorial parts of Gondwanaland, and Patagonia. The difference between those figured specimens along with species from tropical regions was the progression in the secondary venation between unbranched in the smaller and forked in the larger pinnules on the same frond within my specimens. Zodrow (pers. comm.) has pointed out that these terminal pinnae are inadequate to form the basis of a new taxon, hence I have adopted a neutral stance, referring to them as pecopterid fronds.

Figured specimens are NG-1 (Fig. 2b), and specimen NG-2 (Fig. 2c), from locality 80 P 278A, R10 on Table 1. The specimens are held in the School of Natural Resource Sciences, Queensland University of Technology, Brisbane, Australia, pending a decision on their final repository which may be in Indonesia or Australia. Examples also occur at other localiies, see Table 1.

Pecopterid frond sp. B
1940 *Pecopteris* sp. cf. *arcuata* Halle, in Jongmans: 268 - 269, figs 4, 4a.

Jongmans (1940) identified a single fragment of a frond having one complete and three partial pinnules. It is too small to be significant towards any identification other than to say it is completely different from all other pecopterid remains from the Aiduna Formation. In previous tabulations I have called it *Pecopteris* sp. B (Rigby, 1997; 1998a).

Fern frond gen. et sp, indet.
A few fronds from localities 79 SS 7, 79 CP 201 and 79 RY 188C (that is, R1, R2 and R12 on Table 1) are generically indeterminate, being silhouettes (Fig. 2e). They have significantly smaller pinnules than in pecopterid frond sp. A so possibly belong in some other taxon. However, they are sufficiently similar to each other in appearance to be potentially useful in local stratigraphical correlations even though they lack details needed for taxonomical identification.

Genus *Ptychocarpus* Weiss 1869
Type species: *Ptychocarpus hexastichus* Weiss 1869

Ptychocarpus sp.
1940 *Pecopteris unita* Bgt., in Jongmans: 270.
1940 *Ptychocarpus*, in Jongmans; figs 6, 7 - 8.

This identification is based on three fragments. I am unable to comment without seeing Jongmans' specimens.

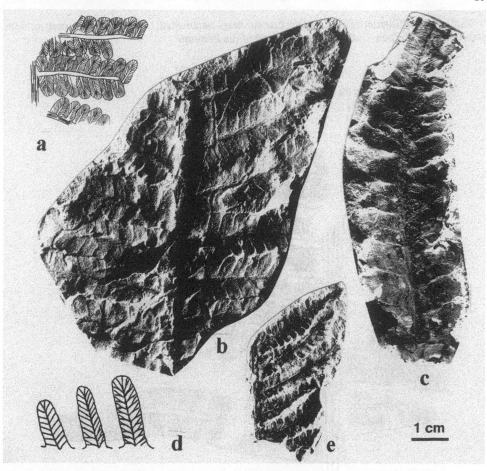

Figure 2. Pecopterid fronds, natural size, except d, freehand sketch. a - d, Pecopterid frond sp. A, from lo-cality 80 P 278A. a, tracing of part of b. b smaller pinnae. c, larger pinnae. d, pinnae showing venation trends from no dichotomies in smallest pinnae, to some dichotomies in middle pinnae, to dichotomies throughout the pinna in the largest. e, Fern frond gen. et sp. indet. from locality 79 SS 7. This indetermi-nate species occurs at 3 localities.

Division Pteridophyta
Order Filicales
Family Osmundaceae
Genus *Cladophlebis* Brongniart 1849
Type species: *Cladophlebis albertsii* (Dunker) Brongniart 1849.

Cladophlebis sp.
1962 *Cladophlebis* cf. *australis* (Morris), in Hopping & Wagner, figs 6, 6a.

The figured specimen by Hopping & Wagner (1962, figs 6, 6a) differs from specimens figured from various parts of Gondwanaland in that the pinnules have at least double the length/width ratio when compared with the type material and other specimens of *Cladophlebis australis* (Morris, 1845; and elsewhere). This feature is sufficiently pronounced to warrant a new species, however I am not able to establish one on the basis of material available to me, namely a photocopy of Hopping & Wagner's figure.

Genus *Fascipteris* Gu & Zhi 1974
Type species: *Fascipteris hallei* (Kawasaki) Gu & Zhi 1974.

The family affiliation of this species has not been established. The genus occurred in both Southern and Northern Cathaysia during much of the Permian.

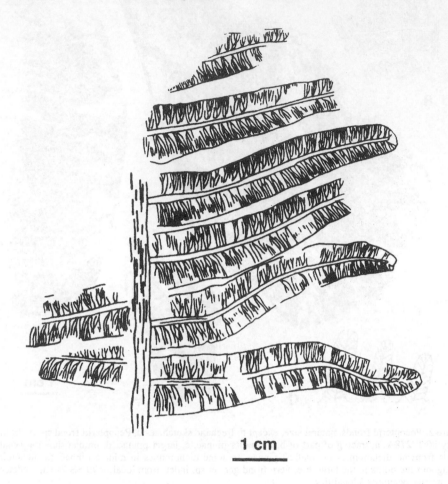

1 cm

Figure 3. *Fascipteris aidunae* Rigby 1997a, natural size. Holotype, locality 79 RY 188C.

Fascipteris aidunae Rigby 1997a
Figure 3
1997a *Fascipteris aidunae* Rigby: 298 - 299; pl. 1, figs 4, 5; text fig. 2I.
   This endemic species is a significant component of the flora.

Division Pteridospermophyta
Genus *Glossopteris* Brongniart (1828) 1831
Type species: *Glossopteris browniana* Brongniart 1828
The classification of the Glossopterids is arbitary as the female fruiting structure is unknown for many species, and for the wide variety of structures where they are known, they appear to have belonged to a number of families. Rigby (1972) was the first to recognize this when he erected the family Arberiaceae for glossopterids with non-compound fertile structures.

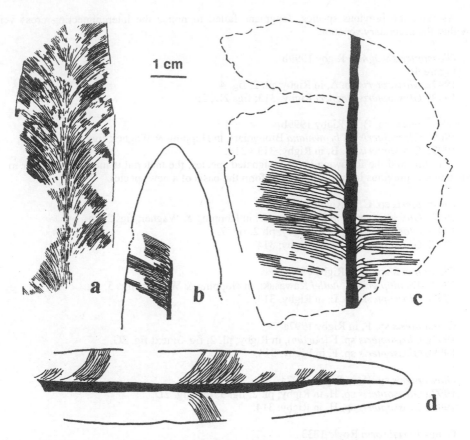

Figure 4. Tracings from photographs of holotypes of *Glossopteris* species, natural size. a, *Glossopteris iriani* Rigby 1997a, locality 80 P 279A. b, *Glossopteris skwarkoi* Rigby 1999b, locality 79 SS 7. c, *Glossopteris wagneri* Rigby 1999b, locality 79 SS 7. d, *Glossopteris jongmansii* Rigby 1999b, locality 79 SS 7.

Glossopteris iriani Rigby 1997a
Figure 4a
1997a *Glossopteris iriani* Rigby: 297 - 298; pl. 2, figs 1, 2; text fig. 2F

Glossopteris jongmansii Rigby 1999b
Figure 4d
1940   *Taeniopteris* sp. cf. *multinervis* Weiss, in Jongmans: 271; figs 10, 10a.
1997a *Glossopteris* sp. F, in Rigby; pl. 2, fig. 5.
1999b *Glossopteris jongmansii* Rigby: 310 - 311; fig. 2H.
The taeniopteroid venation of this species may be seen on the figured specimen. It must be remembered that Jongmans had only a very small collection to study, and there was very little information available on Late Palaeozoic floras from South East Asia at that time.

Glossopteris skwarkoi Rigby 1999b
Figure 4b
1940   *Taeniopteris* cf. *taiyuanensis* Halle, in Jongmans: 270 - 271; pl. 3, figs 9, 9a.
1997a *Glossopteris* sp. G, in Rigby; pl. 2, fig. 3.
1999b *Glossopteris skwarkoi* Rigby: 313; fig. 2I.

As with the previous species, Jongmans failed to notice the interconnecting cross veins within the secondary venation.

Glossopteris wagneri Rigby 1999b
Figure 4c
1997a *Glossopteris* sp. A, in Rigby; pl. 2, fig. 4.
1999b *Glossopteris wagneri* Rigby: 313; figs 2D, 2E.

Glossopteris sp. B, in Rigby 1999b
1962 *Glossopteris* cf. *browniana* Brongniart, in Hopping & Wagner; figs 1, 1a.
1999b *Glossopteris* sp. B, in Rigby: 313 - 314.
    In this and the following letter-designated species, the material is inadequate either to sustain any previous identification, or to form the basis of a new species.

Glossopteris sp. C, in Rigby 1999b
1962 *Glossopteris* cf. *indica* Schimper, in Hopping & Wagner; figs 2, 2a.
1997a *Glossopteris* sp. C, in Rigby; pl. 2, fig. 7.
1999b *Glossopteris* sp. C, in Rigby: 314.

Glossopteris sp. E, in Rigby 1999b
1962 *Taeniopteris* cf. *hallei* Kawasaki, in Hopping & Wagner; figs 5, 5a.
1999b *Glossopteris* sp. E, in Rigby: 314.

Glossopteris sp. F, in Rigby 1997a
1997a *Glossopteris* sp. F (*partim*), in Rigby; pl. 2, fig. 3; text fig. 2G.
1999b *Glossopteris* sp. F, in Rigby: 314.

Glossopteris sp. H, in Rigby 1997a
1997a *Glossopteris* sp. H, in Rigby; pl. 2, fig. 9; text fig. 2D.
1999b *Glossopteris* sp. H, in Rigby: 314.

Genus *Vertebraria* Royle 1833
Type species: *Vertebraria indica* Royle 1833

Vertebraria indica Royle 1833
Previous records from New Guinea:
1940 *Vertebraria*, in Jongmans: 271 - 272;pl. 1, fig. 1.
1962 *Vertebraria* sp., in Hopping & Wagner; pl. 2, fig. 4.

Unclassified Gymnosperms
Order Gigantopteridales
Genus *Gigantonoclea* Koidzumi 1936 *emend.* Gu & Zhi 1974
Type species: *Gigantonoclea lagrelii* (Halle) Koidzumi 1936

Gigantonoclea iriani Rigby 1997a
Figure 5a-b
1997a *Gigantonoclea iriani* Rigby: 298; pl.1, fig. 2, text figs 2A - C

Probable Gymnosperm
Genus *Koraua* Rigby 1997a
Type species: *Koraua hartonoi* Rigby 1997a

Koraua hartonoi Rigby 1997a
Figure 5c
1997a *Koraua hartonoi* Rigby: 300; pl. 1, fig. 1; pl. 2, fig. 6; text fig. 3.
    This is one of a number of monotypic genera of leaves occurring rarely in the then more equatorial, warmer parts of Gondwanaland, including *Benlightfootia* Lacey & Huard-Moine

1966, *Pachwarophyllum* Prasad & Maithy 1990 and *Diphyllopteris* Srivastava 1978. These leaves appear to have been borne in closely adpressed tufts suggesting a gymnospermous affinity.

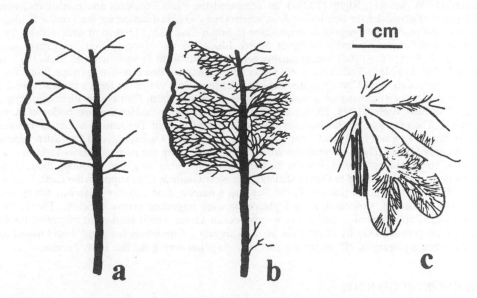

Figure 5, a - b, Holotype of *Gigantonoclea iriani* Rigby 1997a, 1.5 x natural size, locality 80 BH 302D. a, primary and secondary venation. b, tertiary venation superimposed. c, Holotype of *Koraua hartonoi* Rigby 1997a, natural size. The parallel ribbing was part of the stem supporting the leaf tuft. Locality 80 AG 64B.

4 DISCUSSION

The flora from the Aiduna Formation is significant because it is the only Permian land flora known from the island of New Guinea. The report of *Glossopteris* from Papua by Mackay & Little (1911) is discounted as the specimen has been lost, and the locality appears to have been in strata now known to be Tertiary. The flora was dominated by *Glossopteris* species typically related to Gondwanaland, but includes the Cathaysian-related species *Fascipteris aidunae* and *Gigantonoclea iriani* . Both Jongmans (1940) and Hopping & Wagner (1962) seem to have been considering the flora as a predominantly tropical-subtropical Cathaysian-Laurasian flora with an admixture of Gondwanaland elements, whereas my investigations show it to be the other way round as demonstrated by the species now known to be present. There had to be a land connection with both Gondwanaland and Cathaysia as both *Glossopteris* and *Gigantonoclea* were seed plants, and no animal vector for land or sea transport of seed had evolved at that time.

Most of the recent palaeogeographical reconstructions are based using the Plate Tectonics Theory, eg. by Metcalfe (1999). He describes a Late Devonian to Triassic Palaeo-Tethys oceanic basin formed when in part the South and North Chinese plates on one side separated from the Australian plate on the other side of the seaway. This marine basin opened to more than 20° of latitude in South East Asia during the Early Permian when the Aiduna Formation was deposited. His Meso-Tethys opened when a large sliver of Gondwanaland separated from the main continental mass as the Cimmerian continent in the late Early Permian and moved northwards. This block moved too late to have acted as a transporter for *Gigantonoclea* and *Fascipteris* if one assumes these two species originated in New Guinea before colonizing the South and North Cathaysian plates, both of which had many species. Conversely, there was no transporter for these genera to move southwards to populate New Guinea over an ocean 25° to 30° wide. Migration by wind may have been possible for the spores of the fern *Fascipteris* even over the immense distance apart of the Cathaysian blocks and Gondwanaland.

An alternative reconstruction based on the Earth Expansion Theory has been proposed by Carey (1996, fig. 44). Rigby (1999a) has compared the Plate Tectonics and Earth Expansion Theories, in particular for South East Asia, where Carey's reconstruction for the Triassic is identical to the post Early Permian reconstruction in South East Asia. Instead of wide oceanic basins in the Permian, a series of thrust faulted blocks moved along the various suture lines mapped in Metcalfe (1998) letting narrow oceanic basins form at various times. Both the Palaeo-Tethys and Meso-Tethys were still present, but as narrow Red Sea-like seaways. Slight tectonic movements would have exposed the sea floor from time to time to form a passage way for the migration of *Gigantonoclea* and *Fascipteris* from the South Cathaysian Plate during the Early Permian. Rigby (1998, 1999a) used this reconstruction to explain the distribution of these genera. The presence of *Glossopteris* has been reported in both Thailand in a shallow marine sequence by Kon'no (1963) and Anatolia in a flora containing Cathaysian and Laurasian genera by Wagner (1962). These occurrences are to be expected along the narrow Palaeo-Tethyan seaway as suggested by Carey (1996) in his reconstruction where they then lay along the narrow seaway in close proximity to Gondwanaland. The New Guinean flora supports the Earth Expansion Theory which holds that the Tethys Sea was a narrow, Red Sea-like waterway which had its floor exposed at various times and places allowing migration across its width. The Tethys Sea of the Plate Tectonics model was a broad ocean across which seed-plant migration might have been possible only by island rafts. A raft carrying a *Gigantonoclea* plant would have had to have had a journey of 20° of arc in a southerly direction during the late Early Permian.

ACKNOWLEDGEMENTS

D. B. Dow and C. J. Pigram of the Australian Geological Survey Organisation, Canberra, who were engaged in the Irian Jaya Mapping Project, 1979 and 1980, are thanked for allowing me to describe the collection, as are Li X.-x., Nanjing Institute of Geology and Palaeontology, Academia Sinica, and R. H. Wagner, Jardin Botanico, Córdoba, Spain, for useful discussion. Research was undertaken in the School of Natural Resource Sciences, Queensland University of Technology, Brisbane, Queensland.

REFERENCES

Andrews, H. N. 1970. Index of Generic Names of fossil plants. *U. S. Geological Survey Bulletin* 1300: 1-354.

Asama, K. 1970. Evolution and classification of Sphenophyllales in Cathaysia Land. *Bulletin National Science Museum* 13(2): 291-317; pls 1-7.

Ball, L. C. 1912. Mount Mulligan coalfield. *Geological Survey Queensland Publication* 237: 1-39; 7 pls.

Brongniart, A., 1822. Sur la classificacion et la distribution des végétaux fossiles en général, et ceux des terrains de sédiment supérieur en particulier. *Mémoires Museum national d'Histoire naturelle* 8: 203-248; pls 12-15.

Brongniart, A. 1828. Prodrôme d'une histoire des végétaux fossiles, ou recherches botaniques et géologiques sur les végétaux renfermés dans les diverse couches du globe. *Dictionnaire des Sciences naturales* 57: 16-212.

Brongniart, A. 1831. *Histoire des Végétaux fossiles ou Recherches botaniques et géologiques sur les Végétaux renfermes dans les diverses couches du Globe.* Dufour & D'Ocagne, Paris. v. 1, pt. 5; pp. 209-248; 14 pls.

Brongniart, A. 1849. Tableau des genres de végétaux fossiles considérés sous le point de vue de leur classification botanique et de leur distribution géologique. *Dictionnaire universale histoire natural* 13: 52-176.

Carey, S. W. 1996. *Earth Universe Cosmos.* University of Tasmania, Hobart. 231 pp.

Feistmantel, O. 1880. The flora of the Damuda and Panchet Division. Part 1. *Memoirs Geological Survey India. Palaeontologia Indica.* (12)3(2): 1-77; pls 1A-16A*bis.*

Gu & Zhi. 1974. *Chinese Index Fossils. Chinese Plant Fossils.* vol. 1. *Chinese Palaeozoic Plants.* Nanjing Institute of Geology and Palaeontology, Science Press, Beijing. 277 pp.; 130 pls.

Hopping, C. H. & Wagner, R. H. 1962. Plant fossils. In Visser, W. A. and Hermes, J. J. Geological results of the exploration for oil in Netherlands New Guinea. *Koninklijk Nederlands geologisch mijnbouwkundig genootschap. Geologisch serie, Speciaal nummer* 20, Enclosure 17: 1-11.

Jongmans, W. J., 1940. Beiträge zur Kenntnis der Karbonflora von Niederländisch Neu-Guinea. *Geologische Stichting, Mededelingen* 1938-1939: 263-274; pls 1-3.

Koidzumi,G., 1936. *Gigantopteris* flora. *Acta Phytotaxonomica et Geobotanica* 5(2): 130-144.

Kon'no, E., 1963. Some Permian plants from Thailand. *Japanese Journal of Geology and Geography* 34(2-4): 139-159; pl. 8.

Lacey, W. S. & Huard-Moine, D., 1966. Karoo floras of Rhodesia and Malawi - Part 2. The Glossopteris flora in the Wankie District of Southern Rhodesia. *Symposium on Floristics and Stratigraphy of Gondwanaland*: 13-25. Birbal Sahni Institute of Palaeobotany, Lucknow.

Lehner, P., van der Sijp, de Rijke, F., Bilardie, J. & Hermes, J. J., 1955. Geological survey of the Omba - Aidoena area (south coast). Nederlandsche Nieuw Guinee Petroleum Maatschappij Report 26380. (Not seen, cited in Pigram & Panggabean, 1983. There is an inconsistency in listing the author's names.)

Mackay, D. & Little, W. S., 1911. The Mackay - Little expedition in southern New Guinea. *Geographical Journal* 38: 483-487; 2 pls; 1 map.

M'Clelland, J. 1850. *Report of the Geological Survey of India for the season of 1848-49, 2 - Geognosy*. Military Orphan Press, Calcutta. pp. 52-57; pls 12-17.

Metcalfe, I., 1998. Palaeozoic and Mesozoic geological evolution of the SE Asian region: multidisciplinary constraints and implications for biogeography. In Hall, R. & Holloway, J. D. *Biogeography and Geological Evolution of SE Asia*. Backhuys, Leiden. pp. 25-41.

Metcalfe, I. 1999. The Tethys: How many? How old? How deep? How wide? *International symposium Shallow Tethys (ST) 5*. Chiang Mai: 1-15.

Morris, J., 1845. Fossil plants. In Strzelecki, P. E., *Physical description of New South Wales and Van Diemen's Land*. Longmans, Green, Brown & Longmans, London. pp. 245-253; pls 6-7.

Pigram, C. J. & Panggabean, H., 1983. *Geological Data Record. Waghete (Yapekopra) 1:250,000 sheet area Irian Jaya*. Geological Research & Development Centre, Bandung. 126 pp.; 42 figs; 1 map.

Prasad, B. & Maithy, P. K., 1990. *Pachwarophyllum santhalensis* gen. et sp. nov. from the Lower Gondwana of Bihar, India. *Review of Palaeobotany and Palynology* 63: 183-187.

Rigby, J. F. 1972. On *Arberia* White, and some related Lower Gondwana female fructifications. Palaeontology 15(1): 108-120.

Rigby, J. F. 1983. Plant fossils from the Aiduna Formation. Appendix 3, in Pigram, C. J. & Panggabean, H. *Geological data record. Waghete (Yapekopra) 1:250,000 sheet area Irian Jaya*: 91-92, 3 figs.

Rigby, J. F. 1997a. The significance of a Permian flora from Irian Jaya (West New Guinea) containing elements related to coeval floras of Gondwanaland and Cathaysialand, *Palaeobotanist* 45: 295-302.

Rigby, J. F. 1997b. *Glossopteris* occurrences in the Permian of Irian Jaya. *International Conference on stratigraphy and tectonic evolution of Southeast Asia and the South Pacific*. Department of Mineral Resources, Bangkok: 780.

Rigby, J. F. 1998. Upper Palaeozoic floras of SE Asia. In Hall, R. H. & Holloway, J. D. (eds). *Biogeography and geological evolution of SE Asia*. Backhuys, Leiden. pp. 73-82.

Rigby, J. F. 1999a. Permian floras and palaeogeography with particular emphasis on Gondwanaland and the Tethyan margin. *International symposium. Shallow Tethys (ST) 5*. Department Geological Science, Chiang Mai University, Chiang Mai: 185-196.

Rigby, J. F. 1999b. *Glossopteris* occurrences in the Permian of Irian Jaya (West New Guinea). *Proceedings Royal Society Victoria*. 110: 309-315.

Royle, J. F. 1833. *Illustrations of the botany and other branches of natural history of the Himalayan Mountains and of the flora of Cashmere*. vol. 1. Allen, London. (Note the reference to *Vertebraria* is on page xxix*, pl. ii)

Shi G. R., Archbold, N. W. & Zhan L.-P. 1995. Distribution and characteristics of mixed (transitional) mid-Permian (Late Artinskian - Ufimian) marine faunas in Asia and their palaeogeographical implications. *Palaeogeography, Palaeoclimatology, Palaeoecology* 114: 241-271.

Srivastava, A. K., 1978. Studies in the Glossopteris Flora of India - 43. Some new plant fossils from the Lower Gondwana sediments of Auranga Coalfield, Bihar. *Palaeobotanist* 25: 486-495.

Sternberg, G. K., 1825. *Versuch einer geognosischen botanischen Darstellung der Flora der Vorwelt*. Leipzig & Prague. v. 1, pt. 4; pp. 1-48.

Taylor, T. N. & Taylor, E. L., 1993. *The Biology and Evolution of Fossil Plants*. Prentice Hall, Englewood Cliffs. pp. xxii + 982.

Wagner, R. H. 1962. On a mixed Cathaysia and Gondwana flora from SE Anatolia (Turkey). *Congrès Avancement Étude Stratigraphie et Géologie Caronifère, 4ᵉ, Heerlen, Compte rendu* 3: 745-752.

Weiss, C. E., 1869. *Fossile der jüngsten Steinkohlenformation und das Rotliegenden im Saar- Rhein- Gebiete*. Bonn. pt. 1; pp. 1-100; pls 1-12. (not seen, cited in Andrews, 1970)

Yabe, H. 1922. Notes on some Mesozoic plants from Japan, Korea and China, in the collection of the Institute of Geology and Palaeontology of the Tohoku Imperial University. I. *Science Reports Tôhoku Imperial University, Sendai, Japan. Second series (Geology)* 7: 1-28; pls 1-4.

A biogeographic comparison of the dinosaurs and associated vertebrate faunas from the Mesozoic of Australia and Southeast Asia

J. A. Long
Western Australian Museum, Francis St., Perth, Western Australia, 6000

E. Buffetaut
CNRS, 16 cour du Liegat, 75013 Paris, France

ABSTRACT: Dinosaurs, and associated vertebrate faunas, are known from several sites in Australia and Southeast Asia, but the majority of them come from deposits of Early Cretaceous age (Valanginian-Albian). Comparisons between the faunas of these two regions show almost nothing in common of biogeographic significance, a scenario which supports the close proximity of Southeast Asia with mainland Asia since the start of the Mesozoic Era.

1 INTRODUCTION

Dinosaur remains are known from Australia and Southeast Asia in sedimentary depositsranging in age from the latest Triassic through to the end of the Cretaceous Period (Late Cretaceous only in Australia), although nearly all of the well-preserved and taxonomically identifiable material comes mostly from terrestrial deposits of the late Early Cretaceous in both regions. Reviews of the dinosaur faunas for both regions, with maps showing localities can be found in Buffetaut (1997, Buffetaut *et al.* 1989) and Long (1998). In this paper we review the known occurrences of dinosaurs from these two regions and compare their biogeographic affinities.

2 TRIASSIC FAUNAS

Triassic dinosaurs are known from Australia only by trackways (Bartholomai 1966). These represent large theropods having footprints up to 46 cm in length. A recent analysis of the geochemistry of the bones of the prosauropod *Agrosaurus macgillvrayi*, thought originally to have come from a Late Triassic- Early Jurassic site somewhere in northern Queensland (Seeley 1891), actually showed that it is most likely a mislabeled specimen collected from the south of England (Vickers-Rich *et al.* 1999). In Southeast Asia the oldest known dinosaur remains are of a prosauropod pelvis fragment from the Late Triassic Nam Phong Formation in Khon Kaen Province, Thailand. The specimen is similar to other prosauropods common in China and Europe at this time (Buffetaut 1997).

3 JURASSIC FAUNAS

Jurassic dinosaur remains are extremely rare in Australia and known only in southeast Asia by footprints, so there is little basis for biogeographic comparison. The partial skeleton of a moderately large sauropod, *Rhoetosaurus brownei,* was found in the early1920s from the Injune Creek beds (Bajocian age), near Roma in Queensland (Longman 1926, 1927). The Colalura Sandstone, of similar age, exposed near Geraldton, Western Australia, has produced some isolated dinosaur bones including the partial tibia of a small theropod named *Ozraptor subotaii* (Long & Molnar 1998), and a possible sauropod caudal vertebra (Long 1992). Jurassic dinosaur trackways representing ornithopods and a possible stegosaur have been recorded from Queensland by Thulborn (1990, 1994).

The only record of Jurassic dinosaurs from southeast Asia is from the Mid-Late Jurassic Phra Wihan Formation at Hin Lat Pa Chad, in Khon Kaen Province, Thailand. Dinosaur trackways occur here, mostly with individual footprints under 10cm in length, representing both small theropods and ornithischians (Buffetaut 1997).

4 CRETACEOUS FAUNAS

Early Cretaceous dinosaurs are known principally from sites ranging in age from Valanginian-Albian stages in southern Victoria (Otway Ranges and Strzlecki Ranges); the Aptian-Albian opal fields in Lightning Ridge, New South Wales; Andamooka and Coober Pedy, South Australia; and from several shallow marine and terrestrial deposits in the Aptian-Albian of Queensland (see Long 1998 for details of all sites). The faunas, as represented by skeletal remains, appear to be dominated by ornithischians, particularly hypsilophodontians (Rich & Vickers-Rich 1999) of which at least four genera are known *(Fulgurotherium, Leaellynasaura, Atlascopcosaurus, Qantassaurus)*, as well as large ornithopods (e.g. *Muttaburrasaurus,* Molnar 1996a) and thyreophoran remains (e.g. *Minmi;* Molnar 1980, 1996b). Ornithopods are also well-represented by footprints of small-medium sized animals at Winton, Queensland (*Wintonopus latimorum,* Thulborn & Wade, 1979, 1984) and by medium and large-sized animals at Broome, Western Australia (Long 1998). Sauropods are rare apart from some isolated bones from Queensland of *Austrosaurus* (Longman, 1933; Coombs and Molnar 1981), and footrpints at Broome (Thulborn *et al* 1996).

Theropods appear to be a rare part of the fauna, known from scant skeletal remains although their footprints are abundant at Broome (*Megalosauropus broomensis*) and Winton (*Skartopus australis* and cf. *Tyrannosauropus*). *Allosaurus*, a genus well-known from the Late Jurassic of North America, was recorded from the east coast of Victoria by Molnar *et al.* (1981), although a more recent study of the specimen, an astragalus, has pointed to it being outside the generic range of *Allosaurus* and the family Allosauridae, but definitely having allosauroid affinities (Chure 1998).

Isolated bones of theropods include an enigmatic slender coelurosaurian named *Kakuru kujani* from Andamooka (Molnar & Pledge 1980), an isolated metacarpal from Lightning Ridge named as *Rapator ornitholestoides* (Huene 1932), and from the Victorian deposits isolated bones of an ornithomimosaur *Timimus hermani* (Rich & Vickers-Rich 1994), and other bones of oviraptorosaurs (Currie *et al.* 1996) and teeth of dromaeosaurids (Long 1998). Two bones, an ulna and a partial pedal claw suggest the presence of a very large theropod (figured in Tomida 1998). The only large theropods in Gondwana at this time are the carcharodontosaurids *(Giganotosaurus, Carcharodontosaurus)*, the largest members of the allosauroid clade, *Deltadromeus* from Africa (Sereno *et al.* 1996), and the spinosaurids (eg *Suchomimus,* Sereno *et al.* 1998; and *Irritator,* Martill *et al.* 1996), so its quite possible that these bones could represent a taxon from any of these clades. Other theropods are well-represented by their trackways at Broome, Western Australia, from Dinosaur Cove, Victoria, and near Winton, Queensland.

The most enigmatic occurrences of Australian dinosaurs are the records of a possible neoceratopsian having similarities to *Leptoceratops* (Rich & Vickers-Rich 1994), the presence of the early ornithomimosaur *Timimus hermani* (Rich & Vickers-Rich 1994), and the presence of possible oviraptorosaurs (Currie *et al.* 1996). All of these groups are elsewhere known from the Late Cretaceous of North America and Asia, so, if present their early appearance in Australia suggests that they may have had origins in Gondwana before migrating northwards.

The best preserved specimens of Early Cretaceous dinosaurs from Southeast Asia come from the Sao Khua Formation in Phu Wiang National Park near the city of Kalasin in Thailand. The sauropod *Phuwiangosaurus sirindhornae,* known from a partial skeleton, would have been about 15 metres long. Its closest affinities appear to lie with the titanosauroids from India, Africa and South America (Upchurch 1999), so it has no immediate relationship to any of the Chinese or Mongolian sauropods. Theropods from the same site include bones of a small compsognathid, isolated teeth possibly belonging to a spinosaurid (named *Siamosaurus suteethorni)* and other theropod teeth of indeterminate affinity. Recently a partial skeleton of a primitive tyrannosaurid named *Siamotyrannus isanensis* was described by Buffetaut *et al.* (1996), the oldest known member of that family, which is elsewhere restricted to the late Cretaceous of Asia and North America. In addition bones of a fairly advanced small ornithomimosaur are known (Buffetaut 1997). Dinosaur footprints from the Phu Phan Formation, which overlies the Sao Khua Formation indicate the presence of larger theropods, about 1.8 metres high at the hip. The youngest dinosaur fauna from Thailand come from the Khok Kruat Formation, of late Early Cretaceous age (Aptian-Albian). These include theropod teeth, bones of psittacosaurids, including the local species *Psittacosaurus sattayaraki*.

Dinosaurs are also known from the extension of the Sao Khua Formation over in Laos, around the Muong Phalane district. Hoffett (1942, 1943a, b) described isolated large bones of a sauropod (he referred to as a new species"*Titanosaurus falloti*") and of a hadrosaur which he called *Mandschurosaurus laosensis*. New material of a sauropod has recently been described from Tang Vay, Savannakhet Province, Laos, as *Tangvayosaurus hoffetti* (Allain *et al.*, 1999) and it is regarded as a titanosaurid closely allied to *Phuwiangosaurus* from Thailand. The ornithopod bones collected by Hoffet are currently regarded as being indeterminate. Buffetaut (1997) regards the dinosaur remains from the Savannakhet region as being much younger than the main dinosaur bearing sites of the Khorat Plateau in Thailand, possibly as young as basal Late Cretaceous.

5 BIOGEOGRAPHIC COMPARISONS

The only comparisons that can be made between the two region have to be based on similar kind of dinosaurs from similar age ranges. Australian faunas are dominated by hypsilophodontians, but none are known from Southeast Asia, so only the sauropods, theropods, and associated non-dinosaurian vertebrates are a useful basis for comparing biogeographic affinities. Table 1 shows a comparison between the known dinosaur taxa from Australia and Southeast Asia with those known from mainland Asia (China/Mongolia).

The sauropod *Phuwiangosaurus* from Thailand (Martin *et al.* 1994) is a form closely resembling the Nemegtosauridae from the Upper Cretaceous of Mongolia and China according to Buffetaut & Suteethorn (1999). Allain *et al.* (1999) differ in this respect as they regard this genus to be a closely relative of *Tangvayosaurus* from Laos, which they consider as having titanosaurid affinities. In Australia the primitive Early Jurassic sauropod *Rhoetosaurus* shows no direct affinity with those from anywhere else. *Rhoetosaurus* has a peculiar, rapidly tapering stout series of anterior caudal vertebrae which we here suggest could be an adaptation for stiffening the tail possibly for bearing the weight of a bony club. This is only a suggestion, but if it proves to be correct then its closest affinities would possibly lie with the club-tailed sauropods *Shunosaurus* and *Omeiosaurus* from the Late Jurassic of China. *Austrosaurus*, from the Middle Cretaceous of Queensland, is now thought to have titanosaurid affinities (R. Molnar, pers.comm 2000) and therefore may potentially have affinities to the newly described titanosaurid *Tangvayosaurus* from Laos, although its remains are presently too incomplete to make meaningful comparisons.

The Tyrannosauridae, as represented by *Siamotyrannus* from Thailand, are a family otherwise known exclusively from North America and Asia during the Late Cretaceous, with no proven record from Australia. Isolated teeth of a possible spinosaurid, *Siamosaurus*, are also known from Thailand. This family has largely Gondwanan (e.g. *Spinosaurus, Irritator, Suchomimus)* and western Laurasian affinities (e.g.*Baryonyx)*.

Both Australia and Thailand have yielded Early Cretaceous remains of ornithomimosaurs, a group better known from the Late Cretaceous of Laurasia. Although its affinities are not clear, a candidate put forward for the oldest possible ornithomimosaur has been *Elaphrosaurus* from the late Jurassic of eastern Africa. *Elaphrosaurus* may possibly represent an early off-shoot leading to ornithomimosaurs, as suggested by Barsbold and Osmolska (1990), which would support an origin for the group in Gondwana by the start of the Cretaceous Period. However, more recent interpretations place *Elaphrosaurus* at the base of the clade Ceratosauria (*sensu* Gauthier 1986), or as a sister group to the other Gondwana theropods including *Xenotarsosaurus, Noasaurus* and the abelisaurids (Holtz 1994, Sampson *et al.* 1998). If this is correct, then it has no bearing on the origin of ornithomimosaurs. The oldest occurrence of true ornithomimosaurids is *Har-*

pymimus from the Aptian-Albian of Mongolia, approximately contemporaneous with both *Timimus* from Australia and the undescribed ornithomimosaur from Thailand. This suggests that the group was widespread by the late Early Cretaceous in both northern and southern hemispheres.

TAXON	Australia	SE Asia	China/Mongolia
THEROPODA			
Allosauridae	?allosaurid	–	*Chilantaisaurus*
Tyrannosauridae	?ichnotaxon*	*Siamotyrannus*	–
Oviraptorosauridae	gen. indet.	–	–
Ornithomimosauria	*Timimus*	gen. indet.	*Harpymimus*
Dromaeosauridae	gen.idet.	–	–
Fam. indet.	*Kakuru*	–	–
	Rapator	–	–
SAUROPODA			
Titanosauridae?	ichnotaxa**	*Phuwiangosaurus*	?*Mongolosaurus*
THYREOPHORA			
Stegosauridae	ichnotaxon**	–	*Wuehosaurus*
Ankylosauridae	*Minmi*	–	*Shamosaurus*
ORNITHOPODA			
Hypsilophodontidae	*Fulgurotherium*	–	indet.taxon***
	Atlascopcosaurus	–	–
	Leaellynasaura	–	–
	Qantassaurus	–	–
Family Indet.	*Muttaburrasaurus*	–	*Probactrosaurus* etc.
CERATOPSIA			
Psittacosauridae	–	*Psittacosaurus*	*Psittacosaurus* spp.
Protoceratopsidae	?*Leptoceratops*	–	–

* *Tyrannosauropus* Winton site, Queensland (Albian-Cenomanian) could be long to any family of large theropod.
** Broome Sandstone sites, Western Australia (Early Cretaceous); stegosaur prints recorded by Long (1998).
*** Early Cretaceous hypsilophodontids are reported from the North China *Psittacosaurus* fauna by Dong (1993).

Table 1. Comparison between known dinosaur taxa from Australia and Southeast Asia and those from mainland Asia (China/Mongolia)

Possible oviraptorosaur remains have been recorded from the Early Cretaceous of Victoria by Currie *et al* (1996) and they are now recorded from the Late Cretaceous of Argentina (Frankfurt and Chiappe 1999) and Brazil (Frey & Martill 1995) demonstrating that the group was widespread through parts of Gondwana and Euramerica by the Early Cretaceous.

Psittacosaurus sattayaraki which occurs in Thailand (Buffetaut *et al.* 1989), is a well known genus from the Early Cretaceous of Mongolia and northern China, but currently unknown from Australia. However the genus is considered to be a sister taxon to the neoceratopsians, so its geographical range has bearing on the origin of the ceratopsians. An early possible neoceratopsian has been reported from an Aptian site near Kilcunda, Victoria (Rich & Vickers-Rich 1994). Elsewhere the oldest known neoceratopsians are the Albian forms *Kulceratops* from central Asia (Nessov 1995) and *Archaeoceratops* from China (Dong & Azuma 1997). The possible presence of this group in Australia this time would suggest that the group was restricted to East Gondwana and central Asia, and that the group differentiated at atime while a land bridge was still joining the two regions.

The occurrence of other Mesozoic terrestrial vertebrates in both Thailand and Australia is worth considering in the light of biogeographic affinities. The appearance of Late Triassic turtle remains from the Huai Hin Lat Fm (Thailand) has distinctly Laurasian affinities (de Broin *et al.* 1982). Similarly Australia's Cretaceous faunas are peculiar in retaining the world's last labyrinthodont amphibians (*Koolasuchus cleelandi* from Victoria; Long 1998). The group survived into the Middle Jurassic in China (*Sinobrachyops*) and the Late Jurassic in Mongolia (Rich 1996). Furthermore, Cretaceous crocodilians like *Goniopholis phuwiangensis* from Thailand (Buffetaut and Ingavat 1983) bear close affinities to forms from Upper Jurassic-E. Cretaceous of Laurasia, not to those known from Gondwana.

All of these data support the concept that the dinosaurs and associated terrestrial vertebrate faunas of Southeast Asia were clearly aligned with those from the northern hemisphere, and show no similarities to those from Gondwana. The distributions shown by dinosaurs in Australia and Asia support the presence of a continuous land connection between Laurasia and Gondwana in the Middle-Late Jurassic in order to explain the dispersal of groups such as ornithomimosaurs, oviraptorosaurs and neoceratopsians (all appearing late in the Early Cretaceous in the fossil record) by the end of the Early Cretaceous.

ACKNOWLEDGEMENTS

We would like to thank Dr Tom Rich, Museum of Victoria, and Dr Ralph Molnar, Queensland Museum, for their helpful comments on the manuscript. The senior author thanks Dr Ian Metcalfe for finacial support to attend the "Where Worlds Collide" meeting in Armidale.

NOTE ADDED IN PROOF

A number of recent discoveries have improved our knowledge of the Mesozoic vertebrate faunas of Thailand and the picture we now have of them is somewhat different from that outlined above. The main new discoveries are the following :

- The Late Triassic Nam Phong Formation has yielded remains of the earliest known sauropod dinosaur, Isanosarus attavipachi (Buffetaut *et al.*, 2000).
- The Phu Kradung Formation, which on the basis of palynology (Racey *et al.*, 1996) is Late Jurassic or possibly basal Cretaceous, has yielded various dinosaurs, including theropods, small ornithopods, probable euhelopodid dinosaurs (Buffetaut & Suteethorn, 1998), and stegosaurs (Buffetaut *et al.*, 2001).
- The Phra Wihan Formation, in which footprints of ornithopods, theropods and sauropods occur, is now firmly placed in the Early Cretaceous by palynology (Racey *et al.*, 1996).
- Cranial remains of Phuwiangosaurus sirindhornae, from the Early Cretaceous Sao Khua Formation, clearly demonstrate its nemegtosaurid affinities.
- The dinosaur-bearing beds of Laos are in all likelihood a lateral extension of the Khok Kruat Formation of Thailand.

REFERENCES

Allain, R., P. Taquet, B. Battail, J. Dejax, P. Richir, M. Veran, F. Limon-Duparcmeur, R. Vacant, O. Mateus, P. Sayarath, B. Khenthavong & S. Phouyavong. 1999. Un nouveau genre de dinosaure sauropode de la formation des Grès supérieurs (Aptien-Albien) du Laos. *Centre Reserches de l' Academie Scientifique, Paris, Sciences de la terre et des planetes* 329: 609-616.

Bartholomai, A. 1966. Fossil footprints in Queensland. *Australian Natural History* 15: 147-150.

Buffetaut, E. 1997. Southeast Asian dinosaurs. In Currie, P. & K. Padian (eds.) *Encyclopedia of Dinosaurs*: 689-691. Academic Press.

Buffetaut, E. & Ingavat, R. 1983. *Goniopholis phuwiangensis* nov.sp., a new mesosuchian crocodile from the Mesozoic of northeastern Thailand. *Geobios* 16: 79-91.

Buffetaut, E., Sattayarak, N & Suteethorn, V. 1989. A psittacosaurid dinosaur from the Cretaceous of Thailand and its implications for the palaeogeographical history of Asia. *Terra Nova* 1: 370-373.

Buffetaut, E. & Suteethorn, V. 1998. Early Cretaceous dinosaurs fromThailand and their bearing on the early evolution and biogeographical history of some groups of Cretaceous dinosaurs. *New Mexico Museum of Natural History and Science Bulletin* 14: 205-210.

Buffetaut, E. & Suteethorn, V. 1999. The dinosaur fauna of the Sao Khua Formation of Thailand and the beginning of the Cretaceous radiation of dinosaurs in Asia. *Palaeogeography, Palaeoclimatology, Palaeoecology* 150: 13-23.

Buffetaut, E. Suteethorn, V. & Tong, H. 1996. The earliest known tyrannosaur from the Lower Cretaceous of Thailand. *Nature* 381: 689-691.

Buffetaut, E., Suteethorn, V. & Tong, H. 2001. The first thyreophorandinosaur from Southeast Asia: a stegosaur vertebra from the Late Jurassic Phu Kradung Formation of Thailand. *Neues Jahrbuch für Geologie und Paläontologie, Monatshefte* 2: 95-102.

Buffetaut, E., Suteethorn, V., Cuny, G., Tong, H., Le Loeuff, J., Khansubha, S. & Jongautchariyakul, S. 2000. The earliest known sauropod dinosaur. *Nature* 407: 72-74.

Buffetaut, E., Tong, H. & Suteethorn, V. 1994. First post-Triassic labyrinthodont amphibian in South East Asia: a temnospondyl intercentrum from the Jurassic of Thailand. *Neues Jahrbuch für Geologie und Paläontologie, Monatshefte,* 7: 385-390.

Buffetaut, E., Suteethorn, V., Cuny, G., Tong, H., Le Loeuff, J., Khansubha, S. & Jongautchariyakul, S. 2000. The earliest known sauropod dinosaur. *Nature* 407: 72-74.

Chure, D.J. 1998. A reassessment of the Australian *Allosaurus* and its implications for the Australian refugium concept. *Journal of Vertebrate Paleontology* 18 (supplement to number 3): 34A.

Coombs, W.P. Jr. & Molnar, R.E. 1981. Sauropods (Reptilia, Saurischia) from the Cretaceous of Queensland. *Memoirs of the Queensland Museum* 20: 351-373.

Currie, P.J., Rich, T.H. & Vickers-Rich, P. 1996. Possible oviraptorosaur (Theropoda, Dinosauria) specimens from the Early Cretaceous Otway Group of Dinosaur Cove, Australia. *Alcheringa* 20: 73-79.

De Broin, F., Ingavat, R., Janvier, P. & Sattarayak, N. 1982. Triassic turtle remains from northeastern Thailand. *Journal of Vertebrate Paleontology* 2: 41-46.

Dong, Z-M. 1992. *Dinosaurian Faunas of China.* Beijing: China Ocean Press.

Dong Z-M. & Azuma, Y., 1997. On a primitive neoceratopsian from the Early Cretaceous of China. In Z.-M. Dong (ed.), *The Sino-Japanese Silk Road Dinosaur Expedition*: 68-89. Beijing: China Ocean Press.

Frankfurt, N.G. & Chiappe, L.M. 1999. A possible oviraptorosaur from the Late Cretaceous of Argentina. *Journal of Vertebrate Paleontology* 19: 101-105.

Frey, D. & Martill, D. M. 1995. A possible oviraptosaurid theropod from the Santana Formation (Lower Cretaceous, ?Albian) of Brazil. *Neues Jarhbuch für Geologie und Paläontologie Monatshefte* 1995: 397-412.

Gauthier, J.A. 1986. Saurischian monophyly and the origin of birds. In K. Padian (ed.) *The Origin of Birds and the Evolution of Flight. Memoirs of the California Academy of Sciences* 8: 1-55.

Hoffet, J. H. 1942. Description des quelques ossements du Sénonien du Bas-Laos. *C. R. Cons Rechereches Scientifiques Indochine* 1942: 43-57.

Hoffet, J.H., 1943a. Description des quelques ossements du de Titanosauriens du Sénonien du Bas-Laos. *C. R. Cons Rechereches Scientifiques Indochine* 1943: 1-8.

Hoffet, J.H., 1943b. Description des quelques ossements les plus caracteristiques appartenant à des avipelviens du Sénonien du Bas-Laos. *C. R. Cons Rechereches Scientifiques Indochine* 1944: 179-186.

Holtz Jr., T.R. 1994. The phylogenetic implications of the Tyrannosauridae: implications for theropod systematics. *Journal of Paleontology* 68: 1100-1117.

Huene, F. von. 1932. Die fossile Reptil-Ordnung Saurischia, ihre Entwicklung und Geschichte. *Monographs in Geologie und PalÑontologie* 1: 1-361.

Long, J.A. 1992. First dinosaur bones from Western Australia. *The Beagle, Records of the Northern Territory Museum of Arts and Sciences* 9: 21-8.

Long, J.A. 1998. Dinosaurs of Australia and New Zealand and other animals of the Mesozoic Era. Sydney: University of New South Wales Press; Cambridge, USA: Harvard University Press.

Long, J.A. & Molnar, R.E. 1998. A new Jurassic theropod dinosaur from Western Australia. *Records of the Western Australian Museum* 19: 121-129.

Longman, H. 1926. A giant dinosaur from Durham Downs, Queensland. *Memoirs of the Queensland Museum* 8: 183-194.

Longman, H. 1927. The giant dinosaur *Rhoetosaurus brownei*. *Memoirs of the Queensland Museum* 9: 1-18.

Longman, H. 1933. A new dinosaur from the Queensland Cretaceous. *Memoirs of the Queensland Museum* 13: 133-44.

Martill, D. M., Cruickshank, A. R. I., Frey, E., Small, P. G. & Clarke, M. 1996. A new crested maniraptoran dinosaur from the Santana Formation (Lower Cretaceous) of Brazil. *Journal of the Geological Society* 153: 5-8.

Martin, V., Buffetaut, E. & Suteethorn, V. 1994. A new genus of sauropod dinosaur from the Sao Khua Formation (Late Jurassic or Early Cretaceous) of northeastern Thailand. Comptes Rendus l'Acadamie des Sciences, Paris II, 319: 1085-1092.

Molnar, R.E. 1980. An ankylosaur (Ornithischia: Reptilia) from the Lower Cretaceous of southern Queensland. *Memoirs of the Queensland Museum* 20: 77-87.

Molnar, R.E. 1996a. Observations on the Australian ornithopod dinosaur *Muttaburrasaurus*. *Memoirs of the Queensland Museum* 39: 639-52.

Molnar, R.E. 1996b. Preliminary report on a new ankylosaur from the Early Cretaceous of Queensland, Australia. *Memoirs of the Queensland Museum* 39: 653-68.

Molnar, R.E. , Flannery, T.F. and Rich, T.H., 1981. An allosaurid theropod dinosaur from the Early Cretaceous of Victoria, Australia. *Alcheringa* 5: 141-146.

Molnar, R.E. & Pledge, N.S. 1980. A new theropod dinosaur from South Australia. *Alcheringa* 4: 281-7.

Nessov, L.A. 1995. Dinosaurs of northern Eurasia: new data about assemblages, ecology and palaeobiogeography. *University of St. Petersberg, Institute of Earth Crust Publication*: 25-34.

Racey, A., Love, M.A., Canham, A.C., Goodall, J.G.S., Polachan, S. & Jones, P.D. 1998. Stratigraphy and reservoir potential of the Mesozoic Khorat Group, NE Thailand. Part 1 : Stratigraphy and sedimentary evolution. *Journal of Petroleum Geology* 19: 5-40.

Rich, T.H. 1996. Significance of polar dinosaurs in Gondwana. *Memoirs of the Queensland Museum* 39: 711-7.

Rich, T.H. and Vickers-Rich, P. 1994. Neoceratopsians and ornithomimosaurs: dinosaurs of Gondwana origin? *Research and Exploration* 10: 129-131.

Rich, T.H. and Vickers-Rich, P. 1999. The Hypsilophodontidae from southeastern Australia. In Y.Tomida, T.H. Rich, & P.Vickers-Rich (eds.) *Proceedings of the Second Gondwana Dinosaur Symposium*: 167-180. National Science Museum Monographs, Tokyo (no. 15).

Sampson, S. D., Witmer, L.M., Forster, C.A., Krause, D.W., O'Connor, P.M, Dodson, P. & Ravoavy, F. 1998. Predatory dinosaur remains from Madagascar: implications for the Cretaceous biogeography of Gondwana. *Science* 280: 1048-1051.

Seeley, H.G. 1891. On *Agrosaurus macgillvrayi*, a saurischian reptile from the NE-coast of Australia.*Quarterly Journal of the Geological Society of London* 47: 164-5.

Sereno, P.C., Duthiel, D.B., Iarochene, M., Larsson, H.C.E., Lyon, G.H., Magwene, P.M., Sidor, C.A., Varrichio, D.D. & Wilson, J.A. 1996. Predatory dinosaurs from the Sahara and Late Cretaceous faunal differentiation. *Science* 272: 986-991.

Sereno, P. C., Beck, A.L., Duthiel, D.B., Gado, B., Larsson, H.C.E., Lyon, G.H., Marcot, J.D., Rauhut, O.W.M., Sadlier, R.W., Sidor, C.A., Varrichio, D.D., Wilson, G.P. & Wilson, J.A. 1998. A long-snouted predatory dinosaur from Africa and the evolution of spinosaurids. *Science* 282: 1298-1302.

Thulborn, R.A. 1994. Ornithopod dinosaur tracks from the Lower Jurassic of Queensland. *Alcheringa* 18: 247-258.

Thulborn, R.A., Hamley, T. & Foulkes, P. 1996. Preliminary report on sauropod dinosaur tracks in the Broome Sandstone (Lower Cretaceous) of Western Australia. *Gaia* 10: 85-96.

Thulborn, R.A. & Wade, M., 1979. Dinosaur stampede in the Cretaceous of Queensland. *Lethaia* 12: 275-79.

Thulborn, R.A. & Wade, M., 1984. Dinosaur trackways in the Winton Formation (Mid-Cretaceous) of Queensland. *Memoirs of the Queensland Museum* 21: 413-518.

Tomida, Y. (ed.) 1998. *Dinosaurs of Gondwana*. Japan: The Yomiuri Shimbun.

Upchurch, P. 1999. The phylogenetic relationships of the Nemegtosauridae (Saurischia, Sauropoda). *Journal of Vertebrate Paleontology* 19: 106-125.

Vickers-Rich, P., Rich, T.H., McNamara, G. & Milner, A. 1999. Is *Agrosaurus macgillivrayi* Australia's oldest dinosaur?. *Records of the Western Australian Museum Supplement 57*: 191-200.

Early Midle Jurassic (Aalenian) radiolarian fauna from the Xialu chert in the Yarlung Zangbo Suture Zone, southern Tibet

Atsushi Matsuoka, Kenta Kobayashi & Toru Nagahashi
Department of Geology, Niigata University, Niigata, Japan

Qun Yang & Yujing Wang
Nanjing Institute of Geology and Palaeontology, Chinese Academy of Sciences, Nanjing, China

Qinggao Zeng
Institute of Geological Science of Tibet, Lhasa, China

ABSTRACT: An early Middle Jurassic (Aalenian) radiolarian fauna is reported from the Xialu chert in the Yarlung Zangbo Suture Zone. This is the oldest known fossil record from the chert. The fauna comes from a chert sample which contains no terrigenous elements other than clay minerals. This indicates that the Ceno-Tethys was already a rather wide ocean with a pelagic environment far from continents by that time. The rifting and separation of the Lhasa Block from the northern Gondwana margin must have occurred well before early Middle Jurassic time.

1 INTRODUCTION

The Ceno-Tethys is a southern branch of the Tethyan ocean and existed in the Mesozoic and early Cenozoic times (e.g. Metcalfe 1999). The evolution of the Ceno-Tethys has been discussed based mainly on geological and paleontological data obtained from shallow marine sediments on both sides of the ocean. Pelagic sediments deposited in a deep ocean basin of the Ceno-Tethys are commonly incorporated in suture zones. However, geological and paleontological data from the pelagic sediments, which enable us to elucidate the paleoceanography and tectonic evolution of the Ceno-Tethys, are still limited.

The Xialu chert (Wu 1993) is distributed along the southern margin of the Yarlung Zangbo Suture Zone (Fig. 1) and represents deep marine sediments between the Indian Block and Lhasa Block. Wu (1993) reported Late Jurassic and Early Cretaceous radiolarians from the Xialu chert and discussed the geological history of the east Tethys.

We carried out a field survey in southern Tibet in 1998 and visited the type locality of the Xialu chert. As a result of radiolarian biostratigraphic research along two continuous sections, we identified seven different aged radiolarian assemblages from pelagic and hemipelagic sediments ranging from early Middle Jurassic to Early Cretaceous (Matsuoka et al. 1999a). This paper reports the oldest known (Aalenian) radiolarian fauna from the Xialu chert with scanning electron microscopic images and discusses the evolution of the Ceno-Tethys. This can give a direct constraint on the timing of rifting and separation of the Lhasa Block from the Gondwana margin.

2 GEOLOGICAL SETTING

Figure 1 illustrates the general geology around the Yarlung-Zangbo Suture Zone. Cretaceous forearc sediments called the Xigaze Group conformably rests on or are in fault contact with ophiolitic rocks in the northern part of the suture zone. To the north of the suture zone, the Transhimalaya plutonic complex of Cretaceous age crops out. This complex belongs to the Lhasa Block. The Tethys Himalayas, widely distributed to the south of the suture zone, are characterized by sedimentary sequences of Paleozoic-Tertiary age which had accumulated on the northern margin of the Indian Block.

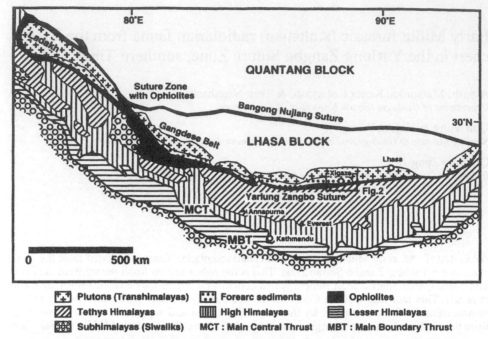

Figure 1. Index map of the study area. After Wang et al. (1996).

The Xialu chert (Wu 1993) crops out about 30 km south of Xigaze and occupies the southern marginal part of the Yarlung Zangbo Suture Zone (Fig. 2). It is in fault contact with an ophiolitic complex to the north and with the Paleogene conglomerate called the Liuqu Formation to the south (Wang et al. 1996). The Xialu chert is composed mainly of bedded cherts associated with subordinate siliceous mudstone. The apparent total thickness of siliceous deposits exceeds 1 km. We have examined two continuous sections, Xialu-W and Xialu-E, which are exposed along either side of a broad valley located south of the village Xialu. The Xialu-W section corresponds to a part of A-B section of Wu (1993) where the occurrence of Late Jurassic radiolarians has been reported (Fig. 2).

Seven different aged radiolarian zones are identified and are assignable to early Middle Jurassic to Early Cretaceous (Matsuoka et al. 1999a). These are assigned to JR 3 (Aalenian), JR 5 (late Bathonian-early Callovian), JR 6 (late Callovian-Oxfordian), JR 7 (Kimmeridgian), JR 8 (early-middle Tithonian), KR 1 (late Tithonian-early Valanginian), and KR 2-3 (late Valanginian-Barremian) of the western Pacific-Japan zonation (Matsuoka 1995). Middle and Late Jurassic assemblages are obtained from the Xialu-W section, while latest Jurassic to Early Cretaceous assemblages are found in the Xialu-E section. Tectonic repetition of siliceous deposits is recognized based on the distribution pattern of radiolarian zones and structural analysis (Matsuoka et al. 1999b, Kobayashi et al. 1999). Of 31 samples collected from the Xialu-W section in our 1998 field work, only one sample (98061235) is assigned to early Middle Jurassic age. The sample is a single layer of bedded chert and contains no terrigenous elements other than clay minerals.

3 RADIOLARIAN FAUNA AND ITS AGE ASSIGNMENT

Sample 98061235 contains abundant and moderately preserved radiolarian tests. Faunal composition with abundance data is presented in Table 1. Representative forms are illustrated in Figure 3. Nassellarians are more abundant than spumellarians in this sample. Abundant genera include

Table 1. List of radiolarians in Sample 98061235 from the Xialu chert.

Radiolarian species	Abundance	Fig. 3-#
Archaeodictyomitra gifuensis Takemura	Few	10
Archaeospngoprunum sp.	Few	
Dictyomitrella (?) sp.	Rare	
Eucyrtidiellum disparile Nagai and Mizutani	Few	16
Encyrtidiellum quinatum Takemura	Few	15
Guexell a sp.	Few	
Higmastra sp.	Few	
Hsuum altile Hori and Otsuka	Rare	8
Hsuum hisuikyoense Isozaki and Matsuda	Abundant	2, 3
Hsuum matsuokai Isozaki and Matsuda	Abundant	4
Laxtorum (?) *hichisoense* Isozaki and Matsuda	Common	1
Mesosaturnalis hexagonus (Yao)	Few	
Napora sp.	Rare	
Palinandromeda sp.	Few	11
Pantanellium sp.	Rare	
Parahsuum levicostatum Takemura	Rare	9
Parahsuum officerense (Pessagno and Whalen)	Few	7
Parahsuum simplum Yao	Few	6
Parahsuum sp.	Abundant	
Parahsuum (?) *magnum* Takemura	Rare	5
Parares sp.	Rare	12
Paronaella sp.	Common	
Parvicingula sp.	Few	
Protunuma sp. cf. *P. fusiformis* Ichikawa and Yao	Few	13
Pseudocrucella sp.	Few	
Sitchocapsa convexa Yao	Common	
Sitchocapsa sp.	Abundant	
Spongotripus (?) sp.	Rare	
Stichomitra (?) sp.	Few	
Tricolocapsa ruesti Tan	Common	14
"*Tripocyclia*" sp.	Few	
Tritrabs simplex Kito and De Wever	Common	
Unuma sp. cf. *U. echinatus* Ichikawa and Yao	Rare	
Unuma sp. cf. *U. typicus* Ichikawa and Yao	Common	
Xiphostylus sp.	Few	

Hsuum, Parahsuum, Laxtorum(?), *Protunuma, Eucyrtidiellum, Tricolocapsa, Stichocapsa, Tritrabs*, and *Paronaella*.

The sample contains *Laxtorum*(?) *hichisoense* Isozaki and Matsuda of which occurrence is restricted to the *Laxtorum*(?) *jurassicum* Zone or JR 3 Zone established in the western Pacific and Japan (Matsuoka 1995). Co-occurrence of *Hsuum matsuokai* Isozaki and Matsuda, *Hsuum hisuikyoense* Isozaki and Matsuda, *Encyrtidiellum quinatum* Takemura, *Eucyrtidiellum disparile* Nagai and Mizutani, *Archaeodictyomitra gifuensis* Takemura, and *Mesosaturnalis hexagonus* (Yao) is characteristic of the zone. However, *Laxtorum*(?) *jurassicum* Isozaki and Matsuda, the nominal species of the zone, has not been found yet.

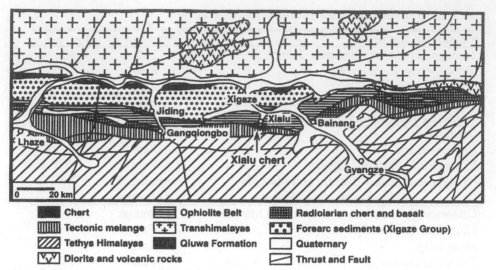

Figure 2. Geological map around the Yarlung Zangbo Suture Zone near Xigaze, southern Tibet. After Wu (1993).

Coeval radiolarian faunas have been reported from Japan (e.g. Isozaki & Matsuda 1985, Takemura 1986), North America (Carter et al. 1988), and the western Tethys regions (e.g. Kito et al. 1990). Matsuoka (1995) discussed the age of the lower and upper limits of the *Laxtorum*(?) *jurassicum* (JR 3) zone and gave an Aalenian age for the zone. A similar age is given when UAZone 95 (Baumgartner et al. 1995), which was established by International joint work on Middle Jurassic-Lower Cretaceous radiolarian biochronology of Tethys, is applied.

4 DISCUSSION

Micropaleontological analysis on the Xialu chert (Matsuoka et al. 1999a) has revealed that a pelagic deposition is recorded from Middle Jurassic to Early Cretaceous. This indicates that the Ceno-Tethys between the Lhasa Block and the northern Gondwana margin was an ocean with a long depositional history of more than 50 m.y. The absence of calcareous sediments suggests the basin was deeper than the CCD throughout the sedimentation period. The Aalenian radiolarian fauna reported herein is the oldest known record so far for the Xialu chert. It is noteworthy to point out that the fauna came from a chert sample which contains no terrigenous elements other than clay minerals. This indicates that the Ceno-Tethys was already a rather wide ocean with a pelagic environment far from continents by that time. The rifting and separation of the Lhasa Block from the northern Gondwana margin must be dated much before early Middle Jurassic time. There is a possibility that radiolarian faunas older than Aalenian age will be discovered from the Xialu chert in the future.

Metcalfe (1999) stated that the Ceno-Tethys was born sometime between Late Triassic and Jurassic times. This work can give a direct constraint on the timing of rifting and separation of the Lhasa block and the northern Gondwana margin. Metcalfe (1999) also pointed out that the rifting started earlier in the west part of the Ceno-Tethys and propagated to the east. To assess this hypothesis much work is required in the suture zones where pelagic sediments of the lost Ceno-Tethys ocean are incorporated.

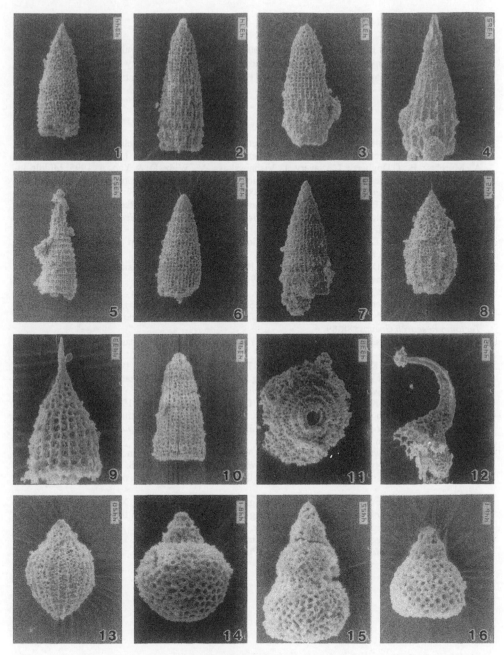

Figure 3. Scanning electron microscopic images of Aalenian radiolarians from the Xialu chert. One interval of the scale put in the margin of each photograph indicates 0.1 mm.

1. *Laxtorum*(?) *hichisoense* Isozaki and Matsuda. 2. *Hsuum hisuikyoense* Isozaki and Matsuda. 3. *Hsuum hisuikyoense* Isozaki and Matsuda. 4. *Hsuum matsuokai* Isozaki and Matsuda. 5. *Parahsuum*(?) *magunum* Takemura. 6. *Parahsuum simplum* Yao. 7. *Parahsuum officerense* (Pessagno and Whalen). 8. *Hsuum altile* Hori and Otsuka. 9. *Parahsuum levicostatum* Takemura. 10. *Archaeodictyomitra gifuensis* Takemura. 11. *Palinandromeda* sp. 12. *Parares* sp.. 13. *Protunuma* sp. cf. *P. fusiformis* Ichikawa and Yao. 14. *Tricolocapsa ruesti* Tan. 15. *Encyrtidiellum quinatum* Takemura . 16. *Eucyrtidiellum disparile* Nagai and Mizutani.

ACKNOWLEDGEMENTS

This paper was greatly improved by reviews from J. Aitchison and K. Wakita. This research has been supported by Grant-in-Aid for Scientific Research from the Ministry of Education, Science, Sports and Culture of Japan (no. 10041114) and Heiwa Nakajima Foundation.

REFERENCES

Baumgarter, P.O., Bartolini, A., Carter, E.S., Conti, M., Cortese, G., Danelian, T., De Wever, P., Dumitrica, P., Dumitrica-Jud, R., Gorican, S., Guex, J., Hull, D.M., Kito, N., Marcucci, M., Matsuoka, A., Murchey, B., O'Dogherty, L., Savary, J., Vishnevskaya, V., Widz, D. & Yao, A. 1995. Middle Jurassic to Early Cretaceous radiolarian biochronology of Tethys based on Unitary Associations. *Mémoires de Géologie (Lausanne)*, no. 23: 1013-1048.

Carter, E.S., Cameron, B.E.B. & Smith, P.L. 1988. Lower and middle Jurassic Radiolarian biostratigraphy and systematic paleontology, Queen Charlotte Islands, British Columbia. *Geological Survey of Canada, Bulletin* 386: 1-109.

Isozaki, Y. & Matsuda, T. 1985. Early Jurassic radiolarians from bedded chert in Kamiaso, Mino Belt, central Japan. *Earth Science* 39: 429-442.

Kito, N., De Wever, P., Danelian, T., & Cordey, F. 1990. Middle to Late Jurassic Radiolarians from Sicily (Italy). *Marine Micropaleontology* 15: 329-349.

Kobayashi, K., Matsuoka, A., Yang, Q., Nagahashi, T., Zeng, Q. & Wang, Y. 1999. Cataclasite series rocks derived from cherts in the Yarlung Zangbo Suture Zone, Xizang (Tibet), China. In H. Takagi, S. Otoh & K. Kimura (eds.), *Abstract of the second Joint Meeting of Korean and Japanese Structure and Tectonic Research Groups*: 88-89.

Matsuoka, A. 1995. Jurassic and Lower Cretaceous radiolarian zonation in Japan and in the western Pacific. *The Island Arc* 4: 140-153.

Matsuoka, A., Yang, Q., Kobayashi, K., Nagahashi, T., Zeng, Q., & Wang, Y. 1999a. Radiolarian dating of siliceous sediments in and around the Yarlung Zangbo Suture Zone, Xizang (Tibet), China. *Terra Nostra – 14th Himalaya-Karakoram-Tibet Workshop*: 96-97.

Matsuoka, A., Yang, Q., Kobayashi, K., Nagahashi, T., Zeng, Q., & Wang, Y. 1999b. Tectonic repetition of siliceous deposits in the Yarlung Zangbo Suture Zone, Xizang (Tibet), China. In H. Takagi, S. Otoh & K. Kimura (eds), *Abstract of the second Joint Meeting of Korean and Japanese Structure and Tectonic Research Groups*: 90-91.

Metcalfe, I. 1999. Gondwana dispersion and Asian Accretion: An overview. In I. Metcalfe (ed.), *Gondwana dispersion and Asian Accretion, IGCP 321 Final Results Volume*: 9-28. Rotterdam: Balkema.

Takemura, A. 1986. Classification of Jurassic nassellarians (Radiolaria). *Palaeontographica Abt A* 195: 29-74.

Wang, C., Xia, D., Zhou, X., Chen, J., Lu, Y., Wang, G., He, Z., Li, X., Wan, X., Zeng, Q., Pubu, C. & Liu, Z. 1996. Geology between the Indus-Yarlung Zangbo Suture Zone and the Himalaya Mountains, Xizang (Tibet), China. *30th IGC Field Trip Guide T121/T387*: 1-72.

Wu, H. 1993, Upper Jurassic and Lower Cretaceous radiolarians of Xialu chert, Yarlung Zangbo ophiolite belt, southern Tibet. *Micropaleontology Special Publ.* no. 6: 115-136.

Section 3

Wallace's Line

Why Wallace drew the line: A re-analysis of Wallace's bird collections in the Malay Archipelago and the origins of biogeography

D. Clode
Department of Zoology, University of Melbourne, Parkville 3010, Australia

R. O'Brien
Museum Victoria, 72 Victoria Crescent, Abbotsford 3067, Australia

ABSTRACT: The significance of Wallace's Line is often misrepresented in modern scientific literature. Many assume that it was a demarcation between Australian and Asian faunas. However, Wallace drew his line to illustrate that the distribution of animals reflects underlying geological history. Arguments over biogeographic boundaries are rarely relevant to Wallace's hypothesis. His work is also often assumed to lack empirical support. We reassess Wallace's Line using 2,517 bird specimens he collected from the Malay Archipelago, as well as modern avifaunal lists. We classified families as Asian, Australian or cosmopolitan and calculated the proportion of Asian to Australian species for each island. The islands fall into three groups, separated by Wallace's Line and Lydekker's Line. Our results suggest that Wallace's Line is supported by the data collected and that it demarcates a major faunal break for birds, concordant with modern geological evidence on the origin of the region.

1 INTRODUCTION

Alfred Russel Wallace (1823-1913) was a remarkable biologist. Self-educated and with no private income, he spent years collecting and preparing immaculate specimens under the most arduous and isolated conditions (van Oosterzee, 1997; Wallace, 1905). In addition, Wallace was one of the finest biological theorists of his time, independently devising a theory of evolution by natural selection and founding the discipline of biogeography (Brooks, 1984). Yet despite such achievements, his key contribution to biology, Wallace's Line, has fallen into disrepute in modern scientific literature (Mayr, 1944; Simpson, 1977; Lincoln, 1975).

Wallace noted a declining proportion of Asian to Australian species as he traveled from west to east through the Malay Archipelago (modern Indonesia, Malaysia, and Brunei). This pattern was consistent with dispersal theories then thought to explain animal distributions (Lyell, 1830; Wallace, 1855). However, when Wallace traveled east from Bali to Lombok, he noticed an unexpectedly large shift in the proportion of Asian to Australian species which was disproportionate to the mere 25 km gap between the islands (Wallace, 1860). Similarly, he regarded the differences between Borneo and Sulawesi as being greater than might be expected from a simple dispersal pattern (Wallace, 1860). The location of this unexpected change in species distributions was subsequently dubbed Wallace's Line (Huxley, 1868: 313). Not content with merely observing or defining such a disjunct distribution, Wallace sought a theoretical explanation for it. His efforts were encouraged by Charles Darwin who wrote 'I am extremely glad to hear that you are attending to distribution in accordance with theoretical ideas. I am a firm believer, that without speculation there is no good and original observation.' (Darwin, 1857, quoted in Burkhardt, 1996:183).

In Wallace's time the frontier of geological theory was defined by Lyell's *Principles of Geology* (1830). Apart from rising and falling sea-levels, volcanic eruptions, subsidence and erosion, continents were generally assumed to be fixed, and geological history was very much a mystery. Wallace considered that animal distributions might reveal past land connections and offer new insights into geological history. To Wallace, the sudden unexpected change in animal distributions across the Macassar and Lombok Straits reflected the barrier these waterways presented to overland dispersal even during periods of lower sea level. As Darwin had suggested, sea-depth was a measure of length of isolation. The lands to the west of Wallace's Line all lie on the shallow Sunda Shelf. Wallace concluded that the pattern of animal distributions he observed suggested 'the western part [of the Malay Archipelago] to be a separated portion of continental Asia, the eastern the fragmentary prolongation of a former Pacific continent' (Wallace, 1869). In essence, Wallace had identified vicariance (or the importance of barriers) as a significant factor in animal distributions, which until then had been considered to be a function entirely of dispersal.

Wallace's Line was much disputed, with alternative lines proposed often based on different taxonomic groups (e.g. Sclater, 1858; Murray, 1866; Lydekker, 1896; Sclater & Sclater, 1899; Weber, 1902 cited in Mayr, 1944). In reviewing this proliferation of lines, Mayr (1944) proposed a line of faunal balance where the fauna was evenly distributed between Asian and Australian species. Simpson (1977) argued that there should be no single line at all, rather a transition zone, commonly termed Wallacea. Wallace's Line appeared to have retained its significance for only a few taxonomic groups (Keast, 1983) and to have lost its value as a general biogeographic principle.

Part of this loss of recognition stems from the assumption that Wallace's Line was primarily intended to demarcate Australian from Asian fauna. 'From the first, a preoccupation of such studies was to draw a line as a boundary between the two' (Simpson, 1977:107). Such an assertion is manifestly incorrect in Wallace's case. Wallace used his line as an illustrative device supporting his argument that the distribution of animals reflects the underlying geological history of the region (Wallace, 1876; 1877; 1881). This concept may seem self-apparent to biogeographers today but was quite novel then. Thus, Mayr's line of faunal balance, where the distribution of Asian to Australian species is roughly 50:50, would hold little interest for Wallace, since it is a statistical construct with no underlying geological foundation. Similarly, while Wallace acknowledged the transition in the proportion of Asian to Australian species from west to east (Wallace, 1860: 175), he felt it was not as smooth as expected through dispersal alone. Wallace wanted to explain the unexpected leap in the transition from Asian to Australian species from Borneo to Sulawesi and from Bali to Lombok, which he interpreted as an echo of geological history.

Another interpretation of Wallace's Line is that he observed the change from the rich continental fauna of Asia to the depauperate island fauna of Wallacea (Mayr, 1944; Keast, 1983). Wallace knew of the depauperate nature of the intermediate islands (referring to Seram as a 'perfect desert of zoology'), but the total number of species does not alter the proportion of Asian or Australian birds on each island. He stated, for example, that Sulawesi had only 220 bird species compared to Borneo's 386, but he also noted that nearly all of Borneo's are Asian, compared to only three quarters of Sulawesi's birds (Wallace, 1869). Since collectors were more interested in species, rarities and types than the numbers of individuals, it seems unlikely (as suggested by Lincoln, 1975) that Wallace would confound a difference in abundance with a difference in proportion.

The smaller central islands of Wallacea are also drier (with an average annual rainfall <2000 mm) than the larger western islands (and New Guinea) where rainfall often exceeds 4000 mm annually. Although such environmental differences have been proposed as an explanation for Wallace's Line (Lincoln, 1975), Wallace argued that these 'physical and geological differences do not coincide with zoological differences' (1860: 174) as both arid and fertile areas are found on both sides. Lincoln (1975) also proposed that Wallace was misled by the presence of conspicuous Australian birds like the sulphur-crested cockatoo on Lombok. It is difficult, however, to understand how Wallace published several books and articles on biogeography if it was based on such a misunderstanding and, indeed, his writings clearly demonstrate that it was not. We can only conclude that modern researchers are not always fully conversant with the original literature they cite.

Misrepresentations of this nature assume that Wallace's work was not based on empirical evidence and that he over-emphasised species supporting his arguments. Simpson, for example, suggests that Wallace established his line 'simply on the observation that there seemed to be an abrupt faunal change between Bali and Lombok' (1977: p113). Certainly scientific literature has changed dramatically in the decades since Wallace's publications. Today, he would need rigorous statistical evidence to support his line. In 1860, his arguments relied as much upon his descriptions of the noisy Australian birds of Lombok as upon his tabulations of the relative proportions of Australian versus Asian birds.

Wallace's written works may not contain all the raw data on which he based his arguments (although they do contain significant summaries for some islands). But much of Wallace's data is still available, as specimens collected during his time in the Malay Archipelago. Many of these are in museum collections, allowing us to re-assess Wallace's claim using his own evidence. By analysing his collections we evaluate whether Wallace's Line is apparent as a faunal break in the proportion of Asian to Australian birds in the region. Furthermore, if Wallace's Line is apparent in his original data set, we will assess whether it is an artefact of his collections or is also apparent in modern, more extensive, island species lists for the area.

2 METHOD

Since Wallace's specimens are dispersed across various museums and usually catalogued taxonomically, rather than by collector, it would be impossible to reconstruct his complete data set of 125,660 specimens. Fortunately, the British Museum of Natural History acquired 2,474 of Wallace's bird specimens in 1873. These specimens (species, sex and provenance) appear in the acquisitions register (volume 23, pp 26-74). To this list, we can add 45 Wallace specimens in the ornithology collection of Museum Victoria.

This provides a sample of the data on which Wallace based his work. Wallace primarily collected birds (8,050) and insects (109,700). However, insects were (and perhaps remain) too poorly understood to allow an adequate biogeographic analysis (Wallace, 1857: 483) which may be why Wallace based his initial arguments on bird distributions (Wallace, 1859). Although Wallace later expanded his theory across taxonomic groups (Wallace, 1860; 1876; 1881) he still relied heavily on birds to illustrate his arguments.

Wallace originally defined species as either Oriental or Australian depending on the predominant distribution of the family. Mayr (1944) did likewise, arguing that 'a specialist of a given group usually has no difficulties in deciding which species are Indo-Malayan and which Australian'. Fossils might provided a more objective basis for defining origins but avian fossils are rare in Australia and insufficient to determine patterns of origin (Rich & van Tets, 1982).

Molecular taxonomy offers a means of defining origins which is independent of current geographic distributions and can be used to determine which of Wallace's birds are Australian or Asian. Molecular taxonomy has altered significant areas of traditional Australasian bird taxonomy (Sibley & Alquist, 1985; 1990; Sibley & Monroe, 1990; Christidis & Schodde, 1991; Nunn & Cracraft, 1996). Many Australian families (like the Pardalotidae warblers and Petroicidae robins) have been separated from Old World families (Sylviidae and Muscicapidae, respectively) as examples of convergent evolution (Cracraft, 1999). Kingfishers previously regarded as Old World are split by molecular studies into the Australian kookaburras (Dacelonidae) and Old World Alcedinidae. Such differences may have added error to earlier analyses, although the differences between traditional and molecular taxonomies are not so great as to be likely to alter the overall pattern of our results. On the basis of molecular taxonomies, we classified all species as belonging to families of Australian (Gondwanan), Asian (Laurasian) or cosmopolitan origin (Table 1).

We then calculated the overall proportion of Asian to Australian species Wallace collected from each island. These data were then graphed against the ranked distribution of islands by their western-most latitude (Table 2). Only islands with species lists exceeding 15 were included. We compared these data with the proportion of Asian to Australian species from modern island species lists summarised in Keast, 1983 (Table 2). It should be noted that Keast (1983) used traditional taxonomies to allocate families to Asian or Australian origin and includes fewer islands. Both data sets were examined for evidence of discontinuities in the transition from Asian to Australian species.

Table 1. The allocation of families to regions of origin (following Sibley & Monroe, 1990)

Origin	Families
Asian	Aegithalidae, Alaudidae, Alcedinidae, Batrachostomatidae, Bucerotidae, Cisticolidae, Hemiprocidae, Irenidae, Laniidae, Melanocharatidae, Meropidae, Muscicapidae, Nectariniidae, Paridae, Passeridae, Phasianidae, Picidae, Pittidae, Pycnonotidae, Sittidae, Sturnidae, Sylviidae, Turdidae, Zosteropidae
Australian	Aegothelidae, Columbidae, Corvidae, Cuculidae, Dacelonidae, Eopsaltridae, Eurostopidae, Jacanidae, Maluridae, Megapodidae, Meliphagidae, Orthonychidae, Pardalotidae, Podargidae, Psittacidae, Ptilonorhynchidae
Cosmopolitan	Accipitridae, Anatidae, Anhingidae, Apodidae, Ardeidae, Burhinidae, Butaroidae, Caprimulgidae, Centropidae, Charadridae, Ciconiidae, Dendrocygnidae, Eurylaimidae, Falconidae, Formicariidae, Fregatidae, Glareolidae, Hirundindae, Laridae, Megalaimidae, Phalacrocoridae, Podicipedidae, Rallidae, Scolopacidae, Strigidae, Sulidae, Threskiornithidae, Turnicidae

Table 2. Island breakdown of species collected by Wallace and modern species list (from Keast, 1983)

Island	N° specimens	N° species	N° Australian species	N° Asian species	N° worldwide species	Modern species list
Ambon	32	31	13	5	13	
Aru Islands	117	95	63	11	20	
Australia	1	1	1	-	-	
Ayu Islands	7	7	5	2	-	
Bacan	78	65	31	12	20	
Bali	9	9	4	4	1	129
Banda	6	6	3	2	1	
Borneo	173	128	26	67	29	386
Buru	89	61	29	13	17	115
Flores	124	66	25	23	17	119
Gorong	13	12	8	1	3	
Halmahera	87	71	29	15	25	120
Java	239	167	52	83	23	287
Kai Islands	12	10	5	2	3	
Lombok	80	59	26	23	7	101
Malaysia	215	172	35	92	36	429
Misool	101	76	51	12	11	
Morotai	57	44	14	9	18	
New Guinea	143	105	69	11	22	
Numfoor	1	1	1	-	-	
Salwati	26	24	11	2	10	
Sangihe	1	1	1	-	-	
Seram	74	61	25	14	18	117
Singapore	15	15	3	8	2	
Sula Islands	65	38	22	8	7	
Sulawesi	176	120	39	38	39	220
Sumatra	203	145	44	74	23	359
Tanimbar	1	1	1	-	-	
Ternate	26	21	9	4	8	
Timor	200	114	48	28	37	113
Waigeo	79	57	43	6	6	
Watubela	8	7	3	2	2	

3 RESULTS

The proportions reveal a gradual decline in the proportion of Asian to Australian species from west to east (Fig. 1). More that 70% of the species Wallace collected in Borneo, Malaysia, Java and Singapore were from Asian families, while fewer than 20% of the species from New Guinea and neighbouring islands were Asian. The question Wallace would have asked, however, is whether or not this transition is smooth or if there is evidence of discontinuities. Figure 1 reveals that the largest change in distribution between islands on a west to east axis divides Java and Borneo in the west from Lombok and Sulawesi in the east. Since Wallace did not collect sufficient specimens from Bali (a fact he later regretted), we are unable to determine whether or not this discontinuity lies precisely on Wallace's Line, but it is certainly in the general area. The second largest discontinuity occurs between the New Guinean islands of Misool, Waigeo etc and the nearby Moluccan Islands of Halmahera, Morotai, Seram and Ambon (essentially along Lydekker's line: Lydekker, 1896). These two breaks in the distribution lie along the edges of the Sahul and Sunda shelves.

Wallace collected fewer than 15 specimens from a number of other islands. Proportions derived from such small sample sizes are prone to significant variation from the addition of single data points, and are therefore not robust enough to be considered reliable indicators of distributions. Some of these islands are nevertheless of interest since they fall close to Wallace's or Lydekker's lines (their data are included in Table 2). Of particular interest is Bali, where Wallace observed birds typical of the Asian region and collected little, unaware that this was their easternmost distribution. This limited data set reveals an equal number of Asian and Australian species. If Bali does fall west of the Wallace Line (rather than east as suggested by his collections) Wallace must have used his personal observations, rather than the birds he collected, to judge it so. The eastern islands with small sample sizes yield similarly ambiguous results and may obscure the apparent discontinuity at Lydekker's Line.

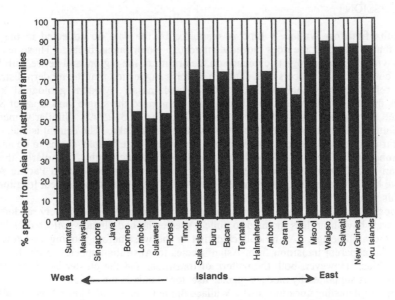

Figure 1: The proportion of bird species from Asian (blank) and Australian (filled) families in a sample of Wallace's collection on landmasses from west to east.

The same discontinuities can be observed in the transition from west to east using a modern species list for the islands (Keast, 1983: Fig. 2). In this data set, Bali clearly falls to the west of the faunal discontinuity, as predicted by Wallace. The ambiguity surrounding the smaller islands near Lydekker's Line, however, is not resolved as Sahul shelf landmasses are only represented by New Guinea in this data set.

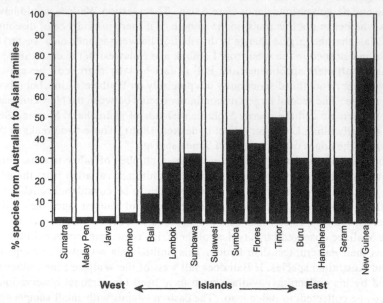

Figure 2: The proportion to bird species from Asian (blank) to Australian (filled) families on landmasses from west to east (from Keast, 1983)

4 DISCUSSION

Our results find that Wallace's Line is supported by the data he collected in the field and suggest that Wallace's Line does indeed demarcate a major faunal break. These results are in keeping with modern geological evidence on the origins of the region, and hence with Wallace's original contention. Wallace's data conform with his suggestion that the modern distribution of species reflects the geological history of the land masses. Modern geological knowledge indicates that the islands west of Wallace's Line comprised the single land mass of Sundaland connected to mainland Asia until the Eocene (Metcalfe, 1993; 1999). Similarly, many islands on the Sahul shelf were also connected to New Guinea/Australia. The central islands, however, have a far more complex and isolated history. Sulawesi, for example, seems to be an amalgam of a number of different islands with different biogeographic origins. Similarly, the northern Moluccan islands seem to have been very recent arrivals from the eastern Pacific Arc which may have had closer contact with Australia and New Guinea than their present location suggests (Hall, this volume).

It is possible of course, that Wallace's sampling technique was not entirely random. In fact, he may well have been less inclined to collect species he had already collected as he traveled through the archipelago. As he collected first in the west before moving east towards New Guinea, the apparent transition from Asian species to Australian species might be a result of sampling bias. However, both the pattern of transition, and the clustering of the islands into three groups is supported by modern species lists for islands in the region.

The more interesting question is why Wallace did not notice Lydekker's Line. Wallace spent a considerable period of time based on Ternate, a small island on the western side of Halmahera which lies close to Lydekker's Line. He collected extensively from New Guinea and surrounding islands and was aware of the very close similarities between Sahul Shelf islands and New Guinea (Wallace, 1857a), even noting the general absence of some families, like the birds of paradise (sub-family Paradisaeidae) on islands west of the shelf and what was subsequently termed Lydekker's Line (Wallace, 1860: 180; Wallace, 1857b). But closer inspection of Wallace's accounts of his travels reveals that many collecting trips across Lydekker's Line were undertaken by his assistant and other specimens were probably obtained through trading. Wallace himself spent an increasing amount of time recovering from malaria. The vast majority of his time in the Malay Archipelago was spent either on the Asian islands or on the intermediate islands of Wallacea. If Wallace made greater use of trading networks and assistants in the east than he did in the west, his opportunities for making a direct comparison between the two sides of Lydekker's Line may have been reduced. Among the birds he did collect (in our data set), fewer than 500 come from the New Guinean islands, while over 800 come from the Asian islands and over a thousand from the islands in between. The fact that Wallace only had small collections from a number of critical islands, such as the Kai and Aru islands (which lie less than 150 km apart but are separated by the Sahul continental shelf), may also have blurred the discontinuity.

Another confounding factor came, not from the birds Wallace observed, but from the mammals he used to extend his theory to a general zoological phenomenon. The typical Asian mammal families of primates, carnivores and deer etc. do, for the most part, stop at Wallace's Line, with some exceptions in Sulawesi, and in other areas where they had clearly been carried by humans. The islands of Wallacea, by contrast, generally lack such large, charismatic creatures. Instead Wallace encountered marsupials like cus-cus and wallabies; creatures clearly of Australian evolutionary origin. But some of these, such as the widespread Common Cuscus (*Phalanger orientalis*) and Common Spotted Cuscus (*Spilocuscus maculatus*) may have been carried between the islands by humans, as pets, for food and for trade (Heinsohn, this volume). Perhaps these early human introductions of marsupial mammals to islands west of Lydekker's Line blurred the distinction for Wallace.

More importantly, Wallace had no reason to expect the existence of a second line. Wallace was not simply recording distributional changes, but was concerned with explaining them in relation to prevailing geological theory. He was primarily interested in demonstrating the existence of differing continental origins for the islands, not a completely isolated oceanic origin for some of them. Wallace was already at the radical end of the geological spectrum in accepting that sea levels had risen and fallen and land eroded and erupted over the course of earth's history. Unlike today, Wallace had no feasible model with which to anticipate either a second faunal break, nor to interpret the anomalous data he recorded on the intermediate islands of Sulawesi and Borneo, for example. At the time there was no geological theory which would allow Wallace to conjecture any more complex 'African' origins for some of the remarkable species he found on Sulawesi, for example: 'lemurs' (tarsiers), 'antelopes' (anoa, *Bubalus* spp.), 'baboons' (*Macaca nigra*) and 'warthogs' (*Babyrousa babyrussa*) (Wallace, 1860: 176-7). It would be many years before geology provided a partial explanation for some of the mysteries Wallace had uncovered.

With the complex geological history of this region increasingly being understood, we now stand a far better chance of assessing Wallace's real legacy—the extent to which species distributions are limited by underlying geological history. This is a far more interesting question than arguing over the placement of arbitrary and illustrative lines. Different taxonomic groups (with different histories and different dispersal abilities) will undoubtedly differ in the extent to which they adhere to different biogeographic boundaries (as foreshadowed in Wallace, 1877) including Wallace's Line. Such variations merely reflect our expanding knowledge of both the species and the effect their geographical history has had on them. Biogeography is much more complex than Wallace thought, simply because geology is more complex than he could have imagined. Re-analysis of Wallace's data presents even stronger evidence supporting of his argument that the geographic patterns of animal distributions reflect the former geological history of the lands they occupy. This principle has now achieved such universal acceptance amongst biologists that modern biogeographers may tend to forget this was the primary purpose of Wallace's argument. Far from being tarnished by inaccuracy, the value of the Wallace Line has been overshadowed by the very success of the theory it was put forward to support.

ACKNOWLEDGEMENTS

We would like to thank Mark Adams and Charles Hussey from the Natural History Museum, London, UK, for arranging access to scanned images of the 1873 acquisitions register. This research was conducted and presented with the support of the Thomas Ramsay Science and Humanities fellowship, Museum Victoria.

REFERENCES

Brooks, J. L. 1984. *Just before the origin: Alfred Russel Wallace's theory of evolution*, New York: Columbia University Press.
Burkhardt, F. 1996. (ed.) *Charles Darwin's Letters: A Selection 1825-1859*, Cambridge: Cambridge University Press.
Christidis, L. and Schodde, R. 1991. Relationships of Australo-Papuan songbirds-protein evidence, *Ibis*, 133: 277-85.
Cracraft, J. 1999. The history of Austrolasian avifauna through space and time, *Where Worlds Collide: Faunal and floral migrations and evolution in SE Asia-Australasia*, Armidale 29 Nov-2 Dec.
Hall, R. this volume. Cenozoic reconstructions of SE Asia and The SW Pacific: changing pattersn of land and sea, *Where Worlds Collide: Faunal and floral migrations and evolution in SE Asia-Australasia*, Armidale 29 Nov-2 Dec.
Heinsohn, T. this volume, Human influences on vertebrate zoogeography: animal translocation and biological invasions across and to the east of Wallace's Line', *Where Worlds Collide: Faunal and floral migrations and evolution in SE Asia-Australasia*, Armidale 29 Nov-2 Dec.
Huxley, T. H. 1868. On the classification and distribution of the *Alectoromorphae* and *Heteromorphae*. *Proceedings of the Zoological Society, London*, 294-319.
Keast, J. A. 1983. In the steps of Alfred Russel Wallace: Biogeography of the Asian-Australian Interchange Zone, In Sims, R. W., Prince, J. H., Whalley, P. E. S. (eds.) *Evolution, Time and Space: The emergence of the biosphere*, London: Academic Press.
Lincoln G. A. 1975. Bird counts either side of Wallace's Line, *Proceedings of the Zoological Society of London*, 177: 349-361.
Lydekker, R. 1896. *A Geographical History of Mammals*, Cambridge: Cambridge University Press.
Lyell, C. 1830. *Principles of Geology*, Vol 1-3 London: J. Murray.
Mayr, E. 1944. Wallace's line in the light of recent zoogeographic studies, *Quarterly Review of Biology*, 19: 1-14.
Murray, A. 1866. *The Geographical Distribution of Mammals*, London: Day and Son.
Metcalfe, I. 1993. South-east Asian terranes: Gondwanaland origins and evolution, In *Gondwana eight: Assembly, evolution and disperal*, Proceedings of the Eighth Gondwana Symposium, Hobart, 21-9 June 1991: Rotterdam: Balkema.
Metcalfe, I. 1999. Geological Origins and natural Resourtces. In A. Kaur & I. Metcalfe (eds), *The Shaping of Malaysia*: 11-41. New York: St. Martins Press.
Nunne, G. B. & Cracraft, J. 1996. Phylogenetic relationships among the major lineages of the birds of paradise (Paradisaeidae) using mitochondrial DNA gene sequences, *Molecular Phylogenetics and Evolution*, 5: 485-59.
Rich, P. V. and van Tets, G. F. 1982. Fossil birds of Australia and New Guinea; phylogenetifc and biostratigraphic input. In Rich, P. V. and Thompson, E. M. (eds.) *The Fossil Vertebrate Record of Australasia*, 235-384, Melbourne: Monash University.
Sclater, P. L. 1858. On the general geographical distribution of the Class Aves, *Zoological Journal of the Linnean Society*, 2, 130-145.
Sclater, W. L. and Sclater P. L. 1899. *The Geography of Mammals*, London: Kegan, Paul, Trench and Trübner.
Sibley, C. G. and Alquist, J. E. 1985. The phylogeny and classification of the Australo-Papuan passerine fauna, *Emu*, 85: 1-14.
Sibley, C. G. and Alquist, J. E. 1990. *Phylogeny and classification of birds: A study in molecular evolution*, New Haven: Yale University Press:.
Sibley, C. G. and Monroe, B. L. 1990. *Distribution and Taxonomy of Birds of the World*, New Haven: Yale University Press:
Simpson G. G. 1977. Too many lines; the limits of the Oriental and Australian zoogeographic regions, *Proceedings of the American Philosophical Society*, 121: 107-120
van Oosterzee, P. 1997. *Where Worlds Collide: The Wallace Line*, Melbourne: Reed Books.
Wallace, A. R. 1855. On the law which has regulated the introduction of new species, *The Annals and Magazine of Natural History*, Ser. 2, Vol 16: 184-196
Wallace, A. R. 1857a. On the natural history of the Aru Islands, *The Annals and Magazine of Natural History*, Ser. 2, Vol 20 (Supp), 475-85
Wallace, A. R. 1857b. On the great Bird of Paradise, Paradisea apoda, Linn.; 'Burong mati' (Dead bird) of the Malays; 'Fanehan' of the Natives of Aru, *Annals and Magazine of Natural History*, Ser. 2, Vol 20, 411-16
Wallace, A. R. 1859. The geographical distribution of birds, *Ibis*, 1: 449-54
Wallace, A. R. 1860. On the zoological geography of the Malay Archipelago, *Journal of the Proceedings of the Linnean Society (Zoology)*, 4: 172-184
Wallace, A. R. 1869. *The Malay Archipelago*, London: Macmillan.
Wallace, A. R. 1876. *The Geographical Distribution of Animals*, London: Macmillan.

Wallace, A. R. 1877. On the comparative antiquity of continents, as indicated by the distribution of living
    and extinct animals, *Proceedings of the Royal Geological Society*, Sept.
Wallace, A. R. 1881. *Island Life*, London: Macmillan.
Wallace, A. R. 1905. *My life: a record of events and opinions*, London: Chapman and Hall.
Weber, M. 1902. Der Indo-australische Archipel und die Gestichte seiner Teirwelt. 46 pp. *Jena*. cited in
    Mayr (1944).

The linear approach to biogeography: Should we erase Wallace's Line?

W.R. Erdelen

Dept. of Biology, Institute of Technology, Jl. Ganesha 10, Bandung 40132, Indonesia

ABSTRACT: Various attempts have been made to subdivide the earth into biogeographic regions. Textbooks of modern biogeography treat the problem of identifying biogeographic units to different extents. Some biogeographers explore sophisticated systems of subdividing the earth into realms and provinces while others omit the issue altogether. Naturally, the basis of such subdivisions are patterns of geographic distribution of species. For large-scale subdivisions of the earth, however, in most cases supraspecific systematic categories such as orders, families or genera have been compared. The different systems of subdividing the earth into biogeographic units reflect our knowledge of the Earth's biota. Accordingly, relatively showy species or, in terms of their geographic distribution or systematic relationships, well known groups such as many of the higher plant species, birds, and mammals, formed the basis of the earlier subdivisions. Although the subdivisions presented to date reflect real patterns, independent of the underlying causal processes, these subdivisions have repeatedly been criticized. Criticisms were based on exceptions which did not fit the patterns that defined realms, regions or provinces. Lines drawn, the most famous certainly being Wallace's Line, were redrawn, replaced or even omitted altogether. A particular problem occurred in zones such as Wallacea where lines blurred and scientists working on different groups of organisms reached different conclusions about biogeographic boundaries. If scientists working on different groups of organisms do not agree on the delimitation of a particular biogeographic unit, it might seem futile to attempt to draw lines. In this paper, however, it is argued that lines, however 'false' or unrepresentative for large arrays of taxonomic groups these may be, should be used as heuristic tools to facilitate and stimulate further research into the biogeography of a particular area. Even lines based on the geographic distribution of only a few taxa may indicate that these share common dispersal characteristics or, more generally, a common biogeographic history. The delimitation of the respective line, and its discussion within the context of other evidence available for, or against, a particular biogeographic hypothesis, might shed further light on the evolution of certain elements of a fauna or flora. This paper focuses on the heuristic values and biogeographic implications of the various attempts to draw biogeographic boundaries in the Wallacean region. Recent developments, problems and needs for a better understanding of the biogeography of the Indo-Australian region are outlined and discussed.

'The western and eastern islands of the [Malay] archipelago belong to regions more distinct and contrasted than any other of the great zoological divisions of the globe. South America and Africa, separated by the Atlantic, do not differ so widely as Asia and Australia.'

Wallace (1860)

1 INTRODUCTION

One of the founders of modern biogeography or, more specifically zoogeography, was Alfred Russel Wallace. Wallace had studied the geographic distribution patterns of animals in the Southeast Asian region, in particular in what is nowadays, from a plant geographer's view, referred to as Malesia and, from a zoogeographer's view, the Indo-Malayan Realm. His most important works in this respect were *The Malay Archipelago* (Wallace 1869) and *The Geographical Distribution of Animals* (Wallace 1876), the latter being one of the benchmark publications that marked the foundation of modern zoogeography. The famous Wallace Line, a term coined by Huxley (1868), is still quoted as one of the major biogeographic boundaries for the subdivision of animal distribution (e.g. Cox & Moore 1993, van Oosterzee 1997). Wallace's Line was meant to be the boundary between the Oriental and Australian faunal regions. Until the end of the last century this line was adopted by virtually all zoogeographers. After 1890, however, first doubts on the validity of Wallace's Line were expressed. Early in this century it was vigorously defended by some scientists and considered imaginary by others (Mayr 1944). Even today opinions differ. Humphries (1990), for instance, stated that '... *although concepts such as "Wallace's Line" and "Wallacea" have been important touchstones in biogeography they are now as much a part of history as outdated concepts such as "centres of origin", filters, sweepstakes and bioaccountancy"* (p. 7). Whitmore (1981b), on the contrary, was more in favor of the usefulness of Wallace's Line: '*The revolution in the earth sciences has led to a total reappraisal of the biogeography of the Malay archipelago but Wallace's Line remains today, as for the past 120 years, a cogent influence, powerfully able to generate hypotheses subject to further test*' (p. 80).

Modern biogeographers rate the importance of subdividing the earth into biogeographic units rather differently. Some explore sophisticated systems of subdividing the earth into realms and provinces or even further, others omit the issue altogether. For instance, Udvardy (1975, 1984) elaborated on a detailed world-wide scheme of biogeographic provinces; Pielou (1979) started her book on biogeography with a chapter on the biogeographic subdivision of the earth; Brown & Lomolino (1998) treated biogeographic subdivisions in their section on the history of biogeography; Huggett (1998) discussed biogeographic regions only in a short section in his first chapter on the definition of biogeography. Based on their work on mammals, Corbet & Hill (1992), subdivided the Indo-Malayan region into Sundaic, Philippine, and Wallacean subregions comprising six, one, and three divisions, respectively.

Naturally, such biogeographic subdivisions of the Earth's biota are essentially based on patterns of geographic ranges of species. For large-scale subdivisions, in most cases supraspecific systematic categories such as orders, families or genera have been compared. The different systems used for subdividing the earth into biogeographic units had been closely linked to our knowledge of biodiversity and its distribution. Accordingly, relatively showy species or, in terms of their geographic distribution or systematic relationships, well known taxa such as many of the higher plant species, birds, and mammals, formed the basis of the earlier subdivisions (overview in George 1981).

Although the subdivisions presented to date reflect real patterns, independently of the underlying causal processes, these subdivisions have repeatedly been criticized. Criticisms were based on exceptions which did not fit the overall patterns that defined realms, regions or provinces. Previously drawn lines, the most famous certainly still being Wallace's Line, were redrawn, replaced or even omitted altogether. A particular problem occurred in zones such as Wallacea where lines blurred and scientists working on different groups of organisms reached different conclusions about biogeographic boundaries.

This paper focuses on the heuristic values and biogeographic implications of the various attempts in drawing biogeographic boundaries in the Wallacean Region. Recent developments, problems and needs for a better understanding of the biogeography of the Wallacean Region are outlined and discussed.

2 A SHORT HISTORY OF LINE DRAWING AND BIOGEOGRAPHIC DYNAMICS IN THE WALLACEAN REGION

The early focus on faunal barriers in the Indo-Australian region was based on the facts that (1) typical Oriental species seemed mostly restricted in their eastward distributional limits to Borneo, Java, and Bali, and (2) that most Australian faunal elements do not extend far westwards beyond the Sahul Shelf (Keast 1983). This is reflected in Wallace's line which runs along the Sunda Shelf and Lydekker's Line (Lydekker 1896) coinciding with the limit of the Sahul Shelf (Fig. 1). A line based on rather different criteria was later drawn, derived from work by Weber (1902). As pointed out by Simpson (1977), the line was modified and named 'Weber's Line' by Pilseneer (1904). Mayr (1944) redefined the line as the line of 'faunal balance'. West of the line more than 50% of the fauna is considered Oriental, east of the line more than 50% are of Australian or Australo-Papuan origin (for details, see Simpson 1977).

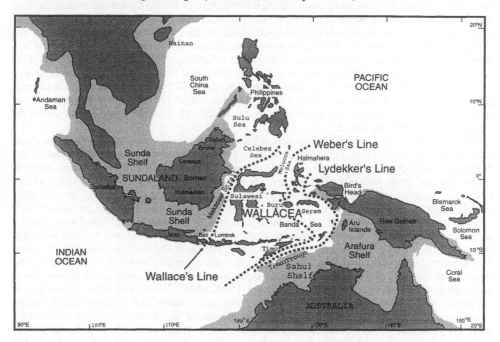

Figure 1. The Wallacea region of SE Asia showing the positions of Wallace's Line, Weber's Line and Lydekker's Line. Figure drawn by I. Metcalfe, after Fig. 1 of Hall (this volume).

The distribution of plants in the region is strikingly different from that of animals. Plant geographers consider the whole region one more or less continuous unit, commonly referred to as Malesia, a term coined by van Steenis (1950). Malesia extends from the Malay Peninsula east to New Guinea, including the Philippines, the island of Borneo, Indonesia, and New Guinea. Van Steenis (1950) distinguished three divisions or provinces within Malesia, viz. a western, a southern, and an eastern division (province). Western Malesia comprises peninsular Malaysia, Sumatra, Borneo, and the Philippines. Southern Malesia is composed of Java and the Lesser Sunda Islands. Eastern Malesia includes Sulawesi, the Moluccas, and New Guinea (for more details see van Steenis 1979).

Explanations for the differences in floristic and faunistic subdivison of the Indo-Australian region are still a matter of debate. Van Steenis (1979) has compiled interesting evidence for explaining the floristic composition of the Lesser Sundas. First, the northern Lesser Sunda Islands which are of volcanic origin and the southern non-volcanic islands such as Sumba and Timor do not differ significantly in their floristic composition. Second, the character and relative species poverty of the flora of the Lesser Sunda Islands may be explained by four factors: the influence of seasonal drought, relatively (compared to the Greater Sunda Islands) small land surface, composition being of many small islands rather than a few big ones, and geological instability, especially for the volcanic islands. Obviously, however, 'stabilizing processes' such as re-immigration or re-establishment of floras after volcanic activities must have been of importance. Otherwise, we would expect that Sumba and Timor, for instance, should be floristically richer than islands within such tectonically unstable regions. If, as assumed by Flenley (1979), floras and faunas in the Indo-Australian region have essentially been shaped in the Tertiary, these may have changed rather differently during the Quaternary. Quaternary changes of plant communities and concomitant dynamics of the animal communities or certain segments thereof are, however, only poorly understood. Our lines of argument still circle around gross approaches such as comparing rainforest and savanna mosaics and their dynamics with animal taxa supposedly being 'adapted' to either of these environments. For instance, the existence of a savanna corridor on the subaerial Sunda Shelf during the cool phases in the Pleistocene is a matter currently discussed as a means for explaining major distributional puzzles shown by several mammals. The search for the causal mechanisms underlying such parallel changes in plant and animal communities certainly has to focus both on ecological characteristics of the species as well as on the relevant Quaternary scenario. Molecular techniques such as DNA analyses allow for new approaches for analyzing divergence times amongst taxa. New light might be shed on the question of homogeneity of the underlying processes, both in space and time, which shaped, for instance, mammal distribution patterns in the Sunda region (Meijaard 2000). In addition, studies of geographic distribution patterns and intraspecific variation might provide useful evidence for or against particular biogeographic hypotheses as the studies of Brandon-Jones (1996, 1998) on *Presbytis* monkeys have shown.

Biogeographic lines or demarcations may reflect different scales in delimiting floral and faunal boundaries, respectively. On the same scale as the Indonesian archipelago, for instance, the Philippine archipelago can be subdivided into several regions. Such major subdivisions were already proposed at the end of the last century (for an overview, see Vane-Wright 1990). A closer look on a smaller scale, however, already indicated that distribution patterns may be extremely complex and that the faunistic composition of particular islands or island groups may be very heterogeneous. For example, the Sulu Islands, located between Borneo and Mindanao, may be an unusual area of endemism rather than 'just' stepping stones between these two major islands (Vane-Wright 1990). Within an extensive region such as Wallacea, the Lesser Sundas and Moluccas may be subdivided into 10 biogeographic units (details in Monk *et al.* 1997). The large island Sumatra and neighboring islands along its westerly coast comprise different phytogeographic and zoogeographic units (Laumonier 1990, 1997). We find striking differences in the degree of isolation of the islands surrounding Sumatra. Islands like Simeulue and Enggano were probably never connected to Sumatra and the Mentawai Islands may have been isolated for more than 500,000 years, whereas the islands to the east of Sumatra have had quite recent connections with Sumatra and neighboring peninsular Malaysia (Whitten *et al.* 1987). Even within Sumatra, related to river and mountain barriers, biogeographic units may be distinguished. Whitten *et al.* (1987) listed 26 zoogeographic regions within Sumatra, comprising major and minor boundaries, referring to separation of species and subspecies, respectively. Possibly the smallest scale for 'biogeographic' units might be seen in individual mountains with distinct floral composition and dominance patterns. As discussed by Flenley (1979) for the upper and lower mountain forests of New Guinea, this may have resulted from extinction and re-formation processes during the Pleistocene. Perhaps dynamics during the repeated climatic cycles of the Pleistocene were much more extreme than our classical perception of gradual depression and re-elevation suggests. If such processes have created 'individuality' of forest communities on very small spatial scales, nowadays seemingly distributed independently of edaphic or present ecological factors, the question arises whether we could find a similar phenomenon in animal distribution patterns. The recently suggested extreme diversity of the Sri Lankan amphibian fauna, estimated to comprise over 250 species, most of them having extremely restricted distributional ranges (<0.5 km^2) might be an example (Pethiyagoda & Manamendra-Arachchi 1998).

In summary, the different scales discussed above include:
➤ the major biogeographic boundaries which led to the subdivision into major biogeographic units as reflected in the various lines drawn within the region,
➤ smaller-scale units comprising island groups,

> biogeographic units such as individual islands (e.g. Borneo or Sumatra, within larger units such as Sundaland),
> biogeographic units that occur within such major islands,
> finally, extremely small-scale biogeographic units such as individual mountain massifs and/or particular habitat types which cover only extremely small areas.

The search for the boundary between the Oriental and Australian realms resulted in the 'major lines' as discussed by Simpson (1977; Fig. 1). The basic approach was the search for a well-defined line which divides these biogeographic regions in an unequivocal way. This search for a general pattern revealed that it was difficult, if not impossible, for three different reasons: inadequate data, comparisons among taxa with rather different dispersal capabilities, and the possibility that such a general pattern, covering all plant and animal taxa within such an enormously large and complex region, might simply not exist. Weber (1902) showed the complexity of the distribution patterns in the region by comparing taxa like the banteng (*Bos javanicus*) which occurs only on Java and Borneo but not on Sumatra, the Malay bear (*Ursus malayanus*) which is found on the Malay peninsula, Sumatra, and Borneo but not on Java, and the leopard (*Panthera pardus*) and tiger (*Panthera tigris*), both presently occurring on Sumatra and Java but not on Borneo. Even Weber's Line, the line of faunal balance (Mayr 1944), is based on a concept that has several flaws. As pointed out by Simpson (1977) there may be no single line of faunal balance for faunas as a whole but taxa-specific balance lines; the assumption of a continuous change towards more Oriental elements in the west and more Australian in the east may not hold; the identification of taxa as being either Oriental or Australian may be difficult, especially in the case of endemics; and finally it may be questioned whether a line of faunal balance should be used to define a zoogeographic, or generally biogeographic, boundary at all.

Nevertheless, the general view held today is that Wallacea, a term coined by Dickerson *et al.* (1928) for the region comprising the Philippines, Sulawesi, the Moluccas and the Lesser Sundas, is a transitional zone between the Asian and Australian region (Monk *et al.* 1997), both in geological and biogeographic or, more precisely, zoogeographic terms. This view is held even in more recent publications (Vane-Wright 1991, Whitmore 1981a, 1987a). Authors like Simpson (1977) and Kitchener & Suyanto (1996b) consider Wallacea not to be a transitional but a unique zone. The major argument in favor of their view is the high level of endemism in this region.

In the case of animal distribution patterns, some of the major uncertainties are clearly a result of our poor knowledge of biodiversity of the region, in particular of eastern Indonesia (see papers in Kitchener & Suyanto, 1996a), and of a general lack of sound biogeographic analyses that cover a broader range of taxa. As in other tropical island systems as well, the question still is as to what extent taxa-specific processes or processes that were invariant over larger taxonomic groups have shaped extant distribution patterns in the Indo-Australian archipelago. As a consequence, the biogeographic grouping of the islands of the huge Indo-Australian archipelago in general and of the Lesser Sunda Islands and the Moluccas in particular, is still a matter of debate (e.g. Ladiges *et al.* 1991, Michaux 1991). In addition, increasing human impacts on remaining natural ecosystems in the region and concomitant changes in the distribution patterns of indigenous flora and fauna or even species extinction underscore the need for more detailed studies on the biogeography of the eastern Indonesian region.

3 MODERN VIEWS - MODERN FRUSTRATIONS

3.1 *Progress and ignorance*

The Indo-Australian and Indo-Pacific regions have been focal areas for biogeographic studies for many years (e.g. Walker 1972). The last decade, however, has seen enormous progress in understanding biogeographic patterns and processes in these regions (e.g. Knight & Holloway 1990, Keast & Miller 1996, Hall & Holloway 1998, Metcalfe 1998b, this volume). This has mainly been the result of a number of conferences that attempted to bring geologists and biologists together to discuss biogeographic issues and problems of these regions. Despite the recent methodological development in the disciplines of geology and biogeography our still fragmentary knowledge has hardly allowed for successful matching of information to create sound hypotheses on the biogeography of the Indo-Australian region or parts thereof (overview in Holloway & Hall 1998). Even despite the enormous body of knowledge that has accumulated in both disciplines quite recently, some facts crucial for understanding biogeographic relationships are still poorly known. We are far from a situation where geologists 'set the stage' and biogeographers interpret their data within a well-established geological framework. A few of our 'hurdles of ignorance' are briefly discussed below.

From the perspective of studying individual taxa or taxonomic groups several problems arise. First, the group under study should be well understood in terms of species composition and their respective geographic ranges. Especially in tropical regions, like the Indo-Australian archipelago, only a few taxonomic groups are relatively well known in terms of species numbers. Second, for many plant and animal groups we lack sound hypotheses on their phylogenetic relationships. Third, the age of many groups and their fossil record are not known or are poorly documented. Even amongst the better studied groups the patterns emerging reflect the complexity of the underlying processes. For instance, rather well known distribution patterns of mammal groups like the Tapiridae (George 1987), Rhinocerotidae, Tupaiidae, Suidae, Soricidae, Phalangeridae, and Paramelidae (listed in west-east arrangement of their distributional ranges, see Fig. 10.8 in George 1987) have not been explained unequivocally to date. Another example is Musser's (1987) study of the mammal fauna of Sulawesi. This study clearly indicates the diversity of mechanisms which have shaped Sulawesi's highly unbalanced and depauperate fauna which consists of ancient endemic non-volant species, forty percent bat species, and one third rat and mice species. In addition, the restriction of a rich fossil fauna comprising a giant land tortoise, species of stegodonts, and pygmy elephants, to Sulawesi's south-western arm has remained a puzzle. Perhaps, as suggested by Musser (1987), these Plio-Pleistocene taxa belonged to an isolated island fauna which had little or no relation to the extant fauna of Sulawesi.

3.2 *From patterns to causes*

Studying the biogeography of a particular group or groups of plants and animals involves several steps. The first, as just outlined, is to compile sound information on species composition and geographic distribution patterns of the taxonomic group in question. Second, knowledge of species ecology is a first prerequisite for understanding these distribution patterns. The study of the Krakatau islands, for instance, has contributed to understanding establishment of a flora and fauna in a comparatively simple geological or, more specifically, volcanic setting (e.g. Thornton *et al.* 1988, Thornton 1996a, 1996b). Species dispersal power, prevalent currents, neighboring species source pools, and niche differentiation and species turnover (just to name a few factors) may be essential for understanding the species complement we finally find on a given island. The more complex the island structure and its geological past, the more difficult it becomes to interpret species occurrences. Species ecology may be linked in varying extent to climatic factors or, in the case of fauna, to vegetation patterns. These themselves have been far from static in the region. One of the most profound changes in the recent geological past have been repeated land connections in the Sunda and Sahul regions during lowered sea levels in the Pleistocene. Enormous areas now under shallow water have repeatedly become dry and huge land masses formed by the exposed Sunda and Sahul shelves were created. Rain forests may have been reduced to refuges (e.g. Flenley 1979) and repeatedly extended their ranges again during the moister interglacials. 'Horizontal' shifts in the arrangement of the mosaics formed by the major vegetation types in the region, i.e. mixed dipterocarp forest, semi-evergreen rain forests, monsoon forests, and savanna (Ashton 1972) most likely have occurred. These may well have been paralleled by vertical shifts in vegetational belts.

These 'Quaternary events' may have shaped the composition and triggered differentiation amongst extant flora and fauna to a significant extent. Many interpretations of extant distribution patterns in the tropics have been attributed to these dynamics of land/sea distribution and changes in patterns of climate and vegetation during the Quaternary. For some of the phylogenetic older taxa this may, however, only have meant fine tuning of patterns and processes which had acted in previous geological periods and shaped flora and fauna independent of Quaternary fluctuations in sea levels and climate. In biogeographic analyses, the further we go back in time the more difficult it becomes to elaborate on the phylogenetic and biogeographic history of a group of organisms. Time elapsed may be proportional not only to diversification but also to extinction rates in most groups of plants and animals; the fossil record is generally only a poor indicator of the spatio-temporal patterns of extinction events. In view of the constant, drastic and complex geological, climatic and vegetational changes that have taken place in the Indo-Australian region it may indeed be questioned whether generally valid zoogeographic boundaries can be drawn (e.g. Walker 1982).

4 BIOGEOGRAPHIC LINES AND BOUNDARIES

With all these difficulties in biogeographic analyses, the usefulness of biogeographic lines or boundaries may be questioned. Such lines or boundaries, however, are at the heart of any biogeographic analysis which aims at understanding ecological and historical processes which have shaped the geographic ranges of species. More specifically, the interspecific (sometimes possibly also intraspecific) variation in the sizes, shapes, boundaries, and internal structures of geographic ranges are the focus of comparative and quantitative biogeographic studies (e.g. Brown *et al.* 1996). Wallace's Line, essentially based on information about birds and larger mammals, with later attempts to delimit the Oriental from the Australian realm, has had enormous heuristic value and may even have triggered much of the biogeographic research which has been carried out in the region. Even today, as exemplified by the conference from which this volume arose, interest in the Wallacean region persists and may even have increased. With the availability of new data in geology and new methods, in particular for phylogeny reconstruction and for the measurement of genetic distinctiveness, the area has become even more interesting for biogeographers.

In summary:
➤ Biogeographic lines (or boundaries) are drawn at rather different spatial scales.
➤ Their explanatory power varies with regard to taxa and spatial scales.
➤ Lines may indicate common properties shared by certain taxa (dispersal power, vicariance events etc.).
➤ In general, lines can be used as heuristic tools to stimulate further research into the biogeography of an area.
➤ 'Fits' and 'non-fits' may lead to more refined biogeographic hypotheses.

5 WHAT WE NEED

The complex geological evolution of the Indo-Australian archipelago involved not only phases during which larger land masses had split up into smaller units, but also times of coalescence of different geological units (Metcalfe 1998a, 1998b), e.g. in the formation of Sulawesi (Audley-Charles 1981, 1987). Depicting the geological history of this region would accordingly require a 'geological area cladogram' which allows for fusion as well as separation processes. Such a novel cladogram would depict the reticulated nature (sensu Holloway & Hall 1998) of the geological evolution of what nowadays constitutes the Indo-Australian archipelago. Unfortunately, the dynamics of the region can at present only be interpreted in the context of a plate tectonic model. For our biogeographic questions, the major hurdle is that we still have no clear picture about the periods during which the different geological units were above or below sea level. This problem was realized by Weber (1902) and re-stated in modern terms by Hall (1998) who discussed the difficulties in drawing detailed paleogeographic maps based on tectonic reconstructions. As also pointed out by Hall (1998), knowledge of distribution of land and sea, bathymetric information, and extension of seas as well as topography of land would be essential for understanding patterns of species distribution.

In theory, after having set the stage in terms of the dynamics of land/sea distribution and break-up and fusion of the various island systems in the Indo-Australian archipelago, we would need to address some of our shortcomings in view of the 'biological components' of a biogeographic analysis of the region. Factors like species composition and geographic ranges of the taxonomic group(s) under study have been discussed above already. For many taxa we are faced with the problem that we do not have any sound hypotheses on the phylogenetic age of the group we are studying and its cladogenetic history. For instance, Pleistocene scenarios may be irrelevant for groups which have mainly been shaped by speciation and extinction events in the Tertiary and, on the contrary, Pleistocene extinctions and speciation patterns may be crucial for understanding extant taxonomic composition of groups which may have an old evolutionary history, but which have been significantly affected by changes in Pleistocene climates. In other words, we would need a timed scenario of the evolutionary history of the organism groups we are studying that we could relate to a geological framework. If both data sets could be obtained independently, we could even test for congruence or discordance of evolutionary and geological processes and further refine our hypotheses or models of biogeographic evolution for a particular time level. By extending this type of analysis to organism groups of different phylogenetic age, we might even develop scenarios or hypotheses for different geological periods and thus obtain a much better understanding of biogeographic patterns and processes in the Indo-Australian archipelago.

6 CONCLUSIONS

If scientists working on different groups of organisms do not agree on the delimitation of a particular biogeographic unit, it might be argued that it would be futile to attempt to draw lines. In this paper, however, it is suggested that lines, however 'false' or unrepresentative for larger arrays of taxonomic groups they may be, should be used as heuristic, hypothesis-generating tools to facilitate and stimulate further research into the biogeography of a particular area. Even lines based on the geographic distribution of only a few taxa may indicate that these share common dispersal characteristics or, more generally, even a common biogeographic history. As suggested earlier (Walker 1982), for an understanding of causal mechanisms the most promising approach might be to study animal groups occurring in areas with comparatively well understood geological and vegetational histories. Lines or boundaries drawn from the study of such groups simply would summarize information on geographic ranges of a given set of taxa and relate it to their phylogenetic histories. The location of such a line and its interpretation within the context of other evidence available for, or against, a particular biogeographic hypothesis, might further our understanding of the evolution of such elements of a fauna or flora. From comparisons of studies on different taxonomic groups we might also formulate new hypotheses on the question of how general the processes have been which have shaped the biota in the Indo-Australian region. Accordingly, the question whether we should erase Wallace's Line, as asked in the subtitle of this paper, should be answered with a clear 'No'. To quote Whitmore (1987b): '*The biogeographical evolution of the Malay archipelago still has its mysteries. Wallace's Line still has us in its thrall.*'

ACKNOWLEDGMENTS

I thank Ian Metcalfe for having invited me to the conference on 'Faunal and floral migrations and evolution in SE Asia - Australasia'. I am grateful to Jeremy Smith for assistance during the editorial process and to Alfred Gramstedt, Tom Heinsohn, Eric Pianka and Nigel Wace for comments on earlier drafts of the manuscript.

REFERENCES

Ashton, P.S. 1972. The quaternary geomorphological history of western Malesia and lowland forest phytogeography. In P. Ashton & M. Ashton (eds), *The Quaternary era in Malesia*: 35-49. Trans. 2nd Aberdeen-Hull symposium on Malesian ecology. Dept. of Geography, University of Hull. Misc. Series No. 13.
Audley-Charles, M.G. 1981. Geological history of the region of Wallace's Line. In T.C. Whitmore (ed.), *Wallace's Line and plate tectonics*: 24-35. Oxford: Clarendon Press.

Audley-Charles, M.G. 1987. Dispersal of Gondwanaland: relevance to evolution of the angiosperms. In T.C. Whitmore (ed.), *Biogeographical evolution of the Malay archipelago*: 5-25. Oxford: Clarendon Press.

Brandon-Jones, D. 1996. *Presbytis* species sympatry in Borneo versus allopatry in Sumatra: an interpretation. In D.S. Edwards & W.E. Booth, S.C. Choy (eds), *Tropical rainforest research - current issues*: 71-76. Dordrecht: Kluwer Academic Publishers.

Brandon-Jones, D. 1998. Pre-glacial Bornean primate impoverishment and Wallace's Line. In R. Hall & J.D. Holloway (eds), *Biogeography and geological evolution of SE Asia*: 393-404. Leiden: Backhuys Publishers.

Brown, J.H. & Lomolino, M.V. 1998. *Biogeography*. Sunderland, Mass: Sinauer Assoc. Publ.

Brown, J.H. & Stevens, G.C., Kaufman, D.M. 1996. The geographic range: size, shape, boundaries, and internal structure. *Annu. Rev. Ecol. Syst.* 27: 597-623.

Corbet, G.B. & Hill, J.E. 1992. *The mammals of the Indomalayan region: a systematic review*. Oxford: Oxford University Press.

Cox, B.C. & Moore, P.D. 1993. *Biogeography - an ecological and evolutionary approach*. London: Blackwell.

Dickerson, R.E. & Merrill, E.D., McGregor, R.C., Schultze, W., Taylor, E.H., Herre, A.W.C.T. 1928. Distribution of life in the Philippines. *Philipp. Bureau Sci. Monogr.* 21: 1-322.

Flenley, J.R. 1979. *The equatorial rain forest: a geological history*. London: Butterworths.

George, W. 1981. Wallace and his line. In T.C. Whitmore (ed.), *Wallace's Line and plate tectonics*: 3-8. Oxford: Clarendon Press.

George, W. 1987. Complex origins. In T.C. Whitmore (ed.), *Biogeographical evolution of the Malay archipelago*: 119-131. Oxford: Clarendon Press.

Hall, R. 1998. The plate tectonics of Cenozoic SE Asia and the distribution of land and sea. In R. Hall & J.D. Holloway (eds), *Biogeography and geological evolution of SE Asia*: 99-131. Leiden: Backhuys Publishers.

Hall, R. & Holloway, J.D. (eds) 1998. *Biogeography and geological evolution of SE Asia*. Leiden: Backhuys Publishers.

Holloway, J.D. & Hall, R. 1998. SE Asian geology and biogeography: an introduction. In R. Hall & J.D. Holloway (eds), *Biogeography and geological evolution of SE Asia*: 1-23. Leiden: Backhuys Publishers.

Huggett, R.J. 1998. *Fundamentals of biogeography*. New York: Routledge.

Humphries, C.J. 1990. The importance of Wallacea to biogeographical thinking. In W.J. Knight & J.D. Holloway (eds), *Insects and the rain forests of South East Asia (Wallacea)*: 7-18. London: The Royal Entomological Society.

Huxley, T.H. 1868. On the classification and distribution of the Alectoromorphae and Heteromorphae. *Proc. Zool. Soc. London*: 294-319.

Keast, J.A. 1983. In the steps of Alfred Russel Wallace: biogeography of the Asian-Australian interchange zone. In R.W. Sims & J.H. Price, P.E.S. Whalley (eds), *Evolution, time and space: the emergence of the biosphere*: 367-407. London: Academic Press.

Keast, J.A. & Miller, S.E. (eds) 1996. *The origin and evolution of Pacific island biotas, New Guinea to eastern Polynesia: patterns and processes*. Amsterdam: SPB Academic Publishing.

Kitchener, D.J. & Suyanto, A. (eds) 1996a. *Proc. first intern. conf. on eastern Indonesian-Australian vertebrate fauna*. Perth: The Western Australian Museum.

Kitchener, D.J. & Suyanto, A. 1996b. Intraspecific morphological variation among island populations of small mammals in southern Indonesia. In D.J. Kitchener & A. Suyanto (eds), *Proc. first intern. conf. on eastern Indonesian-Australian vertebrate fauna*: 7-13. Perth: The Western Australian Museum.

Knight, W.J. & Holloway, J.D. (eds) 1990. *Insects and the rain forests of South East Asia (Wallacea)*. London: The Royal Entomological Society.

Ladiges, P. & Humphries, C., Martinelli, L. (eds) 1991. *Austral biogeography*. Canberra: CSIRO.

Laumonier, Y. 1990. Search for phytogeographic provinces in Sumatra. In P. Baas & K. Kalkman, R. Geesink (eds), *The plant diversity of Malesia*: 193-211. Dordrecht: Kluwer Academic Publishers.

Laumonier, Y. 1997. *The vegetation of Sumatera*. Bogor: SEAMEO-BIOTROP Regional Center for Tropical Biology.

Lydekker, R. 1896. *A geographical history of mammals*. Cambridge: Cambridge University Press.

Mayr, E. 1944. Wallace's Line in the light of recent zoogeographic studies. *Quart. Rev. Biol.* 19: 1-14.

Meijaard, E. 2000. *A reconstruction of the Tertiary and Quaternary distribution of selected mammals and their habitats in Sundaland and Sulawesi, based on their ecology, genetically derived separation lines, and the fossil record*. Proposal for Ph.D. research.

Metcalfe, I. 1998a. Palaeozoic and Mesozoic geological evolution of the SE Asian region: multidisciplinary constraints and implications for biogeography. In R. Hall & J.D. Holloway (eds), *Biogeography and geological evolution of SE Asia*: 25-41. Leiden: Backhuys Publishers.

Metcalfe, I. 1998b (ed.). Gondwana dispersion and Asian accretion. Rotterdam: Balkema Publishers.

Michaux, B. 1991. Distributional patterns and tectonic development in Indonesia: Wallace reinterpreted. *Austral. Syst. Bot.* 4: 25-36.

Monk, K.A. & de Fretes, Y., Reksodiharjo-Lilley, G. 1997. *The ecology of Nusa Tenggara and Maluku.* The ecology of Indonesia series vol. V. Singapore: Periplus Editions.

Musser, G.G. 1987. The mammals of Sulawesi. In T.C. Whitmore (ed.), *Biogeographical evolution of the Malay archipelago*: 73-93. Oxford: Clarendon Press.

Oosterzee, P. van 1997. *Where worlds collide - the Wallace Line.* Ithaca, New York: Cornell University Press.

Pethiyagoda, R. & Manamendra-Arachchi, K. 1998. Evaluating Sri Lanka's amphibian diversity. *Occ. Pap. Wildl. Heritage Trust* 2: 1-12.

Pielou, E.C. 1979. *Biogeography.* New York: John Wiley & Sons.

Pilseneer, P. 1904. La "Ligne de Weber", limite zoologique de l'Asie et de l'Australie. *Bull. Classe Sci. Acad. R. Belgique* 1904: 1001-1022.

Simpson, G.G. 1977. Too many lines; the limits of the Oriental and Australian zoogeographic regions. *Proc. Amer. Philos. Soc.* 121: 107-120.

Steenis, C.G.G.J. van 1950. The delimitation of Malaysia and its main plant geographical divisions. *Flora Malesiana* 1: lxx-lxxv.

Steenis, C.G.G.J. van 1979. Plant-Geography of East Malesia. *Bot. J. Linn. Soc.* 79: 97-178.

Thornton, I.W.B. 1996a. *Krakatau: the destruction and reassembly of an island ecosystem.* Cambridge, Mass: Harvard University Press.

Thornton, I.W.B. 1996b. The origins and development of island biotas as illustrated by Krakatau. In J.A. Keast & S.E. Miller (eds), *The origin and evolution of Pacific island biotas, New Guinea to eastern Polynesia: patterns and processes*: 67-90. Amsterdam: SPB Academic Publishing.

Thornton, I.W.B. & Zann, R.A., Rawlinson, P.A., Tidemann, C.R., Adikerna, A.S. 1988. Colonization of the Krakatau islands by vertebrates: equilibrium, succession, and possible delayed extinction. *Proc. Natl. Acad. Sci. USA* 85: 515-518.

Udvardy, M.D.F. 1975. A classification of the biogeographical provinces of the world. *IUCN Occ. Pap.* 18: 1-48.

Udvardy, M.D.F. 1984. A biogeographical classification system for terrestrial environments. In J.A. McNeely & K.R. Miller (eds), *National parks, conservation, and development*: 34-38. Washington: Smithsonian Institution Press.

Vane-Wright, R.I. 1990. The Philippines - key to the biogeography of Wallacea? In W.J. Knight & J.D. Holloway (eds), *Insects and the rain forests of South East Asia (Wallacea)*: 19-34. London: The Royal Entomological Society.

Vane-Wright, R.I. 1991. Transcending the Wallace Line: do the western edges of the Australian region and the Australian plate coincide? *Austral. Syst. Bot.* 4: 183-198.

Walker, D. (ed.) 1972. *Bridge and barrier: the natural and cultural history of Torres Strait.* Canberra: Australian National University.

Walker, D. 1982. Speculations on the origin and evolution of the Sunda-Sahul rain forests. In G.T. Prance (ed.), *Biological diversification in the tropics*: 554-574. New York: Columbia University Press.

Wallace, A.R. 1860. On the zoological geography of the Malay archipelago. *J. Linn. Soc. (Zool.) London* 2: 1104-1108.

Wallace, A.R. 1869. *The Malay archipelago.* Stenhouse, Scotland: Tynron Press.

Wallace, A.R. 1876. *The geographical distribution of animals.* London: Macmillan.

Weber, M. 1902. Der Indo-australische Archipel und die Geschichte seiner Tierwelt. Jena: Gustav Fischer Verlag.

Whitmore, T.C. (ed.) 1981a. *Wallace's Line and plate tectonics.* Oxford: Clarendon Press.

Whitmore, T.C. 1981b. Wallace's Line and some other plants. In T.C. Whitmore (ed.), *Wallace's Line and plate tectonics*: 70-80. Oxford: Clarendon Press.

Whitmore, T.C. (ed.) 1987a. *Biogeographical evolution of the Malay archipelago.* Oxford: Clarendon Press.

Whitmore, T.C. 1987b. Introduction. In T.C. Whitmore (ed.), *Biogeographical evolution of the Malay archipelago*: 1-4. Oxford: Clarendon Press.

Whitten, A.J. & Damanik, S.J., Anwar, J., Hisyam, N. 1987. *The ecology of Sumatra.* Yogyakarta: Gadjah Mada University Press.

Faunal exchange between Asia and Australia in the Tertiary as evidenced by recent butterflies

R. de Jong

National Museum of Natural History, Dep. of Entomology, Leiden, Netherlands

ABSTRACT: Based on a cladistic study of the genus *Taractrocera* and of the *Taractrocera* group of genera (Lepidoptera: Hesperiidae) it is concluded that in the evolution of the genus and of the group the crossing of the water barrier between Asia and Australia in both directions played an important part. In view of the differentiation between and within the genera at least part of the crossings cannot be of recent age but must date back to the Tertiary. The evolution of *Taratrocera* and its allies is discussed in relation to the geological evolution of the area. It is concluded that the first dispersal event in *Taractrocera*, from Australia to Asia, occurred not later than 10-12 Ma. The early differentiation of the group of genera may have taken place on island arcs now embedded in the Papuan Region from where some taxa dispersed to Australia and from there to Asia, and more recently, directly to Asia.

1 INTRODUCTION

The spread of Asian taxa to the east is well documented. In particular the Asian ancestry of part of the Papuan biota has been the subject of a number of studies (e.g. de Boer & Duffels 1996, Gressit 1982, Schot 1998, Ziegler 1982). Documentation of the spread of Asian taxa into Australia has been more anecdotal. Colonization in the reverse direction is hardly documented. Apart from the observation that Australian taxa do not cross Wallace's Line, the spread of Australian taxa to the northeast has hardly received attention from neontological researchers. There is, however, limited palynological evidence that before the mid-Miocene collision of Australia and Asia, Casuarinaceae, *Dacrydium* and certain Proteaceae migrated northwards (Truswell *et al.* 1987), and one may suggest that some animal species did the same. The present paper deals with a genus of butterflies that seems to indicate that, within the same genus, there has been exchange between Asia and Australia in both directions, apparently starting in the Miocene. The biogeography of related genera is dealt with briefly, and the requirements for discovering further examples are discussed.

2 GEOLOGICAL BACKGROUND

When Pangaea started to break up and North America drifted away from Africa some 180 Ma ago, or perhaps even earlier when an epicontinental sea reached from the Tethys Sea to the Arctic Sea through Europe, the land connection between Asia and Australia (through other parts of Laurasia and Gondwana) was severed once and for all. Since then, faunal exchange of terrestrial organisms between Asia and Australia could only take place by crossing extensive stretches of oceans. For taxa restricted to Asia and Australia long-distance dispersal is the only plausible explanation of their distribution, unless we suppose or have evidence that they once had a wider distribution and became extinct in the rest of their range. Obviously, the latter explanation implies that such taxa were present in Pangaea.

The distance between Asia and Australia has been prohibitively far for any conceivable successful ocean crossing for most of the time during which Australia drifted from its southern Gondwana position to its present position. At some time in the past, however, Australia must have come close enough for the first organisms to bridge the gap, and the chances of successful crossing must have increased with decreasing distance. The reconstruction is mainly based on Hall (1998) (see also de Boer 1995 for a compilation of geological reconstructions up to that time).

In the mid-Oligocene (30 Ma) the northward drift of Australia had brought the fragments of Gondwana to the northwest of Australia, such as the Vogelkop and the southern Moluccas, as well as the entirely isolated eastern arms of Sulawesi, in a line between the southeastern point of continental Asia and the northwestern point of the Australian shelf. However, these fragments were still submerged: there is no geological evidence for any piece of dry land at the time between present-day Java and Australia, although there was a shallow sea over much of the distance. There was a sinusoidal chain of volcanoes extending from the southeastern end of continental Asia (± present-day Java) eastward in an island arc including the (well separated) southwest and north arms of Sulawesi, parts of the proto-Philippine islands and parts of proto-Halmahera. Daly *et al.* (1991) also included Eastern Sulawesi in this arc, but according to Hall (1998), Eastern Sulawesi was situated much further south. This arc, at least the part from the north arm of Sulawesi eastward, could be called the South Philippine Arc, as it fringed the southern edge of the Philippine Sea Plate. The east end of the arc became connected to another volcanic island arc, the South Caroline Arc, moving southward, and through it to the Melanesian Arc, which was pushed away from Australia by sea floor spreading in the Solomon Sea. Along the eastern edge of the Philippine Sea Plate and connected to the South Philippine Arc, there was an arc system, the Izu-Bonin Arc (also known as Marianne Arc plus Bonin Arc). In the Late Miocene or in the Pliocene this arc system reached Japan. The arc systems are mentioned here since some of the volcanoes probably were emergent and possibly formed a dispersal route. If so, it must have been a dead end route for a long time from Asia eastward, as New Guinea (including the still separated Vogelkop) was still completely submerged and Australia, therefore, far away.

Five million years later, at the end of the Oligocene (25 Ma), land may have started to emerge in the fragments now embedded in Eastern Sulawesi and in the Vogelkop region. The distance between Australia and the South Philippine Arc decreased, but there was still hardly any land, although a part of the arc known as the Sepik Arc collided with the Australian plate causing uplift and dry land in what is now central North New Guinea. In the Early Miocene (20 Ma), West and Southeast Sulawesi started to collide and more land emerged, as did the Tukang Besi Platform and a large part of the Arafura Sea, while there was now a continuous shallow sea between Asia and Australia. Practically all of New Guinea, however, was still below sea level. In the north of New Guinea more terranes of the arc system collided with further northward movement of the Australian plate. Their weight pushed down the north part of the plate and a broad epicontinental sea extended between Australia's mainland and the collided terranes. In terms of land proximity, the only feasible dispersion route between Asia and Australia was from Borneo through proto-Sulawesi and Tukang Besi Platform and vice versa, but the stretches of sea to be crossed were still tremendous, of the order of 1000 km.

In the next ten million years conditions did not improve much, but the collision of the Australian plate with the Asian plate caused increasing vulcanism along the Sumatra-Java-Lesser Sunda Islands chain, and in the Late Miocene (10 Ma) there was land along the whole island of Java, over much of Sulawesi, in the area west of (?and including part of) Buru, northern Vogelkop, North and Northeast New Guinea and the greater part of the Arafura Sea including southern New Guinea. By the Early Pliocene (5 Ma), the latter area had become submerged again, but the land area increased considerably in North New Guinea. Moreover, land started to emerge along the Lesser Sunda Islands chain and in the Moluccas (see also Audley-Charles 1993). Halmahera, however, was still to the north of the Vogelkop and could not play a role as stepping stone between Sulawesi and New Guinea. It is uncertain how far the Outer Melanesian Arc, of which Halmahera formed a part, could act as a dispersal route from the Asian continent through Taiwan and proto-Philippines to New Guinea and further east, to the Polynesian islands.

In the last five million years the emergence of the Lesser Sunda Islands and most of New Guinea, and the westward movement of Halmahera to its present position between North Sulawesi and the Vogelkop strongly enlarged the likelihood of faunal exchange between Asia and Australia. Australia has never been so close to the Asian continent as in its present position. The shortest distance between Bali (the southeastern-most subaerial point of the Asian continent) and the coast of Australia is about 1400 km. Surely this is an enormous distance, but it is not insuperable for strong flyers like some butterflies. *Vanessa kershawi* and *Junonia villida* occasionally reach New Zealand (and the former species Norfolk Island as well) in numbers from Australia, a distance of more than 1900 km (Common & Waterhouse 1981). Most organisms are much poorer dispersers, but the Lesser Sunda Islands reduce the longest stretch of water to be crossed, that between Australia and Selaru (the southern island of the Tanimbar Archipelago) to about 350 km, while the shorter route through Timor involves a water gap of about 480 km (Timor, lying on the Australian plate kept about its present distance from Australia during the northward drift). Another possible route leads from Mindanao or Sulawesi through Halmahera to New Guinea and from there to Australia, but the route is longer and involves the colonization of at least half of New Guinea (which, of course, is a much slower process than jump dispersal, although the latter may be a rare event). Lowering of the sea level during the Ice Ages extended the land areas in the region and lowered the minimum distance between Australia and Timor to about 150-200 km, while Australia and New Guinea became broadly connected.

In summary, evidence of dry land is often uncertain and ambiguous, but in an 'optimistic' scenario (from the view point of terrestrial organisms), the evolution of dispersal routes between Asia and Australia could have been as follows:

- 30 Ma – practically no dispersal possibilities, or at most colonization of the South Philippine Arc from Asia;
- 25 Ma – emergent land in East Sulawesi and the Vogelkop offers the first opportunity for faunal exchange; perhaps also exchange possible through the South Philippine Arc;
- 20 Ma – not much change; emergent land in North New Guinea, but South Philippine Arc probably largely submerged;
- 15 Ma – area of exposed land increased, particularly in Northeast New Guinea, but opportunities for faunal exchange not much improved;
- 10 Ma – exposure of most of Arafura Sea and a large part of South New Guinea strongly reduces distance between dry land in North and Northeast New Guinea and Australia; exposed land in Sulawesi area, West Moluccas and the Vogelkop and increasing vulcanism along the Lesser Sunda Island arc enhance opportunities for faunal exchange;
- 5 Ma – emergence of Moluccan islands and Lesser Sunda Islands further enhance the opportunities for dispersal; Philippine islands approaching their present position, to the north more or less connected to Taiwan by a volcanic island arc;
- 5-1 Ma – uplift of most of New Guinea, dramatically increasing 'empty' land surface and opportunities for exchange with Australia;
- 1-0 Ma – lowering of sea level during the Ice Ages connects all islands on the continantal shelves to their respective mainlands, leaving a water gap between the Sunda shelf and the Sahul shelf in which Sulawesi, the Moluccas and the Lesser Sunda Islands are situated; by far best opportunities for exchange in terms of land-water distribution.

3 FAUNAL EXCHANGE

Whatever the land-sea distribution may have been since the mid Oligocene, animals and plants did disperse from Asia to Australia and *vice versa*. It is exactly this dispersal that gave rise to the recognition of a variety of biogeographic lines, from Wallace's Line in the west to Lydekker's Line in the east, and to the recognition of the region between these lines as a region of some biogeographic importance, Wallacea (the name coined by Dickerson 1928). As defined by Moss & Wilson (1998), Wallacea is the area between the East boundary of the strictly Asian fauna (Wallace's Line) and the West boundary of the strictly Australian fauna (Lydekker's Line). Described in this way the region is defined by the absence rather than by the presence of characters. Between these lines Asian and Australian elements occur in varying proportions, with Asian elements dominating in the western islands, and Australian elements dominating in the eastern islands. For this reason, the area could be called a transition zone. In addition to the 'foreign' elements, Wallacea is rich in endemics, generally restricted to one or a few islands only, and certainly not homogeneously distributed throughout Wallacea (see e.g. de Jong 1998, Musser 1987, Vane-Wright 1991, Vane-Wright & Peggie 1994, Whitten et al. 1987). Obviously, these endemics also have a geographic ancestry outside Wallacea. The heterogeneous nature of the Wallacean biota is understandable in view of the complicated and widely different geological histories of the islands concerned (e.g. Hall 1998). Even in cases of similar island faunas, like the butterfly faunas of Seram and Halmahera, the similarity may not be the result of a common history (de Jong 1998).

In addition to the biota of the area not being homogeneous, it is not correct to state that to the west the fauna is strictly Asian, and to the east strictly Australian, whatever is meant by 'strictly'. Species of Asian origin, or at least with Asian ancestry, occur widely in Australia. The 59 endemic rodent species of Australia (nearly a quarter of the native terrestrial mammals) are considered the offspring of several waves of colonization from Asia (Keast 1972). They all belong to the Muridae which are native to the eastern hemisphere (but a few species have been distributed worldwide by human activities). Fossil evidence indicate that murids probably entered Australia for the first time about 6-7 million years ago, although molecular evidence of extant species suggests that they have been in Australia since about 15 million years ago (Archer et al. 1991, 1995). The fossil record of Australian bats goes much further back: 25 million years ago there was already a diverse bat fauna, and the oldest fossil bat known from Australia is 55 million years old. The majority, if not all of these fossils indicate a Laurasian ancestry, not a Gondwana one, but the oldest fossil bat was found in an Early Eocene fauna (near Murgon, southeastern Australia) that also contained a primitive placental mammal showing relationship with South America (Godthelp et al. 1992). Hand et al. (1994: 375) concluded that: 'Early bats probably entered Australia via Asia rather than South America, although their appearance in Australia predates the final breakup of Gondwana.' Members of the bird family Muscicapidae colonized Australia from Asia (Heatwole 1987). At the species level, the hesperiid butterfly Badamia exclamationis occurs from Sri Lanka and Northwest India to East Victoria in Australia, and through the Polynesian islands to Samoa (Common & Waterhouse 1981, Evans 1949). The genus has only one more species, restricted to Polynesia. It belongs to the subfamily Coeliadinae that has three genera in Africa and six in Asia, three of which extend to Australia, but without endemic species in Australia, so that an Asian origin of Badamia is highly likely, even without exactly knowing the phylogeny of the subfamily. The lycaenid butterfly genus Zizeeria contains only two tiny species, Z. knysna occurring from Spain to S Africa, Z. karsandra from Algeria to South Australia. Similar examples can easily be found in other animal groups. Even if we group together as 'strictly Australian' all species that originated in Australia, irrespective of their geographic ancestry, it is clear that faunal elements from west of Wallace's Line occur east of Lydekker's Line.

Dispersal from Australia to mainland Asia is less well documented. There is limited palynological evidence that prior to the mid-Miocene collision of Australia and Asia, Casuarinaceae, Dacrydium and certain Proteaceae migrated northwards (Truswell et al. 1987). According to Sibley & Ahlquist (1985), the group of passerine birds known as Corvi (crows and relatives, ten families in Australia and New Guinea) originated in Australia and dispersed to Asia, where they radiated and from where they colonized other parts of the world. I could not find documentation on other groups. It is possible that such colonization, if it ever occurred, took place so long ago that the traces are difficult to find. Moreover, the distinction of biogeographic lines and regions tends to focus the attention on those groups that agree with the distinction and to neglect the organisms (which may be more abundant) that do not agree. In other words, the pecularities of the islands of Wallacea may have absorbed much attention at the expense of the groups that show different distributional patterns. This paper focuses on a group of butterflies that appear to show dispersal between Asia and Australia in both directions. Although colonization of Asia from Australia may have been much rarer than in the reverse direction, apparently Australian faunal elements did not stop at Wallace's Line, just as Asian elements did not stop at Lydekker's Line.

It is obvious that a region that has no biogeographic characters of its own cannot be called a biogeographic region. In terms of vicariance biogeography, Wallacea is not a centre of endemism. We could also say that the islands constituing Wallacea do not have a shared biogeographic history. The name Wallacea can still be used as a kind of shorthand for the topographic region between the margins of the Sunda shelf and the Sahul shelf, and it will be used in this sense here, but we must realize that it has no more biogeographic meaning than country names.

A few words should be added on the position of New Guinea. Its fauna is closely related to that of Australia, particularly to the fauna of North Queensland. Yet there are many species and also genera restricted to New Guinea and neighbouring islands. In this study we are interested in faunal exchange between Asia and Australia, not between Asia and New Guinea or between Australia and New Guinea. As will be shown below, it is not possible to discuss the faunal exchange without mentioning occurrence in the New Guinea area altogether. For this area the name Papua will be used in a topographic sense. Geologically speaking it is a composite area. It has a very complicated biogeographic history (see e.g. de Boer 1995), quite different from Australia, although it has received an important part of Australia's biota. The region, in its present form, is young, as we have seen above, but part of its fauna may be old. Below, where a geographic origin is called Papuan, it only means that the ancestor probably lived on one (or several) of the fragments that later formed the recent Papuan area, not that it lived on New Guinea as we know it today.

4 METHODS

The method employed here can be called cladistic biogeography in that the phylogeny of organisms is used to deduce relative time of colonization of areas (islands in this case). It does not fully agree with the definition in Humphries & Parenti (1999: 41): 'Cladistic biogeography is about understanding relationships of areas through discovery of biotic patterns' in that it does not search for relationships of areas or biotic patterns, but for individual cases that may throw light on the exchange between Asia and Australia. It may seem a futile difference, but actually it is basally different.

In the above definition of cladistic biogeography the aim is the (understanding of) relationships of areas. The term 'relationship' is confusing in this context. The area relationship is based on the phylogenetic relationships of a number of taxa inhabiting the areas under study. The first step is substituting the names of the taxa at the terminal branches of the found cladograms by their distribution areas. The cladograms are now called area cladograms. Next, the area cladograms are searched for shared speciation patterns (components) (details of the process are irrelevant here) and, based on the shared possession of such patterns, a general area cladogram is calculated. The latter is considered a hypothesis of the historical relationships of the areas and can be compared to what is known of the geological history of the areas.

The equation of areas with species is problematic. Like species, areas can split, but contrary to species they can also merge, leading to a reticulate relationship. Further, phylogenetic relationship is about relative recency of common ancestors. Areas can only have ancestors if they once formed a single area. Thus, the equation of areas with species only makes sense, if this has been the case (like the break-up of Gondwana). While there is a common ancestor of Asia and Australia (Pangaea), there is no common ancestor of Borneo and Sulawesi. In other words, islands that never formed part of another land mass cannot be said to have split and they do not fit in a cladogram. The solution here is, that the so-called area cladogram actually is not a cladogram of areas, but of biotas named by the area where they occur. The definition of cladistic biogeography given by Humphries & Parenti (1999) is based on Croizat's adage that Earth and Life evolve together, but it is one step too far to suppose that this is a one-to-one relationship. Biotas may change and move over areas without the areas changing or moving. Biotas can also merge, so a biotic relationship could as well be reticulate and difficult to compare with species cladograms. Cladistic biogeographers are well aware of this problem. Their solution is to single out repeated speciation patterns and equate them with area history. Disagreement with the pattern is then due to other events (dispersal, extinction). However, not only species with the least dispersive powers tell us something about biotic and geologic change, all species tell us something, be it not the same thing. We do not want to learn only that tiny part of the story that relates to the most sedentary species, we want the whole story. Our aim should not be the elucidation of the history of areas. That field belongs to geology; it must be kept separate so that it can be used to test biotic hypotheses, or to frame biotic hypotheses on. Our aim should be the unraveling of the history of biotas, and since biotas consist of species with different histories, we must start with these histories. Cladistic biogeography, then, is not about understanding relationships of areas, but *about understanding how biotic evolution has been influenced by geologic evolution*, through phylogenetic relationships.

Problems in applying the procedure described in Humphries & Parenti (1999) and earlier publications, particularly Nelson & Platnick (1981), have been aired before. As Polhemus (1996: 64) phrased it: 'Present component and parsimony analyses that rely soley [sic!] on dichotomous branching *ad infinitum* [...] are clearly insufficient tools [...].' Duffels (1994) suggested a search for multiple area cladograms within a given region. It is similar to the suggestion made here that species cladograms should each be judged separately as part of the biotic history. Where these partial histories coincide, a pattern emerges. There is no reason to suppose that pattern is reserved to species that only speciate in response to a 'speciation' (vicariance) event of their distribution area. Absence of pattern is not 'noise', as is the case when one is only interested in the relationships of areas, it is an interesting observation that deserves further examination. Viewed in this light, distribution areas can be seen as characters, and an area cladogram is nothing else but a species cladogram with the areas marked as characters on the terminal branches, complete with character loss (extinction) and homoplasy (dispersal). In that case, however, it is better to avoid the confusing term 'area cladogram' altogether.

Finally, a warning. Although we can treat distribution areas as characters on a species cladogram, we cannot simply trace the evolution of these characters by applying the character optimization method as implemented in MacClade (Maddison & Maddison 1992). As shown in an earlier paper (de Jong 1996), this can lead to incorrect conclusions about ancestral areas. It need not bother us here, since the present use we make of the cladogram is different.

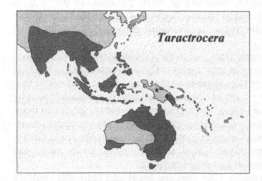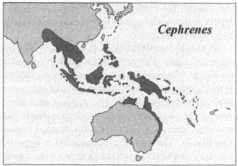

Figure 1. Distribution of the genera *Taractrocera* and *Cephrenes*. Note the absence of *Taractrocera* in the Moluccas and the western part of New Guinea.

5 THE GENUS TARACTROCERA

Taractrocera is a genus of small hesperiid butterflies with a wide distribution, from Sri Lanka and Kashmir to Tasmania (Fig. 1). Surprisingly, in spite of extensive collecting in the Moluccas, the genus has never been encountered there. The 16 species of the genus are nicely divided geographically. Nine species are Oriental, of which four extend into Wallacea, seven species are Australian-Papuan, of which one just reaches Wallacea (Timor). In a preliminary analysis (de Jong 1990), the Oriental species were found to be monophyletic, but this could not be said of the Australian species. Since then two new species have been found to occur at high altitude in New Guinea, and a more extensive analysis showed the situation to be more complex (de Jong, in press). The details of the analysis will be published elsewhere. It is sufficient to observe here that there are two equally parsimonious trees, which differ in the interrelationship of the Australian species at the base (Fig. 2). The difference is not relevant here. The trees show that after an initial radiation in Australia, one species spread to the north, leaving behind an Australian population that evolved into *T. ina*, while the Asian population radiated in the Sunda area (reaching northeast as far as North Burma), after which it extended to India, South China, and, suprisingly, back to Australia (and to New Guinea). Apparently this eastward dispersal was not a single event. Supposing that the sister species of *T. nigrolimbata* was distributed throughout the Oriental Region (i.e. the combined distribution areas of the included Asian species), first dispersal to Australia took place, giving rise to *T. papyria*, later, but from the same Asian population, there was dispersal to New Guinea, giving rise to two New Guinean species, while the Asian population subdivided into three species after the dispersal to New Guinea.

At some unknown date one of the basal species, *T. ilia*, extended its range into New Guinea, where it developed slightly different characters of its own. Since the difference between the Australian and New Guinean populations is so slight, the range extension may have been relatively recent.

Obviously, the correctness of the interpretation of the distribution history of *Taractrocera* depends on the correctness of the reconstruction of its phylogeny. For a discussion of this and of ecological implications, see de Jong (in press).

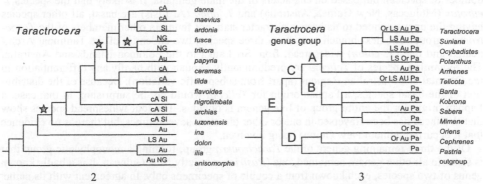

Figures 2-3. 2, One of two equally parsimonious cladograms of *Taractrocera*. The other cladogram has *anisomorpha* and *dolon+ilia* interchanged. The distributions have been roughly indicated as follows: Au = Australia, cA = continental Asia, LS = Lesser Sunda Islands, NG = New Guinea, Sl = Sundaland. Major dispersal events have been marked with a star. 3, Cladogram of the *Taractrocera* group of genera. Distributions have been roughly indicated as follows: Au = Australia, LS = Lesser Sunda Islands, Or = Oriental Region, Pa = Papua (New Guinea with neighbouring and Pacific islands, also including Moluccas). For letters A-E, see text.

6 THE TARACTROCERA GROUP OF GENERA

Taractrocera belongs to an apparently monophyletic group of 13 genera (Evans 1949, Maruyama 1991), widely distributed in the Indo-Australian region and with uncertain affinities. Evans (1955) suggested a link with the American *Phemiades* subgroup of the *Hesperia* group, but there seems to be no apomorphy to support this claim. A preliminary analysis (not publ.) resulted in the cladogram of Figure 3. The genera *Suniana* and *Ocybadistes* form a monophyletic group with *Taractrocera*. *Suniana* has three Papuan-Australian species, one of which (*S. lascivia*) is found on Timor, in addition to New Guinea and Australia. *Ocybadistes* has five Papuan-Australian species, one of which (*O. walkeri*) occurs in a number of Lesser Sunda Islands and Tanimbar. This, together with the early Australian radiation in *Taractrocera*, strongly points to an Australian origin of the three genera. Apparently, they all extended their range to the north, but only *Taractrocera* succeeded in reaching the Asian mainland.

The sister group of the three genera combined is the prolific genus *Potanthus*, with about 30 species in the Oriental region. Six species occur in the Lesser Sunda Islands, of which two are endemics, while one species is found from Sumatra to the Moluccas. Consequently, an Australian origin of this genus is out of the question (but it does not preclude an Australian ancestry).

The genus *Arrhenes*, sister to the four foregoing species combined, is a Papuan genus of six species, two of which extend into Australia. An Oriental origin is out of the question here, as there are no Oriental species. Since there are no Australian endemics, it is likely that the genus is originally Papuan, but a phylogenetic analysis is needed to support this idea. The genus does not occur in the Solomon Islands, but it is found in the Bismarck Archipelago.

Of the apparently monophyletic group of the next five genera that forms the sister group of the five foregoing genera combined, four (*Banta, Kobrona, Sabera* and *Mimene*) form an almost strictly Papuan monophyletic group. Only in *Sabera* and *Mimene* is there some extension into Australia, but there are no Australian endemics. Their sister *Telicota* has a wide distribution, from Sri Lanka and Northwest India to the Solomon Islands and Australia. Although the distribution is similar, the division of the species over the area is quite different from *Taractrocera*. Among the probably more than 30 species there is only one endemic of the Oriental Region, there are three Moluccan endemics (including the Tanimbar Archipelago) and a number of endemics in New Guinea. Of the seven species in Australia, where no endemics occur (if *T. anisodesma* is considered a subspecies of *T. augias*), four are very widespread, occurring westward to Sri Lanka and Northwest India. The phylogeny of *Telicota* is still uncertain (as is the exact number of species), but based on characters of the male genitalia, it is likely that the species *T. eurotas* (Moluccas, New Guinea, Australia) and *T. doba* (Aru Isls) are basal, all other species having the uncus divided to its base, a character state not found in other members of the *Taractrocera* group. The next branch consists of three species, two endemics of the Tanimbar Archipelago and one (*T. colon*) widespread, from Sri Lanka to the Solomon Islands and Australia. The remaining species of *Telicota* are predominantly Papuan with hardly any differentiation in the Oriental and Australian Regions (apart from subspecific variation). Because of the distribution of the basal species, a Papuan origin for *Telicota* would not be surprising. In that case, a Papuan origin for the entire group of five genera is obvious. The four widespread species show that Wallacea has been traversed in one or other direction several times, but there is no evidence that it occurred without New Guinea being involved.

Of the three remaining genera of the *Taractrocera* group, forming a monophyletic group that is sister to all other genera combined, one (*Pastria*) is strictly New Guinean. It is a badly known genus of two species, each known from a couple of specimens only. In agreement with its name, *Oriens* is Oriental, extending to Sulawesi and Flores. There is, however, one species (*O. augustula*) endemic to Fiji and Samoa. It has also been recorded from Australia (one specimen from Darwin, one from Townsville), where it may have been imported inadvertently by boat. This highly disjunct distribution can only be explained by supposing extinction in the intervening area. The ancestral area of the genus could then be either Asia or Papua. Apparently, there has not been exchange between Asia and Australia. The genus *Cephrenes*, finally, deserves some more attention. The distribution is given in Figure 1 and the (preliminary) phylogeny of the five species in Figure 4. If correct, the cladogram indicates a dispersal event between Asia and Australia, leading to the origin of *C. acalle* (Northeast India to Sulawesi and Flores) and *C. trichopepla* (North and East Australia). Since the sister group of *C. acalle* and *C. trichopepla* is strictly Papuan, we can simplify the five-taxon cladogram to a four-taxon cladogram with the following character distribution: (Papua+Australia (Papua (Australia, Asia))). If the ancestral area was Asia, then three dispersal events were needed to arrive at the cladogram as we have found it. If either Papua or Australia was the ancestral area of *Cephrenes*, only two dispersal events were needed. For reasons of parsimony we can thus conclude that the ancestral area of the genus was either Papua or Australia.

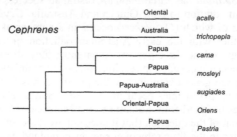

Figure 4. Cladogram of *Cephrenes*, with *Oriens* and *Pastria* as outgroups. Papua = New Guinea with neighbouring and Pacific islands, also including Moluccas. NB – *augiades* has been recorded in few specimens from Sulawesi (Evans 1949) in the same subspecific form as flying in Seram and Ambon, different from the Buru form. This raises some doubt about the reliability of the labels.

7 SUMMARY OF GEOGRAPHIC HISTORY OF THE TARACTROCERA GROUP

Given the monophyly of the *Taractrocera* group, it is clear that there has been exchange between Asia and Australia, but such a general statement, lacking direction and any detail, is unsatisfactory for a group of about 150 species.

The detailed analysis of *Taractrocera* indicates that exchange occurred between Asia and Australia in both directions without New Guinea being involved. Direct dispersal from Australia to Asia seems also to have taken place in *Cephrenes*. Incomplete direct dispersal from Australia to Asia (i.e. the dispersing species not reaching further than the Lesser Sunda Islands) is evident in species of *Taractrocera* (*T. anisomorpha*) and *Suniana* (*S. lascivia*). In the *Taractrocera* group, direct dispersal from Asia to Australia is not evident outside *Taractrocera*. Further dispersal events in the *Taractrocera* group involve exchange between Asia and Papua, and between Australia and Papua.

With regard to the geographic origin of the genera of the *Taractrocera* group, we can draft the following hypotheses (letters referring to the subgroups as marked in Figure 3):

A – We can simplify the cladogram to a three-taxon statement, as follows ('anc' stands for 'ancestor of'): (anc *Arrhenes* (anc *Potanthus* + anc (*Taractrocera+Ocybadistes+Suniana*))). The three ancestors were distributed in Papua, Asia and Australia, respectively. It does not give us a clue as to what the ancestral area of subgroup A may have been. Clearly direct dispersal between Asia (*Potanthus*) and Australia (its sister group) was involved, but on the basis of distribution alone we cannot conclude in which direction dispersal took place. We shall discuss it in the next paragraph in the light of the geological evolution.

B – The phylogeny of *Telicota* needs further attention, but so far a Papuan origin for *Telicota*, and with that for the whole subgroup, is most likely. It implies that there has been dispersal from the Papuan Region in *Telicota*. Whether there has been dispersal in the reverse direction depends on the phylogeny of *Telicota*. There is no evidence of faunal exchange between Asia and Australia without New Guinea being involved.

C – We can simplify the cladogram to a four-taxon statement as follows: (anc B (anc *Arrhenes* (anc *Potanthus* (anc (*Taractrocera+Ocybadistes+Suniana*))))). If the ancesor of subgroup C was Papuan, only two dispersal events were needed to arrive at the four-taxon situation, viz. one to Australia and one to Asia. If either Australia or Asia was the ancestral distribution area, three dispersal events would have been needed (twice to Papua and once to Asia, respecively Australia). So a Papuan origin for subgroup C is the most parsimonious solution.

D – We concluded above that the ancestral area of *Cephrenes* was either Papua or Australia, and that of *Oriens* either Papua or Asia. With their sister group being strictly Papuan, a Papuan origin for subgroup D would be obvious, if *Oriens* also had a Papuan origin. Apparently the genera did not evolve in response to a dispersal event between Asia and Australia, but such an event occurred in *Cephrenes*.

E – On the basis of the above discussion, a Papuan origin for the *Taractrocera* group is a serious possibility.

8 DATING THE FAUNAL EXCHANGE BETWEEN ASIA AND AUSTRALIA

In the absence of fossils, dating the dispersal events in the *Taractrocera* group is a precarious undertaking. Fossil evidence indicates that Australia has not been so splendidly isolated from northern continents as it usually is supposed to have been. The oldest fossil bat from Australia, with an apparent northern affinity, is 55 million years old (Archer *et al.* 1991). Certainly, isolation means different things for different organisms. For butterflies, Australia may have been more isolated than for bats. And the time when fragments of Gondwana that later would become South Tibet, Burma, Malaysia and Sumatra, drifted northward and could have acted as stepping stones, probably was too early for the butterflies anyway. It would seem then that before 25 Ma Australia and Asia simply were too far apart for a successful crossing of the ocean, but we want to be more specific than this general statement for which we did not need to study any butterflies. In three of the thirteen genera of the *Taractrocera* group there are cases in which the same species or sister groups occur in mainland Asia and in Australia (thus excluding dispersal events that ended in Wallacea), namely, in *Taractrocera*, *Telicota* and *Cephrenes* (we exclude *Oriens* here, as it is unlikely that the Australian records refer to autochtonous specimens). In *Telicota* there are four species occurring in mainland Asia as well as in Australia; there are no sister species in the two regions. All four species also occur in New Guinea and there is no need to suppose a direct dispersal between Australia and Asia. Further, in view of the prolific nature of *Telicota* it seems improbable that such widespread species would not have differentiated into a number of species if their distribution areas had been disjunct for millions of years. It would rather seem that the wide distributions are due to relatively recent dispersal, helped by lowering of the sea level in the Pleistocene ice ages.

In *Cephrenes* there is a sister relationship between the Oriental *C. acalle* and the Australian *C. trichopepla*, without involvement of New Guinea. Thus, direct dispersal is likely, and since we concluded above that *Cephrenes* probably had an Australian or Papuan ancestry, the dispersal was from Australia to Asia. Since the sister group of this species pair consists of a pair of Papuan species, one of which (*C. carna*) is restricted to New Guinea and two Moluccan Islands, while the other (*C. moseleyi*) has a wider distribution, from the Moluccas to the Solomon Islands, it would seem that they could only originate when the Moluccas and New Guinea had emerged, i.e. not earlier than 5-6 Ma. Alternatively, they could have been older and have been inhabitants of the South Philippine Arc and the occurrence in the Moluccas may be much younger. So this does not help us to date the dispersal from Australia to Asia.

The degree of differentiation in *Taractrocera* contradicts the idea of a recent faunal exchange such as during the Pleistocene ice ages. Key to dating the dispersal events in this genus may be the two New Guinean endemics. Without these endemics we would have a simple sister group relationship between the Australian *T. papyria* and a monophyletic group of three Asian species, without any involvement of New Guinea or the Moluccas to help dating the dispersal event. As matters stand, the New Guinean endemics seem to be very helpful. Dispersal is easiest for species that live at low altitudes, as for high altitude species the distances to the next high altitude environment are larger. Most *Taractrocera* species are restricted to lower altitudes, although some can occur over a wide range of altitudes. Yet the New Guinean endemics are restricted to high altitude. This suggests that they were trapped by dense rain forest developing on the slopes of the rising mountains and had to adapt themselves or become extinct. In that case they must have arrived there before the mountains were too high to reach the more open spots at higher elevations. At the same time, they became extinct in the lowland. These conditions existed in the late Miocene (10 Ma), when the Arafura Sea and southernmost New Guinea were dry land and several terranes of the Melanesian Arc had already accreted in the north of New Guinea, separated from the dry land in the south by a shallow epicontinental sea. It is suggested here that the ancestor of the New Guinean endemics lived in the lowlands of the dry Arafura Sea, succeeded in reaching dry land in the New Guinea area, and became extinct in the lowland due to gradual submergence of the Arafura Sea. This would also explain the absence of the genus from the Moluccan islands: the last time it dispersed to the east, the Moluccan islands had not yet emerged.

According to the cladogram, there was an earlier dispersal from Asia to Australia giving rise to the Australian *T. papyria*. It would have been simpler if *T. papyria* and the ancestor of the New Guinea endemics had been sister species. In that case the dispersal across the epicontinental sea and the submergence of the Arafura Sea would have caused their separation. As matters stand this is not the case, but this may also be an artefact of the cladogram. Clearly, the dispersal event that gave rise to *T. papyria* could also have taken place in the late Miocene, as soon as the drying up of the Arafura Sea facilitated the crossing from Asia.

Dating the foregoing dispersal events at 10-12 Ma, the dispersal event that brought *Taractrocera* to Asia must have been earlier. We do not have any clue for that, it could be any time between 12 Ma and some time after 25 Ma.

The extreme boundary of 25 Ma for successful crossings could be grossly overestimated. Anyway, if the first dispersal event in *Taractrocera* occurred 12 Ma, the genus must be older, and the same applies to *Ocybadistes* and *Suniana*. In other words, their common ancestor must have lived before 12 Ma, and the same applies for the ancestor of their sister group, *Potanthus*. Consequently, the dispersal event from Australia to Asia that gave rise to *Potanthus* must have taken place around that time, and the first dispersal event in subgroup A of Figure 3 must have occurred still earlier.

In summary, it appears that dispersal between Asia and Australia in the *Taractrocera* group occurred several times, particularly from Australia to Asia, but also in the reverse direction, starting at around 15 Ma. Above we hypothesized that the early differentiation of the *Taractrocera* group (i.e. before 15 Ma) occurred in Papua. However, at that time New Guinea was practically non-existent, and for Papua we should read then 'terranes of the island arcs'. It falls outside the scope of the present study, but it would be interesting to compare the evolution of the *Taractrocera* group with that of the cicadas occurring in the same area, in which the history of the terranes of the island arcs seem to have played an important part (de Boer 1995).

9 PROSPECTS FOR FURTHER RESEARCH

One of the factors determining the success of a colonization is the availability of food in the new area. *Taractrocera* species live on grasses and they do not seem to be very dainty in their choice. The Australian *T. papyria*, for instance, has been found on a variety of common grasses including *Pennisetum*, *Echinopogon*, *Microlaena*, *Paspalum*, *Poa*, *Danthonia* and *Oryza* and moreover it feeds on *Carex* (Cyperaceae). With such a diet, chances of finding food in the new territory are rather high. In addition *Taractrocera* species may be abundant in grassy places along the shore. Thus, *Taractrocera* seems to be well equipped for colonization across the sea. However, most other members of the *Taractrocera* group are grass eaters, yet exchange between Asia and Australia has remained a rare event. It may not be without meaning that *Cephrenes* larvae do not live on grasses but on various palms, including coconut palms, which can be found on every tropical beach.

Members of the *Taractrocera* group are not the only skippers living on grasses. Monocotyledons are the foodplants of practically all Hesperiinae, the subfamily the *Taractrocera* group belongs to, and most species live on grasses. Yet there are few genera in which members are found in both Asia and Australia. Of the 48 hesperiine genera found in Sundaland, only four genera outside the *Taractrocera* group are also found in Australia: *Notocrypta*, *Parnara*, *Borbo* and *Pelopidas*. In none of these genera there are species or pairs of sister species exclusively occurring in Asia and Australia, to the exclusion of New Guinea.

Grass eating is not a prerequisite for a disperser. Three of the 19 genera of dicotyledon-feeding Hesperiidae have species that are found in Sundaland as well as in Australia. In all cases, however, New Guinea is also involved. Further there are a few endemic genera in Papua-Australia of which the relationship is not well known and that could well have sister genera in Asia. A phylogenetic analysis of these genera (*Chaetocneme* – New Guinea, Australia, Bismarcks; *Netrocoryne* – Australia, New Guinea; *Exometoeca* – Southwest Australia) and their relationships with other genera would be highly interesting.

Another group of monocotyledon-feeding Hesperiidae in Australia, the Trapezitinae, must be mentioned here. The subfamily is almost strictly Australian, but there is a slight extension to the Papuan region. It is well developed with 16 genera and about 65 species. The larvae feed on Poaceae, Cyperaceae, Iridaceae and Xanthorrhoeaceae. The relationship of the subfamily is unknown. Evans (1949) suggests that it is related to the *Plastingia* group of the Hesperiinae, an almost exclusively Oriental group of genera. If correct, an Asian-Australian exchange is obvious, but we should better not speculate on this before the phylogenetic relationships have been worked out.

Among the other butterflies, the subfamily Satyrinae (Nymphalidae), which almost exclusively feeds on Poaceae and Cyperaceae, seems to offer particularly good prospects for research on this topic. Miller (1968: 150) wrote about this subfamily (which he considered a family):

'In summary, the satyrid fauna of the Australian region was derived from the Indo-Malayan region, and this derivation very early, resulting in a highly distinctive fauna. If the time schedule for the marsupials is valid for the satyrids, most of the groups reached the region in late Cretaceous to earliest Tertiary times. This indicates that the Hypocystini of the most evolved subfamily, Satyrinae, were in existence by the earliest Tertiary. A few strong flyers, such as *M. leda*, probably arrived over water very recently.'

There is a contradiction in this statement, which was written before general acceptance of plate tectonics. If the Hypocystini (an endemic tribe in the Australian region, with 21 genera of which seven in Australia) are really that old, they cannot well have invaded Australia from the Indo-Malayan region, and a Gondwana origin should be considered. But their supposed age is highly speculative and the scenario is not based on a phylogenetic analysis. A correspondence with the Trapezitinae suggests itself.

10 CONCLUSION

Fossil evidence indicates that Australia has not been as strongly isolated from northern continents on its journey from Antarctica to its collision with Asia as is often supposed. Evidence in extant faunas is scarce, however. This may be partly due to incompletely known phylogenies. The extant faunas of Asia and Australia share a number of species and genera of butterflies. They indicate exchange of faunal elements by dispersal across wide stretches of sea, since there has never been a dry land connection. During the Pleistocene ice ages lowering of the sea level facilitated dispersal. It appears that the Pleistocene dispersal was mainly a one-way route, from Asia to Australia (and New Guinea). Australian elements rarely or never succeeded in reaching further than the Lesser Sunda Islands and Sulawesi. This period is too recent to have led to extensive differentiation in Australia: it is the same species occurring in Asia and Australia. In cases where the colonizer and its source population speciated or even radiated, it would seem that a longer time span is involved. The hesperiid genus *Taractrocera* indicates an early colonization of Asia from Australia, radiation in Asia, and a return colonization of Australia. Since rodents with Asian affinities were present in Australia at least 6-7 million years ago and possibly arrived much earlier, it is supposed that some butterflies could do the same, and the dispersal of the ancestor of *T. papyria* to Australia may date from that period. That places the dispersal of *Taractrocera* from Australia to Asia at a much earlier date, allowing time for radiation in Asia. It is postulated that from the time fragments of dry land emerged between Asia and Australia, about 25 million years ago, opportunities for dispersal improved and strong flyers may have started using that route. Although the geological evidence does not indicate much dry land between Asia and Australia for much of the time since 25 million years ago, the possibility that the evidence got lost, cannot be ruled out. Many more phylogenies of genera and higher taxonomic taxa of butterflies (and other animals, for that matter) need to be examined before we can estimate how frequent dispersal between Asia and Australia has been in the Tertiary.

ACKNOWLEDGMENTS

The author thanks Prof Roger Kitching for discussions and comments on a draft of this paper, and an anonymous reviewer for comments.

REFERENCES

Archer, M., Hand, S.J. & Godthelp, H. 1991. *Riversleigh, The story of animals in ancient rainforests of inland Australia*. Kew, Victoria: Reed Books.

Archer, M., Hand, S.J. & Godthelp, H. 1995. Tertiary environmental and biotic change in Australia. In A.S. Vrba, G.H. Denton, T.C. Partridge & L.H. Burckle (eds), *Paleoclimate and evolution, with emphasis on human origins*. New Haven: Yale University Press.

Audley-Charles, M.G. 1993. Geological evidence bearing upon the Pliocene emergence of Seram, an island colonizable by land plants and animals. In I.D. Edwards, A.A. MacDonald & J. Proctor (eds), *Natural History of Seram*. Andover: Intercept Ltd.

Common, I.F.B. & Waterhouse, D.F. 1981. *Butterflies of Australia*. London etc.: Angus & Robertson Publishers.

Daly, M.C., Cooper, M.A., Wilson, I., Smith D.G. & Hooper, B.G.D. 1991. Cenozoic plate tectonics and basin evolution in Indonesia. *Marine and Petroleum Geology (-MPGD)* 8: 1-21.

de Boer, A.J. 1995. Islands and cicadas adrift in the West-Pacific. Biogeographic patterns related to plate tectonics. *Tijdschrift voor Entomologie* 138: 169-244.

de Boer, A.J. & Duffels, J.P. 1996. Biogeography of Indo-Pacific cicadas east of Wallace's Line. In A. Keast & S.E. Miller (eds), *The origin and evolution of Pacific island biotas, New Guinea to Eastern Polynesia: Patterns and Processes*: 297-330. Amsterdam: SPB Academic Publishing.

de Jong, R. 1990. Some aspects of the biogeography of the Hesperiidae (Lepidoptera, Rhopalocera) of Sulawesi. In W.J. Knight & J.D. Holloway (eds), *Insects and the rain forests of South East Asia (Wallacea)*: 35-42. London: The Royal Entomological Society of London.

de Jong, R. 1996. The continental Asian element in the fauna of the Philippines as exemplified by *Coladenia* Moore, 1881 (Lepidoptera: Hesperiidae). *Cladistics* 12: 323-348.

de Jong, R. 1998. Halmahera and Seram: different histories, but similar butterfly faunas. In R. Hall & J.D. Holloway (eds), *Biogeography and Geological Evolution of SE Asia*: 315-325. Leiden: Backhuys Publishers.

de Jong, R. in press. Phylogeny and biogeography of the genus *Taractrocera* Butler, 1870 (Lepidoptera: Hesperiidae), an example of Southeast Asian-Australian interchange. *Zool. Med. Leiden.*

Dickerson, R.E. 1928. *Distribution of life in the Philippines*. Philippine Bureau of Sciences Monograph 21.

Duffels, J.P. 1994. Abstract: Biogeographic patterns in the Indo-Pacific. Biodiversity and Phylogeny. *XIIth Meeting of the Willi Hennig Society, Copenhagen.*

Evans, W.H. 1949. *A Catalogue of the Hesperiidae from Europe, Asia and Australia in the British Museum (Natural History)*. London: Trustees of the British Museum.

Evans, W.H. 1955. *A Catalogue of the American in the British Museum (Natural History)*, Part IV. London: Trustees of the British Museum.

Godthelp, H., Archer, M., Cifelli, R., Hand, S.J. & Gilkeson, C.F. 1992. Earliest known Australian Tertiary mammal fauna. *Nature* 356: 514-516.

Grant, P.R. (ed.) 1998. Evolution on islands. Oxford, New York & Tokyo: Oxford University Press.

Gressit, J.L. 1982. Zoogeographical summary. In J.L. Gressit (ed.), *Biogeography and ecology of New Guinea*: 897-918. The Hague: Junk.

Hall, R. 1998. The plate tectonics of Cenozoic SE Asia and the distribution of land and sea. In R. Hall & J.D. Holloway (eds), *Biogeography and Geological Evolution of SE Asia*: 99-131. Leiden: Backhuys Publishers.

Hand, S., Novacek, M., Godthelp, H, & Archer, M. 1994. First Eocene bat from Australia. *Jornal Vertebrate Paleontology* 14: 375-381.

Heatwole, H. 1987. Major components and distributions of the terrestrial fauna. In G.R. Dyne (ed.), *Fauna of Australia*, vol. 1A, General Articles: 101-135. Canberra: Australian Government Publishing Service.

Humphries, C.J. & Parenti, L.R. 1999. *Cladistic Biogeography*. 2nd edition. Oxford: Oxford University Press.

Keast, A. 1972. Australian Mammals: Zoogeography and Evolution. In A. Keast (ed.), *Evolution, Mammals, and Southern Continents*: 195-246. Albany: State University of New York Press,.

Maddison, W.P. & Maddison, D.R. 1992. *MacClade: Analysis of phylogeny and character evolution*. Version 3.0. Sunderland, MA: Sinauer Ass.

Maruyama. K. 1991. *Butterflies of Borneo*, vol. 2, No. 2, Hesperiidae. Tokyo: Tobishima Corp.

Miller, L.D. 1968. The higher classification, phylogeny and zoogeography of the Satyridae (Lepidoptera). *Memoirs American Entomological Society*, nr 24.

Moss, S.J. & Wilson, M.E.J. 1998. Biogeographic implications of the Tertiary palaeogeographic evolution of Sulawesi and Borneo. In R. Hall & J.D. Holloway (eds), *Biogeography and Geological Evolution of SE Asia*: 133-163. Leiden: Backhuys Publishers.

Musser, G.G. 1987. The mammals of Sulawesi. In T.C. Whitmore (ed.), *Biogeographical Evolution of the Malay Archipelago*: 73-93. Oxford: Clarendon Press.

Nelson, G. & Platnick, N.I. 1981. *Systematics and biogeography; cladistics and vicariance*. New York: Columbia University Press.

Polhemus, D.A. 1996. Island arcs, and their influence on Indo-Pacific biogeography. In A. Keast & S.E. Miller (eds), *The origin and evolution of Pacific Island biotas, New Guinea to Eastern Polynesia: patterns and processes*: 51-66. Amsterdam: SPB Academic Publishing.

Schot, A.M. 1998. Biogeography of *Aporosa* (Euphorbiaceae): testing a phylogenetic hypothesis using geology and distribution patterns. In R. Hall & J.D. Holloway (eds), *Biogeography and Geological Evolution of SE Asia*: 279-290. Leiden: Backhuys Publishers.

Sibley, C.G. & Ahlquist, J.E. 1985. The phylogeny and classification of the Australo-Papuan passerine birds. *The Emu* 85: 1-14.

Truswell, E.M., Kershaw, A.P. & Sluiter, I.R. 1987. The Australian—Southeast Asian connection: evidence from the palaeobotanical record. In T.C. Whitmore (ed.), *Biogeographical Evolution of the Malay Archipelago*: 32-49. Oxford: Clarendon Press.

Vane-Wright, R.I. 1991. Transcending the Wallace line: do the western edges of the Australian region and the Australian plate coincide? In P.Y. Ladiges, C.J. Humphries & L.W. Martinelli (eds), *Austral Biogeography*. Canberra: CSIRO.

Vane-Wright, R.I. & Peggie, D. 1994. The butterflies of Northern and Central Maluku: diversity, endem-
    ism, biogeography, and conservation priorities. *Tropical Biodiversity* 2: 212-230.
Whitten, A.J., Mustafa, M. & Henderson, G.S. 1987. *The Ecology of Sulawesi*. Gadjah Mada University
    Press, Yogyakarta.
Ziegler, A.C. 1982. An ecological check list of New Guinea Recent mammals. In J.L. Gressit (ed.), *Bio-
    geography and ecology of New Guinea*: 863-893. The Hague: Junk.

Why does the distribution of the Honeyeaters (Meliphagidae) conform so well to Wallace's Line?

Hugh A Ford

Zoology, University of New England, Armidale, NSW 2351, Australia

ABSTRACT: Honeyeaters are the most diverse family of birds in Australia, and occupy many habitats from Bali, through New Guinea and Australia, to Hawaii. However, only one species has crossed Wallace's Line. Five hypotheses could explain the failure of honeyeaters to penetrate the Oriental region. 1. Failure to disperse across sea barriers seems unlikely, because honeyeaters have reached many isolated islands in the Pacific. 2. Sunbirds and bulbuls are potential competitors of small and large honeyeaters, however honeyeaters are aggressive and dominate most other birds. 3. Honeyeaters have co-evolved with the Australian flora, notably the Myrtaceae and Proteaceae, which decline west of Wallace's Line. However, they also occupy rainforest in New Guinea, similar to that in South East Asia, and are flexible and generalised, using fruit and alternative carbohydrates in Australia. 4. Bird-eating raptors occur throughout Wallacea, so they should not prevent the spread of honeyeaters. Also, while Java and Borneo have numerous placental carnivores and primates, there is some geographical overlap between honeyeaters and these potential arboreal nest predators. 5. Avian malaria probably exterminated native birds in Hawaii, including some honeyeaters, but mosquitos must regularly be blown between islands and birds migrate between South East Asia and Wallacea and Australia, both able to carry malaria between these sub-regions. Overall, none of the possible hypotheses explains the nearly total failure of honeyeaters to penetrate west of Wallace's Line. We are unlikely to be able to explain the distribution of honeyeaters and other groups in this region until we understand better their ecology and behaviour.

1 INTRODUCTION

The honeyeaters are the most successful family of birds in the Australasian region, in terms of number of species (c. 180, including the Australian chats). They occupy the full range of habitats from deserts to rainforests and mangroves to sub-alpine shrubland and are frequently among the most abundant birds in these habitats. Some 72 species live in Australia (Christidis & Boles 1994) and 61 species in New Guinea (Beehler & Finch 1985, but excluding *Oedistoma* and *Toxorhampus*, which are now regarded as flowerpeckers - Christidis *et al.* 1993). Species richness declines sharply east and west of these centres. New Zealand has three species, with three to five species occurring in New Caledonia, Samoa and Fiji. The Hawaiian Islands had 5 species, though all but one have probably gone extinct. Despite their success, honeyeaters have almost totally failed to make any impact on South East Asia, in contrast to other typically Australian passerine birds, such as woodswallows and whistlers, which have spread as far as India.

All large islands east of Wallace's Line have several species of honeyeaters, but only a single species, *Lichmera limbata*, occurs west of the line, on Bali (Figure 1; White & Bruce 1986). Wallacea has a distinct honeyeater assemblage, as is true of its birds generally (Clode & O'Brien , this volume). Two genera are endemic to Wallacea: *Myza* consists of two species endemic to Sulawesi, probably related to New Guinea genera (White & Bruce 1986) and *Melitograis*, a distinctive friarbird, occurs in Morotai, Halmahera and Bacan. This endemism indicates that the family has been in the region for a long time and agrees with the relatively early appearance of what is now Sulawesi (Hall, this volume). Although there is a rapid drop in the number of species westwards from New Guinea, two genera have undergone radiation in the Lesser Sundas and Moluccas (Coates & Bishop 1997). *Lichmera* ('brown' honeyeaters) have a total of 8 species and *Philemon* (friarbirds) have 6 species distributed among these islands, which in both cases is more species of these genera than occur in Australia and New Guinea combined. The most distinct members of *Lichmera* occur in Timor and Wetar, possibly indicating where the radiation of this genus started.

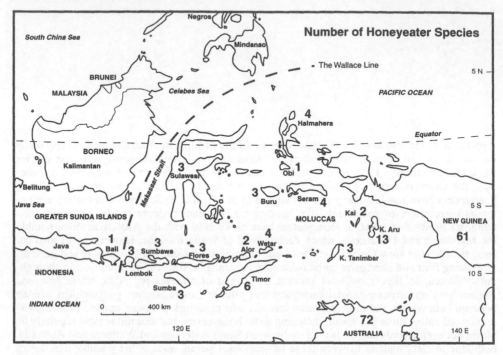

Figure 1. The numbers of species of honeyeaters on each of the islands of Wallacea, and on New Guinea, Australia and Bali (Figure drawn by I. Metcalfe).

2 REASONS FOR LACK OF HONEYEATERS IN ORIENTAL REGION

A number of hypotheses could explain the failure of honeyeaters to penetrate the Oriental region. Honeyeaters may have difficulty crossing sea barriers, they may suffer from competition with ecologically similar birds of Asian origin, they may be unable to find suitable foods, they or their nests may suffer from excessive predation from animals of Asian origin, or new parasites or diseases may exclude them. Each of these will be discussed in turn.

2.1 *Dispersal*

At first sight, failure to disperse across sea barriers seems highly unlikely to be the reason why honeyeaters have almost completely failed to cross Wallace's Line. Members of the family have spread north through Micronesia as far as Saipan (but note that the Bonin Islands 'Honeyeater' is a White-eye - Springer *et al.* 1995), nearly 2000 km from New Guinea. They have also reached Samoa, Fiji, Tonga and Hawaii to the east. Colonisation of the latter would have involved a sea crossing of at least 4000 km and probably further as the Hawaiian honeyeaters appear only distantly related to the other species in Polynesia. One genus *Myzomela* has been particularly successful in reaching numerous, isolated, small islands throughout the western Pacific.

Almost all sizeable islands in Wallacea have one or two *Myzomela* species, yet the genus has failed to colonise Bali, or many other Lesser Sundas (White & Bruce 1986). Its colonisation of Wallacea may be recent.

Many other genera of honeyeaters, however, have been much less successful in crossing even quite modest sea barriers. Eight of the 15 genera of honeyeaters that occur in Victoria (including *Myzomela* and *Philemon*) do not also occur in Tasmania or the Bass Strait Islands (Abbott 1973). Other islands off southern Australia have few honeyeaters (Abbott 1974, 1978), although the lack of suitable habitat may play a part in this (Ford & Paton 1975). Also, only a single species of Australian honeyeater has been recorded in New Zealand, indicating that long-distance dispersal is an infrequent, even freak, occurrence (Falla *et al.* 1981).

2.2 *Interspecific Competition*

Biogeographers tend to emphasise the importance of dispersal across barriers. However, the ability to survive and reproduce once a new island has been reached is also necessary for colonisation to occur.

Honeyeaters are highly aggressive (Longmore 1991) and interspecific competition is believed to play a major role in their community structure (e.g. Ford 1979). Furthermore, the checkerboard patterns of distribution of species, for instance of *Myzomela*, among small islands near New Guinea suggest that competitive exclusion may occur (Diamond 1975). Honeyeaters dominate aggressively many other birds (e. g. Dow 1977), excluding from their habitats a wide range of other, mostly insectivorous, passerines (e. g. Grey *et al.* 1998).

The most obvious potential Asian competitors of small honeyeaters are the sunbirds, which are remarkably similar in morphology and ecology to the honeyeaters of the genus *Myzomela* (Keast 1976). They occur throughout Wallacea, New Guinea and have even colonised northeastern Australia. Ripley (1959) made some interesting observations of co-occurring sunbirds and myzomelas on the island of Batjan (Bacan) in the Moluccas. He noted that both were very similar in size and took similar food, and presumably occupied similar niches. However, the slightly larger Dusky Honeyeaters (*Myzomela obscura simplex*) dominated Black Sunbirds (*Nectarinia sericea*, now *N. aspasia auriceps*), even driving them out of their territories. He proposed that this excessive aggressiveness caused the honeyeaters to neglect their young, leading to low breeding success (an idea originating from Udvardy 1951). However, it is hard to believe that selection would not act against such a disadvantageous level of aggressiveness. Indeed, the behaviour shown by Dusky Honeyeaters on Bacan is characteristic of the high levels of aggression displayed by many honeyeaters elsewhere. There is a need for further studies comparing the foraging behaviours, interspecific interactions and reproductive success of honeyeaters and sunbirds where they are sympatric and allopatric. The *Lichmera* honeyeaters are rather larger than sunbirds and may differ ecologically, though little is known about their foraging behaviour or aggressive interactions.

It is possible that the largely frugivorous bulbuls (Pycnonotidae) are close ecologically to the medium-sized and larger honeyeaters (Keast 1976), such as the friarbirds (*Philemon*). Watling (1982) noted that on Fiji the introduced Red-vented Bulbul (*Pycnonotus cafer*) competed aggressively with the native Wattled Honeyeater (*Foulehaio carunculata*) at nectar sources. He also suggested that they might compete for nest sites. Freifeld (1999) found that the Wattled Honeyeater was abundant in native forest on the island of Tutuila in American Samoa, but scarce in villages and plantations where introduced species such as the Red-vented Bulbul were common. Bulbuls, interestingly, show a distribution that almost mirrors that of honeyeaters: prolific in the Greater Sundas, with 9 species in Bali, but with only three species east of Wallace's Line (Coates & Bishop 1997). Two of these are probably derived from escaped cagebirds. The other species, the Golden Bulbul (*Ixos affinis*) occurs on islands around Sulawesi and in the Moluccas from Morotai to Ambon. Most of these latter islands have one species of friarbird.

The orioles (*Oriolus*) are a widespread Old World group that are similar to friarbirds in morphology, diet, nest type and habitat, and the two genera overlap from Lombok and Halmahera south eastwards to Australia (Keast 1976, Diamond 1982). On many of the islands, orioles and friarbirds also resemble each other to the extent that comparative studies of their ecology and behaviour are hindered. Wallace (1869) suggested that the orioles resembled the large, pugnacious friarbirds to gain protection from hawks. Certainly, friarbirds are notoriously belligerent (14 references in Diamond 1982, also Ford & Debus 1994). However, Diamond has another explanation for the mimicry by orioles of friarbirds - resemblance allowed the orioles to reduce the risk of attack by the friarbirds. Whatever the reason for the evolution of this mimicry, it does seem unlikely that interspecific competition, at least of the interference type, is a reason for the failure of friarbirds to cross Wallace's Line.

2.3 Food and Habitat

The honeyeaters certainly have a close relationship with the Australian flora, notably the families Myrtaceae and Proteaceae (Keast 1968, Ford et al. 1979). The dominance of these plant families falls away towards and beyond Wallace's Line, with the progressively wetter and less seasonal climate. However, most honeyeaters are highly flexible and generalised feeders (Pyke 1980, McFarland & Ford 1991), using a wide range of carbohydrate sources other than nectar (Paton 1980). Within Australia they also visit a range of exotic nectar sources (Ford et al. 1979). It seems unlikely that they would not be able to exploit new nectar sources in South East Asia. Indeed, Rhododendron flowers (a typically Asian genus) are popular with some species (Coates & Bishop 1997) and coconut flowers are visited in New Guinea (Beehler 1980) and Wallacea (Coates & Bishop 1997).

Most of the honeyeaters in New Guinea occupy rainforest, which is floristically and structurally similar to that in South East Asia. Many of the rainforest honeyeaters are more frugivorous and less nectarivorous than those of more open habitats (Beehler 1981) and the Oriental forests would surely provide a wide range of resources that they could exploit. Apparently, an extensive area of Sundaland was covered in sclerophyll or savanna woodland or other open habitats during recent glacials (Anshari et al., this volume, Brandon-Jones, this volume, Kershaw et al., this volume). Honeyeaters of Australian origin should have been able to survive in such habitat, as long as they reached it.

2.4 Predation

The bird-eating falcons (6 species in Australia) and accipiters (7 species in New Guinea) have successfully spread through Wallacea from Asia to Australia and beyond. So, there would appear to be no new avian predators to prevent the spread of honeyeaters into South East Asia. Java and Borneo have a rich array of placental carnivores (especially Mustelidae, Felidae, Viverridae - Ewer 1973). Three viverrids occur on Sulawesi, and the Common Palm Civet (*Paradoxus hermaphroditus*) occurs through the Lesser Sundas and in Seram and the Kai Islands (although Wallace (1869) suggested and Heinsohn (this volume) confirms that it has been introduced to them). Civets include small vertebrates and eggs in their diet, and are to some degree arboreal. A few primates have also crossed into Wallacea as far as Sulawesi and Flores (Napier & Napier 1985), again possibly with human assistance (Heinsohn, this volume). Thus, a number of honeyeaters overlap with these potentially important arboreal nest predators.

The Stitchbird (Hihi - *Notiomystis cincta*), one of New Zealand's unique honeyeaters has disappeared from the North Island, at least partly due to the impact of introduced carnivores (mustelids, cats and rodents - Mills & Williams 1979). However, small birds in Australia have always had to contend with the comparable dasyurids, and there is no strong evidence that arboreal birds have declined due to introduced foxes and cats, although ground-feeding and breeding birds have been adversely affected. Interestingly, the Micronesian Honeyeater (*Myzomela rubrata*) and several other landbirds have apparently gone extinct on Guam due to an exotic predator, the Brown Tree Snake (*Boiga irregularis*), which is from Australia, not Asia (Savidge 1987).

2.5 Disease

A final hypothesis is that a disease that occurs east and north of Wallace's Line has prevented honeyeaters from crossing. There is strong evidence that avian malaria was responsible for the extinction of a number of species of native birds on Hawaii (Van Riper et al. 1986). Both the *Plasmodium* parasite and its *Culex* vector were introduced during the last century. However, three of the honeyeaters went extinct before malaria was believed to have become widespread early in the 20th century. A fourth lingered on until the 1980s. Whether or not malaria has been a contributor to extinctions of Hawaiian honeyeaters, it seems unlikely that it has kept honeyeaters out of South East Asia. Mosquitos are probably regularly blown between islands. A number of bird species migrates between South East Asia and the Lesser Sundas and Moluccas, and a few passerines even reach Australia (e. g. some swallows and wagtails). There seems a high probability that either mosquitos or birds would have carried malaria and many other pathogens between the sub-regions.

3 CONCLUSIONS

Overall, none of the possible reasons for the failure of honeyeaters to penetrate far west of Wallace's Line appears convincing alone, at least when we look for patterns. To a significant degree this may be due to our paucity of knowledge of the food, life history, interspecific interactions of honeyeaters (and indeed most birds) that live in Wallacea. I suggest tentatively that the diverse mammal and bird faunas of Borneo and Java may have together imposed sufficient predation and competition to withstand invading honeyeaters. Only when we understand the ecological processes better is the pattern of distribution of the honeyeaters, and many other groups in the changeover zone between Asia and Australia, likely to be understood.

REFERENCES

Abbott, I., 1973. Birds of Bass Strait. Evolution and ecology of the avifaunas of some Bass Strait Islands, and comparisons with those of Tasmania and Victoria. *Proceedings of the Royal Society of Victoria* 85: 197-223.
Abbott, I., 1974. The avifauna of Kangaroo Island and causes of its impoverishment. *Emu* 74: 124-134.
Abbott, I., 1978. Factors determining the number of land bird species on islands around south-western Australia. *Oecologia* 33: 221-233.
Anshari, G., Kershaw, P., Penny, D., Thamotherampillai, A. & van der Kaars, S. , this volume. Glacial lowland vegetation of Sundaland.
Beehler, B. M. 1980. A comparison of avian foraging at flowering trees in Panama and New Guinea. *Wilson Bulletin* 92: 513-519.
Beehler, B. M. 1981. Ecological structuring of forest bird communities in New Guinea. In J. L. Gressitt (ed.), *Monographiae Biologicae* No 42: 837-861. The Hague: Junk.
Beehler, B. M. & Finch, B. W. 1985. *Species - Checklist of the Birds of New Guinea*. Australasian Ornithological Monograph, No 1. Melbourne: RAOU.
Brandon-Jones, D., this volume. Borneo as a biogeographic barrier to Asian-Australasian migration.
Christidis, L. & Boles, W. E. 1994. *The Taxonomy and Species of Birds of Australia and its Territories*. Royal Australasian Ornithologists Union Monograph, No. 2. Melbourne: RAOU.
Christidis, L., Schodde, R. & Robinson, N. A. 1993. Affinities of the aberrant Australo-Papuan honeyeaters, *Toxorhamphus*, *Oedistoma*, *Timeliopsis* and *Epthianura*: protein evidence. *Australian Journal of Zoology* 41: 423-432.
Clode, D. & O'Brien, R., this volume. The beginning of Biogeography: what made Wallace draw the line?
Coates, B. J. & Bishop, K. D. 1997. *A Guide to the Birds of Wallacea*. Alderley, Queensland: Dove.
Diamond, J. M. 1975. Assembly of species communities. In M. L. Cody and J. M. Diamond (eds.), *Ecology and Evolution of Communities*: 342-444. Harvard: Belknap Press.
Diamond, J. M. 1982. Mimicry of friarbirds by orioles. *Auk* 99: 187-196.
Dow, D. D., 1977. Indiscriminate interspecific aggression leading to almost sole occupancy of space by a single species of bird. *Emu* 77: 115-121.
Ewer, R. F. 1973. *The Carnivores*. Ithaca: Cornell University Press.
Falla, R. A., Sibson, R. B. & Turbott, E. G. 1981. *The New Guide to the Birds of New Zealand and outlying Islands*. Auckland: Collins.
Ford, H. A. 1979. Interspecific competition in Australian honeyeaters - depletion of common resources. *Australian Journal of Ecology* 4: 145-164.
Ford, H. A. & Debus, S. 1994. Aggressive behaviour of Red Wattlebirds *Anthochaera carunculata* and Noisy Friarbirds *Philemon corniculatus*. *Corella* 15: 141-147.
Ford, H. A. & Paton, D. C. 1975. Impoverishment of the avifauna of Kangaroo Island. *Emu* 75: 155-156.
Ford, H. A., Paton, D. C. & Forde, N. 1979. Birds as pollinators of Australian plants. *New Zealand Journal of Botany* 17: 509-520.
Freifeld, H. B. 1999. Habitat relationships of forest birds on Tutuila Island, American Samoa. *Journal of Biogeography* 26: 1191-1213.
Grey, M. J., Clarke, M. F. & Loyn, R. H. 1998. Influence of the Noisy Miner *Manorina melanocephala* on avian diversity and abundance in remnant Grey Box woodland. *Pacific Conservation Biology* 4: 55-69.
Hall, R., this volume. Cenozoic reconstructions of South East Asia and the South West Pacific: changing patterns of land and sea.

Heinsohn, T., this volume. Human influences on vertebrate zoogeography: animal translocation and biological invasions across and to the east of Wallace's Line.

Keast, A. 1968. Seasonal movements in the Australian honeyeaters (Meliphagidae) and their ecological significance. *Emu* 67: 159-210.

Keast, A. 1976. The origins of adaptive zone utilizations and adaptive radiations, as illustrated by the Australian Meliphagidae. *Proceedings of the 16th International Ornithological Congress*: 71-82.

Kershaw, P., Penny, D., van der Kaars, S., Anshari, G. & Thamotherampillai, A. , this volume. Implications of continental collision for late Quaternary climates and vegetation in the Maritime Continent region.

Longmore, W. 1991. *Honeyeaters and their Allies of Australia*. North Ryde, Australia: Angus and Robertson.

McFarland, D. C. & Ford, H. A. 1991. The relationship between foraging ecology and social behaviour in Australian honeyeaters. *Acta XX Congressus Internationalis Ornithologici*: 1141-1155.

Mills, J. A. & Williams, G. R. 1979. The status of endangered New Zealand birds. In M. J. Tyler (ed.), *The Status of Endangered Australasian Wildlife*: 147-168. Royal Zoological Society of South Australia: Adelaide.

Napier, J. R. & Napier, P. H. 1985. *The Natural History of the Primates*. London: British Museum (Natural History).

Paton, D. C. 1980. The importance of manna, honeydew and lerp in the diet of honeyeaters. *Emu* 80: 213-226.

Pyke, G. H. 1980. The foraging behaviour of Australian honeyeaters: a review and some comparisons with hummingbirds. *Australian Journal of Ecology* 5: 343-369.

Ripley, S. D. 1959. Competition between sunbird and honeyeater species in the Moluccan Islands. *American Naturalist* 93: 127-132.

Savidge, J. A. 1987. Extinction of an island forest avifauna by an introduced snake. *Ecology* 68: 660-668.

Springer, M. S., Higuchi, H., Ueda, K., Minton, J. & Sibley, C. G. 1995. Molecular evidence that the Bonin Islands "Honeyeater" is a White-eye. *Journal of the Yamashina Institute of Ornithology* 27: 66-77.

Udvardy, M. D. F. 1951. The significance of interspecific competition in bird life. *Oikos* 3: 98-123.

Van Riper, C., Van Riper, S. G., Goff, M. L. & Laird, M. 1986. The epizootiology and ecological significance of malaria in Hawaiian Land Birds. *Ecological Monographs* 56: 327-344.

Wallace, A. R. 1869. *The Malay Archipelago*. New York: Dover.

Watling, D. 1982. *Birds of Fiji, Tonga and Samoa*. Millwood Press: Wellington, New Zealand.

White, C. M. N. & Bruce, M. D. 1986. *The Birds of Wallacea*. London: British Ornithologists' Union.

Human influences on vertebrate zoogeography: animal translocation and biological invasions across and to the east of Wallace's Line

T.E. Heinsohn

People and the Environment Section, National Museum of Australia, GPO Box 1901, Canberra, ACT 2601, Australia

ABSTRACT: In this paper a combination of literature analysis, historical research and the results of original fieldwork is used to review the impact of animal translocation from ancient times to the present on the vertebrate zoogeography of Wallacea, New Guinea and the Northern Melanesian Islands, referred to collectively as the Circum New Guinea Archipelago. The focus of the paper is on translocated wild rather than domesticated vertebrates. The processes through which these species have been either unwittingly, accidentally or deliberately translocated to new areas are reviewed. Three principal categories of introduced wild species are recognised. These include stowaways, ethnotramps and incidentals. Ethnotramps include economically and culturally favoured animals such as Rusa Deer (*Cervus timorensis*), Long-tailed Macaque (*Macaca fascicularis*), and various civets, cuscuses, wallabies, cassowaries and wild-caught cage birds that are commonly carried around with humans. Stowaways tend to be smaller species such as shrews, rats, or herpetofauna that conceal themselves or their eggs in vessels and cargo. While the Oriental zoogeographic region has furnished parts of the Circum New Guinea Archipelago with around 55 introduced wild terrestrial vertebrates, translocation from the latter region has had negligible impact on the zoogeography of the former. Some Wallacean islands and New Guinea, however, have been a significant source of species introduced within the Circum New Guinea Archipelago. While vertebrate translocation appears to be largely a Holocene phenomenon, several New Guinea marsupials appear to have been exported to surrounding islands in the Late Pleistocene, with the earliest suspected translocation being that of the Northern Common Cuscus (*Phalanger orientalis*) to New Ireland at circa 20 000 BP. It is postulated that ancient human influences may be under-recognised due to the ambiguity of the evidence.

1 INTRODUCTION

The distribution of wild terrestrial vertebrates across archipelagos of continental and oceanic islands is typically explained in terms of natural processes including vicariance and dispersal. The latter includes: active dispersal by moving across land-bridges, swimming and flying across water barriers; or passive dispersal on natural rafts, unaided floating in currents of water, or carriage by strong winds. When dealing with Quaternary zoogeography, an often overlooked extra possibility is that of passive dispersal by ancient human agency, and there is now a growing body of evidence that humans may have been both deliberately and unwittingly carrying species around with them much earlier than previously thought (Flannery & White 1991; Spriggs 1997; Heinsohn 1997, 1998a, 1998b). This paper explores the theme of humans as an ancient force that has led to a violation of the relative natural zoogeographic integrity of Wallace's Line, the transition zone of Wallacea, and the islands of Northern Melanesia. The five principal means by which humans influence zoogeography are: (1) human-induced habitat modification; (2) direct or indirect human-induced decline / extinction of animal species (eg. through predation, competition or disturbance); (3) translocation of pathogens; (4) co-evolution and domestication; and (5) unintentional or deliberate animal translocation. The focus of this paper is on the ancient and

recent translocation of non-domesticated terrestrial vertebrates into and within the chain of is-
lands that extends to the immediate east of Wallace's Line. This is referred to as the Circum
New Guinea Archipelago and includes the Lesser Sundas, Sulawesi Subregion, Moluccas, New
Guinea (and its immediate satellites), and the Bismarck Archipelago and Solomon Islands of
Northern Melanesia (Figure 1).

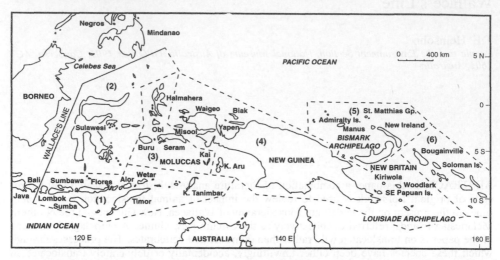

Figure 1. Map of the Circum New Guinea Archipelago. This chain of islands is comprised of the following zoogeographic subregions: (1) Lesser Sundas; (2) Sulawesi; (3) Moluccan; (4) Papuan; (5) Bismarck
Archipelago; and (6) Solomon Islands. Wallacea is comprised of (1), (2) and (3). Figure drawn by I. Metcalfe.

2 STUDY AREA

The dominant natural vegetation of the Circum New Guinea Archipelago, as defined above, is
Malesian and Papuasian rainforest and monsoon forest. A combination of climate and human
impacts such as frequent firing and deforestation has also led to sizeable areas of either natural
or anthropogenic savanna and savanna-woodland, particularly in drier or more densely populated parts of the archipelago. Apart from the legacy of continental drift and long-distance dispersal, the biogeography of the region has also been influenced by the dramatic sea level
changes of the Pleistocene. With sea levels falling by about 200 metres (Chappell & Thom
1977; Bellwood 1997), this led to a significant narrowing of many permanent water barriers,
created contiguous landmasses out of many contemporary archipelagos and led to the formation
of land-bridges between New Guinea and Australia, and continental satellites such as Aru.

3 METHODS

In this paper a combination of literature review, historical research and the results of original
zoogeographic survey and ethnozoological fieldwork in Wallacea and the New Guinea/North
Melanesian region (Heinsohn 1998a; 1998b) is used to provide an up-to-date review of wild
vertebrate translocation and anthropogenic biological invasions in the Circum New Guinea Archipelago. Some of the evidence considered for translocation of wild vertebrates by humans includes: (1) patchy or improbable distributions; (2) genetic similarity over wide inter-island distributions that may indicate recent introduction; (3) sudden appearances of species in
archaeological records; (4) a lack of fossil or subfossil material in palaeontological or archaeological sequences that may indicate recent introduction; (5) ethnozoological evidence of long
associations between humans and certain wild species that are known to be carried around in

watercraft as pets or for trade, food or ceremonial purposes; (6) island assemblages of vertebrates that favour human-associated species over natural dispersal ability; and (7) direct written
or oral historical records of translocation (Heinsohn 1997, 1998a, 1998b, 1999). In the face of
the potential ambiguity of single strands of evidence, it is often a combination of some of the
aforementioned indicators that tips the scales of probability in favour of considering translocation by human agency.

4 COLONISATION BY HUMANS

The initial human invasion of a pristine Wallacea and Meganesia (the combined New Guinean
and Australian continental landmass of the Pleistocene) is thought to have occurred somewhere
between 60 000 and 40 000 years ago, probably by a predominantly Australoid people (Bellwood 1997; Thorne et al. 1999). Parts of the Lesser Sundas may have been colonised in the
Middle Pleistocene by Homo erectus (Bergh et al. 1996; Morwood et al. 1997, Morwood, this
volume). These initial Pleistocene colonisation episodes may have been significantly aided in
places by a narrowing of permanent water barriers due to sea level fluctuations.

Following the relative stabilisation of sea levels in the early Holocene, a further major phase
of human colonisation was that of the Austronesian agricultural and maritime peoples, who expanded into Wallacea, around coastal New Guinea, and into the Pacific islands between 5 000
and 1 000 years ago (Bellwood 1997; Kirch 1997). This was a period that saw the spread of
various plant cultivars and the introduction of the domesticated triumvirate of dogs, pigs and
chickens into Oceania (Spriggs 1997; Kirch 1997). In Wallacea and Western New Guinea the
Austronesian expansion was followed by ongoing economic and cultural influences from Asia
and finally by the impact of the 'modern' spice trade in the last millennium, which involved the
mercantile empires of the Chinese, Arabs, Portuguese, Spanish, Dutch and British (Swadling
1996).

The eastern half of New Guinea and Northern Melanesia remained as scattered indigenous
realms until the late 19th century when colonial rivalries and the lure of copra and gold led the
Germans and British to establish territories. All of these historical phases, including later Australian administration of the Territory of Papua and New Guinea, wartime occupation by the
Japanese, and post-war independence for Indonesia, Papua New Guinea and the Solomon Islands have left salient influences on the biotas of the Circum New Guinea Archipelago. The
Dutch, British and Germans, for example, contributed to the spread of Rusa Deer (Cervus
timorensis) into Melanesia. Just before the Pacific War, the Australians and British introduced
the Cane Toad (Bufo marinus) to Papua New Guinea and the Solomon Islands for biological
control of agricultural pests (Tyler 1994), while during the war the Japanese widely introduced
the Giant African Snail (Achatina fulica) in Melanesia as an emergency food source for hungry
troops (Heinsohn 1998a; Lever 1994). Furthermore, Indonesia's ambitious transmigration program of recent decades has probably contributed to the spread of new commensals to far provinces, as has recent land and resource development and shipping in Papua New Guinea and the
Solomon Islands.

5 THE TRANSLOCATION OF WILD VERTEBRATES

The three principal means by which non-domesticated vertebrates are translocated to new landmasses are: (1) unwittingly as stowaways in vessels or cargo; (2) accidentally as escaped captives or pets; and (3) deliberately as species introduced as part of a 'game-park' or game en-
hancement strategy. Heinsohn (1998a, 1998b) recognises three categories of introduced species.
These include 'stowaways', 'ethnotramps' and 'incidentals'. 'Stowaways' are usually small
animals such as frogs, lizards, snakes, passerine birds, rodents and shrews; they are typically
commensal species that live in and around human structures or they may be littoral species that
enter canoes and praus that have been beached amidst the vegetation of the sea shore. Beached
canoes and praus with thatch shelters on their decks may also be particularly predisposed to
stowaways given the propensity of thatch to be colonised by small vertebrates and their eggs

(Loveridge 1946; Heinsohn 1998a). Alternatively, stowaways may also include commensal or non-commensal species or their eggs caught up in cargoes carried from inland areas. Potted plants, containers of rice, bunches of bananas, other agricultural produce and traditional building materials such as bundles of timber, bark and thatch are ancient cargoes that are particularly prone to infestation with stowaways. Some examples of stowaways include: commensal frogs such as the Asian Painted Frog (*Kaloula pulchra*) and Asian Commensal Tree Frog (*Polypedetes leucomystax*); commensal lizards such as the House Gecko (*Hemidactylus frenatus*); the commensal fossorial Flowerpot Snake (*Ramphotyphlops braminus*); the House Sparrow (*Passer domesticus*); Tree Sparrow (*Passer montanus*); commensal murid rodents such as the Pacific Rat (*Rattus exulans*) and Black Rat (*Rattus rattus*); and shrews such as the House Shrew (*Suncus murinus*).

'Ethnotramps' are economically and culturally favoured wild animals that are commonly captured live and carried around with humans as pets or for food, trade or ceremonial purposes (Heinsohn 1998b). This strong association with humans often leads to greatly expanded geographical ranges as captives escape to establish new populations or are deliberately released as part of a 'game park' strategy. Such strategies may enable new animal products such as meat, pelts, fibre, plumes, bone, horn, antler, and musk to be produced on oceanic islands that are lacking in game. Deliberately introduced predators may also provide a valuable service as rat-catchers. Furthermore, in addition to many commensals that may be captured in the vicinity of settlements, ethnotramps also include species such as cassowaries that are typically captured in the deep bush beyond fields and gardens. Infrequently transported captives that have formed minor rather than widespread introduced populations are referred to as 'incidentals'. Some examples of ethnotramps include: the Javan Rusa Deer (*Cervus timorensis*); Long-tailed Macaque (*Macaca fascicularis*); Malay Civet (*Viverra tangalunga*); Little Civet (*Viverricula indica*); Palm Civet (*Paradoxurus hermaphroditus*); Northern Common Cuscus (*Phalanger orientalis*); Common Spotted Cuscus (*Spilocuscus maculatus*); Admiralty Cuscus (*Spilocuscus kraemeri*); Northern Pademelon (*Thylogale browni*); Dusky Pademelon (*Thylogale brunii*); Agile Wallaby (*Macropus agilis*); and cassowaries such as the Double-wattled Cassowary (*Casuarius casuarius*) and Dwarf Cassowary (*Casuarius bennetti*). Several species of volant birds within the finch family (Estrildinae), myna family (Sturnidae), pigeon family (Columbidae), and parrot family (Psittacidae) could also be regarded as ethnotramps due to their popularity as traded wild-caught cage birds within the Circum New Guinea Archipelago and their occurrence as introduced species in a number of localities.

6 A SURVEY OF RECORDS OF WILD VERTEBRATE TRANSLOCATION

In the Circum New Guinea Archipelago, three principal centres of origin are recognised for introduced wild vertebrates. These include: (1) the Oriental zoogeographic region islands of Western Indonesia, and in particular, Java and Bali; (2) the Wallacean island of Sulawesi; and (3) the large Australasian island of New Guinea (Northern Meganesia) and its continental satellites (collectively this comprises the Papuan zoogeographic subregion). Other minor centres of origin for introduced species include Australia (Southern Meganesia) and various Australasian region oceanic islands with their own endemic species. Certain species, however, such as some widespread or pantropical geckos and skinks, are difficult to attribute to any particular centre of origin. A survey of records regarding actual and postulated translocations of non-volant or weakly volant species from various centres of origin is presented in Table 1. Because strongly volant vertebrates conform less to traditional zoogeographic boundaries and have a more complex pattern of distribution, they are given a separate treatment in Table 2.

Table 1: Records of translocation of non-volant or weakly volant wild terrestrial vertebrates in the Circum New Guinea Archipelago			
Species	Original Source	Where Introduced	Context
Eurasian & Oriental exports			

Cervidae			
Rusa Deer (*Cervus timorensis*)	Java / Bali	Lesser Sundas; Sulawesi; Moluccas; New Guinea; Bismarck Archipelago	Mostly Middle Holocene through to historic times. Carried as captives and deliberately introduced to many areas as a venison and game animal (1, 2, 4, 5, 6, 7, 8, 9, 11, 22, 23, 26, 27, 43, 44, 45, 49)
Barking Deer (*Muntiacus muntjak*)	Western Indonesia / Java / Bali	Lombok	Probably introduced into Lombok in protohistoric / historic times by the Balinese Rajas as a game species (1, 6, 7, 9, 48)
Suidae			
Indonesian Wild Pig (*Sus scrofa vittatus*)	Western Indonesia	Lombok; Sumbawa; Komodo	Probably introduced prehistorically during the Holocene to its patchy island occurrences in the Lesser Sundas as a food or trade animal (9, 17)
Cercopithecidae			
Long-tailed Macaque (*Macaca fascicularis*)	Western Indonesia	Lesser Sundas	Middle Holocene to protohistoric and historic period. Probably carried as pets and food animals (1, 9, 17, 23, 43, 44, 45, 49)
Silvered Leaf Monkey (*Trachypithecus auratus*)	Western Indonesia / Java / Bali	Lombok	Probably introduced into Lombok in protohistoric / historic times by the Balinese Rajahs (1, 7, 9, 48)
Viverridae			
Common Palm Civet (*Paradoxurus hermaphroditus*)	Western Indonesia	Lesser Sundas; Sulawesi Subregion; Moluccas; Kai; Aru	Probably introduced from the latter part of the Holocene to the protohistoric period as a rat-catcher (1, 2, 4, 5, 6, 7, 8, 9, 11, 17, 21, 22, 23, 26, 43, 44, 45, 49)
Malay Civet (*Viverra tangalunga*)	Western Indonesia	Sulawesi; Moluccas	Probably introduced from the Late Holocene to protohistoric / early historic period as a trade animal valued as a source of 'civet' a musk like substance used in perfume (1, 2, 4, 5, 6, 7, 8, 17, 21, 22, 23, 26, 49)
Little Civet (*Viverricula indica*)	Western Indonesia	Lombok; Sumbawa	Probably introduced from the Late Holocene to protohistoric period as a trade animal valued as a source of 'civet' a musk like substance used in perfume (1, 4, 6, 7, 9, 17, 49)
Herpestidae			
Small Asian Mongoose (*Herpestes javanicus*)	South & Southeast Asia / Java / Bali	Ambon; ?Komodo	Probably introduced in protohistoric or historic times as a rat-catcher (6, 26, 38, 49)
Felidae			
Leopard Cat (*Felis bengalensis*)	Western Indonesia / Java / Bali	Lombok	Probably introduced during Holocene prehistoric, protohistoric or historic times as pets or traded captives (9, 49)
Manidae			
Malayan Pangolin (*Manis javanica*)	Western Indonesia / Java / Bali	Lombok	Probably introduced during Holocene prehistoric, protohistoric or historic times as captives traded for their meat, and the medicinal value of their scales (1, 2, 9)
Soricidae			
House Shrew (*Sun-	Asia / Western	Lesser Sundas; Su-	Probably introduced to various islands

cus murinus)	Indonesia	lawesi Subregion; Moluccas; Micronesia	from prehistoric to historic times as a stowaway in cargo, canoes, sailing praus and ships (1, 4, 5, 6, 7, 8, 11, 17, 21, 49)
Javan Shrew (*Crocidura maxi*)	Southern Indonesian Islands	Ambon; Kai; Aru	Probably introduced as a stowaway in cargo or watercraft (11)
Sunda Shrew (*Crocidura monticola*)	Western & southern Indonesian islands	Ambon; Obi	Probably introduced as a stowaway in cargo or watercraft (11)
Hystricidae			
Javan Porcupine (*Hystrix javanica*)	Java; Madura; Bali	Sulawesi; (?I)Tanahdjampea; (?I)Lombok; (?I)Sumbawa; (?I)Flores	Probably introduced during Holocene prehistoric, protohistoric or historic times as captives carried for food or trade (1, 2, 6, 9, 21)
Sciuridae			
Prevost's Squirrel (*Callosciurus prevostii*)	Western Indonesia	Sulawesi	Probably introduced in Holocene prehistoric, protohistoric or historic times as captives carried as pets or for food, or possibly as stowaways in agricultural cargoes (1, 2, 6, 21)
Plantain Squirrel (*Callosciurus notatus*)	Western Indonesia	Lombok; Selayar	Probably introduced in Holocene prehistoric, protohistoric or historic times as captives carried as pets or for food, or possibly as stowaways in agricultural cargoes (1, 2, 9, 21, 49)
Muridae			
Pacific Rat (*Rattus exulans*)	Southeast Asia	Indonesian islands; New Guinea; Melanesia; Polynesia; Micronesia	Probably mostly introduced in the Holocene as a stowaway in canoes and praus and possibly also as a deliberately transported food animal (1, 4, 5, 6, 11, 12, 15, 17, 28, 29, 43, 46)
Himalayan Rat (*Rattus nitidus*)	Himalayas / Asia	Sulawesi; Seram; New Guinea; Palau	Holocene. Probably introduced as stowaways in cargo and captives kept as food (6, 17, 21)
Asian House Rat (*Rattus tanezumi*)	Southeast Asia	Micronesia; Moluccas; New Guinea	Holocene. Probably introduced as stowaways in cargo or watercraft and captives kept as food (11, 49)
Rice-field Rat (*Rattus argentiventer*)	Southeast Asia	Lesser Sundas; Sulawesi Subregion; Moluccas; New Guinea	Holocene. Probably introduced as stowaways in cargo and captives kept as food (4, 5, 6, 9, 11, 12, 17, 21, 29)
Black Rat (*Rattus rattus*)	East Asia	Throughout the world and study area	Holocene. Introduced mostly as stowaways in cargo and watercraft, with a major expansion during historic times (1, 4, 5, 6, 7, 11, 12, 17, 21, 29)
Brown Rat (*Rattus norvegicus*)	East Asia	Scattered throughout the world and study area, preferring ports and urban centres	Holocene. Introduced mostly as stowaways in cargo and watercraft, with a major expansion during historic times (1, 4, 5, 6, 7, 11, 12, 17, 21, 29)
House Mouse (*Mus musculus*)	Eurasian continent	Throughout the world including many parts of the study area	Holocene. Introduced mostly as stowaways in cargo and watercraft, with a major expansion during historic times (1, 4, 5, 6, 7, 9, 11, 12, 21, 29)
Asian House	Southern Asia	Philippines; Indone-	Introduced from Holocene prehistoric

Mouse (*Mus castaneus*)		sian islands	times to the present, mostly as stowaways in cargo and watercraft (9, 17)
Ryukyu Mouse (*Mus caroli*)	Southeast & East Asia	Western Indonesian islands; Flores	Holocene. Probably mostly introduced in protohistoric or historic times as a stowaway in cargo (17)
Phasianidae			
Red Jungle Fowl (*Gallus gallus*) (Wild form)	Southeast Asia / Western Indonesia	(?I)Wallacea; (Domesticated form widely dispersed beyond Southeast Asia into Oceania)	Probably introduced in prehistoric times as a meat and egg bearing species (25, 33)
Green Jungle Fowl (*Gallus varius*)	Southwestern Indonesian islands	Cocos-Keeling Islands; (?I)Some Lesser Sunda islands	??Possibly introduced to parts of Lesser Sundas in prehistoric or protohistoric times as escaped or liberated captives (33)
Boidae - Pythoninae			
Reticulated Python (*Python reticulatus*)	Asia / Indonesian islands	(?I)Some Wallacean islands	This species is a good swimmer and natural disperser, however, as it is carried around as a food animal and rat-catcher some of its occurrences on Wallacean islands may possibly be due to human agency (1)
Elapidae			
Monocled Cobra (*Naja naja sputatrix*)	West Malaysia / Indonesian islands	(?I)Some Wallacean islands	As captives are transported for their skins, medicinal products and for ceremonial purposes, some occurrences on Wallacean islands may be due to human agency (1)
Typhlopidae			
Flowerpot Snake (*Ramphotyphlops braminus*)	Widespread, occuring in Asia; Africa; Mexico; Oceania	Many islands of the Circum New Guinea Archipelago	Probably a comparatively recent introduction in many places as stowaways in potted plants (34, 35, 36, 37, 38, 39, 50)
Varanidae			
Water Monitor (*Varanus salvator*)	Sri Lanka; India; SE Asia; Southern China; Philippines; Indonesia.	(?I)Some Wallacean islands	This species is a good swimmer and natural disperser, however, as it is carried around as a food animal and is heavily exploited for its valuable skin, some of its occurrences on Wallacean islands may possibly be due to human agency (1)
Gekkonidae			
Asian House Gecko (*Hemidactylus frenatus*)	Widespread, occurring in Asia; Africa; Central America; Oceania; & some Indian Ocean islands	Many islands of the Circum New Guinea Archipelago	Probably a recent introduction in many places such as the Solomon Islands where brought in as a stowaway in vessels or their cargo (34, 35, 36, 37, 38, 40, 50)
Wiegmann's Gecko (*Gehyra mutilata*)	Widespread occurring in Asia; Mexico; Madagascar; New Guinea; Oceania	Many islands of the Circum New Guinea Archipelago	Probably a recent introduction in many places such as the Solomon Islands where brought in as a stowaway in vessels or their cargo (36, 38, 40, 50)
Schneider's Gecko (*Cosymbotus platyurus*)	Oriental Region; Indonesian islands	(?I)Some islands in the Circum New Guinea Archipelago	This commensal species may have been introduced to parts of its range in the Circum New Guinea Archipelago as a stow-

			away in vessels or their cargo (38, 40)
Scincidae			
Asian Commensal Skink (*Mabuya multifasciata*)	Oriental Region; Indonesian islands	(?I)Some islands in the Circum New Guinea Archipelago	This commensal species may have been introduced to parts of its range in the Circum New Guinea Archipelago as a stowaway in vessels or their cargo (38, 40)
Ranidae			
Indonesian Ranid Frog (*Rana cancrivora*)	Western Indonesian islands	Sulawesi; (?I)Some other Wallacean islands	Probably introduced as a stowaway in cargo (21, 38)
Indonesian Ranid Frog (*Rana erythraea*)	Western Indonesian islands	Sulawesi; (?I)Some other Wallacean islands	Probably introduced as a stowaway in cargo (21)
Oriental Ranid Frog (*Rana limnocharis*)	Oriental Region; Indonesian islands	(?I)Some Wallacean islands	Probably introduced as a stowaway in cargo (38)
Rhacophoridae			
Asian Commensal Tree Frog (*Polypedates leucomystax*)	Western Indonesian islands	Sulawesi; Timor; (?I)Some other Wallacean islands; New Guinea	Probably introduced as a stowaway in cargo (21, 30, 38, 41)
Microhylidae			
Asian Painted Frog (*Kaloula pulchra*)	Southeast Asia / Indonesian islands	Sulawesi; Flores; (?I)Some other Wallacean islands	Probably introduced as a stowaway in cargo (21, 38)
Sulawesi exports			
Suidae			
Babirusa (*Babyrousa babyrussa*)	Sulawesi / Sula Islands	Buru; (?I)Sula Islands	Probably introduced to Buru in the Central Moluccas in Holocene prehistoric or protohistoric times as a game species or ceremonial trade animal. Possibly also introduced to the Sula Islands, although more likely to be due to natural dispersal (4, 6, 18, 49)
Celebes Wild Pig (*Sus celebensis*)	Sulawesi	Halmahera; Flores; Timor; Simaleue near Sumatra	Probably introduced to its patchy distribution in the North Moluccas and Lesser Sundas via traded captives (6, 17)
Cercopithecidae			
Sulawesi Crested Black Macaque (*Macaca nigra*)	Sulawesi	Bacan	Probably introduced to Bacan in the North Moluccas in Holocene prehistoric or protohistoric times as pets carried on trading praus (1, 7, 11, 23, 49)
Phalangeridae			
Sulwesi Bear Cuscus (*Ailurops ursinus*)	Sulawesi	(?I)Salebabu Island in the Talaud Group	The remote and patchy occurrence on Salebabu Island in the Talaud Group may possibly be due to prehistoric introduction as animals carried for food or trade (1, 3, 11)
New Guinea / Papuan exports			
Muridae			
Large Spiny Rat (*Rattus praetor*)	New Guinea	Parts of the Bismarck Archipelago and Solomon Islands	Probably introduced prehistorically as stowaways in watercraft or captives kept as food (11, 12, 15, 16, 46)

Phalangeridae			
Common Spotted Cuscus (*Spilocuscus maculatus*)	New Guinea	St. Matthias Group; New Ireland; Selayar; Kai; Buru; Ambon; (?I)Seram; Banda; Pandjang; Tioor	Probably introduced from prehistoric to recent times as escaped or liberated captives carried for food, trade and stock purposes (1, 2, 3, 11, 15, 21, 43, 46, 49)
Admiralty Cuscus (*Spilocuscus kraemeri*)	New Guinea	Admiralty Islands; Hermit Islands; Ninigo Group; and Wuvulu	Probably introduced in prehistoric times, possibly as early as the Late Plcistocene, as escaped or liberated captives carried for food trade and stock purposes (1, 2, 3, 11, 46)
Northern Common Cuscus (*Phalanger orientalis*)	New Guinea	Solomon Islands; Bismarck Archipelago; Timor; Leti; Wetar; Babar; Kai; Sanana; Buru; Seram; Ambon; Saparua; Gorong; Banda	Probably introduced from Late Pleistocene to Holocene times as escaped or liberated captives carried for food, trade and stock purposes (1, 2, 3, 11, 15, 16, 17, 19, 43, 44, 45, 46, 49)
Woodlark Cuscus (*Phalanger lullulae*)	Woodlark Island	(?I)Alcester Island	Possibly transported for food or trade purposes (1, 3, 11)
Petauridae			
Sugar Glider (*Petaurus breviceps*)	New Guinea	(?I)Halmahera; (?I)Some other New Guinea satellites	Possibly introduced in prehistoric or protohistoric times as escaped pets or captive food animals (1, 3, 11, 13, 49)
Macropodidae			
Huon Tree-kangaroo (*Dendrolagus matschiei*)	New Guinea	Umboi	Probably introduced in Holocene times as escaped or liberated captives carried as pets or for food or trade purposes (1, 3, 5, 19, 20)
Northern Pademelon (*Thylogale browni*)	New Guinea	Bagabag; Umboi; New Britain; New Ireland; New Hanover	Probably introduced in prehistoric times as escaped or liberated captives carried for food or trade purposes (1, 3, 11, 14, 15, 16, 20, 46)
Dusky Pademelon (*Thylogale brunii*)	New Guinea / Aru	Kai Islands	Probably introduced in prehistoric or protohistoric times as escaped or liberated captives carried for food or trade purposes from Aru (1, 3, 20)
Brown Dorcopsis (*Dorcopsis muelleri*)	New Guinea / Misool	(?I)Halmahera; (?I)Gebe. Now extinct on both islands	Possibly introduced in prehistoric times as escaped or liberated captives carried for food or trade purposes (3, 10, 13)
Agile Wallaby (*Macropus agilis*)	New Guinea	Goodenough; Fergusson; Normanby; Kiriwina	Probably introduced in prehistoric times as escaped or liberated captives carried for food or trade purposes (1, 3, 11, 20)
Peroryctidae			
Common Spiny Bandicoot (*Echymipera kalubu*)	New Guinea	Admiralty Islands	Possibly introduced to the Admiralty Islands in the Late Pleistocene through escaped or liberated captives carried as food (1, 3, 11, 46)
Rufous Spiny Bandicoot (*Echymipera rufescens*)	New Guinea / Aru	(?I)Kai Islands	Possibly introduced in prehistoric times as escaped or liberated captives carried for food or trade purposes from Aru (1, 3, 11)
Casuariidae			

Double-wattled Cassowary (*Casuarius casuarius*)	New Guinea / Aru	Seram	Probably introduced in Holocene prehistoric or protohistoric times as traded captives valued for their meat and plumes (1, 2, 3, 31, 32, 33)
Dwarf Cassowary (*Casuarius bennetti*)	New Guinea	New Britain	Probably introduced in Holocene prehistoric times as traded captives valued for their meat and plumes (1, 2, 3, 31, 32)
Boidae - Pythoninae			
Amethystine Python (*Morelia amethistinus*)	New Guinea / Meganesia and satellites	(?I)Some Circum New Guinea islands	This species is a good swimmer and natural disperser, however, as it is carried around as a food animal, some of its occurrences on Circum New Guinea islands may possibly be due to human agency (1)
Varanidae			
Mangrove Monitor (*Varanus indicus*)	New Guinea and surrounding islands	(?I)Some Circum New Guinea islands	This species is a good swimmer and natural disperser, however, as it is carried around as a food animal, and is extensively exploited for its skin, some of its occurrences on Circum New Guinea islands may possibly be due to human agency (1, 47)
Scincidae			
Northern Skink (*Carlia fusca*)	New Guinea / New Britain	Bougainville; (?I)Timor	Probably a comparatively recent introduction into Bougainville as stowaways in cargo (36, 40)
American exports			
Bufonidae			
Cane Toad (*Bufo marinus*)	South & Central America	New Guinea; Bismarck Archipelago; Solomon Islands	Deliberately introduced in the 1930s to control agricultural pests (24, 30, 41, 42)

Sources: (1) Heinsohn 1998a; (2) Heinsohn 1998b; (3) Heinsohn 1997; (4) Nowak 1991; (5) Honacki *et al.* 1982; (6) Corbet & Hill 1992; (7) Laurie & Hill 1954; (8) Carter *et al.* 1945; (9) Kitchener *et al.* 1990; (10) Flannery *et al.* 1998; (11) Flannery 1995a; (12) Flannery 1995b; (13) Flannery *et al.* 1995; (14) Flannery 1992; (15) Flannery & White 1991; (16) Flannery *et al.* 1988; (17) Groves 1984; (18) Groves 1980; (19) Koopman 1979; (20) Maynes 1989; (21) Whitten *et al.* 1987; (22) Ellen 1993; (23) Wallace 1869; (24) Lever 1994; (25) Lever 1987; (26) Lever 1985; (27) Bentley 1998; (28) Atkinson 1985; (29) Menzies & Dennis 1979; (30) Menzies 1975; (31) del Hoyo *et al.* 1992; (32) White 1975; (33) Coates & Bishop 1997; (34) Loveridge 1946; (35) Gibbons 1985; (36) McCoy 1980; (37) Edgar & Lilley 1993; (38) Auffenberg 1980; (39) Welch 1988; (40) Welch *et al.* 1990; (41) Zweifel & Tyler 1982; (42) Tyler 1994; (43) Bellwood 1997; (44) Glover 1986; (45) Glover 1971; (46) Spriggs 1997 (47) Green & King 1993; (48) Hartert & Everett 1896; (49) Monk *et al.* 1997; (50) Allison 1996.

(?I) = Possibly introduced

Table 2: Some recorded translocations of volant wild vertebrates in the Circum New Guinea Archipelago			
Species	Main Range	Where introduced	Context
Eurasian & Oriental Exports			
Pycnonotidae			
Sooty-headed Bulbul (*Pycnonotus aurigaster*)	Southern China to Southeast Asia as far as Java and Bali	Sulawesi	Introduced as escaped or liberated cage birds (1, 2, 3)
Yellow-vented Bulbul	Southeast Asia, in-	Sulawesi; (?I)Lombok	Introduced as escaped or

(*Pycnonotus goiavier*)	cluding Philippines, Greater Sundas & Lesser Sundas		liberated cage birds (2, 3)
Passerinae			
Tree Sparrow (*Passer montanus*)	Eurasian continent; Sumatra; Java; Bali	Sulawesi; Lesser Sundas; Moluccas	Dispersed as cage bird, and stowaway in the cargo holds of ships (1, 2, 3, 10)
House Sparrow (*Passer domesticus*)	Widespread (originally Eurasian & Mediterranean)	New Guinea	Dispersed as cage bird, and stowaway in the cargo holds of ships (1, 4)
Estrildinae			
Java Sparrow (*Padda oryzivora*)	Java; Bali	Lesser Sundas; Sulawesi; Moluccas	Introduced as escaped or liberated cage birds (1, 2, 3)
Chestnut Munia (*Lonchura malacca*)	South & southeastern Asia, including Philippines & Greater Sundas	Ambon; (?I)Halmahera	Introduced as escaped or liberated cage birds (1, 2)
Red Avadavat (*Amandava amandava*)	Southeast Asia including Greater & Lesser Sunda islands	(?I)Populations east of Wallace's Line may be due to early introduction	Introduced as escaped or liberated cage birds (2)
Sturnidae			
Common Starling (*Sturnus vulgaris*)	Eurasia; Introduced population in Australia	New Guinea	Vagrants from historically introduced population in Australia (4, 6)
Common Myna (*Acridotheres tristis*)	South & Southeast Asia	Solomon Islands	Acclimatisation effort, or as escaped or liberated cage birds (1, 4, 5, 8, 9)
White-vented Myna (*Acridotheres cinereus*)	South & Southeast Asia, including Sumatra, Java, Bali, Flores & Sumba	Sumba; (?I)Sulawesi	Introduced as escaped or liberated cage birds (1, 2, 3)
Hill Myna (*Gracula religiosa*)	South & Southeast Asia including Greater & Lesser Sundas	(?I)Some Wallacean islands	Introduced as escaped or liberated cage birds (1)
Columbidae			
Rock Pigeon (*Columba livia*)	Europe, Asia, Mediterranean	Many towns and areas in the Circum New Guinea Arcipelago	Introduced as escaped or liberated cage and table birds (1, 2, 5, 10)
Red Collared Dove (*Streptopelia tranquebarica*)	South and Southeast Asia	(?I)Sulawesi	Introduced as escaped or liberated cage birds (2, 3)
Spotted Dove (*Streptopelia chinensis*)	South-eastern Asia as far as Western Indonesia & Lesser Sundas	Sulawesi; Moluccas; small islands of Flores Sea; New Britain	Successfully exported from Java to eastern Indonesian islands from 1835 (1, 2, 5)
Zebra Dove (*Geopelia striata*)	Malay Peninsula; Greater Sundas; Western Lesser Sundas	Ambon; Sulawesi	Introduced as escaped or liberated cage birds (2)
Wallacean & Papuan exports			
Columbidae			

Barred Dove (*Geopelia maugei*)	Southeast Moluccas; Lesser Sundas	Tomea Island of Tukangbesi Islands in Sulawesi Subregion	Introduced as escaped or liberated cage birds (2)
Psittacidae			
Blue-streaked Lory (*Eos reticulata*)	Barbar Island; Tanimbar Islands	(?I)Kai Islands; (?I)Damar	Introduced as escaped or liberated captives trapped for pet trade (2, 7)
Purple-naped Lory (*Lorius domicella*)	Seram; Ambon	(?I)Buru	Introduced as escaped or liberated captives trapped for pet trade (2, 7)
Ornate Lorikeet (*Trichoglossus ornatus*)	Sulawesi Subregion	(?I)Sangihe	Introduced as escaped or liberated captives trapped for pet trade (2)
Blue-naped Parrot (*Tanygnathus lucionensis*)	Talaud Islands; North Bornean Islands; Philippines	(?I)Sangihe Islands	Introduced as escaped or liberated captives trapped for pet trade (2)
Eclectus Parrot (*Eclectus roratus*)	Moluccas; Lesser Sundas; New Guinea & satellites; Bismarck Archipelago; Solomon Islands; Australia's Cape York Peninsula	Gorong Island; Banda Islands	Widely captured and translocated for pet trade (1, 2, 7)
Palm Cockatoo (*Probosciger aterrimus*)	New Guinea; Aru; Australia's Cape York Peninsula	Kai Islands	Introduced as escaped or liberated captives trapped for pet trade (2, 7)
Tanimbar Corella (*Cacatua goffini*)	Tanimbar Islands	Kai Islands	Introduced as escaped or liberated captives trapped for pet trade (1, 2, 7)
Yellow-crested Cockatoo (*Cacatua sulphurea*)	Sulawesi Subregion; Lesser Sundas	Singapore; Hong Kong	Introduced as escaped or liberated captives trapped for pet trade (2)
Sulphur-crested Cockatoo (*Cacatua galerita*)	New Guinea & some of its satellite islands; Australia	Ambon; Gorong Islands; Kai Islands	Introduced as escaped or liberated captives trapped for pet trade (1, 2)
Australian exports			
Cracticidae			
Australasian Magpie (*Gymnorhina tibicen*)	Australia; southern New Guinea	Guadalcanal	Introduced prior to 1945, ? now extinct (1, 9)
Meliphagidae			
Noisy Miner (*Manorina melanocephala*)	Australia	Solomon Islands	Introduced in the 1950s as escaped or liberated cage birds (5, 9)
Estrildinae			
Star Finch (*Neochmia ruficauda*)	Australia	New Guinea	Introduced as escaped or liberated cage birds (5)
Sources: (1) Lever 1987; (2) Coates & Bishop 1997; (3) Holmes & Phillipps 1996; (4) Coates 1985; (5) Peckover & Filewood 1976; (6) Beehler *et al.* 1986; (7) Forshaw 1989; (8) Hadden 1981; (9) Doughty *et al.* 1999; (10) Monk *et al.* 1997			
(?I) = Possibly introduced			

7 TRENDS IN VERTEBRATE TRANSLOCATION

In the above tables a number of trends are apparent. Through faunal translocation, the Oriental zoogeographic region has supplied parts of the Circum New Guinea Archipelago with approximately 40 introduced non-volant or weakly volant wild terrestrial vertebrates including two deer; a pig; two monkeys; three civets; a mongoose; a cat; a pangolin; up to three shrews; a porcupine; two squirrels; nine murid rodents; two jungle fowls; possibly a python and an elapid snake; a typhlopid blind snake; at least three geckos and a skink; and at least five frogs. In parallel with this, the large Wallacean island of Sulawesi with its mixed Oriental and Australasian evolutionary influences has through translocation supplied the Babirusa to the Moluccas; another pig to the Moluccas, Lesser Sundas and Simaleue near Sumatra; a monkey to the Moluccas; and possibly a cuscus to the remote Talaud Group. New Guinea (and its continental satellites), at the centre of the Australasian zoogeographic region has through translocation supplied 17 non-volant or weakly volant wild terrestrial vertebrates to some surrounding islands, including: a murid rat; four cuscuses; a glider; five macropods; up to two bandicoots; two cassowaries; possibly a python; and a skink.

In relation to strongly volant wild terrestrial vertebrates, approximately 15 species of birds with Eurasian or Oriental origins have had their ranges extended into or within the Circum New Guinea Archipelago due to translocation; while ten strongly volant birds including nine parrots and a dove with Wallacean or New Guinea origins have had their ranges expanded within the Circum New Guinea Archipelago through translocation. In addition to this, two Australian birds including the Australasian Magpie and Noisy Miner have been introduced to the Solomon Islands, while the Australian Star Finch has been introduced to New Guinea. For the most part, these translocations are due to escaped or liberated captives, and are a by-product of the fairly recent pet trade, which in Indonesia, is extensive (Coates and Bishop 1997; Jepson 1997). Some anthropogenic distributions however, particularly in Indonesia, whether recognised as such or not, may be due to a much more ancient traffic in cage birds (Coates and Bishop, 1997) that is at least as old as the spice trade (Heinsohn 1998a; Swadling 1996).

Out of all of the above, only two Australasian species have been successfully translocated west across Wallace's Line into the Oriental zoogeographic region. This includes the Sulawesi Wild Pig (*Sus celebensis*) to Simaleue near Sumatra, and the Wallacean Yellow-crested Cockatoo (*Cacatua sulphurea*) to Singapore and Hong Kong. Though extralimital to this study, the Australian Red-necked Wallaby (*Macropus rufogriseus*), with some wild individuals in the British Isles and a former population in Germany (Lever 1985), stands out as the only Australasian marsupial to have been successfully translocated west of Wallace's Line into the Palaearctic zoogeographic region of Eurasia.

A general trend in Indo-Melanesian zoogeography is one of faunal homogenisation, with a strong probability of ongoing range expansion by various stowaways and ethnotramps. This poses a general threat to insular biodiversity as vigorous introduced vertebrates outcompete or prey on vulnerable insular species, or introduced conspecifics begin to hybridise with restricted insular forms. Occasionally, however, there may be some conservation benefits, such as the spread of locally endangered species to surrogate localities that increase that species' chances of survival. Two such cases include the anthropogenic spread of the endangered Sulawesi Crested Black Macaque (*Macaca nigra*) to Bacan, and the endangered Sulawesi Babirusa (*Babyrousa babyrussa*) to Buru, where introduced surrogate populations of significant conservation value now occur.

It should also be added that contemporary vertebrate distributions are far from static, with many historically introduced island populations having not yet reached equilibrium. Rusa Deer (*Cervus timorensis*) in New Guinea, for example, after a century of liberation, are probably still consolidating their range (Bentley, 1998); while the historically introduced Common Spotted Cuscus (*Spilocuscus maculatus*) in New Ireland, after 60 to 70 years since liberation at the far northwestern end of the long narrow island has, to date, only invaded a fraction of suitable habitat (Heinsohn 1998a, 1998b).

8 LOMBOK LIMITALS

Several typically Javan or Balinese Oriental mammals, including the Barking Deer (*Muntiacus muntjak nainggolani*), Silvered Leaf Monkey (*Trachypithecus auratus kohlbruggei*), Leopard Cat (*Felis bengalensis javanensis*), and Pangolin (*Manis javanica*), only make it as far as Lombok just across the strait that delineates Wallace's Line. Given that these species are not recorded from neighbouring Sumbawa and other Lesser Sunda islands that would have been joined to Lombok during times of low Pleistocene sea levels, it is concluded here that they are probably recent arrivals by human agency. This is supported by the fact that most of the aforementioned mammals are Balinese subspecies with a history of associations with humans. Furthermore, the non-volant Oriental mammals recorded from Lombok in Kitchener *et al.* (1990) show a bias towards commensal or human-associated species, and point more to an anthropogenic than natural dispersal filter into Lombok from neighbouring Bali. There is also some historical evidence that the Balinese Rajas, whose realms and influence extended across the eastern strait, may have introduced some familiar Balinese animals such as the Barking Deer and Silvered Leaf Monkey to Lombok (Hartert & Everett 1896).

9 ANTIQUITY OF TRANSLOCATION

The antiquity of vertebrate translocation in the Circum New Guinea Archipelago has largely been reconstructed from dated archaeological sequences in which the bones of certain species make sudden appearances. Using such evidence, the oldest recorded incidence of animal translocation is that of the Northern Common Cuscus (*Phalanger orientalis*) to New Ireland from sources in New Britain or New Guinea about 20 000 years ago in the Late Pleistocene (Flannery and White 1991; Spriggs 1997). Given that the undated possible New Britain source population may also have been introduced (from New Guinea), then initial translocation from New Guinea into the Bismarck Archipelago may have occurred at an even earlier date. Based on archaeological dates from New Ireland, the Bismarck Archipelago was after all occupied by humans as early as 35 000 years ago (Spriggs 1997).

Similar evidence of sudden appearances in archaeological records indicates that a further two New Guinean marsupials, the Admiralty Cuscus (*Spilocuscus kraemeri*) and the Common Spiny Bandicoot (*Echymipera kalubu*) may have been introduced to the Admiralty Islands in the Terminal Pleistocene about 13 000 years ago (Spriggs 1997). This brings the suspected number of Pleistocene introductions to three. However, as the arrival of some probably introduced ethnotramp marsupials on large islands such as New Britain and Seram has not yet been dated through archaeological investigation, there may be some further Pleistocene translocation events yet to be recorded. Most of the other New Guinea exports recorded in Table 1 were probably translocated prehistorically in the Holocene with some probably being due to the later Holocene movements of the Austronesians (Spriggs 1997; Heinsohn 1998a). Some very recent exceptions to this include the historically recorded introduction of the Common Spotted Cuscus (*Spilocuscus maculatus*) to New Ireland in the 1930s or 1940s (Heinsohn 1998a, 1998b, 1997, 1999); and the recent arrival of the New Guinean Northern Skink (*Carlia fusca*) in Bougainville (McCoy 1980).

In contrast to the New Guinea exports, the Oriental and Sulawesi exports listed in Table 1 appear on current evidence to be a Holocene phenomenon, with many being associated with the prehistoric Austronesian expansion in the second half of the Holocene; or dating from the later protohistoric and colonial period up to the present (Heinsohn 1998a; Bellwood 1997; Glover 1986, 1971; Ellen 1993; Lever 1985; White 1975).

10 CONCLUSION

Superimposed on the natural vertebrate zoogeography of Wallace's Line and the islands to the east are some significant human influences resulting from animal translocation. The most marked effect is the export of around 55 Oriental wild vertebrate species into various parts of the Circum New Guinea Archipelago. In contrast, only two Australasian vertebrates, the Sulawesi Wild Pig (*Sus celebensis*) and the Wallacean Yellow-crested Cockatoo (*Cacatua sulphurea*) have been successfully introduced to locations within the Oriental zoogeographic region. Thus while the Oriental zoogeographic region has through wild vertebrate translocation had a marked influence on the zoogeography of the Circum New Guinea Archipelago, translocation from the latter region has had negligible influence on the former. Both New Guinea and large Wallacean islands such as Sulawesi have, however, served as significant sources for species translocated within the Circum New Guinea Archipelago. New Guinea, for example, has probably exported around 17 vertebrate species to surrounding islands; whereas Sulawesi has through translocation exported up to four endemic vertebrates to various satellites. Furthermore, while the Oriental and Sulawesi exports appear to be largely a phenomenon of the latter half of the Holocene, the export of New Guinea marsupials probably stretches back to at least 20 000 years ago in the Late Pleistocene.

Due to the ambiguity of the evidence there may be many regionally or locally introduced populations of vertebrates, particularly among the herpetofauna (Edgar & Lilley 1993; Allison 1996), that have not yet been recognised as such, because they do not stand out as obvious exotics. Many questions will ultimately only be answered through further biological surveys and genetic studies, coupled with further archaeological excavations and historical research. These are important questions, because they address the antiquity and magnitude of human interactions with the biosphere. Prior to the work of the Lapita Homeland Archaeological Project, who would have thought that the world's most ancient ethnotramp would turn out to be a humble marsupial (*Phalanger orientalis*)? Furthermore, in addition to being recognised as an early alternative centre for plant domestication and agriculture (Spriggs 1997), the New Guinea/Northern Melanesian Region can now also be recognised as an early centre for game management / 'game park' strategies utilising marsupials that were in some instances probably deliberately introduced to oceanic islands previously depauperate in faunal resources. Indeed, having collected contemporary oral records of cuscuses being carried to tiny outer islands on motor boats and aircraft for release into patches of bush (Heinsohn 1998a; 1999), this is an ongoing tradition that in the Circum New Guinea Archipelago has probably stretched from the ancient raft / canoe age, into the jet age.

ACKNOWLEDGMENTS

The author would like to thank the people of Papua New Guinea and Indonesia for their hospitality and for sharing information, with special thanks to Lowrance Nomaris, Lorex Martin and the people of Put Put Village in New Ireland, and Mr Arifin and family and the people of Pariangan Village on Selayar Island. Gratitude is also extended to the Department of New Ireland, the Papua New Guinea Department of Environment and Conservation, and the Institute of Papua New Guinea Studies for their support of my fieldwork, and to the Department of Forests and Directorate for Nature Conservation (PHPA) in Indonesia for allowing access to many areas. The pioneering research of many biologists and archaeologists and that of the Lapita Homeland Project is acknowledged in the list of references.

REFERENCES

Allison, A. 1996. Zoogeography of amphibians and reptiles of New Guinea and the Pacific region. In A. Keast & S.E. Miller (eds), *The origin and evolution of Pacific island biotas, New Guinea to Eastern Polynesia: patterns and processes*: 407-36. Amsterdam: SPB Academic Publishing.

Atkinson, I.A.E. 1985. The spread of commensal species of *Rattus* to Oceanic islands and their effects on island avifaunas. In P.J. Moors (ed), *Conservation of Island Birds: Case studies for the management of threatened island species*: 35-81. Cambridge: International Council for Bird Preservation.

Auffenberg, W. 1980. The Herpetofauna of Komodo, with notes on adjacent areas. *Bulletin of the Florida State Museum, Biological Science* 25(2): 39-156.

Beehler, B., Pratt, T. & Zimmerman, D. 1986. *Birds of New Guinea*. Princeton: Princeton University Press.

Bellwood, P. 1997. *Prehistory of the Indo-Malaysian Archipelago*. Honolulu: University of Hawaii Press.

Bennett, G. 1860. *Gatherings of a Naturalist in Australasia: Being Observations Principally on the Animal and Vegetable Productions of New South Wales, New Zealand, and some of the Austral Islands.* London: John Van Voorst.

Bentley, A. 1998. *An Introduction to the Deer of Australia.* Melbourne: Australian Deer Research Foundation.

Bergh, G. van den, Mubroto, B., Aziz, F., Sondaar, P. & Vos, J. de, 1996. Did *Homo erectus* reach the island of Flores? *BIPPA* 14: 27-36.

Bowler, J. & Taylor, J. 1993. The Avifauna of Seram. In I.D. Edwards, A.A. Macdonald & J. Proctor (eds), *Natural History of Seram, Maluku, Indonesia*: 143-59. Andover: Intercept.

Carter, T.D., Hill, J.E. & Tate, G.H.H. 1945. *Mammals of the Pacific World.* New York: American Museum of Natural History / Macmillan.

Chappell, J. & Thom, B.G. 1977. Sea Levels and Coasts. In J. Allen, J. Golson & R. Jones (eds), *Sunda and Sahul: Prehistoric Studies in Southeast Asia, Melanesia and Australia*: 275-91. London: Academic Press.

Coates, B.J. 1985. *The Birds of Papua New Guinea: Including the Bismarck Archipelago and Bougainville.* Brisbane: Dove Publications.

Coates, B.J. & Bishop, K.D. 1997. *A Guide to the Birds of Wallacea.* Brisbane: Dove Publications.

Cogger, H.G. 1992. *Reptiles and Amphibians of Australia.* Sydney: Reed.

Corbet, G.B. & Hill, J.E. 1992. *The Mammals of the Indomalayan Region: A Systematic Review.* Oxford: Natural History Museum Publications / Oxford University Press.

Doughty, C., Day, N. & Plant, A. 1999. *Birds of the Solomons, Vanuatu and New Caledonia.* London: Helm.

Edgar, P.W. & Lilley, R.P.H. 1993. Herpetofauna Survey of Manusela National Park. In I.D. Edwards, A.A. Macdonald & J. Proctor (eds), *Natural History of Seram, Maluku, Indonesia*: 131-41. Andover: Intercept.

Egloff, B. 1979. *Recent Prehistory in Southeast Papua, Terra Australis 4.* Canberra: Department of Prehistory, Research School of Pacific and Asian Studies, Australian National University.

Ellen, R.F. 1993. Human Impact on the Environment of Seram. In I.D. Edwards, A.A. Macdonald & J. Proctor (eds), *Natural History of Seram, Maluku, Indonesia*: 191-205. Andover: Intercept.

Flannery, T. 1995a. *Mammals of the South-West Pacific and Moluccan Islands.* Sydney: Australian Museum / Reed.

Flannery, T. 1995b. *Mammals of New Guinea.* Sydney: Australian Museum / Reed.

Flannery, T. 1994. *Possums of the World: A Monograph on the Phalangeroidea.* Sydney: Australian Museum / Geo Productions.

Flannery T., 1992. Taxonomic revision of the *Thylogale brunii* complex (Macropodidae: Marsupialia) in Melanesia, with description of a New Species. *Australian Mammalogy* 15: 7-23.

Flannery, T., Bellwood, P., White, J.P., Ennis, T., Irwin, G., Schubert, K. & Balasubramaniam, S. 1998. Mammals from Holocene Archaeological Deposits on Gebe and Morotai Islands, Northern Moluccas, Indonesia. *Australian Mammalogy* 20(3): 391-400.

Flannery, T., Bellwood, P., White, P., Moore, A., Boeadi & Nitihaminoto, G. 1995. Fossil marsupials (Macropodidae, Peroryctidae) and other mammals of Holocene age from Halmahera, North Moluccas, Indonesia. *Alcheringa* 19: 17-25.

Flannery, T.F., Kirch, P.V., Specht, J. & Spriggs, M. 1988. Holocene mammal faunas from archaeological sites in island Melanesia. *Archaeology in Oceania* 23(3): 89-94.

Flannery, T., Martin, R. & Szalay, A. 1996. *Tree Kangaroos: A Curious Natural History.* Sydney: Reed.

Flannery, T. & White, J.P. 1991. Animal Translocation: Zoogeography of New Ireland Mammals. *National Geographic Research and Exploration* 7(1): 96-113.

Forshaw, J.M. 1989. *Parrots of the World.* Sydney: Lansdowne Editions / Kevin Weldon.

Frankenberg, E. & Werner, Y.L. 1981. Adaptability of the daily activity pattern to changes in Longitude, in a colonising lizard, *Hemidactylus frenetus. Journal of Herpetology* 15(3): 373-76.

Gibbons, J.R.H. 1985. The biogeography and evolution of Pacific Island Reptiles and Amphibians. In G. Grigg, R. Shine & H..Ehmann (eds), *Biology of Australasian Frogs and Reptiles*: 125-42. Sydney: The Royal Zoological Society of New South Wales / Surrey Beatty & Sons.

Glover, I.C. 1986. *Archaeology in Eastern Timor, 1966-67, Terra Australis 11.* Canberra: Department of Prehistory, Research School of Pacific Studies, Australian National University.

Glover, I.C. 1971. Prehistoric Research in Timor. In D.J. Mulvaney & J. Golson (eds), *Aboriginal Man and Environment in Australia*: 158-81. Canberra: Australian National University Press.

Green, B. & King, D. 1993. *Goanna: The Biology of Varanid Lizards.* Sydney: New South Wales University Press.

Groves, C.P. 1984. Of mice and men and pigs in the Indo-Australian Archipelago. *Canberra Anthropology* 7 (1&2): 1-19.

Groves, C.P. 1981. *Ancestors for the pigs: taxonomy and phylogeny of the genus Sus*. Canberra: Department of Prehistory, Research School of Pacific Studies, Australian National University.

Groves, C.P. 1980. Notes on the systematics of Babyrousa (Artiodactyla, Suidae). *Zoologische Mededelingen* 55(3): 29-46.

Hadden, D. 1981. *Birds of the North Solomons*. Wau: Wau Ecology Institute.

Haddon, A.C. & Hornell, J. 1975. *Canoes of Oceania*. Honolulu: Bishop Museum Press.

Harding, T.G. 1967. *Voyagers of the Vitiaz Strait: A Study of a New Guinea Trading System*. Seattle: The American Ethnological Society / University of Washington Press.

Hartert, E. & Everett, A. 1896. List of a collection of birds made in Lombok by Mr. Alfred Everett. *Novitates Zoologicae* 3: 591-99.

Heinsohn, T. 1999. Bikbus wokabaut: human influences and wild nature in Australasia. In J. Dargavel & B. Libbis (eds) *Australia's Ever-changing Forests IV*: 17-36. Canberra: Australian Forest History Society / Centre for Resource and Environmental Studies, Australian National University.

Heinsohn, T. 1998a. *The Realm of the Cuscus: Animal Translocation and Biological Invasions East of Wallace's Line*. Canberra: Unpublished Master of Science thesis, Geography Department, School of Resource and Environmental Management, Australian National University.

Heinsohn, T. 1998b. Captive ecology. *Nature Australia* 26 (2): 36-43.

Heinsohn, T. 1997. Phantoms in the foliage: human influences on the rainforests of the New Guinea Archipelago. In Dargavel J. (ed), *Australia's Ever-changing Forests III*: 278-295. Canberra: Australian Forest History Society / Centre for Resource and Environmental Studies, Australian National University.

Heinsohn, T. 1995. The Biological Invasion of New Ireland, Papua New Guinea by the Introduced Common Spotted Cuscus *Spilocuscus maculatus* (Marsupialia: Phalangeridae). Townsville: Paper presented at the Australian Mammal Society Conference, James Cook University, Townsville, September, 1995.

Herington, J.G. 1977. Wildlife introduced and imported into Papua New Guinea. *Wildlife in Papua New Guinea* 77(2): 1-9

Holmes, D. & Phillipps, K. 1996. *The Birds of Sulawesi*. Kuala Lumpur: Oxford University Press.

Honacki, J.H., Kinman, K.E. & Koeppl, J.W. 1982. *Mammal Species of the World: A Taxonomic and Geographic Reference*. Lawrence: The Association of Systematics Collections / Allen Press.

How, R.A. & Kitchener, D.J. 1997. Biogeography of Indonesian Snakes. *Journal of Biogeography* 24: 725-35

Hoyo, del J., Elliott, A. & Sargatal, J. (eds) 1992. *Handbook of the Birds of the World, Volume 1*. Barcelona: Lynx Edicions.

Iredale, T. 1956. *Birds of New Guinea*. Melbourne: Georgian House.

Jepson, P. 1997. *Birding Indonesia: A Birdwatcher's Guide to the World's Largest Archipelago*. Singapore: Birdlife International / Periplus Editions.

Kirch, P.V. 1997. *The Lapita Peoples: Ancestors of the Oceanic World*. Oxford: Blackwell.

Kirch, P.V. 1988. The Talepakemalai Lapita Site and Oceanic Prehistory. *National Geographic Research* 4(3): 328-42.

Kitchener, D.J., Boeadi, Charlton, L. & Maharaddatunkamsi, 1990. Wild Mammals of Lombok Island. *Records of the Western Australian Museum*, Supplement No. 33: 1-129.

Koopman, K.F. 1979. Zoogeography of Mammals from Islands Off the Northeastern Coast of New Guinea. *American Museum Novitates* 2690(2): 1-17.

Laurie, E. & Hill, J. 1954. *List of Land Mammals of New Guinea, Celebes and Adjacent Islands 1758 – 1952*. London: British Museum (Natural History).

Lever, C. 1994. *Naturalized Animals: The Ecology of Successfully Introduced Species*. London: Poyser.

Lever, C. 1987. *Naturalized Birds of the World*. New York: Longman Scientific and Technical / John Wiley and Sons.

Lever, C. 1985. *Naturalized Mammals of the World*. London: Longman.

Lilley, I. 1986. *Prehistoric Exchange in the Vitiaz Strait, Papua New Guinea*. Canberra: Unpublished Ph.D. thesis, Prehistory Department, Australian National University.

Loveridge, A. 1946. *Reptiles of the Pacific World*. New York: Macmillan.

Macdonald, A.A., Hill, J.E., Boeadi & Cox, R. 1993. The mammals of Seram, with notes on their biology and local usage. In I.D. Edwards, A.A. Macdonald & J. Proctor (eds), *Natural History of Seram, Maluku, Indonesia*: 161-90. Andover: Intercept.

Malinowski, B. 1922. *Argonauts of the Western Pacific: An Account of Native Enterprise and Adventure in the Archipelagoes of Melanesian New Guinea*. London: Routledge & Kegan Paul.

Maynes, G.M. 1989. Zoogeography of the Macropodoidea. In G. Grigg, P. Jarman & I. Hume (eds), *Kangaroos, Wallabies and Rat-kangaroos, Volume 2*: 47-66. Sydney: Surrey Beatty & Sons.

McCoy, M. 1980. *Reptiles of the Solomon Islands*. Wau: Wau Ecology Institute.

Menzies, J.I. 1975. *Handbook of Common New Guinea Frogs*. Wau: Wau Ecology Institute.

Menzies, J.I. & Dennis, E. 1979. *Handbook of New Guinea Rodents*. Wau: Wau Ecology Institute.

Monk, K.A., De Fretes, Y. & Reksodiharjo-Lilley, G. 1997. *The Ecology of Nusa Tenggara and Maluku*. Hong Kong: Periplus Editions.

Morwood, M.J., Aziz, F., van den Bergh, G.D., Sondaar, P.Y. & De Vos, J. 1997. Stone artefacts from the 1994 excavation at Mata Menge, West Central Flores, Indonesia. *Australian Archaeology* 44: 26-34.

Nowak, R.M. 1991. *Walker's Mammals of the World*. Baltimore: The Johns Hopkins University Press.

Peckover, W.S. & Filewood, L.W.C. 1976. *Birds of New Guinea and Tropical Australia: The Birds of Papua New Guinea, Irian Jaya, The Solomon Islands and Tropical North Australia*. Sydney: Reed.

Rand, A.L. & Gilliard, E.T. 1967. *Handbook of New Guinea Birds*. London: Weidenfeld & Nicolson.

Scott, F., Parker, F. & Menzies, J.I. 1977. A Checklist of the Reptiles and Amphibians of Papua New Guinea. *Wildlife in Papua New Guinea* 77(3): 1-18.

Singadan, R.K. 1996. Notes on Hybrid Spotted Cuscus, *Spilocuscus maculatus* X *Spilocuscus kraemeri* (Marsupialia: Phalangeridae). *Science in New Guinea* 22(2): 77-82.

Spriggs, M. 1997. *The Island Melanesians*. Oxford: Blackwell.

Swadling, P. 1996. *Plumes from Paradise: Trade cycles in outer Southeast Asia and their impact on New Guinea and nearby islands until 1920*. Brisbane: Papua New Guinea National Museum / Robert Brown and Associates.

Swadling, P. 1981. *Papua New Guinea's Prehistory: An Introducion*. Port Moresby: National Museum and Art Gallery of Papua New Guinea.

Thorne, A., Grun, R., Mortimer, G., Spooner, N.A., Simpson, J.J., McCulloch, M., Taylor, L. & Curnoe, D. 1999. Australia's oldest human remains: age of the Lake Mungo 3 skeleton. *Journal of Human Evolution* 36: 591-612.

Tyler, M.J. 1994. *Australian Frogs: A Natural History*. Sydney: Reed.

Wallace, A.R. 1869. *The Malay Archipelago, The Land of the Orang-utan and the Bird of Paradise: A Narrative of Travel with Studies of Man and Nature*. London: Macmillan.

Welch, K.R.G. 1988. *Snakes of the Orient: A Checklist*. Malabar: Krieger.

Welch, K.R.G., Cooke, P.S. & Wright, A.S. 1990. *Lizards of the Orient: A Checklist*. Malabar: Krieger, Malabar.

White, C.M.N. 1975. The problem of the cassowary in Seran. *Bulletin of the British Ornithologists' Club* 95 (4): 165-70.

Whitten, A.J., Mustafa, M. & Henderson, G.S. 1987. *The Ecology of Sulawesi*. Yogyakarta: Gadjah Mada University Press.

Zweifel, R.G. & Tyler, M.J. 1982. Amphibia of New Guinea. In J.L. Gressitt (ed), *Biogeography and Ecology of New Guinea*: 759-801. The Hague: Dr. W. Junk.

Wallace's line and marine organisms: the distribution of staghorn corals (*Acropora*) in Indonesia

C. C. Wallace

Museum of Tropical Queensland, 78-102 Flinders St. Townsville, Queensland 4810, Australia

ABSTRACT: Historical vicariance factors affect marine distributions differently from terrestrial distributions, land being a barrier to marine dispersal and the water column being a transport medium for larvae. This paper examines distribution patterns of species of the coral genus *Acropora*, which is found to reach its highest worldwide diversity in the Wallacea region. A morphology-based phylogeny and data from the fossil record indicate an origin of the genus around Africa/Southern Europe with the most recent diversification in Indonesia/West Pacific Ocean. This recent diversification includes forms specialised for deep, calm habitats. It is hypothesised that continuity of reef sites through time and the impact of late-Cenozoic eustatic fluctuations in shaping relictual and endemic faunas have contributed to the high diversity found in Wallacea. Indian Ocean and Pacific Ocean *Acropora* faunas overlap within the region, however the Pacific Ocean influence is stronger, possibly due to access of larvae through the Pacific through-flow current.

1 INTRODUCTION

'For marine biologists the terrestrial biologist's land bridge becomes a barrier to dispersal and we must acknowledge the effects of such barriers on the evolutionary history of the marine realm' (Pandolfi 1992).

Distributions of species of the coral genus *Acropora* in the Indonesian archipelago show a duality reminiscent of the Wallace's Line patterns seen in terrestrial animals and plants, rather than the concentric pattern predicted by the 'centre of origin' model (Wallace 1997). This duality is due to an overlap of Indian Ocean species distributions diminishing eastwards and Pacific Ocean species distributions diminishing westwards within the archipelago (Wallace 1997). Additionally, a large number of species with broad Indo-Pacific distribution, as well as some regional endemics, occur within the archipelago. Thus although Indonesia has a high species number overall, this is due to the presence of a composite fauna with strong regional differences. The highest level of endemicity was seen to occur in the Togian Islands, in Central Sulawesi, a location with a particularly unusual *Acropora* fauna (Wallace 1997). The endemic species of the Togian Islands were hypothesised to indicate a possible late Tethyan relictual fauna (Wallace 1999a), a conclusion which is re-assessed here.

Awareness of the well-known dispersal abilities of marine larvae has been a potent influence on the thinking of marine biologists. A prevailing view of the Indo-Australian arc for marine organisms often invokes a 'centre' from which genera and species are said to have dispersed in a concentric pattern, often combined with the concept of origination at this centre (Briggs 1992, 1993, Piccoli et al. 1987; reviewed Wilson & Rosen 1998). One such model for corals (Stehli & Wells 1971), was based on modern distribution patterns and known ages of genera, and proposed that the central Indo-Pacific served as the centre of origination of most coral genera. This model was not able to take into account the distribution patterns of individual species, phylogenetic evidence, the spatial fossil record or tectonic and eustatic changes that might be relevant to an historical biogeographic viewpoint.

The staghorn corals, genus *Acropora*, are by far the most diverse living reef-building coral genus, with 114 species recognised worldwide (Wallace 1999b) and others possibly unnamed. They occur on most tropical reefs throughout the world, with the exception of the eastern Pacific reefs, and the earliest fossil record of the genus is dated at late Paleocene (65-54m.y., Carbone et al. 1994). Because of this, *Acropora* offers an ideal opportunity to follow global colonisation through time in detail.

From the geographic distribution patterns of documented fossil finds, the centre of marine diversity in Indonesia is seen as unlikely to be the centre of origin of many coral genera but rather the latest centre of diversification following origination around S. Europe/N. Africa (Wilson & Rosen 1998). This is the case for *Acropora*, for which the earliest fossil records indicate an origin in either N. Africa or the regions around the present Mediterranean (Wallace 1999b, p. 20). In this paper I further explore the origins of high species diversity of staghorn corals in Indonesia, through the relationship of fossil records, distribution patterns and habitat preferences of *Acropora* species to a phylogeny derived from skeletal characters. I then re-examine species diversity of the genus in regional Indonesia relative to the rest of the Indo-Pacific, adding new species distribution records from central Sulawesi, Eastern New Guinea and Halmahera.

2 METHODS

The database used for this study was the Worldwide *Acropora* Database, which documents approximately 16,000 specimens of *Acropora* from over 800 locations worldwide (130 in Indonesia), stored in the Museum of Tropical Queensland (Wallace & Wolstenholme 1998, Wallace 1999b, p. 60, Fig. 36). Validated species records from the literature were added to this database for the purpose of area analyses. Some new records resulting from an expedition to the Togian Islands in September 1999 (the *Tethyana* expedition, looking at the distribution of several animal groups) are added to the database for this study and referred to as 'new data'.

For biogeographic analyses, including similarity comparisons and histograms of species compositions, areas were delimited as indicated in Figure 1. These areas are the same as those used for biogeographic analysis of the whole genus (Wallace 1999b, p. 63, Fig. 27), each being represented by major collections in the database as well as the literature record. Thus the size of areas was to some extent determined by collecting effort. Comparisons of similarity between areas were based on Jaccard's similarity coefficient (*similarity* = $a/(a+b+c)$), where a=number of species shared by two areas, b=number of species in area 1 and not in area 2, and c=number of species in area 2 and not in area 1). Areas outside Indonesia were grouped together into broader regions (see Fig. 1) for the purpose of formulating histograms of species compositions.

The kinds of distribution shown by individual species were classified as 'endemic' (found only in a single area), ' Indian Ocean' (found only in the Indian Ocean or in this plus some parts of the Indonesian archipelago), 'Pacific Ocean' (found only in the Pacific Ocean or in this plus some parts of the Indonesian archipelago) or 'Indo-Pacific' (found in the Pacific and Indian Oceans).

Fossil records referred to in this paper are summarised in the species descriptions in Wallace (1999b). These include references from the literature and approximately 200 records from fossil specimens in the Worldwide *Acropora* database.

The phylogenetic cladogram against which data on fossil record, distribution and habitat are compared is the result of an analysis of 98 extant species of the subgenus *Acropora* using 23 skeletal characters by Wallace & Scott (submitted). These are the characters by which both fossil and extant species are presently described and identified, although some progress is being made towards a molecular phylogeny (Wallace and Willis 1994, Wallace 1999b, p. 47-48). For the phylogenetic analysis the *A.* (*Isopora*) subgenus was used as an outgroup to *A.* (*Acropora*) due to its position as a sister group to *A.* (*Acropora*) (Wallace 1999b, Fig. 42). Multiple parsimonious trees were generated using *Hennig86v1.5* (Farris 1988) and *PAUP3.1.1* (Swofford 1991) computer programs and these analyses were summarised as strict consensus trees in *PAUP3.1.1*. Full details of the methods used for Figure 2 are given in Wallace (1999b, pp. 72-91).

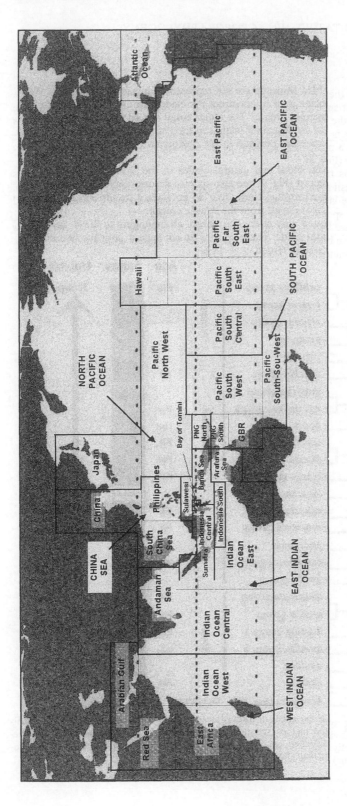

Figure 1. Map of the world indicating divisions used for analysis of species composition. For the purpose of histogram comparisons of parts of Indonesia with the rest of the world (in Figs 4 & 6), areas outside Indonesia are combined as broader regions, delimited by solid lines: dotted lines indicate borders of areas within regions.

3 TIME, SPACE AND HABITAT WITHIN *ACROPORA* PHYLOGENY

3.1 *Time*

Fossil species found up until the Mid Miocene are not represented in the modern fauna. While the fossil record of the extant species used to construct the cladogram is incomplete, it can be used as a guide to the age of the species groups. The oldest fossil record for an extant species is that of *A. humilis* in the Miocene of the Marshall Islands in the Pacific. This species is in '*humilis* group 1', which occupies a near-basal position in the cladogram of the subgenus *Acropora* based on skeletal morphology (Fig. 2).

Each of the species groups from basal and central clades in the *Acropora* phylogeny has species represented in the fossil record (Fig. 2, see also species descriptions in Wallace 1999b). However, of the groups in the large terminal clade, four do not have a fossil record to date and the fossils known (from three species) are all from the Pacific Ocean.

There are three species that are found only in the Caribbean, all belonging to the *A. cervicornis* group, which occurs within the basal clade. These species do not appear until the late Pliocene, following a series of now extinct species (Budd et al. 1994).

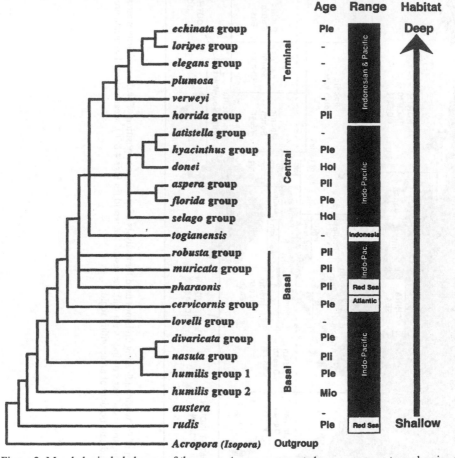

Figure 2. Morphological phylogeny of the genus *Acropora* presented as a consensus tree, showing terminal taxa as species groups or individual species (see methods). Trends in present geographical range, predominant habitat and the currently estimated age of the species groups (oldest fossil record, Ple: Pleistocene, Pli: Pliocene, Mio: Miocene) are also shown.

3.2 *Space*

The cladogram can also be viewed in terms of the distribution patterns of the species groups. The Caribbean species appear in the basal clades, but after some Indian Ocean and broad Indo-Pacific groups, indicating that the genus was quite widespread before the isolation of the Atlantic fauna. All species restricted to the Indian Ocean occur in the older, basal clade. The other species groups in the large basal clade are composed entirely or mostly of species that are widespread throughout the Indo-Pacific. In the terminal clade, however, groups consist mainly of species that are restricted to either the Pacific Ocean, Indonesia, or a combination of both. This suggests a passage of the genus in space, firstly spreading in two directions after its origin around S. Europe/N. Africa, with the final speciation activity taking place around the central Indo-Pacific (as proposed by Wilson & Rosen 1998).

3.3 *Habitat*

There is also a habitat signature in the cladogram. Each group is most characteristically found in particular reef habitats, although most species groups include at least one species with broader habitat ranges. Most species in the terminal clade occur in calm water, either below the zone of tidal activity and wave action on reef slopes, or in lagoons and enclosed bays. This clade includes species from very great depths for *Acropora* (below 50m). The central clade comprises species that usually occur at intermediate to shallow depths, while the basal clade contains most of the species that occupy shallow high energy habitats of the reef top and edge (e.g. the two *humilis* groups and the *robusta* group).

From the habitats occupied by the members of the various groups, it appears that the most recent groups have evolved in calm, subtidal water, whereas earlier clades evolved in shallower depths including high energy, exposed waters. This suggests that events isolating pockets of deeper water have been the most significant in influencing speciation in the central Indo-Pacific region.

4 SPECIES DIVERSITY PATTERNS OF *ACROPORA*

4.1 *Numbers of species*

Five of the 30 areas around the world have a species number of 70 or more (see highlighted numbers under 'Number of species' in Fig. 3). If Indonesia were to be regarded as a single entity, (rather than as six separate areas) its diversity would be highest at 91 (see Table 1). This record could possibly be challenged by further collecting in the south-western Philippines and other parts of the South China Sea. No single area of Indonesia, however, shows such a high species count. The greatest number of species found in an area of Indonesia is 77, in the Bay of Tomini, Central Sulawesi (Fig. 3). This is also the highest species count worldwide. At the same time, the bay is the smallest area included in the study, being the area between the northern and central eastern arms of Sulawesi.

Table 1. Comparison of composition of each of the four distribution categories (see methods) within the Indonesian archipelago.

Distribution Category	Number of Species
Indo-Pacific	64
Endemic to Indonesia	7
Pacific Ocean	16
Indian Ocean	4
Total	91

	Number of species	East Africa	Red Sea	Arabian Gulf	Indian Ocean W.	Indian Ocean C.	Andaman Sea	Indian Ocean E.	Arafura Sea	Sumatra	Indonesia S.	Indonesia C.	Banda Sea	Sulawesi	Bay of Tomini	PNG N.	PNG S.	Philippines	S. China Sea	China	Japan	Pacific N.	GBR	Pacific SW.	Pacific SSW.	Pacific SC.	Pacific SE.	Hawaii	Pacific FSE.	Pacific E.	Atlantic Ocean
		Similarity between areas																													
East Africa	45		0.50	0.14	0.64	0.60	0.59	0.55	0.44	0.55	0.46	0.51	0.50	0.47	0.45	0.49	0.57	0.53	0.58	0.56	0.57	0.55	0.53	0.59	0.53	0.61	0.49	0.07	0.35	0.02	0.00
Red Sea	42	29		0.23	0.54	0.54	0.45	0.49	0.36	0.49	0.48	0.49	0.48	0.43	0.42	0.44	0.50	0.47	0.52	0.47	0.49	0.51	0.44	0.50	0.45	0.47	0.37	0.07	0.29	0.02	0.00
Arabian Gulf	12	7	10		0.19	0.18	0.14	0.15	0.11	0.16	0.15	0.15	0.13	0.12	0.11	0.12	0.15	0.12	0.17	0.13	0.15	0.12	0.13	0.12	0.14	0.14	0.10	0.07	0.09	0.08	0.00
Indian Ocean W.	52	38	33	10		0.66	0.67	0.63	0.39	0.53	0.57	0.56	0.55	0.54	0.48	0.54	0.62	0.56	0.59	0.63	0.65	0.65	0.62	0.62	0.59	0.69	0.58	0.06	0.36	0.02	0.00
Indian Ocean C.	46	34	31	9	39		0.66	0.59	0.41	0.56	0.58	0.59	0.62	0.56	0.52	0.59	0.68	0.61	0.63	0.67	0.66	0.64	0.59	0.65	0.54	0.65	0.51	0.07	0.35	0.02	0.00
Andaman Sea	55	37	30	8	43	40		0.64	0.42	0.58	0.61	0.58	0.61	0.55	0.53	0.58	0.64	0.60	0.61	0.63	0.64	0.60	0.62	0.66	0.56	0.59	0.51	0.05	0.34	0.02	0.00
Indian Ocean E.	70	41	37	11	47	43	49		0.43	0.53	0.70	0.65	0.72	0.69	0.62	0.68	0.71	0.67	0.78	0.74	0.75	0.71	0.74	0.73	0.62	0.66	0.55	0.04	0.31	0.01	0.00
Arafura Sea	30	23	19	4	23	22	25	30		0.43	0.39	0.44	0.41	0.37	0.35	0.37	0.43	0.44	0.47	0.45	0.44	0.46	0.39	0.45	0.44	0.42	0.38	0.06	0.23	0.03	0.00
Sumatra	40	30	27	7	32	31	35	38	21		0.54	0.5	0.51	0.44	0.43	0.45	0.54	0.50	0.56	0.53	0.52	0.52	0.45	0.47	0.49	0.49	0.46	0.08	0.31	0.03	0.00
Indonesia S.	66	35	35	10	43	41	46	56	27	37		0.79	0.81	0.77	0.72	0.74	0.74	0.68	0.68	0.67	0.68	0.66	0.67	0.63	0.54	0.67	0.45	0.05	0.25	0.02	0.00
Indonesia C.	59	35	33	9	40	39	42	51	27	33	55		0.81	0.77	0.74	0.76	0.72	0.70	0.67	0.71	0.72	0.68	0.61	0.65	0.53	0.67	0.42	0.05	0.24	0.02	0.00
Banda Sea	66	37	35	9	42	43	46	57	28	36	59	56		0.84	0.74	0.78	0.77	0.72	0.70	0.71	0.72	0.71	0.70	0.70	0.52	0.69	0.45	0.05	0.27	0.02	0.00
Sulawesi	74	38	35	9	44	43	46	59	28	35	61	58	64		0.80	0.81	0.73	0.70	0.65	0.68	0.70	0.64	0.65	0.65	0.47	0.67	0.41	0.04	0.24	0.01	0.00
Bay of Tomini	77	38	35	9	42	42	46	56	28	35	60	58	61	67		0.81	0.71	0.75	0.61	0.66	0.67	0.60	0.61	0.63	0.45	0.59	0.38	0.04	0.20	0.01	0.00
PNG N.	73	39	35	9	44	44	47	58	28	35	59	57	61	66	67		0.81	0.73	0.68	0.76	0.75	0.67	0.68	0.70	0.49	0.65	0.43	0.04	0.24	0.01	0.00
PNG S.	63	39	35	10	44	44	46	55	28	36	55	51	56	58	58	61		0.75	0.75	0.79	0.78	0.71	0.72	0.73	0.59	0.75	0.49	0.05	0.30	0.02	0.00
Philippines	65	38	34	8	42	42	45	54	29	35	53	51	55	57	61	58	55		0.71	0.84	0.81	0.74	0.75	0.75	0.53	0.68	0.46	0.05	0.25	0.02	0.00
S. China Sea	58	38	34	10	41	40	43	56	28	35	50	47	51	52	51	53	52	51		0.79	0.78	0.69	0.70	0.73	0.63	0.68	0.51	0.05	0.30	0.02	0.00
China	64	39	34	9	45	44	46	57	29	36	52	51	54	56	56	59	56	59	54		0.93	0.83	0.78	0.79	0.58	0.74	0.53	0.05	0.29	0.02	0.00
Japan	65	40	35	10	46	44	47	58	29	36	53	52	55	57	57	59	56	58	54	62		0.79	0.79	0.83	0.59	0.77	0.52	0.05	0.29	0.02	0.00
Pacific N.	62	38	35	8	45	42	44	55	29	35	51	49	53	53	52	54	52	54	49	57	56		0.73	0.76	0.60	0.76	0.59	0.05	0.37	0.02	0.00
GBR	73	41	35	10	48	44	49	61	29	35	56	50	57	58	57	59	57	59	54	60	61	57		0.84	0.62	0.76	0.55	0.04	0.29	0.01	0.00
Pacific SW.	63	40	35	8	44	43	47	56	29	33	50	48	53	54	54	56	53	55	51	56	58	54	62		0.66	0.80	0.58	0.05	0.36	0.02	0.00
Pacific SSW.	45	31	27	7	36	32	36	44	23	28	39	36	38	38	38	39	40	38	40	40	41	40	45	43		0.61	0.57	0.07	0.41	0.02	0.00
Pacific SC.	61	40	33	9	46	42	43	52	27	33	51	48	52	54	51	53	53	51	48	53	55	53	58	55	40		0.63	0.05	0.37	0.02	0.00
Pacific SE.	43	29	23	5	35	30	33	40	20	26	34	30	34	34	33	35	35	34	34	37	37	39	41	39	32	40		0.07	0.46	0.02	0.00
Hawaii	3	3	3	1	3	3	3	3	2	3	3	3	3	3	3	3	3	3	3	3	3	3	3	3	3	3	3		0.13	0.33	0.00
Pacific FSE.	24	18	15	3	20	18	20	22	10	15	18	16	19	19	17	19	20	18	19	20	20	23	22	23	20	23	21	3		0.04	0.00
Pacific E.	1	1	1	1	1	1	1	1	1	1	1	1	1	1	1	1	1	1	1	1	1	1	1	1	1	1	1	1	1		0.00
Atlantic Ocean	3	0	0	0	0	0	0	0	0	0	0	0	0	0	0	0	0	0	0	0	0	0	0	0	0	0	0	0	0	0	
		Number of species shared between areas																													

Figure 3. Species counts and similarity indices (Jaccard's Index) for the Acropora species composition of areas throughout the world, compiled from the Worldwide *Acropora* Database at the Museum of Tropical Queensland. Locations of total species counts of 70 or more are highlighted in black. Results from areas within the Indonesian archipelago are highlighted in grey. Highest similarities for each of the areas within Indonesia are highlighted in black.

The areas with most species numbers within Indonesia (see Fig. 4) are 'Sulawesi' and 'Bay of Tomini', followed by 'Banda Sea' and 'Indonesia South'. These are located totally or mainly in 'Wallacea', the region between the Sunda and Sahul shelves (see Fig. 5). On present records the greatest species diversity (89) of *Acropora* worldwide occurs within Wallacea.

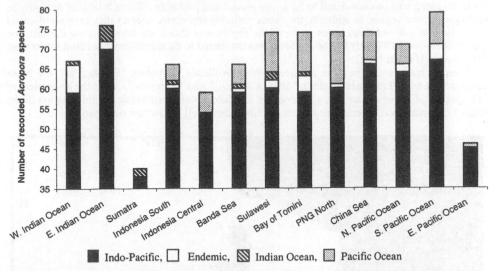

Figure 4. Numerical comparison of *Acropora* species composition in terms of distribution categories in regions of the Indo-Pacific and of the Indonesian archipelago.

What are the special characteristics of this region? The events which led to the docking of the Australian craton in the earliest Miocene (Audley-Charles 1987, Michaux 1991, Hall 1998) also created much of the archipelagic region of SE Asia and gradually closed the late Tethys seaway, creating a partial barrier between the two great ocean basins on each side. At the same time a great variety of habitats would have been created for the development of coral reefs. Following these events came the eustatic events of the Plio-Pleistocene. Low sea stands would have effectively obliterated reef growth on the shallow platforms of the Sunda and Sahul (Asian and Australian) shelves. The centre, however, would have retained its deep-sea channels and the calm, deep seas surrounding its semi-enclosed islands within broad bays such as the Bay of Tomini. During low sea stands, connectivity between reefs may have been less, but sites for coral growth, including sites with more than 60m of reef wall depth, would have remained.

4.2 *Similarity indices*

Most of the areas defined within Indonesia find their greatest similarity (in terms of species) with other areas within the archipelago. Exceptions to this are 'Sumatra', which is most similar to 'Andaman Sea' and 'Indian Ocean Central'; and 'Sulawesi' and 'Bay of Tomini', which are most similar to 'Papua New Guinea North' (Fig. 3). The fauna of southern and western Sumatra includes two species (*Acropora kosurini* and *A. rudis*) which are common on the eastern Asian coastline but not found elsewhere in Indonesia. A small group of islands, the Togian Islands in the Bay of Tomini have amongst their fauna at least six species shared only with New Britain (Papua New Guinea North), and in some cases also with Halmahera and islands in the Molucca Straits.

4.3 *Endemic species*

Two areas of Indonesia, "Sumatra" and "Indonesia Central", do not have endemic species (Fig. 4). These are also the only areas based fully on the Asian continental shelf (Fig. 5) and thus are

likely to consist of reefs that have developed following the last low sea stand. The very small
number of species overall found in "Sumatra" is undoubtedly an underestimate, because many
of the reefs on the Indian Ocean side are difficult to access and because little collecting has been
undertaken (further surveys will be made in 2001). "Indonesia Central", however, includes the
reefs around Java and has been well studied. The small number of species and lack of endemicity in this area can be considered to be a true record and probably reflects a lack of diversity in
reef types in the region. In addition, the fauna probably represents species that have recolonised
since the last low sea stand when much of the region was above sea level. During the last low
sea stand, around 18 000 B.P., the sea level was estimated to be approximately 120m lower than
the present (Hallam 1992).

Endemic species are found in all areas within Wallacea, including "Banda Sea", the Nusa
Tenggara chain ("Indonesia South"), "Bay of Tomini" and "Sulawesi". With the exception of
one species, *Acropora sukarnoi*, which occurs in shallow water just below the tidal region in the
Nusa Tengarra chain, all endemics occur in calm water, well below low tide mark.

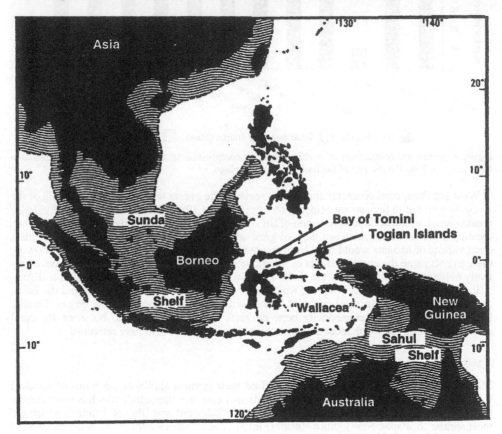

Figure 5. Demarcation of the 200m line for the Indo-Australian arc, indicating the "Wallacea" region
where a continuous presence of deeper water areas would have been maintained throughout the Cenozoic
(modified from Whitmore, 1975).

5 THE DUAL PATTERN OF DISTRIBUTION RE-ASSESSED

From a breakdown of distribution types within the archipelago, based on new data since Wallace (1997), it is clear that the Pacific component of the dual pattern is far stronger than the Indian Ocean component (Table 1). Species shared with the Pacific Ocean form a much larger

proportion of the overall Indonesian fauna than do species shared with the Indian Ocean (Fig. 6).

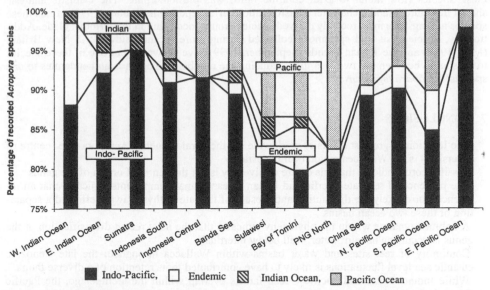

Figure 6. Percentage comparison of *Acropora* species composition in terms of distribution categories in regions of the Indo-Pacific and of the Indonesian archipelago.

The following hypotheses (not necessarily mutually exclusive) are proposed for this pattern:

– That species originating from the Pacific Ocean basin are colonising central Indonesia following the removal of a barrier. The Pacific through-flow current reported by Gordon and Fine (1996) would be the mechanism proposed for this. While this would account for common members of the Pacific fauna that occur more broadly in Wallacea, it does not account for an unusual fauna confined to the Togian Islands within the Bay of Tomini in Central Sulawesi (Wallace 1999a).

– That a 'central Indo-Pacific fauna' or a 'deep-water Indo-Pacific fauna' is retained both in north eastern Papua New Guinea and in parts of Wallacea, most notably the Togian Islands. These islands occur in very calm water, with depths of up to 2000m immediately off shore (Wallace 1999a) with an extensive array of embayments and channels around karst-formed islands. Paulay (1990, 1996) deduced from molluscan data that similar conditions led to the survival of lagoonal faunas in the archipelagos of the western Pacific during glacial sea level falls, whereas oceanic atolls with shallow lagoons lost this fauna. On the basis of new records obtained for a number of marine groups during the *Tethyana* expedition, he proposes this as one possible hypothesis for the origins of the unusual fauna of these islands (Paulay, pers. comm.). Interestingly, the *Acropora* fauna in this area includes at least 4 species that occur within a depth range from 30m to more than 60m depth. All records of these species to date have been from locations that are surrounded by deep water (more than 2000m), for example in northern Papua New Guinea and the South China Sea. It is proposed that these are sites where habitats suitable for these species would have persisted through time.

6 THE TOGIAN ISLANDS FAUNA REASSESSED

In a previous analysis, I proposed that the unusual fauna of the Togian Islands (located in the southern Bay of Tomini, Central Sulawesi) was due to the retention of a Tethyan relictual fauna

(Wallace 1999a). This was based on an indication that a number of endemic species occurred there, and an assumption (from a phylogeny based on a limited number of taxa) that a feature of these species (low radial to axial corallite ratio) was plesiomorphic. The cladistic analysis shown in Figure 2 (and a character transformation series reported in Wallace 1999b) finds these species amongst the most recently derived species groups, occurring in the terminal clade. Additionally, species lists and specimens collected during more recent fieldwork in New Britain (May, 1997) and the Togian Islands (September, 1999) have shown all but one of these species to be shared between the two localities, reducing the number of Togian island endemics to one species, *Acropora togianensis*.

7 CONCLUSIONS

~ The location of greatest diversity worldwide for the coral subgenus Acropora (the 'centre of diversity') is the Wallacea region of Indonesia.
~ Fossil records indicate that this centre of diversity is not the centre of origin of the genus.
~ The presence of separate Pacific and Indian Ocean faunal components indicates that an influence, most likely the diminution and closure of the late Tethys seaway, effected a separation of these two ocean basins.
~ A habitat signature within the phylogeny indicates that the most recent diversification of the genus is occurring in calm water, well below intertidal.
~ Continuity of reef sites and water basins within Wallacea throughout the late Cenozoic eustatic sea level fluctuations is likely to have contributed to its retention of a diverse fauna.
~ While Indian Ocean and Pacific Ocean faunas overlap within the archipelago, the Pacific Ocean influence is stronger.
~ This influence may be due to late Cenozoic retention of lagoonal faunas also retained in the western Pacific, or the introduction of Pacific-derived species via the Pacific through-flow, or both.
~ The Togian Islands fauna is most similar to that of New Britain, and the phylogeny indicates that this shared fauna is not relictual Tethyan, but rather more likely to be the remnants of a fauna derived within the Pacific Ocean or Central Indo-Pacific.

ACKNOWLEDGEMENTS

This research was funded by the Australian Research Council. I thank my colleagues on the *Tethyana* expedition, especially Gustav Paulay, Bert Hoeksema, Mark Erdman and Dave Bellwood, for discussions of the topic, Paul Muir and James True for assistance with development of the manuscript, Barry Goldman for biogeographic analyses, and Jackie Wolstenholme, Bronwen Scott and Denise Seabright for reviewing the text. Revision of the manuscript was greatly facilitated by advice from reviewers Brian Rosen and Moyra Wilson.

REFERENCES

Audley-Charles, M.G. 1987. Dispersal of Gondwanaland: Relevance to evolution of the Angiosperms. In T.C. Whitmore (ed.), *Biogeographical evolution of the Malay Archipelago*: 5-25. Oxford: Clarendon Press.

Briggs, J. C. 1992. The marine East Indies: Centre of origin? *Global Ecology and Biology Letters* 2: 149-156.

Briggs, J. C. 1993. Coincident biogeographic patterns: Indo-West Pacific Ocean. *Evolution* 53: 326-335.

Budd, A.F., Stemann, T.A. & Johnson, K.G. 1994. Stratigraphic distributions of genera and species of neogene to recent Caribbean reef corals. *Journal of Palaeontology* 68(5): 951-977.

Carbone, F., Matteucci, R., Pignatti, J. S., & Russo, A. 1994. Facies analysis and biostratigraphy of the Auradu Limestone Formation in the Berbera-Sheikh area, northwestern Somalia. *Geologica Romana* 29: 213-235.

Farris, J. 1988. *Hennig86 (Version 1.5)* Software manual distributed by the author. 41 Admiral Street, Port Jefferson Station 11776: New York.

Gordon, A.L. & Fine, R.A. 1996. Pathways of water between the Pacific and Indian Oceans. *Nature* 379: 146-149.

Hall, R. 1998. The plate tectonics of Cenozoic SE Asia and the distribution of land and sea. In R. Hall & J.D. Holloway (eds), *Biogeography and Geological Evolution of SE Asia.*: 99-131. Leiden: Backhuys.

Hallam, A. 1992. *Phanerozoic sea-level changes.* New York: Columbia University Press.

Michaux, B. 1991. Distributional patterns and tectonic development in Indonesia: Wallace reinterpreted. In P.Y. Ladiges, C.J. Humphries & L.W. Martinelli (eds), *Austral Biogeography*: 25-36. Melbourne: CSIRO.

Pandolfi, J.M. 1992. Successive isolation rather than evolutionary centres for the origination of Indo-Pacific reef corals. *J. Biogeography* 19:593-609.

Paulay, G. 1990. Effects of late Cenozoic sea-level fluctuations on the bivalve faunas of tropical oceanic islands. *Paleobiology* 16: 415-434.

Paulay, G. 1996. Dynamic clams: changes in the bivalve fauna of Pacific islands as a result of sea-level fluctuations. *American Malacological Bulletin* 12: 45-57.

Piccoli, G., Sartori, S., & Franchino, A. 1987. Benthic molluscs of shallow Tethys and their identity. In K.G. McKenzie (ed.), *Shallow Tethys* 2: 333. Rotterdam: Balkema.

Stehli, F.G. and Wells, J.W. 1971. Diversity and age patterns in hermatypic corals. *Systematic Zoology* 20:115-126.

Swofford, D.L.1991. *PAUP: phylogenetic analysis using parsimony, version 3.1.* Computer program distributed by the Illinois Natural History Survey, Champaign, Illinois.

Wallace, C.C. 1997. The Indo-Pacific centre of coral diversity re-examined at species level. *Proceedings of the 8th International Coral Reef Symposium. Panama,* 1: 365-370.

Wallace, C.C. 1999a. The Togian Islands: coral reefs with a unique coral fauna and an hypothesised Tethys Sea signature. *Coral Reefs* 18: 162.

Wallace, C.C. 1999b. *Staghorn Corals of the World. A revision of the genus* Acropora. Melbourne: CSIRO Publishing.

Wallace, C.C. & Scott, B.J. submitted. Phylogeny and species groupings within the large coral genus *Acropora* (Scleractinia; Astrocoeniinae; Acroporidae) based on skeletal structures *cladistics.*

Wallace, C.C. & Willis, B.L. 1994. Systematics of the coral genus *Acropora*: Implications of new biological findings for species concepts. *Annual Review of Ecology and Systematics* 25: 237-262.

Wallace, C.C, & Wolstenholme, J. 1998. Revision of the coral genus *Acropora* (Scleractinia: Astrocoenina: Acroporidae) in Indonesia. *Zoological Journal of the Linnean Society* 123(3): 199-384.

Whitmore, T.C. 1975. *Tropical rain forests of the Far East.* Oxford: Clarendon Press.

Wilson, E.J. & Rosen, B.R. 1998. Implications of paucity of corals in the Paleogene of SE Asia: Plate tectonics or centre of origin? In Hall, R. and D. Holloway (eds) *Biogeography and Geological Evolution of SE Asia*: 165-195. Lieden: Backbuys Publishers.

Section 4

Plant biogeography and evolution

Why are there so many Primitive Angiosperms in the Rain Forests of Asia-Australasia?

Robert J.Morley

PALYNOVA, Littleport, UK & Royal Holloway, University of London, U.K.

ABSTRACT: The rain forests of tropical and subtropical Asia-Australasia contain significantly greater concentrations of 'primitive' angiosperm families than do those either of the African continent or of the Neotropics. This concentration of primitive taxa has been interpreted as a reflection of their origin in this region, and has led evolutionary botanists to propose that angiosperms actually evolved 'somewhere between Assam and Fiji' (Takhtajan, 1969). A presumed origin in Asia-Australasia has also been supported by the overlap in that area of many family or subfamily pairs, such as the 'northern' Magnoliaceae and the 'southern' Winteraceae, or 'northern' Fagoideae and 'southern' *Nothofagus*. Current views based on the fossil record suggest that angiosperms initially radiated from Western Gondwana or SW Laurasia. This raises the question as to why so many primitive angiosperm families are currently concentrated in the area from NE Australia to southern China. The present paper tries to explain this anomaly by tracing the history of some primitive angiosperm families and some associated taxa by reference to the fossil record, and also in relation to the changing positions of tectonic plates, and changing global climates during the course of the Late Cretaceous and Tertiary. It is suggested that the concentrations of primitive angiosperms in this region are relict, and have nothing to do with angiosperm origins.

1 INTRODUCTION

Many of the early ideas on angiosperm origins were based on biogeography. Takhtajan (1969) brought to attention the concentration of members of angiosperm families which he considered to be 'living fossils' in the rain forests of Asia-Australasia, and on the basis of their representation in this area proposed that the region between 'Assam and Fiji' was the area of origin, or 'cradle' of the angiosperms. His suggestions received widespread support, for example from Thorne (1968, 1976) and Smith (1970).

Subsequent geological evidence revealed that Takhtajan's 'cradle' was a composite region formed by the collision of the Australian and Asian lithospheric plates (e.g. Hamilton, 1977, Hall, 1998), and also that it was very young, relative to the time of origin of angiosperms as revealed from their fossil record elsewhere (Hickey and Doyle, 1977). Undaunted by this, Takhtajan (1987) proposed that this area could still have been the birthplace of angiosperms, with their origin on a 'shard' or terrane of Gondwanan origin which broke from northern Australia during the Jurassic or early Cretaceous and subsequently became embedded within the Asian Plate in southern Asia or Indonesia. In making this proposal, he had in mind Darwin's suggestion (1875) made in a letter to Heer, that angiosperms 'must have largely developed in some isolated area whence owing to geographical changes, they at last succeeded in escaping, and spread quickly around the world'.

The final nail in the coffin with respect to Takhtajan's hypothesis came from the fossil pollen record, which shows much greater pollen differentiation in areas surrounding the Atlantic than have so far been observed anywhere in Asia-Australasia. But the question still remains as to why there are so many primitive angiosperm families concentrated in rain forests in the latter region, especially in southern China, and the region surrounding the Coral Sea. An allied question is why so many plant families (or family groups) occur in bihemispheric pairs, overlapping in the rain forests of Southeast Asia, again raising the question of their area of origin.

An answer to these questions became desirable during a recent attempt to reconstruct the geological history of tropical rain forests in Asia-Australasia (Morley, 2000). From consideration of the fossil record which can be attributed to primitive angiosperm families based on pollen and megafossils, and also by looking at the fossil record and biogeography of some additional families with which fossils of primitive angiosperms are associated, a hypothesis is proposed which appears to reasonably satisfactorily explain concentrations in Asia-Australasia. In doing so, some aspects of angiosperm history are clarified which were previously somewhat cloudy.

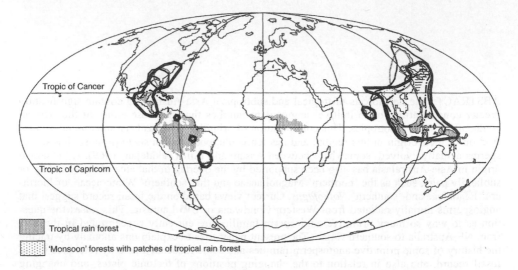

Figure 1 Present day distribution of the 'northern' primitive angiosperm families Calycanthaceae (dotted), Illiciaceae (dot/dash), Magnoliaceae (thick solid), Schisandraceae (thin solid), and Trochodendraceae (horizontal shading).

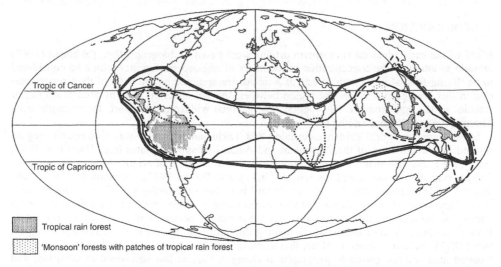

Figure 2 Present day distribution of the 'equatorial' primitive angiosperm families Annonaceae (thick solid), Canellaceae (dotted), Chloranthaceae (dashed) and Myristicaceae (thin solid).

To achieve this, modern distributions of 'primitive angiosperms', will be considered first. These can be divided into three groups as follows:
1- Northern group with Magnoliaceae, Calycanthaceae, Trochodendraceae/Tetracentraceae, Schisandraceae, Illiciaceae (Fig 1)
2- Equatorial group with Annonaceae, Chloranthaceae, Myristicaceae and Canellaceae (Fig 2)
3- Southern group with Amborellaceae, Austrobaileyaceae, Degeneriaceae, Eupomatiaceae, Himantandraceae, Hippuridaceae, Monimiaceae and Winteraceae (Fig 3).

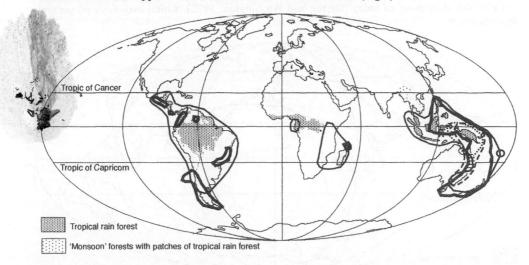

Figure 3 Present day distribution of the 'southern' primitive angiosperm families Amborellaceae, Austrobaileyaceae (horizontal shading), Degeneriaceae (circle), Eupomatiaceae (dashed), Himantandraceae (dotted), Monimiaceae (thin solid) and Winteraceae (thick solid).

Some members within each of these groups have a fossil record. Note that several members of the northern group have disjunct distributions, with occurrences in the Americas and Asia-Australasia, but absence from Europe and Africa. These taxa are members of van Steenis' (1962) amphi-Pacific element, which he proposed necessitated a transpacific land bridge, but which actually represent relict elements of the Boreotropical flora which extended across the northern mid-latitudes during the Late Cretaceous and Early Tertiary (Morley, 2000). The histories of several other amphi-Pacific taxa are used to build up the story, and also the history of Fagaceae (Fig 4) which features prominently in discussions of SE Asian historical biogeography (e.g. van Steenis, 1971; Whitmore, 1981).

The backdrop for this discussion is presented on a latitudinal-time profile, for the Tertiary and Cretaceous (Figs 5-7 and 9), which highlights the time of appearance of monosulcate and triaperturate pollen in the fossil record, and their poleward migration as propounded by Hickey and Doyle (1977). It also incorporates the latitudinal extension through time of moist megathermal forests, and the positions of the subtropical high pressure zones, which were unfavourable zones for rain forests throughout geological time (Morley, 2000), with the exception of narrow coastal zones on east coasts, as seen today in eastern Brazil, Madagascar and eastern Australia. This profile also utilises latitudinal vegetational reconstructions for Australia, with respect to the Southern Hemisphere, Southeast Asia for the equatorial zone, and North America for northern latitudes.

If we trace evidence for the appearance of moist forests during the Cretaceous, it becomes apparent that, although many groups of angiosperms first appeared in tropical latitudes, the first moist, angiosperm-dominated forests were in mid-latitudes, poleward of the sub-tropical high pressure zones (Morley, 2000). Most workers suggest that mid-Cretaceous low latitude climates were arid or semiarid, indicated, for instance by the presence of thick accumulations of salt in central Africa and South America in the area of the opening Atlantic Ocean, the dominance of pollen of Cheirolepidaceae (such as *Classopollis* and *Exesipollenites*, the parent plants of which

are widely believed to have been xerophytes: Vakhmareev, 1981; 1991), and the abundant occurrence and diversity of pollen of Ephedraceae, now restricted to arid locations (e.g. Herngreen
et al.. 1996). However, it is thought that a simple interpretation of widespread tropical aridity is
an oversimplification, for the latitudinal distributions of mid-Cretaceous pollen types demonstrate latitudinal zonations within the tropical zone which parallel the moisture belts suggested
by climate modelers (Morley, 2000). Also, many lithologies from these successions are often
inappropriate for an arid climate. Mid-Cretaceous climate models also propose warmer equatorial climates than those of today (Barron and Washington, 1985), which may have no present
day analogues.

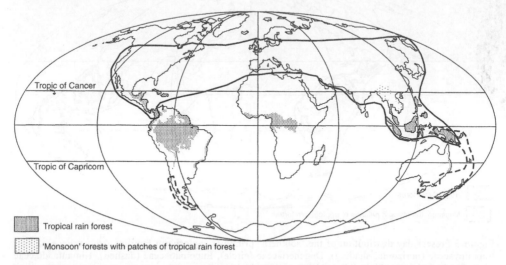

Figure 4 Present day distribution of the subfamilies of Fagaceae (Fagoideae solid, *Nothofagus* dotted)

To explain the numerous suggestions for equatorial aridity, a hot, sub-humid, probably asea-
sonal or possibly monsoonal equatorial climate is proposed. In any case conditions would have
been unfavourable for the development of rain forest. The earliest evidence for rain forest type
settings (based on leaf physiognomy) were in the mid-latitudes, in North America in the Cenomanian (Upchurch & Wolfe, 1987), and in the Southern Hemisphere in the Santonian (part of
the early Senonian), with the development of forests with *Nothofagus*.

2 FOSSIL RECORDS FOR 'PRIMITIVE' ANGIOSPERM FAMILIES

2.1 *Equatorial families*

Three of the four equatorially distributed families (Annonaceae, Chloranthaceae and Myristicaceae) have reasonable fossil records. There is no fossil record for Canellaceae.

The record for Chloranthaceae is based largely on occurrences of the pollen type *Clavatipollenites* (ref C1 on Fig 5) of which the earliest reports are from the Barremian or Hauterivian of
West Gondwanaland, with records from North America, England, Israel and Gabon at about the
same time. The close affinity of this taxon with Chloranthaceae is apparent not only because it
compares closely with pollen of *Ascarina*, but from its association with fossil floral structures,
such as *Couperites* (Friis and Pedersen, 1996). *Clavatipollenites* is first reported somewhat later
from Australia, in the Early Aptian or possibly latest Barremian (Burger, 1991; Dettmann,
1994), and even later in India in the Aptian (Srivastava, 1983). *Clavatipollenites* provides both
the oldest records of angiosperm pollen, and also the oldest record of an existing angiosperm
family. The representation of *Clavatipollenites* after the Cenomanian in the Northern Hemisphere is currently debated. Muller (1981) proposed the extinction of the parent plants produc-

ing *Clavatipollenites* pollen at the end of the Cenomanian (one of the least understood of the angiosperm extinctions), although Krutzsch (1989) suggests that the pollen type may well range into the Tertiary, a suggestion I support. Krutzsch based this suggestion on records of the form-genus *Emmapollis,* which is very similar to *Clavatipollenites,* but was initially thought to be gymnospermous. *Emmapollis* is recorded from the Late Cretaceous and Early Tertiary of Europe, with latest records in the Miocene. Reports of *Clavatipollenites* from the Tertiary of the Southern Hemisphere are much more clear-cut, and it is likely that members of Chloranthaceae were extensive in rain forests of Australia and New Zealand throughout the Tertiary, but became extinct in Australia with the expansion of Neogene aridity. Today, this family has a relict distribution in Central and South America, East Asia, Malesia and New Zealand.

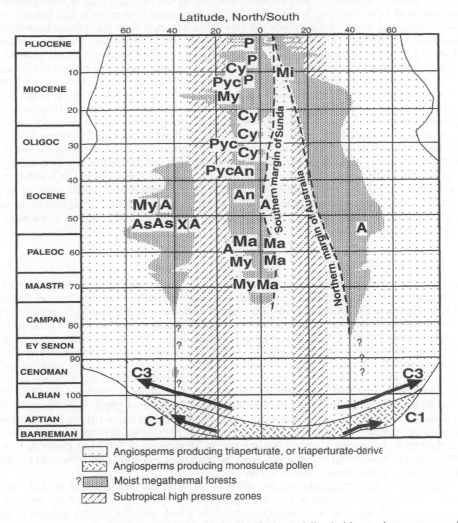

Figure 5 Time versus latitude reconstruction for fossils of 'equatorial' primitive angiosperms, superimposed on the pattern of appearance of the most basic angiosperm pollen types (after Hickey and Doyle, 1977) and the former latitudinal distributions of megathermal rain forest. The latitudinal distribution for the Southern Hemisphere megathermal rain forests is based on a reconstruction for E Australia, that for the low latitudes is based on Asia-Australasia (both from Morley 2000) whereas the Northern Hemisphere reconstruction is based on N American data. Annonaceae: A = Annonaceae undiff; An = *Annona* type; As = *Annonaspermum*; Cy = *Cymbalopetalum* type; Ma = *Malmea* type; Mi = *Milieusia* type; X = *Xylopiae-carpum*. Myristicaceae: My = Myristicaceae undiff; P = *Polyalthia*; Pyc = *Pycnanthus*. C1 = monosulcate; C3 = triaperturate.

Several other mid-Cretaceous pollen types of wide geographical distribution have also been referred to Chloranthaceae, emphasising the importance of this family, or its immediate precursor, during the initial phase of angiosperm radiation. *Asteropollis asteroides* compares with the pollen of modern *Hedyosmum*, and *Stephanocolpites fredricksburgensis* bears similarities with pollen of *Chloranthus*. Pollen of the latter type has been found in floral structures referable to *Chloranthistemon* (Crane *et al.*, 1994). Doyle & Hotton (1991) have also suggested that *Brenneripollis* (syn. *Peromonolites*) *peroreticulatus* was derived from an extinct sister taxon of Chloranthaceae.*Asteropollis asteroides* is reported from Australian Early Albian palynofloras, at which time the only Australian angiosperms appear to have been chloranthaceous (Dettmann, 1994).

The two families Annonaceae and Myristicaceae (Fig 5) are recorded mainly from equatorial localities, with myristicaceous wood being described from the Maastrichtian and Paleocene of North Africa, and pollen of the *Pycnanthus* type being common from the African Late Eocene onward (Jan du Chene *et al.*, 1978, Salard-Chebaldaeff, 1981). However, Walker and Walker (1984) suggest that some mid-Cretaceous pollen types, such as some specimens referred to *Clavatipollenites*, may be derived from Myristicaceae. Annonaceous pollen records are common in all three equatorial regions, with *Malmea* type being recorded as far back as the Maastrichtian in Colombia (Muller 1981). Annonaceous pollen is also recorded widely from the Late Eocene onward in Africa, with tetrads comparable to those of *Annona*, and from the Oligocene and Miocene onward in SE Asia with the *Cymbopetalum* type and *Polyalthia* respectively (Morley, unpublished). Records from outside the tropics are confined to the Late Paleocene/Early Eocene thermal maximum in Europe and N America (Collinson *et al.*, 1993) and the Paleocene of Argentina (Romero, 1986). It thus appears that these families are essentially of equatorial origin, but expanded to the mid-latitudes during periods of particularly warm climate. It is noteworthy that Walker (1971) observed more basic (anasulcate and catasulcate) pollen types in South American extant genera, and proposed the Neotropics, rather than Asia-Australasia, as the centre of origin for the family. Fewer generalisations can be made for Myristicaceae, since its fossil record is much more fragmentary.

2.2 Southern families

The family Winteraceae has a particularly noteworthy fossil record (Fig 6), with occurrences of pollen throughout the Australian Tertiary, and an earliest record of *Drimys*-type in Australia from the Campanian (Dettmann, 1994). The earliest occurrence on a global basis, however, predates the earliest Australian pollen type attributed to this family by 25 Ma, for the first record is from the Aptian of Israel (Walker, Brenner and Walker, 1983). The *Drimys* pollen type is highly distinctive, and cannot be confused with the pollen of any other extant taxon. However, in the equatorial mid-Cretaceous, there are several pollen types, which display some close morphological similarities with pollen of *Drimys*. *Walkeripollis* shows its tetrad and apertural arrangement, but possesses a finer reticulum, and because of this similarity Doyle *et al.* (1990) suggested that there might have been a close affinity between the parent plant of *Walkeripollis* and Winteraceae. *Afropollis* and *Schrankipollis* display similarities of exine structure and aperture, but occur as monads. Nevertheless, the parent plants of both *Afropollis* and *Schrankipollis* are also thought to have been allied to Winteraceae by Doyle *et al.* (1990) and Doyle & Hotton (1991). *Drimys*-type pollen has also been recorded from the Early and Middle Miocene of southernmost Africa by Coetzee and Muller (1984).

Winteraceae therefore appears to have radiated from West Gondwanaland, in the Late Barremian, and subsequently spread to both hemispheres, since there are records of *Afropollis* from North America, China and Australia. Only the clade which includes *Drimys* appears to have survived past the mid-Cenomanian in the Southern Hemisphere, and today it occurs in refugia in South America and Madagascar, and has its greatest representation in terms of both genera and species in the rain forests and related vegetation of New Guinea, New Caledonia and Australia.

The southern family Monimiaceae is well represented today in South America and Asia-Australasia but has few representatives in Africa. The likelihood that it was previously more widely represented in Africa is suggested from the occurrence of numerous fossil woods, for instance from the mid-Tertiary of Libya and Egypt, such as *Xymaloxylon zeltense* and *Atherospermoxylon aegypticum* (Boureau *et al.*, 1983).

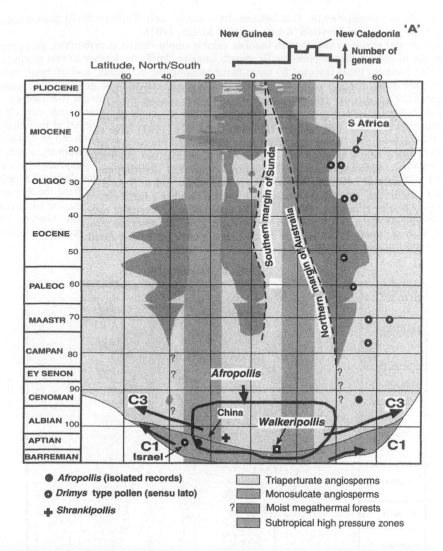

Figure 6 Time versus latitude for fossil Winteraceae, utilising same base as Fig 5. Graph 'A' shows the number of extant winteraceous genera within 4^0 latitudinal belts.

2.3 Northern families

In contrast to its southern sister family Winteraceae, Magnoliaceae has a wholly northern fossil record, with widespread occurrences of megafossils referable to *Magnolia* and *Liriodendron* from North America, Europe and East Asia (Collinson *et al.* 1993) and formed an important component within the Boreotropical Province. Its oldest definitive occurrences are from the Early Eocene. No pollen referable with certainty to Magnoliaceae has been recorded from Cretaceous sediments; *Lethomasites fossulatus* (Ward *et al.* 1989) from the Aptian of Maryland compares closely with some Magnoliaceae, but this pollen type is also found in some Annonaceae and *Degeneria* (Doyle and Hotton, 1991).

All other northern primitive angiosperms that are represented by fossils also appear to have had wholly Northern Hemisphere histories. For instance, Trochodendraceae, currently restricted to maritime East Asia, have widespread early Tertiary macrofossil records from Europe and North America (Crane *et al.*, 1991; Collinson *et al.*, 1993), and Illiciaceae/Schisandraceae, presently exhibiting amphi-Pacific distributions, have many early Tertiary fossil pollen records across the Boreotropical Province (Krutzsch, 1989; Muller, 1981).

Since several primitive angiosperm families exhibit amphi-Pacific distributions, an examination of the history of other amphi-Pacific genera helps to clarify the history of this group (Figs 7, 8). Van Steenis (1962) reported 80 megathermal amphi-Pacific genera, and of these, at least 11 are reported from fossil pollen or macrofossils (Morley, 2000). The genera *Cinnamomum* Lauraceae, *Engelhardtia* (Juglandaceae), *Gordonia* (Theaceae), *Meliosma* (Sabiaceae), *Microtropis* (Celastraceae), *Trigonobalanus* (Fagaceae), and *Saurauia* (Actinidiaceae) were clearly Boreotropical elements, and van der Hammen and Cleef (1983) have suggested that *Hedyosmum* (Chloranthaceae), which exhibits a Late Miocene to Pliocene fossil record in Colombia, may have had a similar origin. Only *Weinmannia* (Cunoniaceae) originated from the south, and *Spathiphyllum* (Araceae) has a poorly defined equatorial distribution. *Symplocos* (Symplocaceae) is clearly of Laurasian origin, but probably dispersed to S America during the earliest Tertiary (when there was a land connection), since it also has a fossil record from the Paleocene of South America (Romero, 1986; 1993). Its appearance in Australia may be due to dispersal from Southeast Asia, for Australian records are from the Miocene onward (Martin, 1994; Macphail *et al.* 1994). Unfortunately *Symplocos* does not yet have a fossil record in Southeast Asia, so this suggestion is speculative.

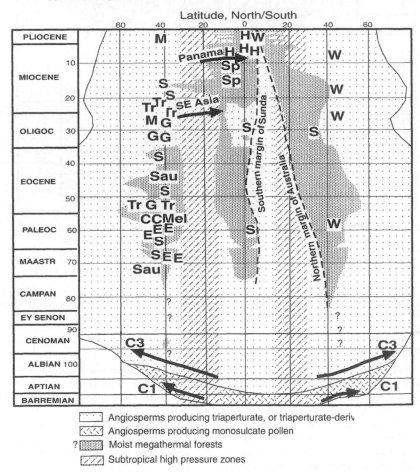

Figure 7 Time versus latitude for amphi-Pacific elements, using the same base as Fig 5. Northern elements: C = *Cinnamomum*; E = *Engelhardia*; G = *Gordonia*; M = *Microtropis*; H = *Hedyosmum*; Tr = *Trigonobalanus*; S = *Symplocos*; Sau = *Saurauia*; Mel = *Meliosma*. Equatorial elements; Sp = *Spathiphyllum*. Southern elements: W = *Weinmannia*.

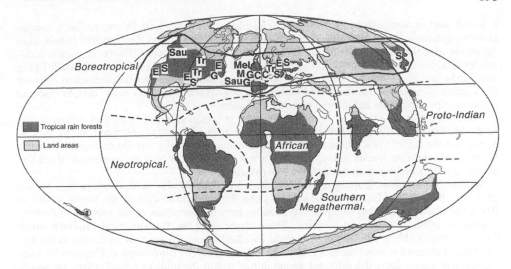

Figure 8 Fossil records for amphi-Pacific elements placed on Early Eocene plate tectonic reconstruction, and with Early Eocene distribution of megathermal rain forests, according to Morley (2000). Notation as for Fig 7.

2.4 Fagaceae

Oaks display a fossil distribution pattern which convincingly suggests wholly separate histories for northern Fagoideae and southern Nothofagoideae (Figs. 8, 9). These subfamilies were brought together at a late stage in their history in New Guinea following the northward dispersal of *Nothofagus* after a phase of Middle Miocene mountain building, and the presumed eastward migration of *Lithocarpus* and *Castanopsis*, either across Wallace's Line during the Neogene, or from SW Sulawesi, where these genera may have found refuge together with other Laurasian taxa, subsequent to the formation of the Makassar Straits in the Late Eocene (Morley, 2000). The fossil record argues strongly against Southeast Asia being the area of origin of oaks; as noted above, *Trigonobalanus*, which displays transitional characters between the two subfamilies, was a Boreotropical taxon. An origin during the mid-Cretaceous, probably from Betulalean ancestors (Jones, 1986) within the Neotropics is most likely.

3 DISCUSSION

Although primitive angiosperm families are more widely represented in rain forests of Asia-Australasia than in other tropical regions, their concentration in this region is due to a succession of historical factors during the Tertiary, and has nothing to with their area of initial radiation during the Cretaceous. Within Asia-Australasia, primitive angiosperm families are concentrated in two main areas: a northern concentration, centred on South China, along the northern margin of the Southeast Asian tropical rain forest block; and a southern concentration, centred around the Coral Sea. The two groups overlap in the region of Wallacea, and immediately surrounding areas.

The fossil record shows that these two groups have developed quite independently of each other since the initial radiation of angiosperms during the mid-Cretaceous, and have been brought into juxtaposition in Asia-Australasia as a result of the Tertiary northern drift of the Australian Plate, and its collision with the Philippine and Asian Plates during the Neogene, and then confined to the low latitudes as mid latitudes became cooler and drier. This pattern is particularly well illustrated by members of the family Fagaceae.

The southern Chinese concentration of primitive elements is relict from the Boreotropical rain forest flora, which had its origins during the mid-or Late Cretaceous within the northern latitudes, and its acme across the Northern Hemisphere during the Late Paleocene/Early Eocene climatic optimum. Many elements of this flora became extinct across the Northern Hemisphere following global climatic deterioration from the Oligocene to Pliocene. However, within Asia-Australasia, numerous Boreotropical elements found refuge in the rain forests of South China and Sunda (see also Tiffney, 1985), taking advantage of the permanent land connection between the equatorial zone and northern mid-latitudes, along which plant taxa were able to disperse between latitudinal zones unhindered by oceanic barriers.

With respect to the European Boreotropical flora, a combination of latitudinal barriers, such as the Mediterranean Sea, the Sahara Desert and the uplift of the Alps prevented southward dispersal into the African low latitudes. For this reason there are hardly any Boreotropical elements relict to African rain forests today, although a few mesothermal taxa are preserved in the Canaries, and the Colchic region of the Black Sea (Morley 2000).

Within the Americas, Boreotropical elements were probably able to find refuge to some degree after the terminal Eocene cooling event along the southern margin of the North American Plate, but could not disperse to equatorial latitudes until the formation of the Isthmus of Panama during the latest Miocene and Pliocene. Those elements of the Boreotropical Province which were eventually able to find refuge in Central and South America are now preserved as the amphi-Pacific element of van Steenis (1962). The concentration of Boreotropical elements in Asia-Australasia thus reflects the different opportunities within the mid- to Late Tertiary for southward dispersal from northern mid-latitudes to the tropical regions (Fig 10).

Equatorial families, such as Annonaceae and Myristicaceae, may have had roots in the mid-Cretaceous (e.g. Walker and Walker, 1984), but appear to have maintained a low latitude distribution throughout the Tertiary, although were able to expand their distributions to the Boreotropical Province and South America during the Early Tertiary climatic optimum. The reduced representation of primitive angiosperm families within the equatorial region compared to the more marginal tropical regions may reflect the somewhat later development of moist megathermal forests in equatorial latitudes (Campanian) compared to mid latitudes (Cenomanian to Turonian).

The earliest elements of the Southern Megathermal Province became established during the mid-Cretaceous, following dispersals from equatorial latitudes from the latest Barremian or Aptian onward, and as with the Boreotropical Province, were in place before the development of moist megathermal vegetation at equatorial latitudes. They did not originate in southern latitudes, but exhibited rapid radiation in that area during the Late Cretaceous (Dettmann, 1994). As with the Boreotropical flora, this flora reached its most extensive development during the Early Tertiary climatic optimum.

The preservation of primitive elements in the rain forests of the Southern Megathermal Province, in Australasia (and also Madagascar) relates to three factors. Firstly, these forests have each experienced continuity since their inception in the Late Cretaceous, and were at no time seriously threatened by the expansion of arid climates (which resulted in extinctions of southern primitive elements in continental Africa) or from invasions by elements of more aggressive megathermal floras, since these regions have been areas of plate disassembly, rather than collision, inhibiting dispersal of outside elements. Secondly, the Tertiary northward drift of the Australian Plate, Madagascar and their detached continental fragments, including New Caledonia, into warmer climatic zones resulted in these rain-forested areas more or less keeping pace with Tertiary global temperature depression (Fig 11). Thirdly, the isolation of fragments of the Southern Megathermal flora on oceanic islands, such as New Caledonia, sufficiently small to have had an oceanic climate even when located within the Southern Hemisphere sub-tropical high pressure zone, and away from other more aggressive floras, created a setting particularly favourable for the preservation of ancient angiosperm lineages.

It is this combination of processes which led to the widespread survival of archaic angiosperm families within NE Australia and on the continental fragments surrounding the Coral Sea, which brought Takhtajan (1969) to the conclusion that the 'cradle of the angiosperms' lay in Asia-Australasia, 'somewhere between Assam and Fiji'. Taking into account data from fossils and plate tectonics, it is more likely that Takhtajan's premise should be reversed - the most archaic angiosperms have in fact survived in those areas furthest from their place of origin in Western Gondwanaland, and thus brings new life to the old 'age and area' theory of Willis (1922) which proposed that the most primitive members of a lineage should occur furthest from their place of origin, having been displaced from their radiation centre by the appearance of subsequent more aggressive lineages.

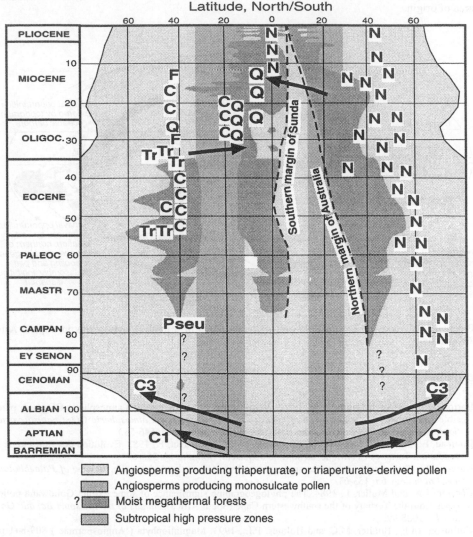

Figure 9 Time versus latitude for Fagaceae using the same base as Fig 5. For simplification, the many records from the northern mid-latitude Neogene are excluded. C = *Castanopsis/Lithocarpus*; F = *Fagus*; N = *Nothofagus*; Pseu = *Pseudofagacea*; Q = *Quercus*; Tr = *Trigonobalanus*.

4 CONCLUSIONS

Primitive angiosperms are concentrated 'between Assam and Fiji' because of:

1- the preferential survival of Boreotropical elements in SE Asian rain forests compared to Africa or the Neotropics;
2- the isolation of Southern Megathermal elements, particularly due to plate disassembly;
3- The northward drift of southern landmasses into warmer latitudes as global climates cooled down during the Neogene.

The concentration of primitive angiosperms in tropical Asia-Australasia is therefore due to a combination of the pattern of movement of the Earth's lithospheric plates during the course of the Tertiary, and to the pattern of global climate deterioration, and has nothing to do with their area of origin.

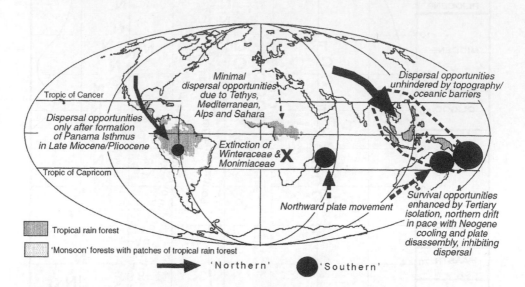

Figure 10 Scenario for the establishment of concentration of primitive angiosperms in tropical Asia-Australasia.

REFERENCES

Barron, E.J. and Washington, W.M., 1985. Warm Cretaceous climates: high atmospheric CO_2 as a plausible mechanism. In: Sundquist *et al.* (eds.), *The carbon cycle and atmospheric CO_2: natural variations, Archaean to the present*. American Geophysical Union 1985: 546-53.

Boureau, E., Salard-Chebaldaeff, M., Koniguer, J. –C., and Louvet, P., 1983. Evolution des flores et de la vegetation Tertiares en Afrique, au nord de l'equateur. *Bothalia* 14: 355-67.

Burger, D., 1991. Early Cretaceous angiosperms from Queensland, Australia. *Review of Palaeobotany and Palynology* 65: 153-63.

Coetzee, J.A. and Muller, J., 1984. The phytogeographic significance of some extinct Gondwana pollen types from the Tertiary of the southwestern Cape (South Africa). *Annals of the Missouri Botanic Garden* 71: 1088-99.

Collinson, M.E., Boulter, M.C. and Holmes, P.L., 1993. Magnoliophyta ('Angiospermae') 809-841 in: *The Fossil Record* 2. Benton, M.J. (ed.) Chapman and Hall, London, 845 pp.

Crane, P.R., Manchester, S.R. and Dilcher, D.L., 1991. Reproductive and vegetative structure of *Nordenskieldia* (Trochendraceae), a vesselless dicotyledon from the Early Tertiary of the Northern Hemisphere. *American Journal of Botany*, 78: 1131-34.

Crane, P.R., Friis, E.M. and Pedersen, K.R., 1994. Palaeobotanical evidence for the early radiation of magnoliid angiosperms. In: Endress, P.K. and Friis, E.M. (eds.) *Early evolution of flowers*. Plant Systematics and Evolution Supplement 8: 51-72.

Darwin C. 1875, quoted in letter to O. Heer, in Takhtajan, 1987.

Dettmann, M.E., 1994. Cretaceous vegetation, the microfossil record. In: Hill, R.S. (ed.), *History of Australian vegetation, Cretaceous to Recent:* 143-170. Cambridge University Press.

Doyle, J.A. and Hotton, C.L., 1991. Diversification of early angiosperm pollen in a cladistic context. In: Blackmore, S. and Barnes, S.H. (eds.) *Pollen and spores, patterns of diversification*. Systematics Association Special Volume 44: 169-195. Clarendon Press, Oxford.

Doyle, J.A., Hotton, C.L. and Ward, J., 1990. Early Cretaceous tetrads, zonasulculate pollen, and Winteraceae. 1: Taxonomy, morphology and ultrastructure. *American Journal of Botany* 77: 1544-67.

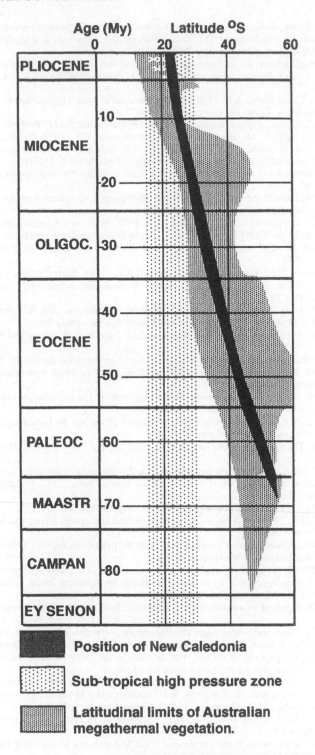

Figure 11 Latitudinal position of New Caledonia during the last 70 Ma superimposed on latitudinal representation of megathermal rain forest in E Australia over the same period.

Friis, E.M. and Pedersen, K.R., 1996. Angiosperm pollen in situ in Cretaceous reproductive organs. 409-426 in: Jansonius, J. and McGregor, D.C. (eds.), *Palynology: principles and applications*, Vol 1, 462 pp. American Association of Stratigraphic Palynologists Foundation.

Hall, R. 1998. The plate tectonics of Cenozoic SE Asia and the distribution of land and sea. In: Hall, R and Holloway, J. (eds), *Biogeography and geological evolution of SE Asia.* 99-124. Bakhuys Publishers, Amsterdam.

Van der Hammen, T. and Cleef, A.M., 1983. *Trigonobalanus* and the Tropical amphi-Pacific element in the north Andean forest. *Journal of Biogeography* 10, 437-40.

Hamilton, W., 1977. *Tectonics of the Indonesian region*. US Geological Survey Professional Paper 1078.

Herngreen, G.F.W., Kedves, M., Rovnina, L.V. and Smirnova, S.B., 1996. Cretaceous palynological provinces: a review. In: Jansonius, J. and McGregor, D.C. (eds.), *Palynology: Principles and Applications*. American Association of Stratigraphic Palynologists Foundation 3, 1157-88.

Hickey, L.J. and Doyle, J.A., 1977. Early Cretaceous fossil evidence for angiosperm evolution. *The Botanical Review,* 43: 1-183.

Krutzsch, W., 1989. Palaeogeography and historical phytogeography (paleochorology) in the Neophyticum. *Plant Systematics and Evolution* 162: 5-61.

Jan du Chene, R.E., Onyike, M.S. and Sowunmi, M.A. 1978. Some new Eocene pollen of the Ogwashi-Asaba Formation, South-West Nigeria. *Revista Espanola de Micropalaeontologia* 10: 285-322.

Jones, J.H., 1986 Evolution of Fagaceae: the implication of foliar features. *Annals of the Missouri Botanical Garden* 73: 228-75.

Macphail, M.K., Alley, N.F., Truswell, E.M. and Sluiter, R.K., 1994. Early Tertiary vegetation: evidence from spores and pollen. In: Hill, R.S. (ed.), *History of Australian vegetation, Cretaceous to Recent.* Cambridge University Press, 262-75.

Martin, H.H., 1994 Tertiary phytogeography, palynological evidence. In: Hill, R.S. (ed.), *History of Australian vegetation, Cretaceous to Recent.* 104-142. Cambridge University Press.

Morley, R.J., 2000. *Origin and Evolution of Tropical Rain Forests*. John Wiley and Sons, London, 362 pp.

Muller, J., 1981. Fossil pollen records of extant angiosperms. *The Botanical Review* 47: 1-142.

Romero, E.J., 1986. Palaeogene phytogeography and climatology of South America. *Annals of the Missouri Botanic Garden* 73: 449-61.

Romero, E.J., 1993. South American paleofloras. In: Goldblatt, P. (ed.), *Biological relationships between Africa and South America.* 62-85. Yale University Press.

Salard-Chebaldaeff, M., 1981. Palynologie maastrichtienne et tertiare du Cameroon: results botanique. *Review of Palaeobotany and Palynology* 28: 401-39.

Smith, A.C., 1970. The Pacific as a key to flowering plant history. *Univ. Hawaii, Harold L. Lyon Arboretum Lecture* 1.

Srivastava, S.K., 1983. Cretaceous geophytoprovinces and palaeogeography of the Indian Plate based on palynological data. In: Maheswari, H.K. (ed.), *Cretaceous of India.* 141-157. Indian Association of Palynostratigraphy, Lucknow.

Van Steenis, C.G.G.J., 1962. The land-bridge theory in botany. *Blumea,* 11: 235-372.

Van Steenis, C.G.G.J., 1971. *Nothofagus*, key genus in time and space, living and fossil, ecology and phylogeny. *Blumea* 19: 65-98.

Takhtajan, A., 1969. *Flowering plants, origin and dispersal* (transl. C. Jeffrey). Oliver and Boyd, Edinburgh, Smithsonian Institution, Washington, 300 pp.

Takhtajan, A., 1987. Flowering plant origin and dispersal: cradle of the angiosperms revisited. In: Whitmore, T.C. (ed.). *Biogeographical Evolution of the Malay Archipelago.* 26-31. Oxford Monographs on Biogeography 4, Oxford Scientific Publications.

Thorne, R.F., 1968. Synopsis of a putatively phylogenetic classification of the flowering plants. *Aliso* 6: 57-66.

Thorne, R.F., 1976. When and where might the tropical angiospermous flora have originated? In: Mabberley, D.J. and Chiang Kiaw Lan, (eds.), *Tropical Botany*. Gardens Bulletin, Singapore, 29, 183-189.

Tiffney B.H., 1985. Perspectives on the origin of the floristic similarity between eastern Asia and eastern North America. *Journal of the Arnold Arboretum* 66: 73-94.

Upchurch, G.R. and Wolfe, J.A. 1987. Mid-Cretaceous to Early Tertiary vegetation and climate: evidence from fossil leaves and woods. In: E.M.Friis, W.G.Chaloner and P.H.Crane (eds.), *The origins of angiosperms and their biological consequences.* 75-105. Cambridge University Press.

Vakhrameev, V.A., 1981. Pollen of *Classopollis*, indicator of Jurassic and Cretaceous climates. *Palaeobotanist,* 229: 301-7.

Vakhrameev, V.A., 1991. Jurassic and Cretaceous floras and climates of the Earth (translated by Ju. V. Litvinov). Cambridge University Press 318 pp.

Walker, J.W., 1971 Pollen morphology, phytogeography and phylogeny of the Annonaceae. *Contr. Gray Herbarium, Harvard University* 202: 1-132.

Walker, J.W. and Walker, A.G., 1984. Ultrastructure of Early Cretaceous angiosperm pollen and the origin and early evolution of flowering plants. *Annals of the Missouri Botanical Garden* 71: 464-521.

Walker, J.W., Brenner, G.J. and Walker, A.G., 1983. Winteraceous pollen in the Lower Cretaceous of Israel: early evidence of a magnolialian angiosperm family. *Science* 22: 1273-5.

Ward, J.V., Doyle, J.A. and Hotton, C.L., 1989. Probable granular magnoliid angiosperm pollen from the Early Cretaceous. *Pollen et Spores* 33: 101-20.

Whitmore, T.C. 1981. *Wallace's Line and plate tectonics*. Clarendon Press, Oxford, 90 pp.

Willis, J.C., 1922. *Age and Area*. Cambridge University Press.

Australian Paleogene vegetation and environments: evidence for palaeo–Gondwanan elements in the fossil records of Lauraceae and Proteaceae

A.J. Vadala & D.R. Greenwood

School of Life Sciences and Technology, Victoria University of Technology, PO Box 14428, Melbourne City MC, Victoria 8001, Australia

ABSTRACT: Tropical rainforests in the northeast of Australia have been interpreted as being either communities largely comprising taxa that 'invaded' newly available environments from Sundaland during and after the Middle Miocene collision of the Australian plate and the Sunda plate, or refugia for humid–mesothermal Gondwanan taxa. Recent biogeographic analyses have suggested four 'tracks' (areas of endemism) that potentially account for some previously hypothesised floristic 'elements' defined by 'tropical' or Malesian origins. Early Cenozoic (Paleogene) macrofloral records of Lauraceae and Proteaceae are informative on these issues. Unequivocal macrofossil evidence for Lauraceae and Proteaceae occurs at least from the Early Paleocene (~ 65 million years, Ma) in Australia. This evidence appears contrary to the suggestion of their past dispersal from Malesia to the Australian plate. Tropical floristic 'elements' defined by Malesian origins are not appropriate for elucidating current biogeographic patterns of these families in Australia.

1 INTRODUCTION

This paper reviews and provides macrofossil evidence supporting the ancient nature and Gondwanan origin of extant floristic elements of the Wet Tropics region of northeastern Australia. Traditional descriptive phytogeographic analyses of the extant Australian flora had identified three floristic elements, defined by hypothesised origin (*sensu* Crisp *et al.* 1999):
1) a Gondwanan element that comprises a rainforest flora with centres of diversity in the temperate south and humid tropical northeast and shares genera with, or has closely related genera in, other Austral landmasses;
2) an autochthonous element characterised by high endemism and represented by the sclerophyllous and dry-climate adapted vegetation of much of Australia (particularly the southwest of Western Australia);
3) a tropical element composed of taxa shared with southeast Asia, largely centred in the humid tropics and monsoonal tropics (Herbert 1932, 1967; Burbidge 1960; Barlow 1981; Schodde 1989; Crisp *et al.* 1999).
Rainforests of the Wet Tropics region of northeastern Queensland mainly comprise taxa belonging to the third of these elements. They have been considered as a vegetation type largely comprising taxa that 'invaded' newly available environments from Sundaland during and after the Miocene collision of the Australian and Eurasian plates. Alternatively, they have been regarded as refugia for humid mesothermal Gondwanan taxa. These narrative analyses have emphasised the role of either continental drift or long–distance dispersal of plant propagules in shaping the modern flora of Australia. However, more recent biogeographic analyses have stressed the 'autochthonous' character of much of the flora of the Wet Tropics (e.g. Webb *et al.* 1984; Webb *et al.* 1986; Truswell *et al.* 1987) that may reflect an ancient Gondwanan heritage that was also proposed by Barlow (1981). Crisp *et al.* (1999) concluded that the lack of success in identifying generalised biogeographic tracks (sharply differentiated areas of endemism)

within the Australian craton (Australia and New Guinea) may reflect the lack of significant barriers to dispersal and consequently possible range expansion for some taxa across Australia.

Analyses of Australian fossil microfloras have demonstrated that many of the 'tropical' elements of the extant flora of the Wet Tropics region were present in Australia prior to its final separation from the remainder of Gondwana (Truswell *et al.* 1987; Drinnan & Crane 1990). These floral elements were once presented as descendants of immigrant Malesian or tropical floral elements (see Crisp *et al.* 1999). Truswell *et al.* (1987) concluded that some exchange of taxa between the Australian craton and lands to the northwest of Australia had occurred, but had not resulted in any major alteration to the structure or compostion of Australian forests. Webb *et al.* (1984) also suggested that the Middle Miocene was the beginning of arid periods that would not have favoured the spread of rainforest immigrants.

The evidence presented here consists of a review of the Australian macrofossil record of Lauraceae and Proteaceae from the Paleogene (~ 65–23.3 million years (Ma): Fig. 3). New data from a Late Paleocene leaf macroflora from Cambalong Creek in the Southern Highlands of New South Wales are also presented. This macroflora may include the oldest known Australian taxa of tribes Laureae and Cryptocaryeae of Lauraceae and of tribes Banksieae, Oriteae, Stenocarpinae, Helicieae and Knightieae (Grevilleoideae) of Proteaceae. These data are used as supporting evidence for the presence of Lauraceae and Proteaceae in eastern Gondwana prior to separation of some Austral landmasses, and clearly precede the Miocene 'contact phase' (Hall 1996, 1997) between Australia and those parts of Malesia from which tropical floral elements may have 'invaded'. Southeastern Australia is important in the evolution of the modern flora due to the interactions between vegetation and the physical environment. The area underwent significant tectonism in the Paleogene during uplift of the Eastern Highlands (Wilford & Brown 1994). Coeval subsidence along a failed rift saw formation of the Gippsland Basin, and the area experienced major sea-level changes associated with initial Antarctic glaciation and the opening of Bass Strait as Australia and Antarctica rifted apart (Crook 1981; Kemp 1981; Powell *et al.* 1981; Blackburn & Sluiter 1994; Wilford & Brown 1994).

1.1 *The plant macrofossil record*

Evidence of the Cenozoic vegetation is abundant throughout southeastern Australia. Microfloras and macrofloras have been recorded from numerous localities spanning the Paleogene and Neogene (Carpenter *et al.* 1994; Christophel 1994; Macphail *et al.* 1994; Greenwood *et al.* in press, and references therein). Paleogene macrofloras in particular constitute a record of the vegetation of Australia prior to its final separation from East Antarctica and preceding the Miocene collision between the Australian craton and the Sunda Arcs.

The most detailed hypotheses regarding palaeovegetation and phytogeography for the Late Cretaceous and Cenozoic of southeastern Australia are currently based on palynology (e.g., Kemp 1981; Martin 1981, 1991, 1994, 1998; Truswell *et al.* 1987; Drinnan & Crane 1990; Truswell 1990, 1993; Kershaw *et al.* 1994; Macphail *et al.* 1994). These have been complemented by systematic taxonomic research on leaf macrofossils (e.g., Christophel 1981, 1989, 1994; Hill 1983, 1992*a* & *b*, 1994; Hill & Jordan 1993; Hill & Pole 1992; Hill & Carpenter 1991; Carpenter & Jordan 1997; Jordan *et al.* 1998; Hill *et al.* 1999). The Murray and Gippsland Basins provide virtually continuous sequences of Late Cretaceous to Pliocene-Pleistocene palynofloras and the highly detailed dinoflagellate, foraminiferal and sequence stratigraphy for these basins acts as an independent control for spore–pollen zonation (e.g. Holdgate & Sluiter 1991; Macphail *et al.* 1994). However, reconstructions of palaeovegetation based on palynology have significant limitations. For example, the family Lauraceae is a major component of southeastern Australian (particularly Victorian) Cenozoic macrofloras, yet it is absent from the palynological record because the thin sporopollenin exine of the pollen preserves poorly (Truswell *et al.* 1987; Drinnan *et al.* 1990; Martin 1994; Hill *et al.* 1999; although see Macphail 1980). Definitive cuticular morphological characters have been identified for and within many significant families in the fossil and extant floras of Australia (e.g., Hill 1986, 1990, 1991, 1992*b*, 1994; Hill & Carpenter 1991; Hill & Christophel 1988; Hill & Read 1991; Carpenter *et al.* 1994 and references therein; Christophel & Rowett 1996; Hill & Christophel 1996; Carpenter & Jordan 1997; Jordan *et al.* 1998). Such a high degree of taxonomic resolution (to generic level for instance) is rarely available from palaeopalynological analyses (Macphail *et al.* 1994).

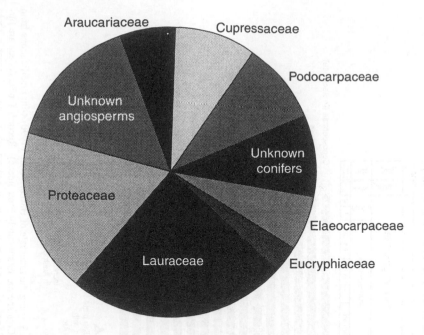

Figure 1: Composition of the Late Paleocene macroflora recovered from Cambalong Creek indicated as percentage of 173 total specimens.

Extant lineage	Australia RF	Australia non-RF	New Guinea	New Caledonia	New Zealand	South America	Malesia	SE Asia	India	Sri Lanka	Japan	Africa
Beilschmiedia	●		●		●	●	●	●	●			●
Cryptocarya	●		●			●	●	●				●
Endiandra	●	●	●				●	●				
Litsea	●	●	●		●	●	●	●			●	
Banksieae (Banksiinae)		●										
Embothrieae (Stenocarpinae)	●	●	●	●			●					
Helicieae (Hollandaeinae/Heliciinae)	●	●	●					●	●	●	●	
Knightieae (Knightiinae)	●											
Oriteae	●	●				●						

Figure 2: Extant distributions of closest living relatives of Late Paleocene Lauraceae and Proteaceae macrofossils recovered from Cambalong Creek. Grey circle for *Endiandra* indicates that some taxa grow in gallery and temperate forests rather than tropical rainforest. Grey circle for *Litsea* indicates that one taxon (*L. glutinosa*) grows throughout coastal northern Queensland, Northern Territory and the Kimberley region of Western Australia. Black circles for Helicieae indicate the distribution of *Helicia*, grey circles for this tribe indicate distribution of *Hollandaea*, a genus of two species both restricted to small areas in the Wet Tropics region of northeastern Queensland. Distribution shown for Knightieae is for *Darlingia*, a genus of two species also restricted to the Wet Tropics region of northeastern Queensland.

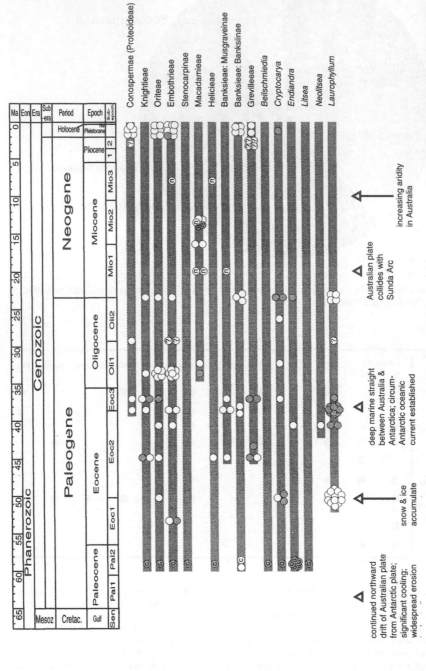

Figure 3: Stratigraphic ranges of Lauraceae and Proteaceae macrofossils listed in Tables 2 and 3, using the timescale of Harland et al. (1990) with modifications from Berggren and Aubry (1996). All Proteaceae tribes belong to subfamily Grevilleoideae except Conospermeae (subfamily Proteoideae). Records for Banksieae are further divided into subtribes. White circles correspond with published records in Tables 2 and 3; grey circles represent unpublished records and possible records (those marked by '?') in Tables 2 and 3. Circles with 'N' indicate New Zealand records; circles with 'C' indicate undescribed taxa from Cambalong Creek; circles with '?' indicate records from localities with an uncertain stratigraphic position.

1.2 Lauraceae and Proteaceae as targets for research

Proteaceae and Lauraceae are comparatively well represented temporally and spatially in Australian macrofloras, where oldest fossils are recorded from the early Paleogene (Fig. 3). Proteaceae and Lauraceae leaf macrofossils with cuticular preservation are particularly amenable to identification with high levels of taxonomic resolution due to relatively recent research into extant tribal and generic limits. The tribal taxonomy of extant Proteaceae was established with the detailed and extensive work of Johnson & Briggs (1963, 1975) and revised by Douglas (1995). General cuticular morphological characteristics of the family have been described from macrofloras spanning the Paleogene and Neogene (Blackburn 1985; Hill & Merrifield 1993; Carpenter & Pole 1995; Carpenter & Jordan 1997; Jordan et al. 1998). Cuticular features of fossil extant taxa have been described at subfamilial, generic, tribal and subtribal levels (e.g. Cookson & Duigan 1950; Lange 1978; Hill & Christophel 1988; Hill & Merrifield 1993; Carpenter 1994; Carpenter et al. 1994; Carpenter & Pole 1995; Jordan 1995; Carpenter & Jordan 1997; Jordan et al. 1998; Vadala & Drinnan 1998).

By contrast, intrafamilial relationships within Lauraceae are poorly understood (Eklund 1999). Suprageneric taxonomy of extant Lauraceae is less well established than for Proteaceae and consequently numerous suprageneric classifications for the family exist (see van der Werff & Richter 1996). Bandulksa (1926, 1928) recognized the strong similarities in cuticular morphology between some extant and fossil genera of Lauraceae. However, the lack of good cuticular morphological data for extant and fossil Australian genera of Lauraceae limited the usefulness of this record for many years (see Hill 1986). The published Cenozoic record of the family in Australia is substantial (Hill 1986, 1988a, b) and dates from the Early Eocene (Table 1 and Fig. 3). Abundant leaf impressions with ascending acrodromous venation led to the gradual development of the concept of a 'Cinnamomum flora' (reviewed by Duigan 1951). Most of the early identifications of these Paleogene and Neogene leaves with Cinnamomum were based only on gross leaf morphology. However, most of the earliest described leaves could not even be accepted as Lauraceae without cuticular morphological evidence (Hill 1988a). Hill (1986, 1988a) reassigned some of these taxa with adequate cuticular preservation to the genus Laurophyllum and the concept of the 'Cinnamomum flora' was gradually rejected. Nevertheless, the fossil record of Lauraceae is impressive compared with that of other taxa that are significant in the modern flora of Australia, such as Acacia, Eucalyptus and Casuarina.

Many Paleogene and Neogene Lauraceae from Australia have been described as Laurophyllum (e.g. Hill 1996; Carpenter & Pole 1995), a genus indicating only general affinity with Lauraceae and limiting the systematic or phylogentic use of the record. Subsequent research has demonstrated that leaf venation and shape (Christophel & Hyland 1993) and cuticular morphology (Christophel & Rowett 1996) can be used as reliable characters in the taxonomy of extant Australian Lauraceae. The potential utility of the Australian macrofossil record of Lauraceae has been increased since the first critical work on the Australian taxa (Hill 1986) by refinement of generic and suprageneric concepts in Lauraceae. This has been based on a wide suite of characters (van der Werff & Richter 1996) including cuticular morphology (Christophel et al. 1996) and foliar morphology (Klucking 1987; Cristophel & Hyland 1993), and a major revision of the arborescent Australian genera of the family (Hyland 1989).

Table 1. Published records of Paleogene and Neogene Lauraceae macrofossils from Australia. The table does not include some fossils of possible lauraceous affinity that were included in the survey of Hill (1988b) and which were described without cuticular detail on the basis of leaf morphology only. Specimens listed with extant affinity as '?Lauraceae' should be considered doubtful (see text).

Locality (Age)	Fossil taxon	Extant affinity
Cobungra River[9] (?Early Eocene)	Cinnamomum polymorphoides[9]	?Lauraceae
Nerriga (Early/Middle Eocene)[20]	Laurophyllum acrocryptocaryoides[20]	Cryptocarya (C. bellendenkerana, C. grandis)[20]
Nerriga (Early/Middle Eocene)[21]	Laurophyllum acrodromum L. conspicuum	?Endiandra: E. pubens group*
	L. acuminatum	?Endiandra : E. jonesii group*

	L. acutum	?*Endiandra* : *E. jonesii* or *E. pubens* group*
	L. angulosum	?*Cryptocarya* : *C. pleurosperma* group*
	L. squamatum	?*Cryptocarya* : *C. pleurosperma* group*
	L. lanceolatum	?*Neolitsea dealbata* or *Litsea* : *L. fawcettiana* group*
	L. brochidodromum	?*Endiandra* : *E. jonesii* group*
	L. intramarginatum	?*Endiandra* : *E. jonesii* group*
	L. sinuatum	
	L. pubescens	?*Endiandra* : *E. jonesii* group*
	L. arcuatum	?*Endiandra* : *E. pubens* group*
Nelly Creek (Middle Eocene)	Parataxon 5[3]	Lauraceae[6]
Anglesea clay lenses (late Middle Eocene)	Parataxon 4[16]	Lauraceae[16]
Anglesea clay lenses (late Middle Eocene)	Lauraceae I[17]	*Endiandra muelleri*, ?*Litsea*[17]
	Lauraceae II[17]	*Neolitsea dealbata*[17]
	Lauraceae III[17]	?*Cinnamomum*/?*Cryptocarya*[17]
	Lauraceae IV[17]	
Lefroy paleodrainage (Middle/Late Eocene)[19]	CUT-L-001[19]	?*Litsea bennettii* group*
	CUT-L-002[19]	
	CUT-L-003[19]	?*Endiandra*: *E. pubens* group*
	CUT-L-004[19]	?*Endiandra*: *E. pubens* group*
	CUT-L-005[19]	?*Endiandra*: *E. jonesii* group*
	CUT-L-006[19]	?*Lindera**
	CUT-L-007[19]	?*Lindera**
	CUT-L-008[19]	?*Endiandra*: *E. pubens* group*
	CUT-L-009[19]	?*Endiandra*: *E. pubens* group*
	CUT-L-010[19]	
Hasties (Late Eocene)	*Laurophyllum* cf. *L. arcuatum*[7]	*Laurophyllum arcuatum* (also *L. acuminatum*)[7]
Jungle Creek (Late Eocene)	*Cryptocaryoxylon gippslandicum*[18]	*Cryptocarya oblata*[18]
Kojonup Sandstone (Late Eocene)	cf. Lauraceae[5]	Lauraceae[5]
Pallinup Siltstone (Late Eocene)	WAM.P88.21[5]	Lauraceae[5]
Vegetable Creek (Late Eocene)	*Cinnamomum nuytsii*[23] (= *Laurophyllum nuytsii*[22])	?Lauraceae
Golden Grove (Eocene)[2]	*Cinnamomum polymorphoides*[9]	?Lauraceae
	Acrodromous primary venation taxon[2]	?Lauraceae
	Pinnate primary venation taxon[2]	
Pascoe Vale (Late Eocene/Early Oligocene)	*Cinnamomum* sp.[15]	?Lauraceae
Sedan Coalfield (Oligocene – Micoene or Late Eocene - Miocene)	Parataxon AA 007: 'aff. *Endiandra*'[1]	?*Endiandra*[1]
	Parataxon AG 005: 'aff. *Cryptocarya*'[1]	?*Cryptocarya*[1]
	Parataxon AA 006: 'aff. *Cryptocarya*'[1]	?*Cryptocarya*[1]
Narracan (Early Oligocene)	*Cinnamomum polymorphoides*[9]	?Lauraceae
	C. polymorphoides[9, 12]	fossil *Cinnamomum burmanni*[12]
	Cryptocarya australis[12]	*Cryptocarya australis*/*C. murrayi*/*C. mackinnoniana*[12]
Dalton (Late Oligocene)	*Cinnamomum polymorphoides*[9]	?Lauraceae
Morwell Open Cut (Late Oligocene)	*Cinnamomum praevirens*[11]	*Cinnamomum virens*, *C. oliveri*[11]
Darlimurla (Late Oligocene)	*Cryptocarya praeobovata*[12]	*Cryptocarya obovata*[12]

Tambellup Siltstone (?Oligocene)	*'Laurophyllum'*[5]	
West Dale (?Oligocene)	*Laurophyllum striatum*[4]	*Laurophyllum arcuatum/L. brochidodromum*[4]
Berwick Quarry (Late Oligocene/ Early Miocene)	*Laurophyllum* sp. 'sinuous'[14] *Laurophyllum* sp. 'thick'[14] *Laurophyllum* sp. 'butterfly'[14] *Laurophyllum* sp. 'smooth'[14]	
Newstead (= Elsmore); (Oligocene/Miocene)	*Cinnamomum leichhardtii*[9]	?Lauraccae
Maddingley (Miocene)	*Cinnamomum polymorphoides*[9,13]	*Cinnamomum polymorphum*[13]
Pitfield (Miocene)	*Cinnamomum polymorphoides*[9]	?Lauraceae
Werribee Ck./Lyalls Ck. (Miocene)	*Cinnamomum polymorphoides*[9,13]	*Cinnamomum polymorphum*[13]
Werribee Ck./Lyalls Ck. (Miocene)	*Laurus werribeensis*[13]	?Lauraceae
Regatta Point (Early/?Middle Pleistocene)	*Laurophyllum australum*[6]	*Cryptocarya novae-anglica/C.* sp.nov (Mt. Bellenden Ker, Queensland)[6]
Mount Bischoff, Tasmania	*Laurus sprentii*[8]	?Lauraceae
Travertine lake deposits near Hobart	*Cinnamomum woodwardi*[10]	?Lauraceae
William Creek, South Australia	*Cinnamomum* sp.[10]	?Lauraceae

*Possible extant affinities of *Laurophyllum* specimens from the Lefroy Palaeodrainage and Nerriga were produced by keying parataxa using the key to extant Australian genera of Christophel & Rowett (1996) where possible from the descriptions and illustrations provided in Carpenter & Pole (1995) and Hill (1986).
[1]Rowett (1991); [2]Christophel & Greenwood (1987); [3]Christophel, Scriven & Greenwood (1992); [4]Hill & Merrifield (1993); [5]McLoughlin & Hill (1996); [6]Jordan (1997); [7]Pole (1992b); [8]Johnston (1886); [9]Chapman (1926); [10]Chapman (1921); [11]Deane (1925); [12]Paterson (1935); [13]McCoy (1876); [14]Pole *et al.* (1993); [15]Douglas (1967); [16]Rowett & Christophel (1990); [17]Christophel, Harris & Syber (1987); [18]Leisman (1986); [19]Carpenter & Pole (1995); [20]Conran & Christophel (1998); [21]Hill (1986); [22]Hill (1988a); [23]Ettingshausen (1888).

2 MACROFOSSIL EVIDENCE FOR LAURACEAE AND PROTEACEAE

2.1 *Lauraceae – Laurasian records*

Recent phylogenies based on DNA sequence analyses have emphasised the antiquity of Laurales (Qiu *et al.* 1999). Drinnan & Crane (1990) indicated that Lauraceae had differentiated early in angiosperm evolution, likely by the Albian, ~ 110 Ma. Indeed, fossils of definite lauraceous affinity are known from the early Cenomanian (~ 97 Ma: Drinnan *et al.* 1990; Eklund & Kvacek 1998). Monosulcate pollen typical of the magnoliid dicotyledon (including Laurales, Winterales and Chloranthaceae) and monocotyledon grade first appear in the fossil record around the Hauterivian (Hughes & McDougall 1987), preceding the appearance of triaperturate pollen typical of the non–magnoliid dicotyledon clade (Crane 1987; McLoughlin *et al.* 1995).

Supporting this antiquity is a good mid–Cretaceous macrofossil record of Lauraceae (Eklund & Kvacek 1998 and references therein). Early Cenomanian (~ 97 Ma) inflorescences and flowers of *Mauldinia mirabilis* were described from the Potomac Group, eastern North America (Drinnan *et al.* 1990), and Cenomanian *M. bohemica* inflorescences are known from the Peruc-Korycany Formation in the Czeck Republic (Eklund & Kvacek 1998). Macrofossils of Lauraceae are common across middle and low palaeolatitudes in the Northern Hemisphere in the Maastrichtian Northern Gondwana and *Normapolles* Provinces (Crane 1987). Pole (1992a) suggested lauraceous affinity for several leaves with pinnate margins, acrodromous primary venation and percurrent secondary venation from the Upper Cretaceous Taratu Formation, Otago, New Zealand. The Late Cretaceous (Santonian/ Campanian; ~ 83 Ma) coincided with increased seafloor spreading between Australia and Antarctica (Veevers *et al.* 1991), separation of India from Australia (Powell *et al.* 1981; Wilford & Brown 1994), and initial spreading to form the Tasman Sea (Crook 1981; Veevers *et al.* 1991). Audley-Charles (1987) suggested that dispersal of land plants between mainland Asia and Australia would have been possible around that time (~ 90 Ma). However, more recent tectonic reconstructions of the region indicate several thousand kilometres still separated the Eurasian continental margin and the leading edge of the Australian craton at the Early/Middle Eocene, ~ 50 Ma (Hall 1996, 1997).

Diversity within the Lauraceae persisted in the Paleogene with leaves, wood and reproductive structures being abundant and diverse in macrofloras from most parts of the world (Eklund & Kvacek 1998). Indeed, Paleocene macrofloras spanning northern America, central and eastern Europe (the Southern Laurasian floristic province of the Cenomanian and *Normapolles* Province of the Santonian–Campanian) are typically dominated by palms, Euphorbiaceae and Laurales (Crane 1987). The London Clay flora contains eight taxa of *Beilschmiedia* (some possibly attributable to the closely related genus *Endiandra*), two taxa of *Cinnamomum*, one taxon of *Litsea*, one taxon of *Crowella*, five taxa of *Laurocalyx* and 30 taxa of *Laurocarpum* (Chandler 1964). This indicates the existence of two of the three extant tribes of Lauraceae recognised by van der Werff & Richter (1996): Perseeae (as *Cinnamomum*) and Cryptocaryeae (as *Beilschmiedia*, *Cryptocarya* and *Endiandra*) in the North Atlantic/European floristic Province during the Paleocene (*sensu* Crane 1986). Paleocene Lauraceae leaves are also represented in macrofloras preserved in nine other Lower Eocene sedimentary beds from southern England (Chandler 1964), and were a dominant component of the vegetation preserved in both the Bournemouth Beds (Bandulska 1928) and London Clay (Chandler 1964). Exact generic relationships of Lauraceae leaf macrofossils from the Eocene London and Bournemouth Clay floras and from Late Cretaceous and Early Paleogene localities in the Unites Stated are uncertain beyond the more general '*Cinnamomum*' and '*Laurophyllum*' types. The presence of Lauraceae macrofossils in these areas strongly suggest that Lauraceae was a prominent component of early angiosperm plant communities in Laurasia, and substantially predates the earliest known records of the family from Gondwana.

2.2 Lauraceae – Australian records

Published Australian macrofossil records of Lauraceae extend from the Early Eocene (Table 1 and Fig. 3). This extensive record connotes the prominence of Lauraceae in the eastern Australian sector of Gondwana before the formation of a deep marine strait between Tasmania and Antarctica in the early Late Eocene/Early Oligocene and the ensuing development of Circum–Antarctic oceanic circulation (Kemp 1981; Martin 1991; Veevers *et al.* 1991; Wilford & Brown 1994). These events clearly predate the 'contact phase' between the Australian craton and the Eurasian plate in the Miocene (Powell *et al.* 1981; Truswell *et al.* 1987; Metcalfe 1990; Hall 1996, 1997: Fig. 3).

Many of the older published records in Table 1 should be considered doubtful. For example Early Eocene and Late Oligocene/Miocene records of *Cinnamomum* (McCoy 1876; Chapman 1921, 1926; Deane 1925), *Cryptocarya* (Paterson 1935), and *Laurus* (McCoy 1876) were made on the basis of leaf morphology, and lack the cuticular morphological information needed for definite placement in Lauraceae (Hill 1988a). Published macrofossils with cuticular preservation enabling secure assignment to Lauraceae date from the Early/Middle Eocene, with Hill (1986) and Conran & Christophel (1998) describing a total of 13 taxa of *Laurophyllum* from the Nerriga locality in New South Wales (Table 1 and Fig. 3). Conran & Christophel (1998) indicated the fossil taxon *L. acrocryptocaryoides* from Nerriga had a combination of cuticular morphological characters such as wide, butterfly-shaped cuticular scales and rounded epidermal cells (Conran & Christophel 1988, figs. 2C, 2D) characteristic of extant *Cryptocarya* (Christophel & Rowett 1996). This reiterates the presence of Lauraceae similar to extant Tribe Cryptocaryeae in southeastern Australia in the Paleogene (early Middle Eocene; ~50 Ma).

Hill (1986) described 12 taxa of Lauraceae from Nerriga on the basis of micromorphological characters and assigned all to *Laurophyllum* (Table 1). The taxa described by Hill (1986) may include *Endiandra*, *Cryptocarya* and either *Neolitsea* or *Litsea* using the key of Christophel & Rowett (1996; Table 1). *Laurophyllum acrodromum*, *L. brochidodromum*, *L. intramarginatum*, *L. acuminatum*, *L. acutum* and *L. arcuatum* descibed by Hill (1986) all appear to have combinations of micromorphological characters typical of extant *Endiandra* (Christophel & Rowett 1996). These include epidermal cells with irregularly thickened and/or granulate periclinal walls and cuticular scales that appear double (Hill 1986, figs. 7C and 7E, 13C and 13E, 14D, 9C, 10C, 17C). *L. acrodromum*, *L. brochidodromum*, *L. intramarginatum* and *L. arcuatum* also have guard cells with polar extensions or rods (Hill 1996, figs. 7E, 13C, 14D, 17D) while *L. acuminatum*, *L. acutum* and *L. intramarginatum* have mostly angular adaxial cell wall outlines (Hill 1986, figs. 9B, 10A, 14E). These characters are also typical of extant *Endiandra* in combination with the other characters described above (Christophel & Rowett 1996). The fossil described as *L. lanceolatum* by Hill (1986) has cuticular ledges that appear single, thin and mainly straight (Hill 1986, figs. 9G and 9H), which are typical of extant *Litsea* (Christophel & Rowett 1996). *L. angulosum* and *L. squamatum* from Nerriga appear to have wide, butterfly-like cuticular scales (Hill 1986, figs. 11F, 12E). This implies a close relationship with extant *Cryptocarya* (Christophel & Rowett 1996). These Nerriga fossils suggest the presence in southeastern Australia of Tribes Laureae (as *Litsea*) and Cryptocaryeae (as *Cryptocarya* and *Endiandra*; van der Werff & Richter 1996) in the Early/Middle Eocene (~ 50 Ma).

Lauraceae also feature prominently in the Middle/Late Eocene (~ 50–35 Ma) of southern Australia (Table 1), at which time there was a substantial seaway in the Indian Ocean, Southern Ocean and Tasman Sea (Fig. 3; Veevers *et al.* 1991), and the Australian craton was still distant from Sundaland and the Eurasian plate (Hall 1996, 1997). Christophel *et al.* (1987) described four late Middle Eocene taxa of Lauraceae from the Anglesea clay lenses in southern Australia (Victoria). These macrofossils have strong affinites with extant *Endiandra/Litsea*, *Neolitsea* and *Cinnamomum/Crytpocarya* (Christophel *et al.* 1987), reflecting a similar range of diversity in Lauraceae as may have been present earlier at Nerriga.

Carpenter & Pole (1995) described ten taxa of *Laurophyllum* from the Middle/Late Eocene Lefroy Palaeodrainage (Pidinga Formation) in Western Australia, some of which may reperesent *Litsea*, *Lindera* and *Endiandra* (Table 1). Four of the fossil cuticle types described by Carpenter & Pole (1995) have combinations of characters including 'double' cuticular scales (Carpenter & Pole 1995, figs. 25, 37-38, 40, 42-44, 54, 56) typical of extant *Endiandra* (Christophel & Rowett 1996). Fossil type 1 has a combination of characters including granular periclinal walls; single, thin and straight cuticular ledges and prominent abaxial papillae (Carpenter & Pole 1995, figs. 24, 28, 25) that are typical of extant *Litsea* (Christophel & Rowett 1996). Fossil cuticle type 7 has highly sinuous abaxial cell outlines and heavily 'beaded' abaxial anticlinal walls (Carpenter & Pole, 1995 figs. 47, 45) that are typical of the extant genus *Lindera* (Christophel & Rowett 1996).

Lauraceae macrofosssils from the Late Paleocene (~ 58–60 Ma) at Cambalong Creek (Taylor *et al.* 1990) grew at a time when Australia had already been separated from India and New Zealand by a long period of development of the southeast Indian Ocean and Tasman Sea (Powell *et al.* 1981; Wilford & Brown 1994), although Australia and Antarctica were still partially joined (Veevers *et al.* 1991). Australia was also still distant from Sundaland, the continental block forming southeast Asia, at approximately 60 Ma (Powell *et al.* 1981; Hall 1996, 1997).

Lauraceous taxa from the Cambalong Creek macroflora will be formally described in a future publication (Vadala & Drinnan, in prep.). Lauraceae comprise approximately 20% of total taxa in the macroflora, as eight species in four genera (Fig. 1: A.J. Vadala, unpubl.). These fossils predate all published records of Lauraceae from Australia (Table 1 and Fig. 3) and have been identified as extant genera using morphological characters preserved by the mummified leaf cuticles. Primary classification and sorting of fossil specimens was achieved by running the fossil specimens through the key to Australian genera of Lauraceae developed by Christophel & Rowett (1996). This identified fossils belonging to *Beilschmiedia*, *Cryptocarya*, *Endiandra* and *Litsea*. Steps in the key were used to derive a binary (qualitative) character set (28-33 characters) and a continuous (numerical) character set (7 characters) for the fossils specimens and for 25 extant Australian species of *Beilschmiedia*, *Endiandra* and *Cryptocarya*. The two character sets were used in conjunction and separately in pattern analyses for the three genera to determine further taxonomic divisions between specimens. These analyses applied sequential-agglomerative-heirarchical-combinatorial strategies using dissimilarity metrics (Belbin 1987) to the data sets.

The Cambalong Creek Lauraceae indicate the presence in southeastern Australia of taxa similar to the extant tribes Laureae (*sensu* van der werff & Richter 1996; as *Litsea*) and Cryptocaryeae (as *Beilschmiedia*, *Cryptocarya* and *Endiandra*) before the breakup of the last fragments of Gondwana (Powell *et al.* 1981; Veevers *et al.* 1991). This precedes the Miocene contact between the Australian continent and Sundaland by ~ 40 Ma (Fig. 3; Powell *et al.* 1981; Truswell *et al.* 1987; Metcalfe 1990). The extant distributions of the nearest living relatives of these taxa are shown in Fig. 2 and Table 3.

Extant *Endiandra* consists of approximately 100 species (Table 3; Hyland 1989) occurring from Australia through New Guinea to Malesia and across broader southeast Asia (Fig. 2). Most of the Australian species grow in rainforest (Table 3). *Beilschmiedia* consists of 200-250 extant species (Table 3; Hyland 1989) found in Africa, Australia, South America, New Zealand and New Guinea, through Malesia and broader southeast Asia to India (Fig. 2). The 11 extant Australian taxa of *Beilschmiedia* are all restricted to rainforest habitats (Hyland 1989: Table 3). *Cryptocarya* consists of 200-250 extant species (Hyland 1989) found in Australia, South America, Africa, New Guinea, Malesia and broader southeast Asia (Fig. 2). All 46 Australian species are restricted to rainforest (Fig. 2 and Table 3; Hyland 1989), with habitats varying from north Queensland seasonal rainforests with *Agathis* to drier rainforests of northern NSW and southern and central Queensland, to monsoon forests in northern Queensland, Northern Territory and the Kimberley region of Western Australia (Hyland 1989). *Litsea* consists of 100 species (Hyland 1989) found in Australia, New Zealand, South America, New Guinea, through Malesia and southeast Asia to Japan (Fig. 2). All but one of the 11 Australian species are rainforest trees (Fig. 2 and Table 3; Hyland 1989).

Five species of *Endiandra* have been identified from sediments at Cambalong Creek. The fossil taxa have rounded abaxial epidermal cell outlines, granular inner periclinal walls (Figs. 4 and 5, p) and 'double' cuticular scales consisting of a narrow inner and outer ridge (Figs. 4 and 5; o, i). These characters are typical of most of the 38 extant species of Australian *Endiandra* (Christophel & Rowett 1996). The Late Paleocene species of *Endiandra* are most closely related to the '*E. pubens* group' of Christophel & Rowett (1996) and some compare very favourably with extant *E. globosa*, *E. wolfei* and *E. cowleyana* (cf. Fig. 4 and Fig. 5).

One of the fossil taxa from Cambalong Creek has been identified as *Beilschmiedia*. The fossil taxon has thick (> 2.5 μm) epidermal anticlinal walls with buttressed thickenings (Fig. 6, b) and prominent inner stomatal ledges. These characters are typical of most of the 11 extant Australian taxa of *Beilschmiedia* (Christophel & Rowett 1996). The fossil is similar in cuticle morphology to extant *B. tooram* and *B. recurva* (cf. Fig. 6 and Fig. 7) from rainforests of northern Queensland (Hyland 1989). Crisp *et al.* (1999) described *Beilschmiedia* as one of the genera exhibiting an 'Equatorial track', with an Afro–Indo–Malesian distribution (Fig. 2). Members of this track had been described as the 'tropical element' of the Australian flora by Burbidge (1960) and as the 'Irian Element' by Schodde (1989). Crisp *et al.* (1999) however alluded to the more likely Late Cretaceous, Gondwanan origins of taxa exhibiting this track. A more ancient origin is supported by the presence of *Beilschmiedia* fossils in the Late Paleocene of southeastern Australia.

Twenty-two fossil cuticle fragments from Cambalong Creek have been identified as one species of *Cryptocarya*. The fossils have rounded epidermal anticlinal walls and wide, butterfly-shaped cuticular scales (Fig. 8, s) that are characteristic of most of the 46 extant Australian species of *Cryptocarya* (Fig. 9, s; Christophel & Rowett 1996). The fossils are similar to extant *C. bidwillii*, *C. clarksoniana* (cf. Fig. 8 and Fig. 9) and *C. cunninghamii*.

Three cuticle fragments from Cambalong Creek have been identified as a single species of *Litsea*. Fossil cuticles CMB 4-22/35/2c-28 feature prominent thickened rings of cuticle encircling the outer surface of the stomates, and papillae on the outer abaxial surface (Fig. 10, pa). These characters compare favourably with the '*L. bennettii* group' of Christophel & Rowett (1996), particularly *L. connorsii* (cf. Fig. 10 and Fig. 11), a taxon that grows in rainforests and forest margins in northern Queensland over an altitudinal range of 600-1200 m asl (Hyland 1989).

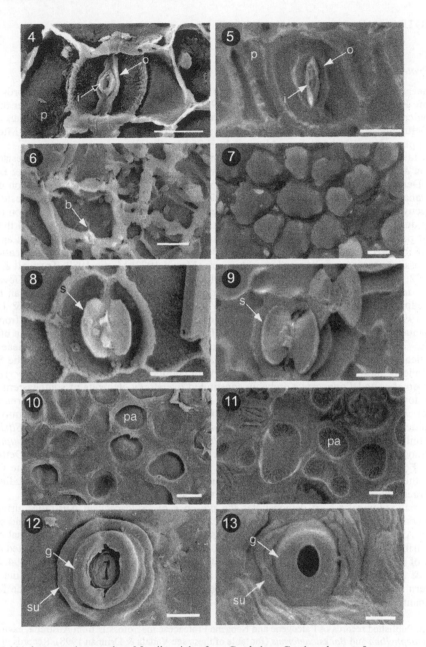

Figs. 4-13: electron micrographs of fossil cuticles from Cambalong Creek and extant Lauraceae and Proteaceae cuticles; all scale bars indicate 10 μm. Fossils are indicated by the prefix CMB; extant taxa followed by accession numbers in parentheses: MEL = Royal Botanic Gardens Melbourne. Figs.4-5, inner
surfaces of stomates; i = inner stomatal ledge, o = outer stomatal ledge, p = granular inner periclinal wall.
Fig. 4: CMB2c-21; Fig. 5: *Endiandra cowleyana* (MEL1604225). Figs 6-7, inner adaxial surfaces of non-
vein-course cells. Fig. 6: CMB4-9, b = buttressed irregular thickening on anticlinal wall; Fig. 7: *Beilschmiedia recurva* (MEL1606701). Figs. 8-9, inner surfaces of stomates; s = cuticular scale. Fig. 8:
CMB4-23; Fig. 9: *Cryptocarya clarksoniana* (MEL1605606). Figs 10-11, inner abaxial surfaces of non-
vein-course cells with papillae; pa = outline of base of papilla. Fig. 10: CMB4-35; Fig. 11: *Litsea connorsii* (MEL1604219). Figs. 12-13, outer surfaces of stomates; su = outer surface of subsidiary cell, g = outer
surface of guard cell. Fig. 12: CMB2c-24; Fig. 13: *Stenocarpus verticis* (MEL669961).

3 PROTEACEAE

3.1 *Proteaceae in Gondwana*

Proteaceae have a long macrofossil record from Australia, with earliest published records from the Late Paleocene (Table 2; Carpenter *et al.* 1994; Vadala & Drinnan 1998). However the palynological evidence for the family in Australia is diverse and extends to the Turonian (~ 90 Ma; Dettmann & Jarzen 1990; Dettmann 1994; Hill *et al.* 1999). This precedes the Miocene contact phase between Australia and Eurasia by ~ 75 Ma (Fig. 3).

The rainforest ancestor to the extant subfamilies ('Proto–Proteaceae') was hypothesised by Johnson & Briggs (1963) to have existed somewhere in northern Gondwana (Johnson & Briggs 1975; Dettmann 1989; Hill *et al.* 1995) prior to the Late Cretaceous and the separation of the gondwanan landmasses (Johnson & Briggs 1975). Dettmann & Jarzen (1991) suggested the phylogeny and ecogeography of extant Proteaceae implied evolution and diversification of the family during the mid–Cretaceous in Gondwana. Johnson & Briggs (1981) proposed that the earliest palynological records of Proteaceae actually post–date the evolution and first appearance of the precursors of extant subfamilies. Despite problems regarding the identification of some fossil proteaceous pollen with extant genera (Martin 1973; Martin 1982; Truswell & Harris 1982; Hill *et al.* 1995) fossil pollen belonging to *Beauprea, Macadamia, Gevuina–Hicksbeachia* and *Knightia* has been reliably identified from Campanian-Maastrichtian (~ 74 Ma) sediments of southeastern Australia, New Zealand and Antarctica (Dettmann & Jarzen 1990, 1991). The oldest palynological record of the family in New Zealand dates from the Campanian or possibly Santonian (Pole 1998). These pollen data reiterate the existence of subfamilies Proteoideae (tribe Conospermae) and Grevilleoideae (tribes Macadamieae and Knightieae) in the southeast Australian/Antarctic sector of Gondwana from at least the Senonian. Indeed, these genera of Proteaceae along with conifers including *Araucaria, Dacrydium, Lagarostrobus* and *Podocarpus* probably constituted overstorey elements of southeastern Australian forests in the Late Cretaceous (Specht *et al.* 1992). *Carnarvonia* (subfamily Carnarvonioideae), *Telopea* (subfamily Grevilleoideae, tribe Embothrieae) and *Persoonia* (subfamily Persoonioideae, tribe Persoonieae) probably formed part of the forest understorey, with *Stirlingia* (subfamily Proteoideae, tribe Conospermae), *Adenanthos* (tribe Franklandieae) and *Beauprea* in scleromorphic communities on the fringes of these forests (Dettmann & Jarzen 1991; Dettmann 1994). Forest and scleromorphic communities in southern Gondwana during the Late Cretaceous, therefore, included representatives of four of the extant seven subfamilies, including the two largest subfamilies, Proteoideae and Grevilleoideae.

The palynological record demonstrates the presence of Proteaceae in Australia during the Cretaceous and prior to any contact with southeast Asia (Fig. 3). Indeed, Antarctica and southeastern Australia have been integral to hypotheses regarding the evolution and dispersal of the extant tribes of Proteaceae and their progenitors. Northern Gondwana has been postulated as the origin of some of these ancestors (Johnson & Briggs 1975, 1981; Dettmann 1989, 1994; Dettmann & Jarzen 1990, 1991; Hill *et al.* 1995; Hill *et al.* 1999) and the opening of the early Southern Ocean and concomitant habitat changes coincided with diversification of the family (Dettmann 1989, 1994; Hill *et al.* 1999).

Table 2. Published records of Paleogene and Neogene Proteaceae macrofossils from Australia, excluding *Banksieaephyllum* and *Banksieaeformis* (for table of these see Vadala & Drinnan 1998). Records of New Zealand Proteaceae are limited to those described by Pole (1998).

Locality (Age)	Fossil taxon	Extant affinity
Mount Somers coal mine, New Zealand (Paleocene)	CUT-P-013[29]	Proteaceae[29]
Regatta Point (Early Eocene)	Unidentified Proteaceae (at least 8 taxa)[28]	Proteaceae[28]
Brooker (Early Eocene)	*Euproteaciphyllum brookeren-*	?Lomatia[28]

	sis[28]	
	E. tasmanicum[28]	Proteaceae[28]
Buckland (Early Eocene)	*E.* cf. *brookerensis*[28]	?*Lomatia*[28]
Livingstone, North Otago, NZ (Early/Middle Eocene)	Grevilleoideae cf. *Orites excelsa*[29]	*Orites excelsa*[29]
Golden Grove (Middle Eocene)	Proteaceae aff. *Neorites*[8]	*Neorites* (immature leaves)[8]
	Musgraveinanthus[8]	*Musgravea* fruits[8]
Maslin Bay (Middle Eocene)	Maslin Bay cf. Proteaceac III[2]	*Darlingia* (?*Knightia*)[2]
	Maslin Bay cf. Proteaceae II[2]	*Helicia* (?*Darlingia*)[2]
	Maslin Bay cf. Proteaceae IV[2]	?*Finschia*[2]/*Helicia*/*Grevillea*[2]
Maslin Bay (Middle Eocene)	*Maslinia grevilleoides*[3]	*Grevillea*[3]
Nelly Creek (Middle Eocene)	Parataxon 2[14]	?*Grevillea*[14]
	Parataxa 7, 9, 12[14]	Unknown [14]
Cowan and Lefroy Paleodrainages (Middle/Late Eocene)	Banksieae[20]	Banksieae[20]
	CUT-P-002[20,29]	Banksieae: *Musgravea*[29]
	CUT-P-003[20,29]	Banksieae: *Musgravea*[29]
	CUT-P-004[20]	*Telopea*[20]
	Darlingia cf. *Ferruginea*[20]	*Darlingia ferruginea*[20]
	Lomatia fraxinifolia[20]	*Lomatia fraxinifolia*[20]
Hasties (Middle/Late Eocene)	*Cenarrhenes nitida*[28]	*Cenarrhenes-Beauprea*[28]
Merlinleigh Sandstone, Kennedy Range, WA (Middle/Late Eocene)	*Banksia archaeocarpa* infructescence[4]	Series *Spicigerae*; *Banksia attenuata* (Series *Crytostylis*)[4]
Anglesea (late Middle Eocene)	Parataxon 12[10] (=Type 12, 'Lobed Prteaceae'[7])	*Orites*[10]
Kalgoorlie (late Middle Eocene, ~39 my)	?Banksieae[18]	Unknown [18]
	?*Darlingia*[18]	Unknown [18]
	?Gevuininae[18]	Unknown [18]
	Lomatia[18]	*Lomatia fraxinifolia*[18]
Anglesea (Late Eocene)	Bivalved fruit[5]	*Cardwellia*[5]
	Follicle[5]	?*Grevillea*[5]
	Leaves[5]	*Orites*/*Darlingia*[5]
	Musgraveinanthus alcoensis[4]	Musgraveinae (inflorescence)[5]
Kojonup Sandstone (Late Eocene)	Deeply-dissected leaf[25]	?*Grevillea*[25]
	Fructification[25]	Proteaceae[25]
Lake Lefroy (Late Eocene)	Lake Lefroy cf. Proteaceae I[2]	*Synaphea*[2]
Nelly Creek (Eocene)	Proteaceae leaves (toothed)[14]	Unspecified 'rainforest taxa'[14]
Cethana (Early Oligocene)	*L. fraxinifolia*[9,19,27]	*Lomatia fraxinifolia*[9,19,27]
	Lomatia xeromorpha[9,19]	*Lomatia tinctoria*[9,19]
Cethana (Early Oligocene)	*Euproteaciphyllum lomatiodes*[27]	cf. *Lomatia*[28]
	E. tridacnoides[27]	cf. *Lomatia*[28]
	E. gevuininoides[27]	?Gevuininae-*Hicksbeachia*[28]
	E. cethanicum[27]	Grevilleoideae[28]
	E. linearis[27]	Grevilleoideae[28]
	E. rugulatum[27]	Grevilleoideae[28]
	E. attenuatum[27]	Grevilleoideae[28]
	E. ornamentalis[27]	cf. *Darlingia*[27,28]
	E. integrifolium[27]	Grevilleoideae[28]
	E. microphyllum[27]	cf. *Orites*[27,28]
	Telopea truncata[27]	*Telopea truncata*[27]
Glencoe (mid Early Oligocene)	*Wilkinsonia glencoensis*[11,24] (syn. *Athertonia glencoesis*)[11,24]	*Athertonia* (?*Heliciopsi*s) endocarp[11,24]
Lea River (Early Oligocene)	*E. papillosum*[28]	*Lomatia polymorpha*[28]
	Orites scleromorpha[28]	*Orites milliganii*[28]
	O. milliganoides[28]	*O. milliganii*[28]
Lemonthyme (Early Oligocene)	*Euproteaciphyllum polymorphum*[28]	Grevilleoideae[28]
	E. microlobium[28]	Grevilleoideae[28]
	E. falcatum[28]	Grevilleoideae[28]
	E. serratum[28]	Grevilleoideae[28]
Leven River	*O. excelsioides*[27]	*O. excelsa*[27]

(Early Oligocene) West Dale (?Oligocene)	Proteaceae cf. *Alloxylon*[17]	*Alloxylon* (e.g., *A. wickhamii*, *A. pinnata*)[17]
	Proteaceae cf. *Stenocarpus*[17]	*Stenocarpus*[17]
	Proteaceae spp. 1-6[17]	Unknown [17]
Yallourn Open Cut (Oligocene)	Proteaceae aff. *Conospermum*[6]	*Conospermum*[6]
Moonpeelyata (Late Oligocene/ Early Miocene)	Proteaceae sp. 1[28]	Grevilleoideae[28]
Morwell Open Cut (Oligocene/Miocene)	Proteaceae aff. *Darlingia*[6]	*Darlingia*[6]
	Proteaceae aff. *Orites*[6]	*Orites*[6]
	Proteaceae aff. *Stenocarpus* or *Oreocallis*[6]	*Stenocarpus salignus*[6]
Sedan Coalfield (Oligocene/Miocene)	Parataxon S1 001 '*Banksiea-ephyllum* aff. *B. laeve*'[13]	*Banksieaephyllum laeve*[13]
	Parataxon LC 004: '*B.* aff. *B. fastigatum*'[15]	*B. fastigatum*[15]
	Parataxon LC 012: '*B.* aff. *B. obovatum*'[15]	*B. obovatum*[15]
Yallourn and Morwell (Oligocene/ Miocene)	Proteaceae taxa 58, 59, 60, 61[6]	Unknown[6]
Manuherikia Group, NZ (Early Micoene)	CUT-P-017[29] CUT-P-003[29]	Macadamieae: *Macadamia*[29] Gevuininae-*Hicksbeachia* (?*Eu-plassa*)[29]
	CUT-P-018[29]	Banksieae: *Musgravea*[29]
Beneree (early/mid Miocene)	*Wilkinsonia bilaminata*[1]	*Athertonia*[24]
Gulgong (early/mid Miocene)	*Wilkinsonia bilaminata*[1]	*Athertonia*[24]
Gulgong (mid Miocene)	*Wilkinsonia bilaminata*[24]	*Athertonia*[24]
Yallourn Formation (mid Miocene)	*Wilkinsonia bilaminata*[16]	*Athertonia*[16]
New Zealand (Miocene)	?*Euplassa*[19,21] ?Gevuininae/*Hicksbeachia*[19,21] *Macadamia*[19,21]	?*T. papuana*[19,21] ?*Turrillia bleasdalei*[19,21] *Macadamia ternifolia/ M. tetraphylla*[19,21]
Mangonui Formation, North Island,NZ (Late Miocene)	CUT-P-015[29] CUT-P-016[29]	Embothrieae[29] Helicieae: *Helicia*[29]
Melville Island (Van Diemen Sandstone) (?Late Pliocene)	*Grevillea* sp. A ('deeply lobed')[26]	*Grevillea whitiana/G. pteridifolia/ G. dryandri/G. rubicunda*[26] *Grevillea longifolia*[26]
	Grevillea sp. B ('serrate')[26]	*Grevillea dryophylla*[26]
	Grevillea sp. C ('deeply serrate')[26]	*Dilobeia thouarsii*[26]
	Proteaceae cf. *Dilobia*[24]	
Regatta Point (Early Pleistocene)	*Agastachys odoorata*[22]	*Agastachys odorata*[22]
	Banksia kingii[22]	*Banksia saxicola - B. canei*[22]
	B. strahanensis[12]	*B. spinulosa*[12]
	Cenarrhenes nitida[22]	*Cenarrhenes nitida*[22]
	Hakea sp.[22]	*Hakea*[22]
	cf. *Lomatia*[22]	?*Lomatia*[22]
	Orites revoluta[22]	*Orites revoluta*[22]
	O. truncata[22]	*O. diversifolia/O. milliganii*[22]
	Proteaceae cf. *Lomatia*[22]	*Lomatia/Knightia/Orites*[22]
	Telopea truncata[22]	*Telopea truncata*[22]
	Telopea cf. *mongaensis*[22]	*Telopea mongaensis*[22]
	Telopea strahanensis[22]	*Telopea*[22]
Marionoak Formation (Early Pleistocene)	*Orites revoluta*[22]	*O. revoluta*[22]
	O. acicularis[22]	*O. acicularis*[22]
Regatta Point	*O. revoluta*[23]	*O. revoluta*[23]

(RPA: Early/Middle	*O. milligani*[23]	*O. milligani*[23]
Pleistocene)	*Telopea truncata*[23]	*T. truncata*[23]
Henty lignites	*Banksia* sp.[28]	*Banksia*[28]
(?Early Pleistocene)		
Regency Formation	*Agastachys odorata*[28]	*A. odorata*[28]
(Middle Pleistocene)	*Cenarrhenes nitida*[28]	*C. nitida*[28]
Melaleuka Inlet	*Agastachys odorata*[28]	*Agastachys odorata*[28]
(Late Pleistocene)	*Banksia kingii*[12]	*Banksia saxicola – B. canei*[12]
	Hakea sp.[28]	*Hakea*[28]
	Lomatia aff. *tasmanica*[28]	*Lomatia tasmanica*[28]

[1]von Mueller (1883); [2]Lange (1978); [3]Blackburn (1981); [4]McNamara & Scott (1983); [5]Christophel (1984); [6]Blackburn (1985); [7]Christophel et al. (1987); [8]Christophel & Greenwood (1987); [9]Carpenter & Hill (1988); [10]Rowett & Christophel (1990); [11]Rozefelds (1990); [12]Jordan & Hill (1991); [13]Rowett (1991); [14]Christophel et al. (1992); [15]Rowett (1992); [16]Rozefelds (1992); [17]Hill & Merrifield (1993); [18]Carpenter (1994); [19]Carpenter et al. (1994); [20]Carpenter & Pole (1995); [21]Hill et al. (1995); [22]Jordan (1995); [23]Jordan et al. (1995); [24]Rozefelds (1995); [25]McLoughlin & Hill (1996); [26]Pole & Bowman (1996); [27]Carpenter & Jordan (1997); [28]Jordan et al. (1998); [29]Pole (1998).

3.2 Proteaceae after the isolation of Australia

The palynological record also indicates a high abundance and diversity of Proteaceae during the Paleogene, particularly during the Late Paleocene/Eocene (~ 56 Ma: Martin 1978; Martin 1982; Hill *et al.* 1985). The macrofossil record corroborates these pollen data (Table 2), and fossils with cuticular preservation provide the most reliable and unambiguous evidence for proteaceous affinity (Carpenter & Jordan 1997; Jordan *et al.* 1998). Records of tribes Grevilleeae and Banksieae, which now dominate the sclerophyllous flora of Australia, are abundant in the Paleogene and Neogene macrofossil records of cuticle (Table 2). The oldest macrofossils able to be attributed with confidence to Proteaceae have been described from the Late Paleocene of the Southern Highlands of New South Wales (Fig. 3). Carpenter *et al.* (1994) described cuticles of *Banksiseaephyllum taylorii* from Lake Bungarby and Vadala & Drinnan (1998) described *B. praefastigatum* from nearby Cambalong Creek. These taxa have simple leaves with serrate margins, superficial stomates, epidermal cells with irregularly thickened anticlinal walls and trichome bases with some degree of thickening (Carpenter *et al.* 1994; Vadala & Drinnan 1997). These characters are typical of extant *Banksia* and *Dryandra*, which are indistinguishable on these criteria alone, and the fossils are consequently attributed to the proteaceous fossil genus *Banksieaephyllum* (tribe Banksieae; Cookson & Duigan 1950).

The published fossil record of Proteaceae is extensive throughout the Cenozoic of southeastern Australia, and comprises mainly subfamily Grevilleoideae (Table 2 and Fig. 3: Hill *et al.* 1995; Carpenter & Jordan 1997; Jordan *et al.* 1998; Vadala & Drinnan 1998). The Paleocene and Early Eocene (~ 65–35 Ma) macrofossil record corresponds to an apparent increase in diversity and abundance of Proteaceae pollen during the Late Paleocene/Eocene (Martin 1978; Martin 1982; Hill *et al.* 1995), prior to the complete separation of the Australian continent from Antarctica in the Late Eocene/Early Oligocene (Crook 1981; Veevers *et al.* 1991; Wilford & Brown 1994). This preceded the Australian craton/Sunda Arc collision in the Late/Middle Miocene (Crook 1981; Powell *et al.* 1981; Hall 1996, 1997) by ~ 20–15 Ma (Fig. 3).

The Paleogene macrofossil record of tribe Banksieae (consisting of subtribes Banksiinae and Musgraveinae) is in concord with the size and diversity of the tribe in the modern flora of Australia (Table 3). The record for Banksiinae (consisting of *Banksia* and *Dryandra*) is extensive: 27 taxa of *Banksieaephyllum* and *Banksieaeformis* have been described from Western Australia, South Australia, Victoria and Tasmania, although the identification of several may be doubtful (Carpenter & Jordan 1997; Jordan *et al.* 1998). These range in age from Late Paleocene (~ 60 Ma) to Early Micoene (~ 23 Ma; Fig. 3; also reviewed in Vadala & Drinnan 1998). Dettmann & Jarzen (1991) hypothesized that both rainforest and sclerophyll members of Proteaceae had evolved by the Campanian/Maastrichtian (~ 74 Ma). Indeed, the oldest described taxa of *Banksieaephyllum* exhibit either sclerophyllous or more mesic characters by the Paleogene (Carpenter *et al.* 1994; Hill *et al.* 1995; Vadala & Drinnan 1998). The record of Musgraveinae is also impressive (Fig. 3 and Table 2). Christophel & Greenwood (1987) recorded Middle Ecoene *Musgravea* flowers from Golden Grove, and at least three taxa of Banksieae have been recovered from the Middle/Late Eocene Lefroy and Cowan Palaeodrainages in Western Australia (Carpenter & Pole 1995). Two of these taxa have been described as *Musgravea* (Pole 1998). Mid- to late Middle Eocene inflorescences of *Musgravea* have also been recovered from Anglesea in Victoria and Golden Grove in South Australia (Fig. 3 and Table 2; Christophel 1984).

Table 3: Australian and worldwide abundance of extant genera and tribes to which lauraceous and proteaceous fossils from Cambalong Creek are most similar.

Taxon	Total species	Species in Australia	Endemic species
Lauraceae			
Tribe Laureae: *Litsea* [1]	~ 100	11	
Tribe Cryptocaryeae: *Beilschmiedia* [1]	200-250	11	
Cryptocarya [1]	200-250	46	
Endiandra [1]	~ 100	38	
Proteaceae			
Tribe Oriteae: *Orites* [2]	9	7	4*
Tribe Stenocarpinae: *Stenocarpus* [3]	~ 25	9	7**
Tribe Helicieae			
Heliciinae: *Helicia* [4]	~ 90	9	8***
Hollandaeinae: *Hollandaea* [5]	2	2	2
Tribe Knightieae			
Knightiinae:*Darlingia* [6]	2	2	2
Tribe Banksieae			
Banksiinae: *Banksia* [7]	76	76	75
Dryandra [8]	93	93	93

* 4 taxa endemic to Tasmania[2]
**2 taxa extend to New Guinea and Aru Is.[3]
***1 taxon extends to New Guinea[5]
[1] Hyland (1989); [2] George & Hyland (1995); [3] Foreman (1995a); [4] Foreman (1995b); [5] Hyland (1995a); [6] Hyland (1995b); [7] George (1999a); [8] George (1999b).

The Cenozoic macrofossil record of tribe Knightieae is less extensive than that of Banksieae (Fig. 3 and Table 2). Lange (1978) identified *Darlingia/Knightia* (tribe Knightieae) from Middle Eocene Maslin Bay and Carpenter & Pole (1995) described *Darlingia* from the Middle/Late Eocene Cowan and Lefroy Palaeodrainages in Western Australia. Late Eocene leaves with a combination of characters similar to either *Darlingia* or *Orites* were also described from Anglesea (Christophel 1984).

The oldest Australian macrofossil record of tribe Oriteae is Middle Eocene *Neorites* from Golden Grove in South Australia (Christophel & Greenwood 1987), although Pole (1998) described leaves with similarities to extant *Orites excelsa* from Early/Middle Eocene sediments in New Zealand (Table 2 and Fig. 3). Late Middle Eocene specimens of *Orites* (Rowett & Christophel 1990) and leaves of either *Orites* or *Darlingia* (Christophel 1984) have been described from Anglesea in Victoria.

Tribe Embothrieae is preserved in the macrofossil record mainly as *Lomatia*, although the oldest fossil may be a Late Paleocene taxon from Cambalong Creek, with affinites to *Stenocarpus* (see above). Jordan *et al.* (1998) described two Early Eocene taxa of *Euproteaciphyllum* from Tasmania that have strong micromorphological similarities with extant *Lomatia* (Table 2). Carpenter & Pole (1995) described a Middle/Late Eocene *Lomatia* macrofossil from the Cowan and Lefroy Palaeodrainages as indistinguishable from the extant northeast Queensland rainforest taxon *L. fraxinifolia*; another late Middle Eocene *Lomatia* fossil from Kalgoorlie probably has the same affinities (Table 2; Carpenter 1994).

Other tribes of Grevilleoideae are less extensively represented in the Paleocene and Eocene macrofossil records (Fig. 3). Blackburn (1981) identified *Maslinia grevilleoides* from the Middle Eocene Maslin Bay locality as closely related to extant *Grevillea* (tribe Grevilleeae; Fig. 3). Many other tribes are represented in the record by fossils with suggested or implied affinities (Table 2). For example, Lange (1978) identified Middle Eocene leaves with possible affinities to *Helicia* (tribe Helicieae) and *Grevillea* (tribe Grevilleeae) from Maslin Bay (Table 2 and Fig. 3). Fossil leaves with possible affinities to *Grevillea* have also been described from Anglesea

(Christophel 1984) and the Late Eocene (~ 36 Ma) Kojonup Sandstone in Western Australia (Table 2 and Fig. 3; McLoughlin & Hill 1996). The only published taxon of subfamily Proteoideae from the Paleocene or Eocene is Middle/Late Eocene *Cenarrhenes nitida* (tribe Conospermeae) from Hasties in Tasmania (Table 2 and Fig. 3; Jordan *et al.* 1998). The Australian macrofossil record of Proteaceae from the Oligocene (~ 35-23 Ma) and the Neogene (~ 23-1 Ma; Table 2) indicates the family was equally diverse following the separation of the Australian and Antarctic continents in the Late Eocene/Early Oligocene and the collision of the Australian plate with the Sunda Arcs in the Middle Miocene (Fig. 3).

Proteaceae macrofossils from Cambalong Creek other than *Banksieaephyllum* (see above) will be formally described elsewhere, but likely represent the earliest macrofossils of tribes Embothrieae, Helicieae and Oriteae described from Australia (subfamily Grevilleoideae; A.J. Vadala, unpubl.). These fossil taxa have cuticular characters described by Carpenter & Jordan (1997) as typical of Grevilleioideae. These include granulation on the inner cuticle surface, hypostomaty, with stomates aligned randomly over the cuticle rather than parallel to the long axis of the leaf, and most have trichome bases associated with at least one epidermal basal cell. Most of the fossil taxa are too fragmentary to allow reliable comparisons of leaf morphology with extant taxa, and most lack enough distinctive cuticular morphological features to be placed with confidence in any extant genus of Proteaceae. However, the fossils all have brachyparacytic stomates and most have trichome bases overlying one or several epidermal cells, characters typical of the fossil proteaceous genus *Euproteaciphyllum* (Carpenter & Jordan 1997; Jordan *et al.* 1998).

Each fossil taxon from Cambalong Creek has a suite of cuticular characters enabling it to be compared favourably with taxa in either one of the extant tribes Banksieae, Embothrieae, Helicieae, Knightieae or Oriteae in subfamily Grevilleoideae. Distributions of the nearest living relatives of the fossils are shown in Fig. 2 and Table 3. Carpenter (1994) described tribal and generic characteristics of extant taxa growing in the Wet Tropics region of north Queensland; certain of these extant taxa share many similarities with proteaceous macrofossils from Cambalong Creek, and contribute to a solid taxonomic framework for the fossils.

Fossil proteaceous cuticles from Cambalong Creek were compared with those of 19 extant taxa representing two subfamilies, six subtribes and five tribes on the basis of a data set of 32 micromorphological characters. Pattern analyses of the data sets were conducted as for the Lauraceae fossils from the locality (described above). Two fossil taxa compare favourably with extant taxa in tribe Embothrieae (including the genus *Stenocarpus*; Table 1) on the basis of thickened bands of cuticle over the outer surface of guard cells and subsidiary cells (Fig. 12; g, su), which are characteristic of most extant taxa in the tribe (cf. Fig. 12 and Fig. 13; Carpenter 1994). These fossils share many micromorphological characters with extant *Stenocarpus sinuatus* and *S. verticis*, including striated or rugulated outer abaxial surfaces, superficial stomates, prominent inner cuticular ledges and epidermal anticlinal walls with irregular thickenings.

One fossil taxon from Cambalong Creek is very similar in cuticular morphology to two extant taxa in tribe Helicieae: *Hollandaea riparia* (subtribe Hollandaeinae) and *Helicia glabriflora* (subtribe Heliciinae: cf. Figs. 14 and 16 with Figs. 15 and 17). The fossil has highly granular inner periclinal surfaces (Fig. 16, p), typical of extant *Helicia* and *Hollandaea* (Fig. 17, p; Carpenter 1994), and all three taxa have vein courses marked by elongate cuticular striations. The fossil taxon has a thickened ring of cuticle over the outer surface of the guard cells and striated thickening over subsidiary cells (Fig. 14; g, su), which are typical of extant subtribe Heliciinae, including *Hollandaea riparia* (Fig. 15; g, su: Carpenter 1994). The inner stomatal structure of the fossil is similar to that of *H. riparia* in terms of prominent inner cuticular ledges (Fig. 16; i), granular periclinal walls over the guard cells and thickened, heavily granular periclinal walls over the subsidiary cells. The fossil taxon differs from extant Helicieae by not having trichomes, which are large and characteristic in extant taxa of the tribe, though very rare in *H. riparia* (Carpenter 1994).

A fourth fossil taxon from Cambalong Creek appears closely related to extant *Darlingia* (tribe Knightieae), a genus of two species endemic to rainforests of north-east Queensland (Table 3 and Fig. 2; Hyland 1995b). Epidermal cell outer surfaces of the fossil are covered in intricate striations (Figs 18 and 20, st) that are typical of extant *Darlingia* (Figs. 19 and 21, st), *Eucarpha* and *Knightia* in subtribe Knightiinae (Carpenter 1994; Carpenter & Pole 1995).

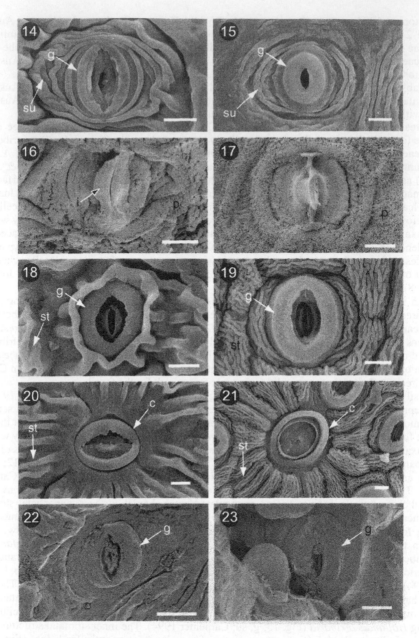

Figs. 14-23: electron micrographs of fossil cuticles from Cambalong Creek and extant Lauraceae and Proteaceae cuticles; all scale bars indicate 10 μm. Fossils are indicated by the prefix CMB; extant taxa followed by accession numbers in parentheses: MEL = Royal Botanic Gardens Melbourne; AWD = A.W. Douglas. Figs.14-15, outer surfaces of stomates, labelled as for Figs. 12-13. Fig.14: CMB4-62a; Fig. 15: *Hollandaea riparia* (MEL712266). Figs 16-17, inner surfaces of stomates; i = inner stomatal ledge, p = granular inner periclinal walls. Fig. 16: CMB4-62a; Fig. 17: *Helicia glabriflora* (MEL232949). Figs 18-19, outer surfaces of stomates; g = outer surface of guard cell, st = cuticular striations. Fig. 18: CMB2c-29; Fig. 19: *Darlingia darlingiana* (AWD629). Figs. 20-21, outer surfaces of trichome bases; c = thickened collar, st - cuticular striations. Fig. 20: CMB2c-39; Fig. 21: D. *darlingiana* (AWD629). Figs. 22-23, outer surfaces of stomates; g = outer surface of guard cells. Fig. 22: CMB4-14; Fig. 23: *Orites diversifolia* (MEL593824).

The thickened ring of cuticle over the outer surface of the guard cells of the fossil (Fig. 18, g) is also characteristic of extant Knightiinae (Fig. 19, g; Carpenter 1994; Carpenter & Pole 1995). The fossil has large trichome bases with a raised ring (collar) of cuticle around the insertion point of the foot cell (Fig.20, c), and pronounced radiating striations (Fig. 20, st). These characters are also typical of extant Knightiinae (Fig. 21; c, st: Carpenter 1994). However, the fossil has cuticular thickenings at the poles of the guard cells, which are not present in extant Knightiinae (Carpenter 1994). This fossil may indicate the existence of either *Darlingia* in particular, Knightiinae generally, or a close relative in the Late Paleocene in the Southern Highlands. This correlates with the Cretaceous occurrence in southeastern Australia of pollen similar to that produced by extant *Knightia* (Dettmann & Jarzen 1991; Specht *et al.* 1992).

A fifth fossil taxon of Proteaceae from Cambalong Creek (Fig. 22) bears many similarities to extant *Orites* and *Neorites*, and may represent a Late Paleocene relative of Oriteae, providing some support for the hypothesis of Johnson & Briggs (1975) that *Orites* must have evolved by the Paleocene. This fossil is characterised by heavy thickening over the outer surface of the guard cells (Fig. 22, g), abaxial trichomes associated with 2–3 basal cells and striations over the vein-courses only. These characters are typical of extant Oriteae (e.g. *Orites diversifolia*: cf. Fig. 22 and Fig. 23; Carpenter 1994). The Late Paleocene fossil is similar to three taxa endemic to north Queensland montane rainforests, *O. megacarpa*, *O.excelsa*, and *O. fragrans* in terms of granular inner cuticular surfaces and abaxial trichome structure (Carpenter 1994). The oldest described example of tribe Oriteae (consisting of extant genera *Orites* and *Neorites*) dates from the Early Oligocene (~ 35 Ma; Carpenter & Jordan 1997), although Christophel *et al.* (1987) suggested a late Middle Eocene (~ 38 Ma) fossil from Anglesea in Victoria may have been related to extant *Orites*.

The proteaceous taxa from Cambalong Creek described above are significant macrofossil evidence for taxa related to tribes Banksieae, Embothrieae, Helicieae, Knightieae and Oriteae in south-eastern Australia approximately 60 Ma. This closes the wide temporal gap that has existed between the earliest occurrence of these tribes in the macrofossil record and the palynological record. The oldest published macrofossils are Middle Eocene taxa possibly related to *Darlingia* (tribe Knightieae: Table 2; Lange 1978), while the palynological record dates from the Campanian-Maastrichtian, including forms with affinities to extant *Adenanthos*, *Beauprea*, *Stirlingia* (Proteoideae), *Persoonia* (Persoonioideae), *Carnarvonia* (Carnarvonioideae) and Grevilleoideae including *Gevuina-Hicksbeachia*, *Grevillea*, *Knightia*, *Macadamia*, *Telopea*, and possibly *Embothrium* (Dettmann 1989; Dettmann & Jarzen 1990, 1991; Specht *et al.* 1992).

4 DISCUSSION

The nature of the palaeobotanical record, particularly the macrofossil record with identifications based on cuticular characters, reiterates the ancient nature of Lauraceae and Proteaceae in Australia. Johnson & Briggs (1981) hypothesised that most of the tribes and subtribes of Proteaceae had evolved by the beginning of the Late Cretaceous, well before any known fossils with the characteristics of extant Proteaceae. The published macrofossil records (Table 2) correspond on a more general level with the hypothesis of Late Cretaceous diversification of Proteaceae in southern high latitudes (Dettmann & Jarzen 1991). Macrofossil and palynological evidence support an ancient presence of Proteaceae in Gondwana (~ 90 Ma; Dettmann 1989, 1992, 1994; Dettmann & Jarzen 1991; Hill *et al.* 1995), notwithstanding the unpublished Late Paleocene taxa from Cambalong Creek. The palaeobotanical record is consistent with the hypothesis of Dettmann & Jarzen (1990) that evolution and initial diversification of several clades within Proteaceae (e.g. subfamilies Proteoideae and Grevilleoideae) occurred in the region of the embryonic Southern Ocean before the separation of Australia from Antarctica. The palynological and macrofossil records are consistent with the hypothesis of Johnson & Briggs (1975) that Proteaceae originated before the Middle Cretaceous as part of a mesic, moist forest flora and dispersed with the breakup of Gondwana. This is reflected in biogeographic data superimposed onto phylogenies based on Proteaceae chloroplast sequences (atpB gene and the *atpB – rbcL* intergenic spacer), which evince divergence of the major groups in the family prior to or during the break-up of Gondwana (Hoot & Douglas 1998). By contrast, many published Cenozoic lauraceous fossils from Australia are younger than those at Cambalong Creek, Nerriga and Angle-

sea described above (Table 1). The more reliable of these descriptions, such as Eo-
cene/Oligocene and Miocene *Cryptocarya* (Leisman 1986; Rowett 1991; Pole *et al.*
1993) and *Endiandra* (Rowett 1991), all nevertheless precede the 'contact phase' or collision between
Australia and southeast Asia in the Miocene (Fig. 3; Truswell *et al.* 1987; Metcalfe 1990; Hall
1996, 1997).

The long Australian fossil record of Lauraceae and Proteaceae demonstrates that the pre-
Cenozoic distribution of these plant taxa throughout the Australian region was significant to the
current biogeography of these groups, as reviewed for the Austral landmasses by Drinnan &
Crane (1990). These distributions arose before the Cenozoic, and prior to any floristic ex-
changes between the Australian Craton and Malesia. The fossil record and extant distribution of
Proteaceae and Lauraceae consequently imply neither family was introduced into the Australian
Craton via Malesia subsequent to the Miocene contact phase. Australia and South America are
centres of diversity of extant Proteaceae. Forty-six of 79 genera and 1100 of approximately
1700 species of Proteaceae are found in virtually all except the most arid habitats in Australia
(Douglas 1995). Most taxa of the sclerophyllous subtribe Banksiinae (the fossil record of which
extends to the Late Paleocene, ~ 58–60 Ma; Fig. 3) are endemic to Australia, and most are restricted in distribution to the Southwest botanical province of Western Australia (George 1999a,
b). This high degree of diversity and endemism reiterates a long evolutionary history for the
Proteaceae on the Australian landmass.

Martin (1981) recognised a disjunction between traditional concepts of 'floristic elements'
defined on the basis of extant distributions, and the Cenozoic fossil record, which indicates that
many extant northern Australian taxa with 'tropical' distributions have been in Australia since
the Paleogene. Martin (1981) also proposed that the Cenozoic fossil record of such taxa exemplified continual floral evolution closely linked to climatic change since the Cretaceous, rather
than indicating recent migrations from the Malesian region. The temporal and spatial extent of
the macrofossil record reviewed here indicates this could be the case for genera with a long fossil history in Australia and even longer records in Europe and North America, specifically *Beilschmiedia*, *Cryptocarya* and *Endiandra*, for example. *Beilschmiedia*, *Cryptocarya*, *Endiandra*,
Litsea and *Neolitsea* are all typical tree components of tropical and subtropical forests in Australia, with all but *Beilschmiedia* also typical of warm temperate forests (Specht 1981). These
closed forests and sclerophyll communities of Australia contain a flora that Specht (1981) suggested must be regarded as of ancient origin. The palynological record indicates that these
closed forests were present over most of southern and central Australia in the Paleogene; however, there is no extant equivalent vegetation for comparison with most of these fossil assemblages (Martin 1981).

The macrofossil record for Lauraceae and Proteaceae presented is contrary to the thesis of
Herbert (1932) that the rainforests of north Queensland are 'essentially Malaysian'. The 'pa-
laeotropic element' was defined on the basis of presumed Malaysian or tropical origin (Herbert
1932, 1967). Proteaceae and Lauraceae constitute part of this northeast Queensland rainforest
flora, but Proteaceae first appeared in the Australian region soon after the early diversification
of the family in Middle or Late Cretaceous. The flora of Australia at that time was probably the
result of a gradual shift in floristic composition of plant communities from the Neocomian to the
Senonian, with mostly deciduous gymnosperm communities dominated by forms with cosmopolitan Jurassic affinities and with no modern analogues (Hill *et al.* 1995) disappearing and being replaced by angiosperms (McLoughlin *et al.* 1995). The Australian palaeobotanical record
of Lauraceae is not as ancient, but nevertheless indicates the presence of the family in southeast
Australia many millions of years prior to contact between the Australian craton and Malesia
(Fig. 3).

The paleobotanical evidence presented in this review reiterates in a specific sense the importance of the Cretaceous flora of Gondwana to the biogeography of some extant Austral angiosperms, discussed in detail by Drinnan & Crane (1990). Barlow (1981) suggested that the temperate and subtropical rainforests of eastern Australia were derived from an ancient Gondwanan
flora. The Gondwanan origins of at least some extant rainforest taxa in northeastern Australia
previously considered to be of Indo–Malayan or Malesian origin suggested by the fossil record
presented here was also indicated by the detailed ecological data of Webb *et al.* (1984). This
plant macrofossil record complicates earlier concepts of the extant flora of Australia being com-

posed of discrete 'elements', some of which were defined by distinct Malesian, Antarctic or 'autochthonous' origins.

ACKNOWLEDGEMENTS

The authors are grateful to Associate Professor Ian Metcalfe for the opportunity to contribute to this volume, to Associate Professor Andrew Drinnan and Dr Stephen McLoughlin for valuable comments and suggestions on the manuscript, to Professor James Ross and the staff of the Herbarium, Royal Botanic Gardens Melbourne for use of the extant Lauraceae and Proteaceae collections, and to Dr Andrew Douglas for access to his collection of extant Proteaceae. Fossil data from Cambalong Creek presented in this report was gathered whilst AJV was a postgraduate student in the School of Botany, The University of Melbourne. Preparation of this report was funded by Australian Research Council Large Grant A39802019 to DRG.

REFERENCES

Audley-Charles, M.G. 1987. Dispersal of Gondwanaland: relevance to evolution of the angiosperms. In T.C. Whitmore (ed), *Biogeographical Evolution of the Malay Archipelago*: 5–25. Oxford: Clarendon Press.

Bandulska, H. 1926. On the cuticles of some fossil and recent Lauraceae. *Botanical Journal of the Linnean Society* 47: 383–425.

Bandulska, H. 1928. A Cinnamon from the Bournemouth Eocene. *Botanical Journal of the Linnean Society* 48: 139–147.

Barlow, B.A. 1981. The Australian flora: its origin and evolution. In A.S. George (ed), *Flora of Australia, Volume 1*: 25–76. Canberra: Australian Government Publishing Service.

Belbin, L. 1987. *PATN Pattern Analysis Package: Reference Manual*. Canberra: CSIRO Division of Wildlife and Rangelands Research.

Berggren, W.A., & Aubry, M.–P. 1996. A Late Paleocene–Early Eocene NW European and North Sea magnetobiochronological correlation network. In R.W.O'B Knox, R.M. Confield, and R.E Dunay (eds), *Correlation of the Early Paleogene in Northwest Europe*. Geological Society Special Publication No. 101: 309–352.

Blackburn, D.T. 1981. Tertiary macrofossil flora of Maslin Bay, South Australia: numerical taxonomic study of selected leaves. *Alcheringa* 5: 9–28.

Blackburn, D.T. 1985. *Palaeobotany of the Yallourn and Morwell coal seams: SECV Palaeobotanical Project, Report No. 3*. Melbourne: State Electricity Commission of Victoria. Unpublished report.

Blackburn, D.T., & Sluiter, I.R.K. 1994. The Oligo–Miocene coal floras of southeastern Australia. In R.S. Hill (ed), *History of the Australian Vegetation: Cretaceous to Recent*: 328–367. Cambridge: Cambridge University Press.

Burbidge, N.T. 1960. The phytogeography of the Australian region. *Australian Journal of Botany* 8: 75–209.

Carpenter, R.J. 1994. Cuticular morphology and aspects of the ecology and fossil history of North Queensland rainforest Proeaceae. *Botanical Journal of the Linnean Society* 116: 249–303.

Carpenter, R.J., & Hill, R. S. (1988). Early Tertiary *Lomatia* (Proteaceae) macrofossils from Tasmania, Australia. *Review of Palaeobotany and Palynology* 56: 141–150.

Carpenter, R.J., Hill, R.S. & Jordan, G.J. 1994. Cainozoic vegetation in Tasmania: macrofossil evidence. In R.S. Hill (ed), *History of the Australian Vegetation: Cretaceous to Recent*: 276–298. Cambridge: Cambridge University Press.

Carpenter, R.J. & Jordan, G.J. 1997. Early Tertiary macrofossils of Proteaceae from Tasmania. *Australian Systematic Botany* 10: 533–563.

Carpenter, R.J., Jordan, G.J., & Hill, R.S. 1994. *Banksieaephyllum taylorii* (Proteaceae) from the Late Paleocene of New South Wales and its relevance to the origin of Australia's scleromorphic flora. *Australian Systematic Botany* 7: 385–392.

Carpenter, R.J., & Pole, M. 1995. Eocene plant fossils from the Lefroy and Cowan Palaeodrainages, Western Australia. *Australian Systematic Botany* 8: 1107–1154.

Chandler, M.E.J. 1964. *The Lower Tertiary Floras of Southern England IV: a Summary and Survey of Findings in the Light of Recent Botanical Observations*. London: Trustees of the British Museum (Natural History).

Chapman, F. 1921. A sketch of the geological history of Australian plants: the Cainozoic flora. *Victorian Naturalist* 37: 127–133.

Chapman, F. 1926. New or little-known fossils in the National Museum. Part XXIX. On some Paleogene and Neogene plant remains from Narracan, South Gippsland. *Proceedings of the Royal Society of Victoria (new series)* 38: 183–191.

Christophel, D.C. 1981. Tertiary megafossil floras of Australia as indicators of floristic associations and palaeoclimate. In A. Keast (ed), *Ecological Biogeography of Australia*, Volume 1: 379–390. The Hague: Dr. W. Junk Publishers.

Christophel, D.C. 1984. Early Tertiary Proteaceae: the first floral evidence for the Musgraveinae. *Australian Journal of Botany* 32: 177–186.

Christophel, D.C. 1989. Evolution of the Australian flora through the Tertiary. *Plant Systematics and Evolution* 162: 63–78.

Christophel, D.C. 1994. The early Tertiary macrofloras of continental Australia. In R.S. Hill (ed), *History of The Australian Vegetation: Cretaceous to Recent*: 262–275. Cambridge: Cambridge University Press.

Christophel, D.C. & Greenwood, D.R. 1987. A megafossil flora from the Eocene of Golden Grove, South Australia. *Transactions of the Royal Society of South Australia* 111: 155–162.

Christophel, D.C., Harris, W.K. & Syber, A.K. 1987. The Eocene flora of the Anglesea locality, Victoria. *Alcheringa* 11: 303–323.

Christophel, D.C. & Hyland, B.P.M. 1993. *Leaf Atlas of Australian Tropical Rainforest Trees*. Melbourne: CSIRO Australia.

Christophel, D.C., Kerrigan, R. & Rowett, A.I. 1996. The use of cuticular features in the taxonomy of the Lauraceae. *Annals of the Missouri Botanical Gardens* 83: 419–432.

Christophel, D.C. & Rowett, A.I. 1996. *Leaf and Cuticle Atlas of Australian Leafy Lauraceae*. Canberra: Australian Biological Resources Study.

Christophel, D.C., Scriven, L.J. & Greenwood, D.R. 1992. An Eocene megafossil flora from Nelly Creek, South Australia. *Transactions of the Royal Society of South Australia* 116: 65–76.

Conran, J.G. & Christophel, D.C. 1998. A new species of triplinerved *Laurophyllum* from the Eocene of Nerriga, New South Wales. *Alcheringa* 22: 343–348.

Cookson, I.C. & Duigan, S.L. 1950. Fossil Banksieae from Yallourn, Victoria, with notes on the morphology and anatomy of living species. *Australian Journal of Scientific Research, Series B (Biological Sciences)* 3: 133–165.

Crane, P.R. 1987. Vegetational consequences of the angiosperm diversification. In E.M. Friis, W.G. Chaloner & P.R. Crane (eds), *The Origins of Angiosperms and Their Biological Consequences*: 107–144. Cambridge: Cambridge University Press.

Crisp, M.D., West, J.G. & Linder, H.P. 1999. Biogeography of the terrestrial flora. In A.E. Orchard & H.S. Thompson (eds), *Flora of Australia Volume 1: Introduction*. Second Edition: 321–367. Melbourne: ABRS/CSIRO Australia.

Crook, K.A.W. 1981. The break-up of the Australian–Antarctic segment of Gondwanaland. In A. Keast (ed), *Ecological Biogeography of Australia*, Volume 1: 3–14. The hague: Dr. W. Junk Publishers.

Deane, H. 1925. Fossil leaves from the Open Cut, State Brown Coal Mine, Morwell. *Records of the Geological Survey of Victoria* 4: 492–498.

Dettmann, M.E. 1989. Antarctica: Cretaceous cradle of austral temperate rainforests? In J.A. Crame (ed), *Origins and Evolution of the Antarctic Biota*, Geology Society Special Publication No. 47: 89–105. London: The Geological Society: London.

Dettmann, M.E. 1992. Structure and floristics of Cretaceous vegetation of southern Gondwana: implications for angiosperm biogeography. *Palaeobotanist* 41: 224–233.

Dettmann, M.E. 1994. Cretaceous vegetation: the microfossil record. In R.S. Hill (ed) *History of the Australian Vegetation: Cretaceous to Recent*: 143–170. Cambridge: Cambridge University Press.

Dettmann, ME. & Jarzen, D.M. 1990. The Antarctic/Australian rift valley: Late Cretaceous cradle of northeastern Australasian relicts? *Review of Palaeobotany and Palynology* 65: 131–144.

Dettmann, M.E. & Jarzen, D.M. 1991. Pollen evidence for Late Cretaceous differentiation of Proteaceae in southern polar forests. *Canadian Journal of Botany* 69: 901–906.

Douglas, A.W. 1995. Morphological features. In P. McCarthy (ed) *Flora of Australia Volume 16: Elaeagnaceae, Proteaceae I*: 14–20. Melbourne: ABRS/CSIRO Australia.

Drinnan, A.N. & Crane, P.R. 1990. Cretaceous palaeobotany and its bearing on the biogeography of Austral angiosperms. In T.N. Taylor & E.L. Taylor (eds) *Antarctic Paleobiology: its Role in the Reconstruction of Gondwana*: 192–219. (New York: Springer—Verlag New York Inc.

Drinnan, A.N., Crane, P.R., Friis, E.M. & Pedersen, K.R. 1990. Lauraceous flowers from the Potomac Group (Mid–Cretaceous) of eastern North America. *Botanical Gazette* 151: 370–384.

Duigan, S.L. 1951. A catalogue of the Australian Tertiary flora. *Proceedings of the Royal Society of Victoria* 63: 41–56.

Eklund, H. 1999. *Big Survivors with Small Flowers: Fossil History and Evolution of Laurales and Chloranthaceae*. Comprehensive Summaries of Uppsala Dissertations from the faculty of Science and Technology 495. Uppsala: Acta Universitatis Upsaliensis.

Eklund, H. & Kvacek, J. 1998. Lauraceous inflorescences and flowers from the Cenomanian of Bohemia (Czeck Republic, Central Europe). *International Journal of Plant Sciences* 159: 668–686.

Ettingshausen, C. von 1888. *Contributions to the Tertiary flora of Australia*. Memoirs of the Geological Survey of New South Wales: Palaeontology No. 2.Sydney: Department of Mines.

Foreman, D.B. 1995a. Stenocarpus. In A.E. Orchard (ed), *Flora of Australia Volume 16, Eleagnaceae, Proteaceae 1*: 363–369. Melbourne: ABRS/CSIRO Australia.

Foreman, D.B. 1995b. Helicia. In A.E. Orchard (ed), *Flora of Australia Volume 16, Eleagnaceae, Proteaceae 1*: 393–399. Melbourne: ABRS/CSIRO Australia.

George, A.S. 1999a. Banksia. In A.E. Orchard, H.S. Thompson & P.M. McCarthy (eds), *Flora of Australia Volume 17B, Proteaceae 3, Hakea to Dryandra*: 175–251. Melbourne: ABRS/CSIRO Australia.

George, A.S. 1999b. Dryandra. In A.E. Orchard, H.S. Thompson & P.M. McCarthy (eds), *Flora of Australia Volume 17B, Proteaceae 3, Hakea to Dryandra*: 251–363. Melbourne: ABRS/CSIRO Australia.

George, A.S. & Hyland, B.P.M. 1995. Orites. In A.E. Orchard (ed), Flora of Australia Volume 16, *Eleagnaceae, Proteaceae 1*: 346–352. Melbourne: ABRS/CSIRO Australia.

Greenwood, D.R., Vadala, A.J. & Douglas, J.G. 2000. Victorian Paleogene and Neogene macrofloras: a conspectus. *Proceedings of the Royal Society of Victoria* 112: 65-92.

Hall, R. 1996. Reconstructing Cenozoic SE Asia. In R. Hall & D. Blundell (eds), *Tectonic Evolution of Southeast Asia*, Geological Society Special Publication No. 106: 153–184.

Hall. R. 1997. Cenozoic plate tectonic reconstructions of SE Asia. In A.J. Fraser, S.J. Matthews & R.W. Murphy (eds), *Petroleum Geology of Southeast Asia*, Geological Society Special Publication No. 126: 11–23.

Harland, W.B., Armstrong, R.L., Cox, A.V., Craig, L.E., Smith, A.G. & Smith, D.G. 1990. *A Geologic Time Scale 1989*. Cambridge: Cambridge Univeristy Press.

Herbert, D.A. 1932. The relationships of the Queensland flora. *Proceedings of the Royal Society of Queensland* 44: 1–23.

Herbert, D.A. 1967. Ecological segregation and Australian phytogeographic elements. *Proceedings of the Royal Society of Queensland* 78: 101–111.

Hill, R.S. 1983. Evolution of *Nothofagus cunninghamii* and its relationship to *N. moorei* as inferred from Tasmanian macrofossils. *Australian Journal of Botany* 31: 453–465.

Hill, R.S. 1986. Lauraceous leaves from the Eocene of Nerriga, New South Wales. *Alcheringa* 10: 327–352.

Hill, R.S. 1988a. A re-investigation of *Nothofagus muelleri* (Ett.) Paterson and *Cinnamomum nuytsii* Ett. from the Late Eocene of Vegetable Creek. *Alcheringa* 12: 221–231.

Hill, R.S. 1988b. Australian Tertiary angiosperm and gymnosperm leaf remains — an updated catalogue. *Alcheringa* 12: 207–219.

Hill, R.S. 1990. Tertiary Proteaceae in Australia: a re-investigation of *Banksia adunca* and *Dryandra urniformis*. *Proceedings of the Royal Society of Victoria* 102: 23–28.

Hill, R.S. 1991. Leaves of *Eucryphia* (Eucryphiaceae) from Tertiary sediments in south-eastern Australia. *Australian Systematic Botany* 4: 481–497.

Hill, R.S. 1992a. *Nothofagus*: evolution from a Southern perspective. *Trends in Ecology and Evolution* 7: 190–194.

Hill, R.S. 1992b. Australian vegetation during the Tertiary: macrofossil evidence. *The Beagle, Records of the Northern Territory Museum of Arts and Sciences* 9: 1–10.

Hill, R.S. 1994. The history of selected Australian taxa. In R.S. Hill (ed), *History of The Australian Vegetation: Cretaceous to Recent*: 390–420. Cambridge: Cambridge University Press.

Hill, R.S. & Carpenter, R.J. 1991. Evolution of *Acmopyle* and *Dacrycarpus* (Podocarpaceae) foliage as inferred from macrofossils in south-eastern Australia. *Australian Systematic Botany* 4: 449–479.

Hill, R.S. & Christophel, D.C. 1988. Tertiary leaves of the tribe Banksieae (Proteaceae) from south-eastern Australia. *Botanical Journal of the Linnean Society* 97: 205–227.

Hill, R.S. & Jordan, G.J. 1993. The evolutionary history of *Nothofagus* (Nothofagaceae). *Australian Systematic Botany* 6: 111–126.

Hill, R.S.& Merrifield, H.E. 1993. An Early Tertiary macroflora from West Dale, southwestern Australia. *Alcheringa* 17: 285–326.

Hill, R.S. & Pole, M. 1992. Leaf and shoot morphology of extant *Afrocarpus*, *Nageia* and *Retrophyllum* (Podocarpaceae) species, and species with similar leaf arrangement, from Paleogene and Neogene sediments in Australasia. *Australian Systematic Botany* 5: 337–358.

Hill, R.S. & Read, J. 1991. A revised infrageneric classification of *Nothofagus* (Fagaceae). *Botanical Journal of the Linnean Society* 105: 37–72.

Hill, R.S., Scriven, L.J.& Jordan, G.J. 1995. The fossil record of the Australian Proteaceae. In P. McCarthy (ed), *Flora of Australia Volume 16: Elaeagnaceae, Proteaceae I*: 21–30. Melbourne: ABRS/CSIRO Australia.

Hill, R.S., Truswell, E.M., McLoughlin, S. & Dettmann, M.E. 1999. Evolution of the Australian flora: fossil evidence. In A.E. Orchard & H.S. Thompson (eds), *Flora of Australia, Volume 1: Introduction*. Second Edition: 251–320. Melbourne: ABRS/CSIRO Australia.

Holdgate, G.R. & Sluiter, I.R.K. 1991. Oligocene–Miocene marine incursions in the LaTrobe Valley depression, onshore Gippsland Basin: evidence, facies relationships and chronology. In M.A.J. Williams, P. DeDeckker & A.P. Kershaw (eds), The Cainozoic in Australia: a Re-Appraisal of the Evidence. Geological Society of Australia Special Publication No. 18: pp. 137–157. Sydney: Geological Society of Australia.

Hoot, S.B. & Douglas, A.W. 1998. Phylogeny of the Proteaceae based on *atp*B and *atp*B–*rbc*L intergenic spacer region sequences. *Australian Systematic Botany* 11: 301–320.

Hughes, N.F. & McDougall, A.B. 1987. Records of angiospermid pollen entry into the English Early Cretaceous succession. *Review of Palaeobotany and Palynology* 50: 255–272.

Hyland, B.P.M. 1989. A revision of Lauraceae in Australia (excluding *Cassytha*). *Australian Systematic Botany* 2: 135–367.

Hyland, B.P.M. 1995a. *Hollandaea*. In A. E. Orchard (ed), *Flora of Australia Volume 16, Eleagnaceae, Proteaceae 1*: 391–393. Melbourne: ABRS/CSIRO Australia.

Hyland, B.P.M. 1995b. *Darlingia*. In A. E. Orchard (ed), *Flora of Australia Volume 16, Eleagnaceae, Proteaceae 1*: 356–357. Melbourne: ABRS/CSIRO Australia.

Johnson, L.A.S. & Briggs, B.G. 1963. Evolution in the Proteaceae. *Australian Journal of Botany* 11: 21–61.

Johnson, L.A.S. & Briggs, B.G. 1975. On the Proteaceae — the evolution and classification of a southern family. *Botanical Journal of the Linnean Society* 70: 83–182.

Johnson, L.A.S. & Briggs, B.G. 1981. Three old southern families — Myrtaceae, Proteaceae and Restionaceae. In A. Keast (ed), *Ecological Biogeography of Australia, Volume 1*: 429-469. The Hague: Dr. W. Junk Publishers.

Johnston, R.M. 1886. Descriptions of new species of fossil leaves from the Tertiary deposits of Mount Bischoff belonging to the genera *Eucalyptus, Laurus, Quercus, Cycadites*, Etc. *Papers and Proceedings of the Royal Society of Tasmania* 1885: 322–325.

Jordan, G.J. 1995. Early–Middle Pleistocene leaves of extinct and extant Proteaceae from western Tasmania, Australia. *Botanical Journal of the Linnean Society* 118: 19–35.

Jordan, G.J. 1997. Evidence of Pleistocene plant extinction and diversity from Regatta Point, western Tasmania, Australia. *Botanical Journal of the Linnean Society* 123: 45–71.

Jordan, G.J., Carpenter, R.J. & Hill, R.S. 1998. The macrofossil record of Proteaceae in Tasmania: a review with new species. *Australian Systematic Botany* 11: 465–501.

Jordan, G.J. & Hill, R.S. 1991. Two new *Banksia* species from Pleistocene sediments in Western Australia. *Australian Systematic Botany* 4: 499–511.

Jordan, G.J., Macphail, M.K., Barnes, R. & Hill, R.S. 1995. An early to middle Pleistocene flora of subalpine affinities in lowland western Tasmania. *Australian Journal of Botany* 43: 231–242.

Kemp, E.M. 1981. Tertiary palaeogeography and the evolution of Australian climate. In A. Keast (ed), *Ecological Biogeography of Australia, Volume 1*: 31–50. The Hague: Dr. W. Junk Publishers.

Kershaw, A.P., Martin, H.A. & McEwen Mason, J.R. 1994. The Neogene: a period of transition. In R.S. Hill (ed), *History of the Australian Vegetation: Cretaceous to Recent*: 299–327. Cambridge: Cambridge University Press.

Klucking, E.P. 1987. *Lauraceae Volume 2: Leaf Venation Patterns*. Berlin: J. Cramer, in der Gebruder Borntraeger Verlagsbuchhandlung.

Lange, R.T. 1978. Some Eocene leaf fragments comparable to Proteaceae. *Journal of the Royal Society of Western Australia* 60: 107–114.

Leisman, G.A. 1986. *Cryptocaryoxylon gippslandicum* gen. et sp. nov. from the Tertiary of eastern Victoria. *Alcheringa* 10: 225–34.

Macphail, M.K. 1980. Fossil and modern *Beilschmiedia* (Lauraceae) pollen in New Zealand. *New Zealand Journal of Botany* 18: 453–457.

Macphail, M.K., Alley, N.F., Truswell, E.M. & Sluiter, I.R.K. 1994. Early Teetiary vegetation: evidence from spores and pollen. In R.S. Hill (ed), *History of the Australian Vegetation: Cretaceous to Recent*: 189–261. Cambridge: Cambridge University Press.

Martin, A.R.H. 1982. Proteaceae and the early differentiation of the central Australian flora. In W.R. Barker & P.J.M. Greenslade (eds), *Evolution of the Flora and Fauna of Arid Australia*: 77–83. South Australia: Peacock Publications.

Martin, H.A. 1973. The palynology of some Tertiary Pleistocene deposits, Lachlan River Valley, New South Wales. *Australian Journal of Botany Supplementary Series, Supplement No. 6*: 1–57.

Martin, H.A. 1978. Evolution of the Australian flora and vegetation through the Tertiary: evidence from pollen. *Alcheringa* 2: 181–202.

Martin, H.A. 1981. The Tertiary flora. In A. Keast (ed), *Ecological Biogeography of Australia: Volume 1*: 391–406. The Hague: Dr. W. Junk Publishers.

Martin, H.A. 1991. Tertiary stratigraphic palynology and palaeoclimate of the inland river systems in New South wales. In M.A.J. Williams, P. DeDeckker & A.P. Kershaw (eds), *The Cainozoic in Auatralia: a Re-Appraisal of the Evidence*. Geological Society of Australia Special Publication No. 18: 181–194. Sydney: Geological Society of Australia.

Martin, H.A. 1994. Australian Tertiary phytogeography: evidence from palynology. In R.S. Hill (ed), *History of the Australian Vegetation: Cretaceous to Recent*: 104–142. Cambridge: Cambridge University Press.

Martin, H.A. 1998. Tertiary climatic evolution and the development of aridity in Australia. *Proceedings of the Linnean Society of New South Wales* 119: 115–136.

McCoy, F. 1876. Prodromus of Palaeontology of Victoria, Decade IV. *Geological Survey of Victoria Special Publication* 1876, 29–32.

McLoughlin, S., Drinnan, A.N. & Rozefelds, A.C.F. 1995. A Cenomanian flora from the Winton Formation, Eromanga Basin, Queensland, Australia. *Memoirs of the Queensland Museum* 38: 273–313.

McLoughlin, S. & Hill, R.S. 1996. The succession of Western Australian Phanerozoic terrestrial floras. In S.D. Hopper, J.A. Chappill, M.S. Harvey & A.S. George (eds), *Gondwanan Heritage: Past, Present and Future of the Western Australian Biota*: 61–80. Chipping Norton: Surrey Beatty and Sons.

McNamara, K.J. & Scott, J.K. 1983. A new species of *Banksia* (Proteaceae) from the Eocene Merlinleigh Sandstone of the Kennedy Range, Western Australia. *Alcheringa* 7: 185–193.

Metcalfe, I. 1990. Allochthonous terrane processes in Southeast Asia. *Philosophical Transactions of the Royal Society of London* A 331: 625–640.

Mueller, F. von 1883. *Observations on New Vegetable Fossils of the Auriferous Drifts*. Melbourne: Geological Survey of Victoria.

Paterson, H.T. 1935. Notes on plant remains from Narracan and Darlimurla, South Gippsland. *Proceedings of the Royal Society of Victoria (new series)* 48: 67–74.

Pole, M.S. 1992a. Cretaceous macrofloras of Eastern Otago, New Zealand: angiosperms. *Australian Journal of Botany* 40: 169–206.

Pole, M. S. 1992b. Eocene vegetation from Hasties, north-eastern Tasmania. *Australian Systematic Botany* 5: 431–475.

Pole, M.S. 1998. The Proteaceae record in New Zealand. *Australian Systematic Botany* 11: 343–372.

Pole, M.S. & Bowman, D.M.J.S. 1996. Tertiary plant fossils from Australia's 'Top End.' *Australian Systematic Botany* 9: 113–126.

Pole, M.S., Hill, R.S., Green, N. & Macphail, M.K. 1993. The Oligocene Berwick quarry flora — rainforest in a drying environment. *Australian Systematic Botany* 6: 399–427.

Powell, c. McA., Johnson, B.D. & Veevers, J.J. 1981. The Early Cretaceous break-up of eastern Gondwanaland, the separation of Australia and India, and their interaction with southeast Asia. In A. Keast (ed), *Ecological Biogeography of Australia, Volume 1*: 15–29. The Hague: Dr. W. Junk Publishers.

Qiu, Y.–L., Lee, J., Bernasconi–Quadroni, F., Soltis, D.E., Soltis, P.S., Zanis, M., Zimmer, E.A., Chen, Z., Savolainen, V. & Chase, M.W. 1999. The earliest angiosperms: evidence from mitochondrial, plastid and nuclear genomes. *Nature* 402: 404–407.

Rowett, A.I. 1991. The dispersed cuticular floras of South Australian Tertiary coalfields, part 1: Sedan. *Transactions of the Royal Society of South Australia* 115: 21–36.

Rowett, A.I. 1992. Dispersed cuticular floras of South Australian Tertiary coalfields, Part 2: Lochiel. *Transactions of the Royal Society of South Australia* 116: 95–107.

Rowett, A.I. & Christophel, D.C. 1990. The dispersed cuticle profile of the Eocene Anglesea clay lenses. In J.G. Douglas & D.C. Christophel (eds), *Proceedings of the Third International Organisation of Palaeobotany Conference*: 115–121. Melbourne: A–Z Printers.

Rozefelds, A.C.F. 1990. A Mid–Tertiary rainforest flora from Capella, Central Queensland. In J.G. Douglas & D.C. Christophel (eds), *Proceedings of the Third International Organisation of Palaeobotany Conference, Melbourne*: 123–136. Melbourne: A–Z Printers.

Rozefelds, A.C.F. 1992. The subtribe Hicksbeachiinae (Proteaceae) in the Australian Tertiary. *Memoirs of the Queensland Museum* 32: 195–202.

Rozefelds, A.C.F. 1995. Miocene *Wilkinsonia* fruits (Hicksbeachiinae, Proteaceae) from the base of the Yallourn Formation, LaTrobe Valley, Victoria. *Papers and Proceedings of the Royal Society of Tasmania* 129: 59–62.

Schodde, R. 1989. Origins, radiations and sifting in the Australasian biota — changing concepts from new data and old. *Australian Systematic Botany Society Newsletter* 60: 2–11.

Specht, R.L. 1981. Evolution of the Australian flora: some generalizations. In A. Keast (ed), *Ecological Biogeography of Australia*: 783–805. The Hague: Dr. W. Junk Publishers.

Specht, R.L., Dettmann, M.E. & Jarzen, D.M. 1992. Community associations and structure in the Late Cretaceous vegetation of southeast Australasia and Antarctica. *Palaeogeography, Palaeoclimatology, Palaeoecology* 94: 283–309.

Taylor, G., Truswell, E.M., McQueen, K.G. & Brown, M.C. 1990. Early Tertiary paleogeography, landform evolution, and paleoclimates of the Southern Monaro, N.S.W., Australia. *Palaeogeography, Palaeocliamtology, Palaeoecology* 78: 109–134.

Truswell, E.M. 1990. Australian rainforests: the 100 million year record. In L.J. Webb & J. Kikkawa (eds), *Australian Tropical Rainforests: Science–Value–Meaning*: 7–22. Melbourne: CSIRO Australia.

Truswell, E.M. 1993. Vegetation changes in the Australian Tertiary in response to climatic and phytogeographic forcing factors. *Australian Systematic Botany* 6: 533–557.

Truswell, E.M. & Harris, W.K. 1982. The Cainozoic palaeobotanical record in arid Australia: fossil evidence for the origins of an arid-adapted flora. In W.R. Barker & P.J.M. Greenslade (eds), *Evolution of the Flora and Fauna of Arid Australia*: 67–76. South Australia: Peacock Publications.

Truswell, E.M., Kershaw, A.P. & Sluiter, I.R.K. 1987. The Australian–South-east Asian connection: evidence from the palaeobotanical record. In T.C. Whitmore (ed), *Biogeographical Evolution of the Malay Archipelago*: 32–42. Oxford: Clarendon Press.

Vadala, A.J. & Drinnan, A.N. 1998. Elaborating the fossil history of the Banksiinae: a new species of *Banksieaephyllum* (Proteaceae) from the Late Paleocene of New South Wales. *Australian Systematic Botany* 11: 439–463.

van der Werff, H. & Richter, H.G. 1996. Toward an improved classification of Lauraceae. *Annals of the Missouri Botanical Gardens* 83: 409–418.

Veevers, J.J., Powell, C.McA. & Roots, S.R. 1991. Review of seafloor spreading around Australia. I. Synthesis of the patterns of spreading. *Australian Journal of Earth Sciences* 38: 373–389.

Webb, L.J., Tracey, J.G. & Jessup, L.W. 1986. Recent evidence for autochthony of Australian tropical and subtropical rainforest floristic elements. *Telopea* 2: 575–589.

Webb, L.J., Tracey, J.G. & Williams, W.T. 1984. A floristic framework of Australian rainforests. *Australian Journal of Ecology* 9: 169–198.

Wilford, G.E. & Brown, P.J. 1994. Maps of late Mesozoic–Cenozoic Gondwana break-up: some palaeogeographical implications. In R.S. Hill (ed), *History of the Australian Vegetation: Cretaceous to Recent*: 5–13. Cambridge: Cambridge University Press.

Vegetation and climate in lowland southeast Asia at the Last Glacial Maximum

A. Peter Kershaw, Dan Penny, Sander van der Kaars, Gusti Anshari & Asha Thamotherampillai
Centre for Palynology and Palaeoecology, School of Geography and Environmental Science, Monash University, Vic. 3800, Australia

ABSTRACT: The regional geography of southeast Asia, and global climate forcing, were very different to today during the Last Glacial Maximum (LGM). In order to determine how these factors influenced the lowland vegetation and climate of the region, pollen data for the LGM are examined from ten recently constructed late Quaternary pollen records derived from terrestrial and marine cores within the southeast Asian region. Interpretation is facilitated by comparison of LGM pollen spectra with those extracted from the same records for the mid-Holocene period, prior to a major influence of people on the vegetation landscape. It is found that precipitation was lower (probably by c. 30-50%) than today, while mean temperatures may have been reduced by as much as 6-7°C, assuming little influence of reduced atmospheric carbon dioxide levels. Rainforest was replaced by grassland to some extent in areas presently experiencing some seasonality in climate but rainfall reduction was insufficient to have a major influence on the extent of evergreeen forest in the core area of tropical rainforest. Montane forest elements descended to low altitudes but augmented rather than replaced existing lowland rainforest plants. Continental shelves, exposed by sea levels at least 120 m lower than today, appear to have been covered largely by rainforest in wetter areas and by grassland, open woodland and sedgeland in drier areas, except for the colonisation by *Artemisia* steppe of the exposed northern part of the South China Sea. Apart from this steppe invasion, there is no other evidence of large scale regional movements of vegetation types. Despite the lower rainfall, the extent of continental shelf exposure within the Indonesian region probably resulted in the presence of a greater area of tropical evergreen rainforest than exists today. The reconstruction has implications for debates over LGM tropical sea surface temperature, rainforest refugia, and the pattern and timing of human colonisation of the southwest Pacific region.

1 INTRODUCTION

There has been a great deal of speculation on the extent and directions of Quaternary climate change in the lowland parts of the southeast Asian region (Heaney, 1991). This speculation has been promoted by interests in the explanation of biogeographic patterns (eg. Brandon-Jones, 1996) and the contributions that the region can make to an understanding of broad regional and global climate change (eg. Wright *et al.*, 1993) and changes in terrestrial carbon storage (eg. Adams et al. , 1990). The Quaternary history of Southeast Asian lowlands also has a bearing on environments confronting the arrival of early people and their colonisation of the region and, subsequently Australasia. However, in line with other humid tropical regions of the world, evidence of Quaternary environments has been slow in forthcoming, is fragmentary in nature and sometimes contradictory. The most substantial data from the region have derived from pollen records in the high mountains of Sumatra and Java (Stuijts *et al.* , 1988, Stuijts, 1993). These records have demonstrated the maintenance of wet conditions at high altitudes through the last 40,000 years or so, but with substantially lower temperatures than today during the last glacial period. However, it is considered that they may not be applicable to extensive lowland areas because estimates of the degree of temperature change between the last glacial period and the present, around 7°C, are much greater than than those of around 1-2°C derived from most (eg. CLIMAP, 1976, Thunell *et al.* , 1994) but not all (see Anderson & Webb, 1994) sea surface temperature estimates from the tropics.

Here we attempt to provide some indication of conditions existing during one critical past period, the Last Glacial Maximum (LGM) centred on 18,000 radiocarbon years before present (BP), from recently constructed pollen records from low altitude sites in the Southeast Asian region. We also provide some indication of the degree of climate and environmental change during the latter part of the Quaternary by comparison of the LGM with pollen data from the height of the present interglacial period, or Holocene, centred on 6000 years BP. This latter period provides evidence of conditions prior to major human impact, generally reflected in pollen records within the last 5000 to 3000 years.

2 THE STUDY SITES AND THEIR REGIONAL SETTING

The location of 10 lowland pollen records from which comparative LGM and mid-Holocene data have been obtained are shown in relation to major environmental variables on Figure 1. The 200 m bathymetric line approximates the edge of the continental shelves that were exposed during the LGM. It also approximates the extent of warm shallow water which, within the Intertropical Convergence Zone (ITCZ) and the Pacific Warm Pool, helps promote the intense convective activity that results in high annual precipitation in the core of the region. The climates of this core region, which exhibits little seasonality, are defined broadly by a lack of an annual water deficit (see Fig. 1).

To the north and south of this core, precipitation gradually decreases and becomes more seasonal due to the reduced influence of the ITCZ and greater exposure to monsoon rainfall variability. In the northern hemisphere winter, dry conditions are promoted by the northerly monsoon emanating from the Tibetan Plateau region, while in the summer, wet conditions result from the formation of the southerly monsoon over the ITCZ and the Indian Ocean. In the southern hemisphere, summer rain is derived from the northwest monsoon which is, partially, a southern extension of the northern Asian monsoon that has acquired moisture from the warm continental seas of southeast Asia. The dry winter period results from the dominance of southeasterly winds that have passed over the dry Australian continent. Local variation in rainfall amounts results largely from orographic effects.

In general terms, evergreen rainforests, including extensive peat and swamp forests, dominate those areas with an annual water deficit of less than 100 mm. More seasonal, semi-evergreen rainforest is predominant in those areas with a water deficit of between 100 and 200 mm while 'dry' rainforest or savanna and savanna woodland vegetation cover those areas with an annual water deficit of greater than 200 mm (see Fig. 1). There is a clear distinction between woodlands north and south of the equator with *Pinus* characteristic of the Indochina region, largely confined to highlands, and *Eucalyptus* extending from Australia to islands on the Sahul Shelf. Also shown on the map are the submarine extensions of major river systems over the continental shelf that would have been exposed during the LGM.

The palynological records are derived from a variety of site situations. Four sites are from the ocean beyond the direct influence of sea level fluctuations. They provide continuous records of changing regional vegetation and environments and are generally well dated through accompanying oxygen isotope records and associated radiocarbon dates. Two records from the South China Sea were constructed from cores collected by the joint German-Chinese Sonne Cruise 95 in 1994. Core 17940 (20°07'N, 117°23'E) is located on the northern continental slope, adjacent to southern China, at a water depth of 1727 m. The pollen record from this site covers the last 37,000 years in 13 m of sediment (Sun & Li, 1999). Core 17964 (6°10'N, 112°13'E) is located on the southern continental slope, northwest of Borneo, and is well within the western portion of Pacific Warm Pool. The pollen record covers the last 26,000 years from 13 m of core sediment (Sun *et al.*, 2000). The other two marine records are from cores taken from the southeastern part of the region. Core G6-4 (10°47'S, 118°04'E) was collected from the Lombok Ridge at a water depth of 3510 m, about 170 m southwest of Sumba in the Lesser Sunda Islands and 850 km off northern Australia, by the Indonesian-Dutch Snellius expedition in 1985. The nine m core provides a condensed pollen record for about the last 300,000 years (van der Kaars, 1991; Wang *et al.,* 1999). Core SHI-9014 (5° 46' S, 126° 58' E) w as taken from a w ater depth of 3163 m in the Banda Sea Basin, between Sulawesi, New Guinea and Australia, as part of a French-Indonesian programme in 1990. The palynological record, from the 7.5 m core, extends through the last 175,000 years (van der Kaars *et al.* , 2000).

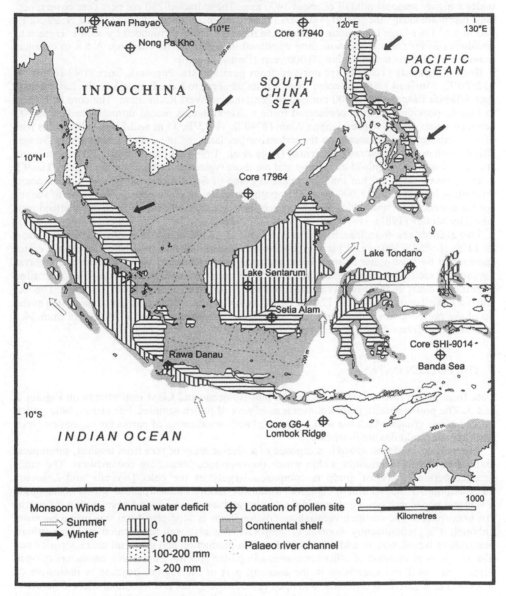

Figure 1. Locations of pollen analytical records in relation to selected environmental features of the southeast Asian region. The annual water deficit pattern is from Kayane (1971) and other environmental data are from Heaney (1991) and W.P.O. (1995).

Of the six terrestrial records covering the late Quaternary period, two are from northern Thailand, two are from Kalimantan, and there is one each from western Java and northern Sulawesi. Nong Pa Kho (17°06'N, 102°56'E; 180 m a.s.l.) is a depression lying at the western edge of the Kumphawapi floodplain on the Khorat Plateau of northeastern Thailand. Remnant patches of native vegetation suggest the predominance of deciduous forest in the region existing under a highly seasonal rainfall of about 1400 mm. The extracted 230 cm peat core covers, perhaps discontinuously, the last 40,000 years (Penny, in press). Kwan Phayao (19°10'N, 99°52'E; 380 m a.s.l.) is a large lake within an illuviated montane basin, surrounded by semi-evergreen to deciduous forest and sub-montane pine woodland in northwestern Thailand. A 5.8 m core has provided a pollen record of the last 20,000 years (Penny, in prep).

Both Kalimantan records are derived from peat forests. Pemerak Core HN3 (0°50'N, 112°00'E; 35m a.s.l.) was collected from forests adjacent to Lake Pemerak in the Lake Sentarum Wildlife Reserve, some 300 km east of Pontianak in West Kalimantan. The core is 120 cm in length, providing what is considered to be a discontinuous record through the last 30,000 years (Anshari *et al.*, in press). Setia Alam (2°40'S, 114°5'E; 35 m a.s.l.) was collected as part of a collaborative programme on the extensive peatlands of the upper catchment of Sungai (River) Sebangau in Central Kalimantan (Page *et al.*, 1999). Unlike Pemerak, these peatlands are coastal rather than inland in location and are more typical of peatlands in the region. The 9.5 m core, taken some 6.5 km into the peatland dome of Setia Alam, provides a pollen record through at least the last 18,000 years (Thamotherampillai *et al.*, in prep.). The value of the record is enhanced by similarities to more generalised and undated records from the peatland produced by Morley (1981).

Two records are from lake basins within lowland to sub-montane rainforest. Rawa Danau (6°11'S, 105°58'E; 90 m a.s.l.) is a largely swamp-covered lake formed within a volcanic crater surrounded by fresh-water swamp forest and evergreen rainforest, at the western end of Java. Twenty-six metres of peat and lake sediment have been extracted to date, providing a detailed record of the last 24,000 years (van der Kaars *et al.*, in press; van der Kaars, in prep.). The extensive Lake Tondano (1°17'N, 124°54'E; 680 m a.s.l.) lies within an intermontane basin at the tip of the northern arm of Sulawesi. The existing record covers the last 33,000 years from 14.5 m of sediment (Dam *et al.*, in press).

3 THE POLLEN DATA

Data from pollen records are shown for the mid-Holocene and LGM respectively on Figures 2 and 3. The pollen spectra represent averages of several pollen samples. For clarity, only pollen taxa or taxon groups which are well represented and/or indicative of particular vegetation types and climatic conditions are illustrated.

The tropical rainforest group is composed of a diverse array of taxa from tropical, subtropical and submontane communities within which the Dipterocarpaceae are conspicuous. The montane/temperate rainforest group is composed largely of the oaks *Quercus* and *Lithocarpus/Castanopsis* and 'southern conifers' within the family Podocarpaceae which characterise lower and upper montane rainforest in the tropics and warm temperate forests within the northern hemisphere. The northern hemisphere conifer *Pinus* is separated from this group because, although it is predominantly montane or temperate, it is also well represented in drier northern hemisphere woodlands. In addition, its high pollen production and dispersal characteristics can distort the representation of other components in pollen spectra. *Eucalyptus* characterises open forest and woodland vegetation in the southern part of the region. Poaceae is indicative of grassland or a grassland understorey to open or disturbed forest and woodland while Asteraceae is best represented in semi-arid environments in Australia to the south of the region and in cool dry steppe vegtation to the north within continental Asia. The distinctive asteraceous pollen type, *Artemisia*, is derived from a dominant of steppe vegetation. All pollen taxa and taxon groups are displayed on the diagrams as percentages of their combined total for each spectrum.

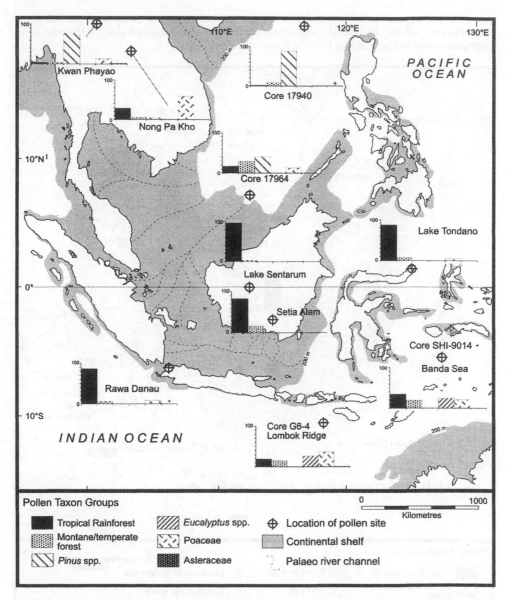

Figure 2. Relative representation of major pollen taxa and taxon groups dated to around 6000 years BP in pollen records from lowland southeast Asia.

4 PATTERNS OF POLLEN REPRESENTATION

Continuous pollen records from the region suggest that the mid-Holocene spectra reflect what would be the major natural vegetation patterns at the present day without substantial human impact. *Pinus* is restricted to the northern sites, extending equatorward to the southern part of the South China Sea while *Eucalyptus* representation is restricted to the Lombok Ridge and Banda Sea cores which are closest to the Australian mainland. Tropical rainforest pollen dominates those sites in the humid equatorial region that are generally surrounded by this kind of vegetation. Montane or temperate forest pollen is present at all sites but percentages are relatively low while Poaceae values are generally significant outside the core rainforest area. Asteraceae is inconspicuous and virtually absent from the humid rainforest sites.

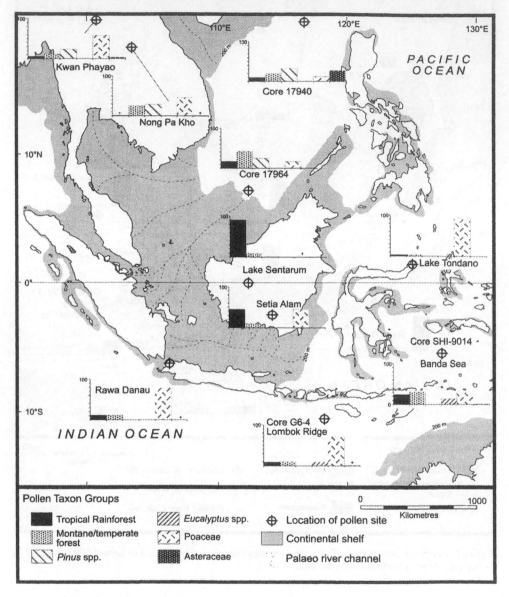

Figure 3. Relative representation of major pollen taxa and taxon groups dated to the Last Glacial Maximum (c 18,000 years BP) in pollen records from lowland southeast Asia.

There are also some marked variations in representation within and between biogeographic regions that can be related to specific site location and pollen dispersal. In the north, the two Thai sites are very different from each other. The presence of pine woodland in the continental highlands, within the vicinity of Kwan Phayao, has resulted in the under-representation of more abundant deciduous and semi-evergreen forests in the pollen record. By contrast, deciduous forest and grassland vegetation is more strongly represented in the Nong Pa Kho record, where pine currently maintains a limited distribution on topographically elevated sites. The significance of pine pollen production and transport is even more evident in the South China Sea cores. At the northern site, pine pollen swamps the spectrum totally distorting vegetation patterns on the adjacent land masses while at the southern site high pine values occur south of any vegetation types containing significant numbers of pine trees.

Differences between marine and terrestrial sites are well illustrated by variation in representation of montane or temperate forest taxa, which are largely derived from tropical mountains. Values are high at all marine sites, with the exception of the pine-dominated South China Sea core 17940. This is most likely because high altitude tropical forests contain a high proportion of these taxa combined with the fact that winds at high altitudes facilitate more regional pollen dispersal. It may be surprising that montane elements are recorded at all in the smaller basins or peat forests within the terrestrial equatorial sites. All sites, though, are in close proximity to mountains and, more importantly, lower montane elements also occur in the lowlands, particularly in poorly drained areas from which pollen records are frequently derived. These taxon occurrences may also be remnants of more extensive past distributions under climatic conditions different to those of today.

Comparison of the Holocene spectra with those from the LGM indicates that it is unlikely that there were major changes in biogeographic distributions at the broad regional scale. Neither *Pinus* or *Eucalyptus,* for example, change their site representation. However, the high values for Asteraceae, predominantly *Artemisia,* in South China Sea core 17964, probably indicate the development of steppe grassland. In the absence of high values of *Artemisia* within terrestrial pollen records in southern China to the north of this site at the LGM, Sun *et al.* (2000) consider that steppe grassland probably colonised the exposed continental shelf. The lack of an *Artemisia* response in the terrestrial north Thailand sites may add support for this proposal.

There certainly were major distributional changes in vegetation within the pollen catchments of individual sites. At the northern South China Sea site, the higher values of all pollen components except for *Pinus* are probably a reflection of contracted woodlands and subtropical forests dominated by or containing pine. A generally more open vegetation is indicated by much increased levels of grasses as well as *Artemisia* relative to those of tropical and montane forests, at the LGM compared with the Holocene. Lower pine values are also recorded at Kwan Phayao where it appears that there was a replacement of pine woodland by grasses. The reverse situation around Nang Pa Kho suggests that pine woodlands may have shifted and that during the LGM they were better represented at lower altitudes The lower grass values during the LGM may be an artefact of the local presence of pine pollen rather than a less open vegetation. Regionally more restricted pine-containing vegetation is also indicated in the southern South China Sea core.

Within the present area of tropical evergreen rainforest, there are higher percentages of montane rainforest taxa at the LGM in most sites and, where percentages are lower than in the Holocene, they are proportionately less reduced than tropical lowland taxa. This indicates that montane vegetation components expanded into or near to the lowlands at the LGM, replacing tropical lowland rainforest taxa to some degree. There was a strong expansion in grasses at most sites indicating that lowland rainforest at least was reduced in surrounding areas. Exceptions to this expansion are Lake Sentarum and South China Sea core 17964 which record a similar rainforest cover to the Holocene.

The marine cores in the south of the region show similar patterns of reduced low land rainforest, expanded montane forest relative to lowland rainforest, and generally more open vegetation both from the Indonesian and Australian regions at the LGM.

5 DISCUSSION

The pollen data indicate that the lowland southeast Asian region has, in common with many other parts of the world, experienced significant change in relation to global climate forcing. While the LGM was generally drier and cooler than the Holocene, the nature of the data make it difficult to provide quantitative estimates of climate change or determine whether or not the degree of change was similar over the whole region.

In relation to temperature, lowland pollen spectra would have been influenced by well-dispersed pollen from more extensive montane vegetation during the LGM. Higher values of montane taxa could largely be a reflection of temperature lowering at high altitudes. However, there are similar changes in representation of montane elements in terrestrial sites which have limited pollen catchment areas and marine records with extensive catchments reflecting broad regional vegetation patterns. Consequently, it is likely that montane elements were present in the lowlands, a conclusion also reached in tropical South America (Colinvaux *et al.*, 2000). Colinvaux *et al.* further suggest that present day lowland and montane elements were intermixed, forming associations with no modern analogues but characteristic of the 'more normal' glacial conditions which have predominated during the later part of the Quaternary era. The southeast Asian pollen data give general support to the proposal of van der Kaars (1998), derived from the pollen analysis of Bandung Basin in west Java, that temperatures at low altitudes may have been some 7°C lower than today, an estimate similar to those from highland areas within the Indonesian region.

In the northern part of the region, the proposed presence of *Artemisia* steppe on the continental shelf also points to substantially lower temperatures during the LGM. The high representation of this taxon, which achieves a wide distribution today only at temperate latitudes, indicates a cooling of some 6°C, similar to other estimates from pollen records in southern China (Zheng & Lei, 1999).

Reduced precipitation during the LGM is evident at most sites. Such a reduction is predictable in light of the emerged continental shelves. The loss of shallow seas would have increased the continentality of the region, resulting in a reduction in the influences of the ITCZ and summer monsoons. Moisture reduction would have been exacerbated by reduced sea surface temperatures. It has been estimated that precipitation was reduced by about 30% in western Java (van der Kaars, 1998). The presence of *Artemisia* on the continental shelf of the northern part of the South China Sea could indicate a reduction in precipitation of around 50%, although the north Thailand sites suggest that, regionally, this would be an overestimate.

Exceptions to the general site pattern are Lake Sentarum and South China Sea core 17964 which show no evidence of drying. In the case of Lake Sentarum, it is possible that, as at montane sites, the vegetation is insensitive to variations in precipitation due to the perhumid nature of the climate. With the South China Sea core, it might be expected that any regional reduction in precipitation would be detected because of its proximity to seasonal as well as ever-wet environments. Sun *et al.* (2000) suggest that wet conditions may have been maintained through the last glacial period because of the sustained or even intensified activity of the winter monsoon. Although causing a reduction in rainfall in the northern part of the South China Sea, moisture absorbed on passage over the sea would have been available to the southern part of the region. However, the influence of the monsoon must have waned before reaching western Java which, as previously mentioned, experienced substantially lower precipitation levels during the LGM. An alternative scenario is that conditions were regionally drier with extensive grassland but that there was a heavy bias towards river-borne pollen derived from gallery rainforests along rivers crossing the continental shelf and discharging into the South China Sea close to the core site (see Figure 1). It is unlikely, however, that there would be no indication in the pollen record if grasslands had been more extensive than in the Holocene.

Further support for the maintenance of high rainfall and rainforest in the southern part of the South China Sea is provided from recent pollen data from peat sediments (van der Kaars, unpublished data) in industry cores taken from the Sunda Shelf in the vicinity of Pulau Tudjah, southeastern Sumatra (Aleva, 1973). Although undated, it is probable that the peat sequences formed during the latest Pleistocene, as the peat is unlikely to have survived subsequent low sea level stands. The pollen spectra are dominated by rainforest or peat swamp forest taxa with a high proportion of *Pandanus*. Values for Cyperaceae are low and Poaceae occur (at a low percentage) in only one of the 14 analysed samples. In four samples there are significant percentages of Rhizophoraceae suggesting increased proximity of mangroves, most likely during the last marine transgression. Other pollen records from the region indicate that the mangrove signal dates from about 15,000 years BP: consequently, much of the sequence would fall within the LGM.

The weight of evidence suggests that a core of rainforest was maintained in the southeast Asian region during the LGM. The presence of a 'dry corridor' which is thought to have extended from Indochina to Australia through the Malayan Peninsula and the continental shelf between Sumatra and Borneo (Morley & Flenley, 1987; Heaney, 1991) is not supported by the data presented here. Although there is geomorphic and pollen evidence for dry conditions within the corridor during the Quaternary, this evidence is undated and most likely related to one or more earlier periods. Consequently, despite the near continuous land connection, there is unlikely to have been a clear migration route for organisms, which may have included *Homo sapiens,* adapted to more open environments during the later part of the Quaternary period. Furthermore, despite generally drier conditions, the exposure of the continental shelf is likely to have resulted in a net increase in the area of rainforest rather than a reduction. Even at those pollen sites where there is evidence for substantial grassland vegetation, rainforest pollen is still represented and rainforest probably survived along streams and other locally moist areas as well as at higher altitudes. The data suggest that, although there were altitudinal changes in the ranges of rainforest types or components and localised changes is the extent of lowland rainforest, there were no broad regional distributional changes in location of rainforest types. Patterns of distribution of evergreen, semi-evergreen and deciduous forests are likely to have developed over a period much longer than the last glacial cycle.

6 CONCLUSIONS

The data presented, predominantly from pollen records produced or published within the last two years, provide a realistic basis for assessment of patterns of vegetation and climate change in the southeast Asian region in the late Quaternary. During the LGM, some elements of montane forest extended into the tropical lowlands suggesting that temperatures were regionally depressed, perhaps by as much as 6-7°C in some places, assuming that lower atmospheric carbon dioxide levels had little influence on plant distributions at this time. Such temperature depression is in conflict with some estimates of tropical sea surface temperatures that have suggested that low altitude temperatures were little different to those of today. There is also evidence for substantial fragmentation of forests during the LGM, implying reductions of at least 30% in mean annual rainfall, or much greater rainfall variability, outside the core area of tropical rainforest. Within the core area, rainforest is considered to have expanded over the continental shelf during the LGM and it is likely that the extent of evergreen rainforest was greater than today. This does not necessarily imply rainfall higher than, or even similar to, present: under the perhumid climate, the forest vegetation may simply have been insensitive to any rainfall change. Apart from colonisation of continental shelves, submerged during interglacial periods, there is little evidence of wholesale migration of vegetation types in response to climate change on this scale. Basic vegetation patterns have probably developed over a much longer period of time.

ACKNOWLEDGMENTS

Our research into the vegetation history of Southeast Asia is supported by an ARC grant, a Monash University Research Fund Fellowship to Dan Penny, a Monash University Logan Fellowship to Sander van der Kaars and an AUSAID scholarship to Gusti Anshari.

REFERENCES

Adams, J.M., Faure, H., Faure-Denard, L., McGlade, J.M. & Woodward, F.I. 1990. Increases in terrestrial carbon storage from the Last Glacial Maximum to the present. *Nature* 348: 711-714.

Aleva, G.J.J. 1973. Aspects of the historical and physical geology of the Sunda Shelf essential to the exploration of submarine tin placers. *Geologie en Mijnbouw* 52: 79-91.

Anderson, D. & Webb, R.S. 1994. Ice-age tropical temperatures revisited. *Nature* 367: 23-24.

Anshari, G., Kershaw, A.P. & van der Kaars, S. in press. A Late Pleistocene and Holocene pollen and charcoal record from peat swamp forest, Lake Sentarum Wildlife Reserve, West Kalimantan, Indonesia. *Palaeogeography, Palaeoclimatology, Palaeoecology.*

Brandon-Jones, D. 1996. The Asian Colobinae (Mammalia: Cercopithecidae) as indicators of Quaternary climate change. *Biological Journal of the Linnean Society* 59: 327-350.

Chappell, J. 1994.Upper Quaternary sea levels, coral terraces, oxygen isotopes and deep-sea temperatures. *Journal of Geography (Japan)*, 103: 828-840.

CLIMAP Project Members 1976. The surface of Ice-Age Earth. *Science* 191: 1131-1137.

Colinvaux, P.A., De Oliviera, P.E. & Bush, M.B. 2000. Amazonian and neotropical plant communities on glacial time-scales: The failure of the aridity and refuge hypotheses. *Quaternary Science Reviews* 19: 141-169.

Dam, R.A.C., van der Kaars, W.A., Suparan, P. & Fluin, J. in press. Palaeoenvironmental developments in the Lake Tondano area (N. Sulawesi, Indonesia) since 33,000 B.P. *Palaeogeography, Palaeoclimatology, Palaeoecology.*

Heaney, L.R. 1991. A synopsis of climate and vegetational change in Southeast Asia. *Climate Change* 19: 53-61.

Kayane, I. 1971. Hydrological regions in monsoon Asia. In M.M. Yoshino (ed.) *Balance of Monsoon Asia - a Climatological Approach.* University of Hawaii Press, Honolulu.

Morley, R.J. 1981. Development and vegetation dynamics of a lowland ombrogenous swamp in Kalimantan Tengah, Indonesia. *Journal of Biogeography* 8: 383-404.

Morley, R.J. & Flenley, J.R. 1987. Late Cainozoic vegetational and environmental changes in the Malay Archipelago. In: T.C. Whitmore (ed.), *Biogeography of the Malay Archipelago* 50-59. Oxford: Oxford Monographs on Biogeography 4, Oxford Scientific Publications.

Page, S.E., Rieley, J.O., Shotyk, O.W. & Weiss, D. 1999. Interdependence of peat and vegetation in a tropical peat swamp forest. *Philosophical Transactions of the Royal Society of London, Series B.* 354: 1-13.

Penny, D. in press. A 40,000 year palynological record from north-east Thailand; implications for biogeography and palaeo-environmental reconstruction. *Palaeogeography, Palaeoclimatology, Palaeoecology.*

Stuijts, I., 1993. *Late Pleistocene and Holocene Vegetation of West Java, Indonesia.* A.A. Balkema, Rotterdam.

Stuijts, I., Newsome, J.C. & Flenley, J.R. 1988. Evidence for late Quaternary vegetational change in the Sumatran and Javan highlands. *Review of Palaeobotany and Palynology* 55: 207-216.

Sun, X. & Li, X. 1999. A pollen record of the last 37 ka in deep sea core 17940 from the northern slope of the South China Sea. *Marine Geology* 156: 227-244.

Sun, X., Li, X., Luo, Y. & Chen, X. 2000. The vegetation and climate of the last glaciation in the emerged continental shelf of the South China Sea. *Palaeogeography, Palaeoclimatology, Palaeoecology.* 160: 301-316.

Thunell, R., Anderson, D., Geller, D. & Miao, Q. 1994. Sea-surface temperature estimates for the tropical Western Pacific during the last glaciation and their implications for the Pacific Warm Pool. *Quaternary Research* 41: 255-264.

van der Kaars, W.A., 1991. Palynology of eastern Indonesian marine piston-cores: A Late Quaternary vegetational and climatic record for Australasia. *Palaeogeography, Palaeoclimatology, Palaeoecology* 85: 239-302.

van der Kaars, S. 1998. Marine and terrestrial pollen records of the last glacial cycle from the Indonesian region: Bandung basin and Banda Sea. *Palaeoclimates: Data and Modelling* 3: 209-219.

van der Kaars, S., Penny, D., Tibby, J., Fluin, J., Dam, R. & Suparan, P. in press. Palynology and palaeolimnology of a tropical lowland swamp: Rawa Danau, West Java, Indonesia. *Palaeogeography, Palaeoclimatology, Palaeoecology.*

van der Kaars, W.A., Wang, X, Kershaw, A.P. & Guichard, F. 2000. A late Quaternary palaeoecological record from the Banda Sea, Indonesia: patterns of vegetation, climate and biomass burning in Indonesia and northern Australia. *Palaeogeography, Palaeoclimatology, Palaeoecology* 155: 135-153.

W.P.O. (1995) *Westpac Palaeogeographic Maps: The Last Glacial Maximum Palaeogeographic Map for the Western Pacific Region.* Marine Geology Laboratory, Tongji Univeristy, Shanghai.

Wang, X., van der Kaars, S., Kershaw, P., Bird, M. & Jansen, F. 1998. A record of fire, vegetation and climate through the last three glacial cycles from Lombok Ridge core G6-4, eastern Indian Ocean, Indonesia. *Palaeogeography, Palaeoclimatology, Palaeoecology* 147: 241-256.

Wright, H.E. Jnr., Kutzbatch, J.E., Webb III, T., Ruddiman, W.E.F., Street-Perrott, F.A. & Bartlein, P.J. (eds.) 1993. *Global Climates Since the Last Glacial Maximum.* University of Minnesota Press, Minneapolis.

Zheng, Z. & Lei, Z.-Q. 1999. A 400,000 years record of vegetational and climate change from a volcanic basin, Leizhou Peninsula, southern China. *Palaeogeography, Palaeoclimatology, Palaeoecology* 145: 339-362.

The restiads invade the north: the diaspora of the Restionaceae

Barbara G. Briggs

Royal Botanic Gardens, Mrs Macquaries Rd, Sydney, Australia 2000

ABSTRACT: Restionaceae are an almost exclusively southern hemisphere plant family, related to the grasses (Poaceae). They show evidence of trans-oceanic dispersals, as well as an ancient Gondwanic link between Australia and southern Africa. The only exclusively tropical genus of Restionaceae is *Dapsilanthus*, recently segregated from *Leptocarpus*, with three species in northern Australia, New Guinea and the Aru Islands and one in south-east Asia (Malaysia, Cambodia, Thailand, Hainan Island). Growing often in saline soils and producing small readily-dispersible nut fruits, *Dapsilanthus* and similarly the related genus *Apodasmia* (in east and west Australia, New Zealand and Chile) have dispersed far beyond the range of other Restionaceae.

Outstanding features of the Restionaceae are their almost exclusively Southern Hemisphere distribution and their close relationship to the grasses (Poaceae) (Linder *et al.* 1998, Briggs & Johnson 1999, Meney & Pate 1999). Most are dioecious so their successful dispersal requires both sexes to be established. All are wind-pollinated and so do not required associated pollinators; most also do not appear to require particular seed dispersal agents although some are myrmechorous (Pate & Meney 1998). Almost all are restricted to habitats of low soil fertility and when eutrophication occurs Restionaceae are generally the first plants to fail, outcompeted or not thriving in the altered conditions (Meney *et al.* 1999, Pate 2000). They are most abundant in seasonally dry 'Mediterranean' climates, the leaves are reduced to small sheaths and the stems, the main photosynthetic organs, have a range of anatomical structures that appear to protect the chlorenchyma during seasonal drought. The disseminules may be seeds, late-dehiscent capsules, nuts, or nuts surrounded by perianth and bracts.

The main diversity of Restionaceae is in South Africa and the south of Western Australia (Table 1); 32 species occur in eastern Australia (including Tasmania), of which three occur also in Western Australia; six are elsewhere in Australia. Four species (three endemics and one shared with eastern Australia) occur in New Zealand, one in Chile, and one in south-east Asia. Two species extend from northern Australia to New Guinea and the Aru Islands (Table 2).

Relationships among the monocotyledons have recently been greatly clarified by analyses based on DNA sequence data (Chase *et al.* 1993, 2000, Duvall *et al.* 1993) or on both DNA and morphological data (Stevenson *et al.* 2000). Conclusions based on DNA data (Briggs *et al.* 2000, Briggs 2000) indicate that the closest relatives of the Restionaceae are the Centrolepidaceae, Anarthriaceae, Hopkinsiaceae, Lyginiaceae, Ecdeiocoleaceae, Joinvilleaceae, Flagellariaceae, and Poaceae. The families Hopkinsiaceae and Lyginiaceae were recently described (Briggs & Johnson 2000), since the distinctiveness of the south Western Australian *Hopkinsia* and *Lyginia* has become apparent from analyses of DNA sequence data and morphological studies. Studies based on morphology (Kellogg & Linder 1993, Linder *et al.* 2000) give general support to this grouping. Apart from the wide ranging Poaceae, these families all have primarily Southern Hemisphere distributions. Especially notable is their high concentration in the south-west of the Australian continent: seven of these nine allied families occur in that region and four are limited to it. They constitute the Poales as recognised by Dahlgren *et al.* (1985) and form a monophyletic subgroup within a wider concept of the Poales that has been used by the Angiosperm Phylogeny Group (1998) and by Chase *et al.* (2000).

Table 1. Distribution of genera and species of Restionaceae.

Region	Genera	Species
South Africa + Madagascar	19	320
New Zealand	3	4
South America	1	1
New Guinea + Aru Islands	1	2
southeast Asia	1	1
Australia	34	147
south Western Australia	21	109
eastern mainland Australia	14	30
Tasmania	12	14
northern Australia	2	4

In addition to the nine families mentioned above, the findings of Chase *et al.* (2000) and Stevenson *et al.* (2000) lead to the inclusion in Poales of a further eleven families: Cyperaceae, Juncaceae, Eriocaulaceae, Xyridaceae, Bromeliaceae, Sparganiaceae, Typhaceae, Thurniaceae, Mayacaceae, Hydatellaceae and Rapataceae. Of these, the last four are wholly or mainly on the southern hemisphere land masses, but the others occur in both hemispheres. The inclusion of Hydatellaceae among these families had been in doubt until recent DNA sequence data linked it with Xyridaceae (Stevenson *et al.* 2000); recent evidence has also led to Prioniaceae being subsumed into Thurniaceae (Chase *et al.* 2000). The concentration of its closest allies in the south has led to the suggestion that the Poaceae itself may have had a Southern Hemisphere origin (Doyle *et al.* 1992). The identity of fossil pollens identified as Restionaceae from the Northern Hemisphere (e.g. Hochuli 1979) has been queried on closer scrutiny (Linder 1987).

Table 2. Genera and species of Restionaceae with distributions spanning different regions.

Region	Genera	Species
Africa + Australasia	0	0
Australia + New Zealand	3	1
southwest + southeast Australia	7	4
Western Australia +eastern Australia + New Zealand + Chile*	1	0
northern + southeast Australia	1	1
northern Australia + New Guinea + Aru Is.**	1	2
northern Australia + New Guinea + southeast Asia**	1	0

* *Apodasmia*, formerly included in *Leptocarpus* R. Br.

'*A. ceramophilus*' B.G. Briggs & L.A.S. Johnson ined.	Western Australia
A. brownii (Hook. f.) B.G. Briggs & L.A.S. Johnson	eastern Australia
A. similis (Edgar) B.G. Briggs & L.A.S. Johnson	New Zealand
A. chilensis (Gay) B.G. Briggs & L.A.S. Johnson	Chile

**Dapsilanthus, formerly included in *Leptocarpus*

D. spathaceus (R. Br.) B.G. Briggs & L.A.S. Johnson	northern Australia, Aru Is.
D. elatior (R. Br.) B.G. Briggs & L.A.S. Johnson	northern Australia, New Guinea
D. ramosus (R. Br.) B.G. Briggs & L.A.S. Johnson	northern Australia, Aru Is., New Guinea
D. disjunctus (Mast.) B.G. Briggs & L.A.S. Johnson	southeast Asia: Malaysia, Cambodia, Thailand, Hainan

The basal division within Restionaceae appears, from molecular data (Briggs *et al.* 2000) to separate the African and non-African groups. Morphological cladistics, although in considerable agreement with the DNA data, places this division slightly higher, above the origin of the plesiomorphic Australian *Lepyrodia* group of genera (Linder *et al.* 2000). Conclusions from both approaches imply an ancient division between African and Australasian groups, consistent with separation accompanying the fragmentation of Gondwana. The African species outnumber the non-African but include less diversity of features. Among them, only *Restio* reaches the tropics with two species as far north as Malawi and one each in Zimbabwe and Madagascar (Linder 1984, 1986, 1995).

The four New Zealand species, representing three genera, are all congeneric with Australian members or, in the case of *Empodisma minus*, conspecific. Since there has been so little evolutionary divergence, it is unlikely that these links date from Gondwanan times, except possibly for *Sporadanthus*, which belongs to an early lineage within Restionaceae and where flowers and fruit of the two New Zealand species show several notable apomorphies not shared by their Australian congeners. It has one species in the North Island of New Zealand (de Lange *et al.* 1999), one on Chatham Island and six divided between east and west Australia (Briggs & Johnson 1998b).

Beyond the main centres of Africa, Madagascar, Australia and New Zealand, the spread of Restionaceae has been achieved by members of the *Leptocarpus* group of genera (Table 2), which includes the most successful putative migrants of the family, *Apodasmia* and *Dapsilanthus*. These are plants of seasonal swamps and produce numerous tiny nuts, 0.5—2 mm long, that are dispersed with the membranous pericarp surrounded by the perianth and usually a bract. The perianths of some genera of the *Leptocarpus* group bear slender awns and dense hairs, facilitating dispersal by wind or floating on water; *Apodasmia* and *Dapsilanthus* lack these and presumably their trans-oceanic movement depended on transport by birds.

Following recent revision (Briggs & Johnson 1998a, 1999) the *Leptocarpus* group of Restionaceae consists of seven genera (Table 3). The main concentration is in the south of Western Australia and the sole eastern Australian species of both *Leptocarpus* and *Hypolaena* occur also in the west. The four species of *Apodasmia* are the most widely scattered, while *Dapsilanthus* is the only wholly tropical genus of Restionaceae. Of these genera, *Stenotalis*, *Apodasmia* and *Dapsilanthus* are newly segregated from *Leptocarpus*, while four species have been transferred to *Meeboldina* from the much depleted *Leptocarpus*.

Table 3. Distribution of the *Leptocarpus* group of genera

Genus	South-west Australia	South-east Australia	North Aust. and extra-Australia
Leptocarpus	3 species	1 species	—
Hypolaena	8 spp.	1 sp.	—
Stenotalis	1 sp.	—	—
Meeboldina	11 spp.	—	—
Chaetanthus	3 spp.	—	—
Dapsilanthus	—	—	North Aust. 3 spp.; N.G. 2 spp.; S-E Asia 1 sp.
Apodasmia	1sp.	1 sp.	N. Z. 1 sp.; Chile 1 sp.

Chloroplast DNA sequence data confirm the monophyly of the *Leptocarpus* group (Briggs *et al.* 2000), but suggest that the bizzare *Alexgeorgea* of south Western Australia may be a close relative. *Alexgeorgea* produces much larger nuts than related genera and the female flowers are borne on subterranean rhizomes with only the style and stigmas protruding above ground—a remarkable set of apomorphies in a wind-pollinated group. In Restionaceae, basal or near basal female flowers are rare but are found also in *Lepidobolus basiflorus* J.S. Pate & K.A.Meney and in the monoecious *Coleocarya*. A few Poaceae and Cyperaceae similarly have basal spikelets, but in some cases these are amphicarpic with the basal fruits developed from cleistogamous flowers (Cheplick 1987, Bruhl 1994).

Apodasmia chilensis, the sole South American Restionaceous species, is minimally distinct from the New Zealand *A. similis*, giving evidence for relatively recent trans-oceanic dispersal rather than an ancient Gondwanic link. *Apodasmia* is the most notably salt-tolerant genus of the family: the eastern Australian *A. brownii* often grows in foredune coastal sites, sometimes below the level of the highest tides, while a rare Western Australian species, known as '*A. ceramophilus*' but not yet formally named, is found inland around hypersaline lakes; such a genus may be notably well-suited to establish successfully after trans-oceanic dispersal.

At the Australia—Asia collision zone, *Dapsilanthus* has three species in northern Australia, of which two occur also in New Guinea and the Aru Islands, as well as *Dapsilanthus disjunctus* which extends from Malaysia, Cambodia and Thailand to the south-eastern Chinese island of Hainan. Species of Restionaceae characteristically occur in heaths, shrublands and woodlands on low-nutrient soils, mostly in sites that are seasonally moist/seasonally arid. The northern Australian and New Guinea species are mostly on seasonally wet open flats in monsoonal woodlands. Such habitats may have been more extensive during past glacial periods, but now scarcity of suitable habitat would restrict occurrence in this region. The single Asian species extends well to the north of all other Restionaceae and grows in sandy areas near the sea and inland on saline soils. Perhaps—as with *Apodasmia*—salt-tolerance has enabled *Dapsilanthus* to establish after crossing sea barriers, and so to invade the Northern Hemisphere.

REFERENCES

Angiosperm Phylogeny Group (APG). 1998. An ordinal classification for the families of flowering plants. *Ann. Missouri Bot. Gard.* 85: 531-553.

Briggs, B.G. (2000) What is significant--the Wollemi Pine or the southern rushes? *Ann. Missouri Bot. Gard.* 87: 72--80.

Briggs, B.G. & Johnson, L.A.S. 1998a. New genera and species of Australia Restionaceae (Poales). *Telopea* 7: 345-373.

Briggs, B.G. & Johnson, L.A.S. 1998b. New combinations arising from a new classification of non-African Restionaceae. *Telopea* 8: 21-31.

Briggs, B.G. & Johnson, L.A.S. 1999. A guide to a new classification of Restionaceae and allied families. In K.A. Meney & J.S. Pate (eds) *Australian Rushes, Biology, Identification and Conservation of Restionaceae and allied families*: 25-56. Nedlands: University of Western Australian Press.

Briggs, B. G. & Johnson, L.A.S. 2000. Hopkinsiaceae and Lyginiaceae, two new families of Poales in Western Australia, with revisions of *Hopkinsia* and *Lyginia*. *Telopea* 8: 477-502.

Briggs, B.G., Marchant, A.D., Gilmore, S. & Porter, C.L. 2000. A molecular phylogeny of Restionaceae and allies. In K.L. Wilson, & D.A. Morrison (eds) *Monocots—Systematics and Evolution* Proc. 2nd int. conf. comparative biol. monocots, Sydney 1998: 661-671. Melbourne: CSIRO.

Bruhl, J.J. 1994. Amphicarpy in the Cyperaceae, with novel variation in the wetland sedge *Eleocharis caespitosissima* Baker. *Aust. J. Bot.* 42: 441-448.

Chase, M. W., Soltis, D. E., Olmstead, R. G., Morgan, D., Les, D. H., Mishler, B. D., Duvall, M. R., Price, R. A., Hills, H. G., Qui, Y-L., Kron, K. A., Rettig, J. H., Conti, E., Palmer, J. D., Manhart, J. R., Sytsma, K. J., Michaels, H. J., Kress, W. J., Karol, K. G., Clark, W. D., Hedren, M., Gaut, B. S., Jansen, R. K., Kim, K-J., Wimpee, C. F., Smith, J. F., Furnier, G. R., Strauss, S. H., Xiang, Q-Y., Plunkett, G. M., Soltis, P. S., Swensen, S. M., Williams, S. E., Gadek, P. A., Quinn, C. J., Eguiarte, L. E., Golenberg, E., Learn, G. H., Graham, S. W., Barrett, S. C. H., Dayanandan, S. and Albert, V. A. 1993. Phylogenetics of seed plants: an analysis of nucleotide sequences from the plastid gene *rbc*L. *Ann. Missouri Bot. Gard.* 80: 528-580.

Chase, M.W., Soltis, D.E., Soltis, P.S., Rudall, P.J., Fay, M.F., Hahhn, W.H., Sullivan, S., Joseph, J., Molvray, M., Kores, P.J., Givnish, T.J., Sytsma, K.J. & Pires, J.C. 2000 Higher-level systematics of the monocotyledons: an assessment of current knowledge and a new classification. In K.L. Wilson, & D.A. Morrison (eds) *Monocots—Systematics and Evolution* Proc. 2nd int. conf. comparative biol. monocots, Sydney 1998: 3-16. Melbourne: CSIRO.

Cheplick, G.P. 1987. The ecology of amphicarpic plants. *Trends Evol. Ecol.* 2: 97-101.

Dahlgren, R.M.T., Clifford, H.T. & Yeo, P.F. 1985. *The Families of Monocotyledons: Structure, Evolution and Taxonomy*. New York: Springer-Verlag.

de Lange, P.J., Heenan, P.B., Clarkson, B.D. & Clarkson, B.D. 1999. Taxonomy, ecology and conservation of *Sporadanthus* (Restionaceae) in New Zealand. *New Zealand J. Bot.* 37: 413-431.

Doyle, J. J., Davis, J.L., Soreng, R.J., Garvin, D. & Anderson, M.J. 1992. Chloroplast DNA inversions and the origin of the grass family (Gramineae). *Proc. Natl Acad., USA* 89: 7722-7726.

Duvall, M.R., Clegg, M.T., Chase, M.W., Clark,W.D., Kress, W.J., Hills, H.G., Eguiarte, L.E., Smith, J.F., Gaut, B.S., Zimmer,E.A. & Learn, G.H., Jr. 1993. Phylogenetic hypotheses for the monocotyledons constructed from *rbc*L sequence data. *Ann. Missouri. Bot. Gard.* 80: 607-619.

Hochuli, P. A. 1979. Ursprung und verbreitung der Restionaceen. *Vierteljahrschr. Naturf. Ges. Zürich* 124: 109-131.

Kellogg, E.A. & Linder, H.P. 1995. Phylogeny of Poales. Pp. 511-542 in P. J. Rudall, P. J. Cribb, D. F. Cutler & C. J. Humphries (eds) *Monocotyledons: Systematics and Evolution*. Royal Botanic Gardens, Kew.

Linder, H.P. 1984. A phylogenetic classification of the genera of the African Restionaceae. *Bothalia* 15: 11-76.

Linder, H.P. 1986. A review of the tropical African and Malagasy Restionaceae. *Kew Bull.* 41: 99-106.

Linder, H. P. 1987. The evolutionary history of the Poales/Restionales - a hypothesis. *Kew Bull.* 42: 297-318.

Linder, H.P. 1995. *Restio mlanjiensis*, a new species from south-central Africa. *Kew Bull.* 50: 623-625.

Linder, H. P., Briggs, B.G. & Johnson, L.A.S. 1998. Restionaceae. In K. Kubitzki (ed.) *The Families and Genera of Flowering Plants IV*: 425-445. Berlin: Springer-Verlag.

Linder, H. P., Briggs, B.G. & Johnson, L.A.S. 2000. Restionaceae—a morphological phylogeny. In K.L. Wilson, & D.A. Morrison (eds) *Monocots—Systematics and Evolution* Proc. 2nd int. conf. comparative biol. monocots, Sydney 1998: 653-660. Melbourne: CSIRO.

Meney, K.A. & Pate, J.S. (eds) 1999. *Australian Rushes, Biology, Identification and Conservation of Restionaceae and allied families*. Nedlands: University of Western Australia Press.

Meney, K.A., J.S. Pate, J.S., Dixon, K.W., Briggs, B.G. & Johnson, L.A.S. 1999. In K.A. Meney & J.S. Pate (eds) *Australian Rushes, Biology, Identification and Conservation of Restionaceae and allied families*: 465-479. Nedlands: University of Western Australia Press.

Pate, J.S. 2000. Fire response and conservation biology of Western Australian species of Restionaceae. In K.L. Wilson, & D.A. Morrison (eds) *Monocots—Systematics and Evolution* Proc. 2nd int. conf. comparative biol. monocots, Sydney 1998: 685-691. Melbourne: CSIRO.

Pate, J.S. & Meney, K.A. 1998. Morphological features of Restionaceae and allied families. In K.A. Meney & J.S. Pate (eds) *Australian Rushes, Biology, Identification and Conservation of Restionaceae and allied families*: 3-23. Nedlands: University of Western Australia Press.

Stevenson, D.W., Davis, J.I., Freudenstein, J.V., Hardy, C.R., Simmons, M.P. & Specht, C.D. 2000. A phylogenetic analysis of the monocotyledons based on morphological and molecular character sets, with comments on the placements of *Acorus* and Hydatellaceae. In K.L. Wilson, & D.A. Morrison (eds) *Monocots—Systematics and Evolution* Proc. 2nd int. conf. comparative biol. monocots, Sydney 1998: 17-24. Melbourne: CSIRO.

Evolutionary history of *Alectryon* in Australia

K.J. Edwards & P.A. Gadek
School of Tropical Biology, James Cook University, Cairns, Queensland, Australia

ABSTRACT: Phylogenetic analysis of nucleotide sequences from *mat*K, upstream and down-stream spacer regions between *trn*K and *mat*K, *rps*16 and the internal transcribed spacer (ITS-1) were conducted for 14 species of *Alectryon* and five related genera in Sapindaceae. Phylogenetic analysis shows a number of nested clades within *Alectryon,* with *A. subcinereus* representing the first diverging lineage of the genus. Two morphological characters, presence/absence of petals and aril patterning, are congruent with the molecular phylogeny. Two major groupings are evident within *Alectryon* - those with petals and smooth arils and those without petals and patterned arils. The phylogeny of *Alectryon* suggests a temperate origin for the genus, with radiation and speciation northwards into highly seasonal habitats and dispersal to New Guinea, New Zealand and further afield.

1 INTRODUCTION

Alectryon Gaertner (Sapindaceae) currently consists of 25 species, occurring in Melanesia, Philippines, Indonesia, New Guinea, Australia, New Zealand, New Caledonia, New Hebrides, Solomon Islands, Fiji, Samoa, Sandwich Islands and Hawaii. Significantly, more than half the total number of species are found in Australia (15 species, of which 13 are endemic). Most species are fairly narrow endemics, although several are more widely distributed. For example, *A. oleifolius* is widely distributed across Australia, and *A. connatus* is found also in Papua New Guinea. The genus is broadly distributed throughout the Malesian region and tends to occur in areas of drier rainforest where the distribution is consistent with the corridor of seasonally dry habitats that occurs through the Malesian-Australian region (Barlow, 1994).

The genus *Alectryon* was initially circumscribed by Radlkofer (1895, 1933) and comprised 26 species grouped into six sections, the Australian species being distributed in four of these. Since then a number of partial revisions have been undertaken whereby new species have been added and taxa combined. Revisions include the taxa occurring in Malesia (Leenhouts, 1988), Australia (Reynolds, 1985), Hawaii (Linney, 1987) and New Zealand (de Lange et al., 1999). Four Australian species previously comprising *Heterodendrum* Desf. were combined with *Alectryon sens. lat.* after a reassessment of variability in the two genera (Reynolds, 1987). These species were previously distinguished from *Alectryon* on the basis of their simple leaves and drier habitat occurrence (*A. oleifolius,* formerly *H. oleifolium,* extends to the margins of deserts) in contrast to *Alectryon* which was considered to be confined mostly to rainforests (Reynolds, 1987).

Alectryon is a genus in Sapindaceae, a reasonably large family of mainly tropical and subtropical trees and shrubs. The centre of diversity for the family is the south-east Asian region, with the family being well-represented in Australia (30 genera and approximately 200 species). Sapindaceae is divided into two well-circumscribed and accepted subfamilies, Sapindoideae and

Dodonaeoideae, with the former being fragmented into 14 tribes. *Alectryon* is currently accepted as a member of the Nephelieae tribe in the Sapindoideae subfamily (Radlkofer, 1933; Muller & Leenhouts, 1976).

Sapindaceae are generally regarded as showing a Gondwanan origin. Most genera in Australia are restricted to the northern tropical region as elements of tropical rainforest although there is some extension down the east coast as elements of subtropical rainforest. *Alectryon* is well-represented in the flora of both the tropical and subtropical rainforests of Australia, especially the dry monsoonal forests of the northern tropics. It is a particularly interesting genus to study since it contains species that occur in both typical rainforest and the dry monsoonal vine thickets of northern Australia. We are particularly interested in the origins of the flora of the monsoonal vine thickets and would like to establish whether taxa within the genus *Alectryon* share common evolutionary histories in these areas.

The aim of this study was to investigate the evolutionary history of the Australian species of the genus by constructing phylogenetic trees based on DNA sequence data from both the chloroplast and nuclear genomes. This is the first in a series of investigations that are proposed to cover Australian tropical genera in the family. The present paper presents results based on molecular phylogenies for *Alectryon* and discusses the distribution patterns and evolutionary history of *Alectryon* in Australia.

2 METHODS

The taxa investigated include 12 of the 15 Australian species, one species from New Zealand and one from New Guinea. Representatives from five of the six subgeneric sections (Radlkofer 1933) are sampled for this study.

2.1 *DNA extraction, amplification and sequencing*

DNA was isolated using a modified CTAB protocol (Doyle & Doyle, 1987). Maceration of plant tissue was carried out at 60°C with the addition of a pinch of sterile sand. DNA amplifications of *mat*K (including the 5' and 3' spacer regions), *rps*16 and ITS-1 were achieved via the polymerase chain reaction (PCR) to obtain sufficient quantities of DNA for sequencing. All PCR reactions employed *Taq* DNA polymerase (Life Technologies) following the manufacturer's suggested concentrations for all reagents. Primers for amplification and sequencing of *mat*K, *mat*K spacers, *rps*16 and ITS-1 will be described elsewhere. PCR amplification products were purified using PCR® SPINCLEAN™ (PROGEN) prior to sequencing.

Sequencing reactions were carried out using the Big Dye terminator Cycle Sequencing Ready Reaction Kit (Perkin Elmer Applied Biosystems) and products sequenced on an ABI-377 Prism Automated DNA sequencer. The nucleotide sequences were aligned and consensus sequences were constructed using DAPSA (Harley, 1996).

2.2 *Phylogenetic Analyses*

Parsimony analyses were performed in PAUP* v4.0b2a (beta version; Swofford, 1999) using the heuristic search option with random taxon addition, TBR (tree bisection reconnection) branch swapping and MULPARS option. Characters were polarised by the outgroup method. A nested hierarchical approach was followed in the analyses of data from nuclear and chloroplast DNA regions. Initially, a dataset comprising *mat*K and its 5' and 3' spacer regions (*trn*K-*mat*K and *mat*K-*trn*K, respectively) was compiled (not shown here) from representatives of Nephelieae, and analysed using representatives from the Dodonaeoideae subfamily as the outgroup (*Dodonaea viscosa, Distichostemon hispidulus* and *Harpullia ramiflora*). For a subset of taxa (all *Alectryon* representatives, *Pappea capensis, Litchi chinensis* and *Dimocarpus australianus*), sequences from the *rps*16 intron were then added and analysed, using the last two taxa as outgroups. Finally, for a subset of *Alectryon* representatives, sequences of ITS-1 were compiled and analysed both separately and in combination with the previous datasets, rooted on *Litchi chinensis* and *Dimocarpus australianus*.

Relative support for clades identified by parsimony analysis was inferred by using bootstrap (PAUP*) and decay analysis using AutoDecay v2.7 (Eriksson, 1995; Bremer, 1988; Donoghue *et al.*, 1992). Indels (insertion/deletion events) were identified by eye from aligned sequences. Gaps were excluded from the parsimony analysis but those that occurred for more than one taxon and were potentially informative were mapped on to trees later.

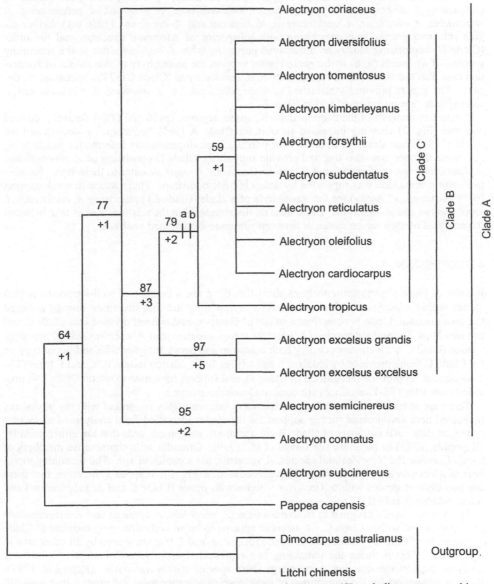

Figure 1. Strict consensus of 90 most parsimonious trees of 141 steps (CI excluding autapomorphies = 0.75, RI = 0.88, RC = 0.78) derived from heuristic analysis of sequence data from *mat*K, spacers and *rps*16. Bootstrap percentages are shown above branches and decay values below. Informative indels a and b are mapped onto the tree.

3 RESULTS

Parsimony analysis of the combined *mat*K, 5' and 3' spacers and *rps*16 sequences (all from the chloroplast genome) gave 90 equally parsimonious trees with the strict consensus tree for the analysis being shown in Figure 1. Analysis of the combined dataset improved support for groupings within *Alectryon* compared to analysis of *mat*K alone (not shown). Within *Alectryon*, there is some support (77% bootstrap, +1 decay) for the monophyly of Clade A sister to *A. subcinereus*

Within Clade A there is moderate support (87% bootstrap, +3 decay) for a clade (Clade B) comprising *A. diversifolius*, *A. tomentosus*, *A. kimberleyanus*, *A. forsythii*, *A. subdentatus*, *A. reticulatus*, *A. oleifolius*, *A. cardiocarpus*, *A. tropicus* and *A. excelsus*. These taxa further divide into two groups, one comprising both subspecies of Alectryon excelsus, and the other (Clade C) comprising mostly an unresolved polytomy with *A. tropicus* sister to the remaining species. Two indels (a, b) in the *rps*16 region support the monophyly of this subset of Australian taxa plus the New Guinea representative, *A. cardiocarpus* (Clade C) (79% bootstrap, +2 decay). The genera previously classified as *Heterodendrum*, i.e. *A. tropicus*, *A. oleifolius* and *A. diversifolius*, are interspersed within Clade C.

Parsimony analysis of the combined *mat*K, spacer regions, rps16 and ITS-1 dataset produced one tree (Fig. 2) showing increased support for Clade A (96% bootstrap, +6 decay) and for Clade C (81% bootstrap, +3 decay). Four potentially phylogenetically informative indels (c, d, e, f) were mapped onto this tree and provide support for Clade D consisting of *A. diversifolius*, *A. kimberleyanus*, *A. forsythii* and *A. subdentatus* with *A. tropicus* sister to these taxa. Resolution within this clade was supported by decay but not bootstrap. There was also weak support (68% bootstrap, +1 decay) for the monophyly of a clade (Clade E) comprising *A. coriaceus*, *A. semicinereus* and *A. connatus*. There were no discrepancies in the relationships of taxa between the analyses of each region alone, or between subsequent combined analyses.

4 DISCUSSION

Results of these phylogenetic analyses show that there are a number of well-supported nested clades within *Alectryon*, with *A. subcinereus* representing the first diverging lineage amongst the taxa sampled. Clade B comprises a subset of Clade A, and is itself divided into Clade C and the two New Zealand subspecies (*A. excelsus* subsp. *excelsus* and *A. excelsus* subsp. *grandis*). Clades B and C are well-supported by both bootstrap and decay analyses with additional support for Clade C being provided by indels a, b and c (Figs 1, 2). Shared indels, d, e, and f, from ITS-1 sequences support the recognition of Clade D, and support for a monophyletic Clade E is provided only after ITS-1 sequences are combined into the analysis.

There are at least two morphological characters that are highly congruent with the phylogeny presented here and provide further support for the clades identified from analysis of molecular sequence data. All members of Clade B are apetalous, and posses arils that are either granular (Reynolds, 1985) or papillose (de Lange et al., 1999). Granular arils characterise members of Clade C, while the New Zealand species, *A. excelsus*, has a papillose aril. The remaining members of *Alectryon* are petaliferous and have smooth arils (Fig. 3). It appears therefore, that there are two distinct groups within *Alectryon* - those with petals (Clade E and *A. subcinereus*) and those without (Clade B).

Species previously assigned to *Heterodendrum* (*A. oleifolius*, *A. tropicus* and *A. diversifolius*) are interspersed within Clade C. *A. tropicus* appears to be an early diverging member of Clade C, supported further by an absence of three indels, d, e, and f, that are shared by all other taxa in this clade. This includes the remaining two representatives of *Heterodendrum*. While our analyses support the decision to submerge these species within *Alectryon* (Reynolds, 1987), there is no support for the recognition of these species as a monophyletic group. It is apparent that the original classification based on the character 'simple leaves' was artificial (Fig. 3).

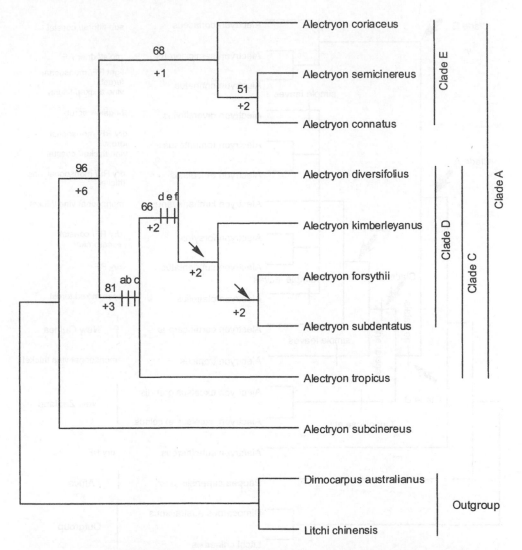

Figure 2. Single most parsimonious tree found from heuristic search of sequence data from *mat*K, spacers, *rps*16 and ITS-1 (length including autapomorphies = 179, CI excluding autapomorphies = 0.75, RI = 0.79, RC = 0.70). Bootstrap percentages are shown above branches and decay values below. Branches that collapse on bootstrap are indicated by arrows. Informative indels c,d,e,f are mapped onto the tree.

A study of distribution patterns of 11 plant taxa within continental Australia (Crisp et al., 1995) suggested that tropical floras appeared to have followed two tracks of speciation with separate histories, resulting in the extant monsoonal/arid flora, and the wet tropical flora. Their preferred area cladogram suggested an ancient vicariance between wet and dry tropical areas. Based on the distribution of extant *Alectryon* our results suggest that *Alectryon* fits into the monsoonal/arid flora clade of Crisp et al. (1995).

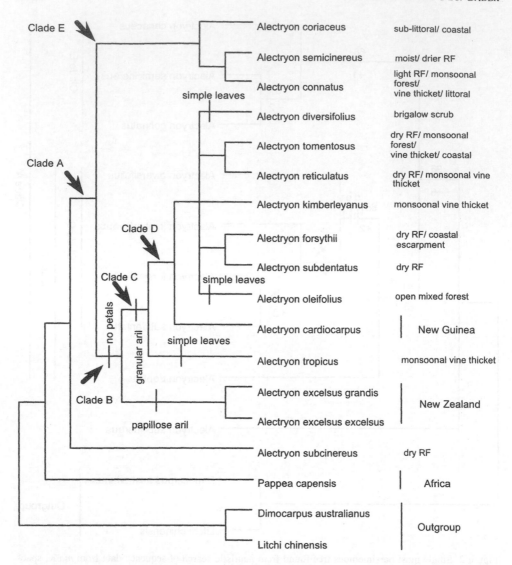

Figure 3. One of the 90 most parsimonious trees derived from heuristic analysis of sequence data from *mat*K, spacers and *rps*16 showing habitat types for Australian *Alectryon*. Support for clades from morphological characters are mapped onto the tree. Simple leaves are shown as a homoplasious state on branches leading to *A. tropicus*, *A. oleifolius* and *A. diversifolius*. All other taxa have compound leaves. Descriptions of habitat affinities were taken from Reynolds (1985) and Hyland & Wiffen (1993).

Both Clades C and E have representatives from monsoonal forests. Table 1 lists the species found in several representative tropical Australian vine thickets and show that *A. connatus* (Clade E) is found together with representatives of Clade C (*A. tomentosus*, *A. diversifolius* and *A. oleifolius*) in the vine thickets at 40 Mile Scrub, Barrabas and Undara. Taken together with the phylogeny presented here, we interpret this to indicate that seasonally dry monsoonal rainforest communities comprise a number of elements that even within the same genus do not share common evolutionary histories. That is, based on the data presented here, the community type (monsoonal vine thickets) does not appear to reflect a common origin for its present constituents.

Table 1. Distribution of *Alectryon* in tropical monsoonal vine thickets (from Kahn & Lawrie, 1987; unpublished species lists).

Species	Clade	Vine thicket locality				
		40 mile scrub	Barrabas	Kinrara	Toomba	Undara
A. connatus	E	x	x			x
A. tomentosus	C			x		x
A. diversifolius	C	x			x	
A. oleifolius	C	x	x	x	x	x

'Dry' rainforests are found in isolated pockets across northern tropical Australia. They are structurally and floristically distinctive (Adam, 1994), show a marked response to seasonality (Gillison, 1987), and predominately comprise taxa usually associated with tropical wet rainforests (Webb & Tracey, 1981). They include such vegetation types as softwood scrubs, araucarian vine thicket and deciduous and semi-evergreen vine thickets. The term 'monsoonal forest' is usually used to encompass the seasonally dry tropical closed-canopy forests (dry rainforests or vine thickets) of the wet-dry tropics of northern Australia.

The origins of these dry monsoonal forests are uncertain, due in part to the lack of fossil evidence, and have been the subject of much debate and controversy. Unlike the tropical and subtropical rainforests of Australia, which have a long fossil record supporting an autochthonous origin, the pre-Holocene fossil record of dry monsoonal plant communities is limited and restricted to anecdotal macrofloral accounts and limited palynological evidence (Greenwood, 1996). Discussion of the evolutionary history of monsoonal forest communities has therefore been restricted to analyses of phytogeographical patterns (Greenwood, 1996). Phytogeographic analyses have suggested an immigrant origin for some floral elements of present-day monsoonal forests (Barlow & Hyland, 1988), while other elements appear to have a common history with the tropical rainforests (Webb et al., 1986).

The phylogeny of *Alectryon* presented here reflects a Gondwanan origin of the genus, the first two diverging lineages comprising species endemic to Australia or New Guinea. Subsequent lineages appear to be the result of dispersal to New Zealand and Papua New Guinea. Within Australia, recent range expansions into new habitats may explain the distribution pattern of the widely distributed *A. oleifolius*. The New Zealand endemic *A. excelsus* is interpreted here to be an example of recent dispersal events rather than the result of vicariance.

Ultimately the genus appears to have had a temperate origin, radiating and speciating northwards into highly seasonal habitats as well as dispersing to New Guinea, New Zealand and further afield. The presence of the Australian taxon *A. connatus* in Papua New Guinea further supports this interpretation and may reflect a dispersal event between Australia and Papua New Guinea across the Torres Strait, possibly during periods of low sea levels during the Pliocene (Barlow, 1994).

An analysis of the geographic affinities of the monsoonal forest taxa of northern Australia (Liddle et al., 1994) found that, by contrast to the ancient endemicity shown by taxa in rainforest and sclerophyll communities, most species extended overseas (to Malesia) and many as far as Africa and America. This suggested that monsoonal rainforest communities were highly dynamic and comprised pantropical taxa that had greater ability to disperse than most of the Australian flora. This also lends support to our suggestion that *Alectryon* has dispersed from Australia into Asia and the Pacific region. Although no fossil evidence is available for *Alectryon* to interpret previous distributions, Vadala & Greenwood (this volume) attribute some taxa appearing in the Cenozoic macrofloras of southeastern Australia to extant *Heterodendrum*. Since our work supports the submersion of *Heterodendrum* in *Alectryon*, this observation suggests that some elements of *Alectryon* have had a long history on the Australian continent.

Further studies are needed to increase the sampling of species within *Alectryon* (particularly extra-Australian members) and to establish congruence in distribution patterns and evolutionary histories for other Australian taxa that have elements in both rainforest and monsoonal forest. These investigations are currently underway.

ACKNOWLEDGEMENTS

We are indebted to the following organisations and people: Deborah Edwards at the Australian National Botanic Gardens, Canberra; Peter Wilson at the Royal Botanic Gardens, Sydney; Karen Sheath at the University of Auckland, New Zealand and the CSIRO Arboretum, Atherton for fresh leaf material; the Australian National Herbarium, Canberra for permission to sample their herbarium collections and Angela Higgins at the UNSW Automated DNA Sequencing Facility. We also thank Barbara Briggs and Hubert Turner for their helpful comments and suggestions that improved the manuscript. This project was supported by a James Cook University Merit Research Grant and a James Cook University Postdoctoral Fellowship to KJE.

REFERENCES

Barlow, B.A. 1994. Phytogeography of the Australian region. In R.H. Groves (ed.), *Australian Vegetation*: 3-35. Cambridge: Cambridge University Press.

Barlow, B.A. & Hyland, B.P.M. 1988. The origins of the flora of Australia's wet tropics. *Proc. Ecol. Soc. Aust.* 15: 1-17.

Bremer, K. 1988. The limits of amino acid sequence data in angiosperm phylogenetic reconstruction. *Evolution* 42: 795-803.

Crisp, M.D., Linder, H.P. & Weston, P.H. 1995. Cladistic biogeography of plants in Australia and New Guinea: congruent pattern reveals two endemic tropical tracks. *Syst. Biol.* 44: 457-473.

de Lange, P.J., Cameron, E.K. & Murray, B.G. 1999. *Alectryon excelsus* subsp. *grandis* (Sapindaceae): a new combination for an uncommon small tree endemic to the Three Kings Islands, New Zealand. *New Zealand J. Bot.* 37: 7-16.

Donoghue, M.D., Olmstead, R.G., Smith, J.F. & Palmer, J.D. 1992. Phylogenetic relationships of Dipsacales based on *rbc*L sequences. *Annals Miss. Bot. Gard.* 79: 333-345.

Doyle, J.J & Doyle, J.L. 1987. A rapid isolation procedure for small quantities of fresh leaf tissue. *Phytochemical Bulletin* 19: 11-15.

Eriksson, T. 1995. Auto Decay v2.7. Harvard University Herbaria: Cambridge, Mass.

Gillison, A.N. 1987. The 'dry' rainforests of *Terra Australia*. In: *The Rainforest Legacy*, 1: 305-321. Canberra: Australian Government Publishing Service.

Greenwood, D.R. 1996. Eocene monsoon forests in central Australia? *Aust. Syst. Bot.* 9: 95-112.

Harley, E.H. 1996. DAPSA: A program for DNA and Protein Sequencing Analysis 3.8. Department of Chemical Pathology, University of Cape Town.

Hyland, B.P.M. & Whiffen, T. 1993. Australian tropical rain forest trees - an interactive identification system, vol 2: East Melbourne: CSIRO Publications.

Kahn, T.P. & Lawrie, B.C. 1987. Vine thickets of the inland Townsville region. In *The Rainforest Legacy*. 1: 159-199. Canberra: Australian Government Publishing Service.

Leenhouts, P.W. 1988. A revision of *Alectryon* (Sapindaceae) in Malesia. *Blumea* 33: 313-327.

Liddle, D.T., Russell-Smith, J., Brock, J., Leach, G.J. & Connors, G.T. 1994. Atlas of the vascular plants of the Northern Territory. Canberra: Australian Biological Resources Study.

Linney, G. 1987. Nomenclature and taxonomic changes in Hawaiian *Alectryon* (Sapindaceae). *Pacific Sci.* 41: 68-73.

Muller, J. & Leenhouts, P.W. 1976. A general survey of pollen types in Sapindaceae in relation to taxonomy. In I. K. Ferguson & J. Muller (eds), *The evolutionary significance of the exine*: 407-445. London: Academic Press.

Radlkofer, L. 1895. Sapindaceae. In A. Engler & K. Prantl (eds), *Die naturlichen Pflanzenfamilien*, 3(5). Leipzig: Wilhelm Engelman.

Radlkofer, L. 1933. Sapindaceae tribus I-VIII, In A. Engler (ed.) *Das Pflanzenreich* 98: 983-1002.

Reynolds, S.T. 1985. Sapindaceae. *Flora of Australia* 25: 4-101.

Reynolds, S.T. 1987. Notes on Sapindaceae V. *Austrobaileya* 2: 328-338.

Swofford, D.L. 1999. PAUP*. Phylogenetic analysis using parsimony (*and other methods). Version 4. Sunderland, Massachusetts: Sinauer Associates.

Vadala, A. J., Greenwood, D. R., Drinnan, A. N. & Banks, M. A. 1999. Late Paleocene and Early Eocene Vegetation in Southeastern Australia. In Metcalfe, I. (ed.), *Where worlds collide: Faunal and floral migrations and evolution in SE Asia-Australasia; Proc. intern. conf.*, Armidale, 29 November - 2 December 1999. Armidale: University of New England.

Webb, L.J. & Tracey, J.G. 1981. Australian Rainforests: patterns and change. In A. Keast (ed.) *Ecological biogeography of Australia*: 605-694. The Hague: Junk.

Webb, L.J., Tracey, J.G. & Jessup, L.W. 1986. Recent evidence for autochthony of Australian tropical and subtropical rainforest floristic elements. *Telopea* 2: 575-589.

Section 5

Non Primates

Australasian distributions in Trichoptera (Insecta) - a frequent pattern or a rare case?

Wolfram Mey

Museum für Naturkunde, Humboldt-Universität, Invalidenstr. 43, D - 10115 Berlin, Germany

ABSTRACT: Australasian distributions are defined as intercontinental ranges with areas in Asia and Australia. The aquatic insect order Trichoptera is reviewed within the context of obtaining figures on the abundance of Australasian distributions at different taxonomic levels. Continuous and disjunct patterns are distinguished. Asia and Australia have 17 families in common. Ten families are widespread and regarded as uninformative. Seven families exhibit Australasian distribution patterns, which are depicted on a global scale and discussed in some detail as to age and origin. Australasian distributions carry a different weight at different taxonomic levels: they make up 60 % in families, about 20 % in genera and 1% or less in species. It is concluded, that only a minor part of the Asian and Australian groups are involved in the Austral-Asian interchange.

1 INTRODUCTION

The term "Australasia" has been used with a different meaning in the biogeographic literature. It denotes a major faunal region, comprising South Asia and Australia, and is thus synonymous with Indo-Australia (e.g. Parsons 1999), or defines the tropical areas of South and Southeast Asia to New Guinea and North Australia or is applied to the Australian biogeographic region including Wallacea (e.g. Brown & Lomolino, 1998; Heads, 1999). Here the term is used strictly in its geographical sense, regarding Asia and Australia together as a combined area including the intervening archipelagos. An Australasian distribution of taxa is therefore an intercontinental range crossing continental boundaries between Asia and Australia.

Contact between the Asian and Australian biotas occurred when the northward moving Australian plate collided with island arcs in the midwestern Pacific in the Late Oligocene to Mid Miocene (Burett et al., 1991; Hall,1996; Polhemus, 1996). These islands were in close proximity to other archipelago systems (e.g. Halmahera and Banda arcs), which provided a stepping stone dispersal route system to the West. The development of this peculiar route allowed Oriental as well as Australian species of caddisflies to penetrate towards the Australian or Asian continent respectively. Since dispersal abilities differ considerably among the expanding taxa different distances from their original ranges were reached. Eventually, some groups entered the opposite continent and became an allochthonous part of the continental fauna. According to this basic dispersal scenario a variety of Australasian distribution ranges might represent origins from either the Asian side or from the Australian side. Of course, not all groups occurring in Australasia are involved. A certain amount of the fauna remains stationary. But what are the real figures? How big is is the mobile fraction of the fauna? Is it only a minor part or are many more species involved? The relationship between expansive and stationary elements is an aspect which defines the faunal composition of the contact zone. Since that relationship is a relevant feature of faunal borders in general, it can be used to quantify the "faunal jump" and to compare the figures with other transition zones.

2 THE TRICHOPTERA AS EXAMPLE GROUP

Quantitative aspects of the faunal exchange have scarcely been estimated. It is a difficult task because it requires the availability of complete species inventories on both Asian and Australian continents. In addition, the full range of phylogenetic analyses of widespread genera have to be considered to determine their probable origin. In insects these prerequisites are only exceptionally met. This applies to caddisflies or Trichoptera too. However, the caddisfly fauna of the Australian Region is comparatively well known. The identification atlas of Neboiss (1986) provides a fine synopsis of all taxa. The Sundaland fauna is extensively documented in Ulmer (1930, 1951). Although the inventory is by no means complete, most of the currently recognised supraspecific taxa are treated allowing a first and rough comparison.

During the last years the phylogenetic relationships among Trichoptera families have been studied intensively (Schmid 1980, Weaver 1984, Wiggins and Wichard 1989). The most elaborate and best documented hypothesis was proposed by Frania and Wiggins (1997). Their cladogram can be used to combine phylogeny with biogeographical pattern.

On this data base the Trichoptera appears to be a suitable candidate group to address the question posed in the title of the article at different taxonomic levels.

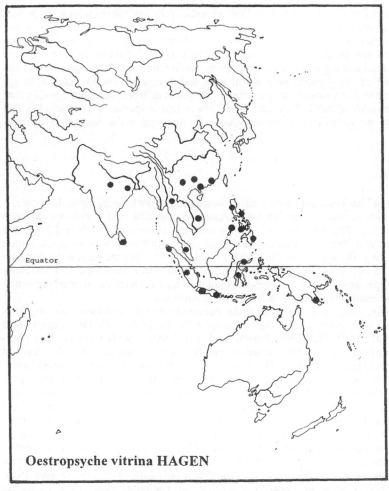

Oestropsyche vitrina HAGEN

Fig. 1: Distribution of Oestropsyche vitrina Hag.(Hydropsychidae: Macronematinae) in Australasia.

3 DISTRIBUTION TYPES

3.1 *Continuous ranges*

Species ranges made up of parts on both Asian and Australian continents are of different size varying from a few islands to large areas. In some cases a single species has conquered a vast region ranging from one continent to the other (fig. 1). The movement could have been in both directions. However, this leaves us with the question whether the species originated in Asia or in Australia? A decision can be made by looking at the range of the sisterspecies or analysing the taxonomic diversity center of the genus or related genera.

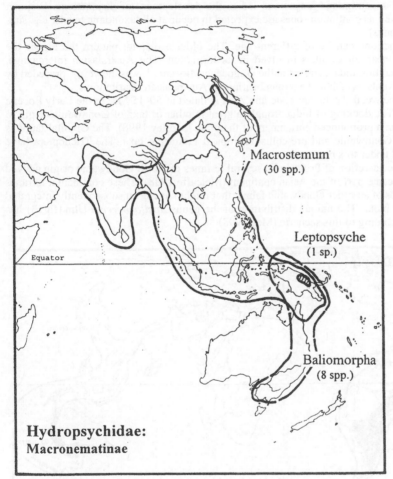

Fig. 2: The ranges of three related genera of Hydropsychidae in Australasia. Species of *Macrostemum* occur in Africa too.

The expansion of species might have been accompanied by speciation processes (= isolation dispersal). It means, that the population in the newly conquered range undergoes speciation. The newly developed species continues the expansion at some time and arrives at the adjacent or next island within the chain of islands between SE Asia and Australia. The resulting congruence between phylogenetical and chorological pattern allows to discern the direction of the range expansion. This evolutionary process was described by Hennig (1950: 212), and later named by himself the "Progression rule". According to this concept an Australasian distribution is composed of ranges of two or more species, which exhibit sister group relationships. The genus *Macrostemum* Kol. provides an excellent example of a taxon widespread throughout the region (fig. 2). Only one species, however, reaches Australia. The majority of species live on continental Asia and Sundaland, which gives an Asian bias to this genus. This type of distribution is supposed to be the most frequent pattern in the Australasian interchange.

3.2 *Disjunct ranges*

The complicated geological evolution and the climatic changes during the Tertiary and Pleistocene had a strong impact on the distribution of the biotas. They triggered new speciations, radiations, migrations and of course also extinctions. The latter two processes caused the disintegration of more or less continuous ranges and led to the development of disjunctions. Disjunct range patterns as well as continuous ones are expected to occur at all taxonomic levels(e.g. subspecies, species, genus).

However, disjunctions can be of different age. The older a disjoint pattern the higher the probability that a vicariant event is involved in its formation. The Australasian interchange, which is still in operation today, started in the Oligocene/Miocene. It was, however, preceded by three much older events, which had a strong bearing on Australasian distributions:

- The collision between the Indian plate and Asia is dated at 50-45 Ma, in the Early Eocene (Packham, 1996). The docking of India brought a great number of taxa of Gondwanic origin to Asia and resulted in a pronounced enrichment of the biota (Morley 1998). The Oriental fauna is thus a mixture of Gondwanic and pre-collision tropical Laurasian taxa. The Trichoptera have not been examined under this perspective.

- Long before the accretion of Proto-India several terranes of North Gondwanic origin arrived at Laurasia and became part of the Asian continent (Metcalfe 1998). These continental terranes have carried species of an older Gondwanic fauna, that might have been subsequently integrated into the Laurasian biota. The unique distribution of the genus *Apsilochorema* Ulm.(fig. 3) has been explained according to this scenario (Mey 1998).

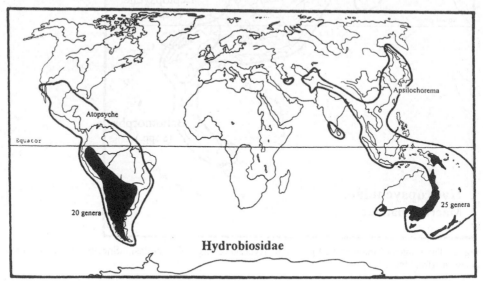

Fig. 3: Distribution area of Hydrobiosidae. Only the range of *Apsilochorema* Ulm. has an Australasian pattern.

A Pangaean configuration of continents existed until the end of the Triassic (Metcalfe 1998). Accordingly, the fossil record provides much evidence for a quite homogenous fauna and flora distributed over that supercontinent (Hallam 1994). After the splitting of this landmass and the separation of the Gondwanic continents from Laurasia an independent evolution of the shared species and groups began. After more than 100 Ma of isolation and divergent evolution on northern and southern continents the old Pangaean groups went extinct or had been transformed, and at the same time gave rise to new evolutionary lines. Most of the groups, which have their roots in the Triassic or Jurassic, are treated today as families or subfamilies. However, not all of them underwent a recognizesable change. Some clades remained unchanged in their morphology and in their environmental preferences, and most interestingly, kept its global distribution, e.g. Philopotamidae.

Besides ancient disjunctions one may expect to find disjunctive patterns of a more recent age. Islands are notorious for plant and animal extinctions (MacArthur & Wilson 1967). Since the formation of the dispersal route between Asia and Australia more than 20 Ma years elapsed. This appears to be long enough to assume that a certain number of dispersal and subsequent extinction events have taken place in the region. A distinction between ancient and young disjunctiv patterns is possible on the basis of a phylogenetic analysis. The distribution of the family Odontoceridae (fig. 4) is an example for a disjunctive Australasian pattern with the open question about its age.

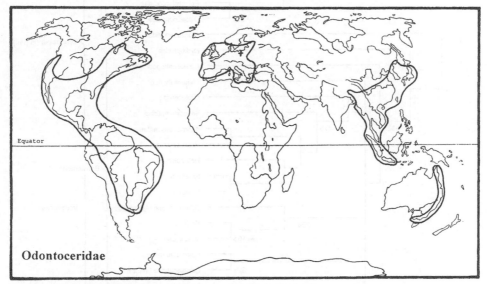

Fig. 4: Distribution area of Odontoceridae.

4 FAMILIES WITH AUSTRALASIAN DISTRIBUTIONS

At present 45 families are recognised in the Trichoptera (Tab. 1). Twenty-eight families are known overall from Asia and from Australia. Seventeen families occur in both continents and thus exhibit an Australasian distribution. However, 10 of these families are widespread and may be regarded as uninformative. They belong to the suborders Spicipalpia and Annulipalpia, which are almost wholly consistent in their representation on continents of the both northern and southern hemispheres. They have a near-cosmopolitan distribution. A pattern of marked contrast is reavealed in the suborder Integripalpia, where most families are endemic to continents of either Laurasia or Gondwana. The contrast suggests that a different arrangement of continents governed the distribution of families of Spicipalpia and Annulipalpia, than was available to most families of Integripalpia. Wiggins (1984) has presented the idea, that family level diversification in Spicipalpia and Annulipalpia preceded the break-up of Pangea, and that most of the Integripalpia families evolved after the break-up. The fossil record corroborates this view. The age of the Trichoptera dates back at least to the Triassic. The family Philopotamidae is known from deposits of this formation. According to the fossil record the caddisflies were already a diverse group in mid-Jurassic (Novokshonov & Sukatcheva 1993). Even the Integripalpia clade is an old branch of the order. Recently, the families Calamoceratidae, Phryganeidae and Plectro-tarsiidae were identified from fossil deposits of the Lower Cretaceous in South England (Su-katcheva 1999).

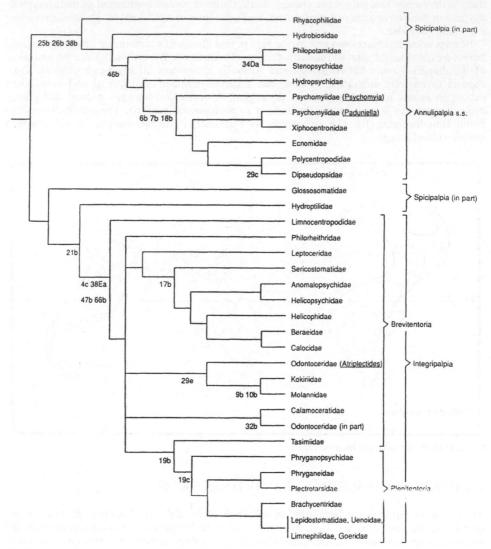

Fig. 5: Strict consensus tree for families of Trichoptera. Figures denote uniquely derived character states (after Frania & Wiggins 1997).

We have two further biogeographic arguments to roughly adjust the family cladogram (fig. 5) to the geological time scale: Relationships within the Limnephilidae have been analysed recently (Gall 1997)(cf. fig. 6). The distributions of the subfamilies seem to be compatible with an origin preceding the break-up of Pangea, because the subfamily Dicosmoecinae, the most primitive group within the family, is represented in both northern and southern hemispheres (fig. 7), whereas the other subfamilies are almost entirely restricted to the northern hemisphere (fig. 8). The second example comes from the Kokiriidae-Molannidae branch in fig. 5. The Molannidae have a continuous northern hemisphere range including the Oriental region, while the species of

Kokiriidae are confined to southern continents. This points again to a Pangean distribution of the ancestors of this clade.

Table 1: The families of Trichoptera and its distribution in the Old World.

Taxon	Asia	Australia	Africa	Europe
Spicipalpia				
1. Rhyacophilidae	x	-	-	x
2. Hydrobiosidae	x	x	-	-
3. Hydroptilidae	x	x	x	x
4. Glossosomatidae	x	x	x	x
Annulipalpia				
5. Philopotamidae	x	x	x	x
6. Stenopsychidae	x	x	x	-
7. Hydropsychidae	x	x	x	x
8. Polycentropodidae	x	x	x	x
9. Xiphocentronidae	x	-	x	-
10. Psychomyidae	x	x	x	x
11. Dipseudopsidae	x	x	x	-
12. Ecnomidae	x	x	x	x
Integripalpia				
13. Phryganopsychidae	x	-	-	-
14. Phryganeidae	x	-	-	x
15. Plectrotarsiidae	-	x	-	-
16. Brachycentridae	x	-	-	x
17. Lepidostomatidae	x	x	-	x
18. Pisuliidae	-	-	x	-
19. Uenoidae	x	-	-	x
20. Goeridae	x	x	x	x
21. Apataniidae	x	-	-	x
22. Rossianiidae	-	-	-	-
23. Limnephilidae	x	x	-	x
24. Oeconesidae	-	x	-	-
25. Odontoceridae	x	x	x	x
26. Limnocentropodidae	x	-	-	-
27. Philorheithridae	-	x	-	-
28. Calamoceratidae	x	x	x	x
29. Atriplectididae	-	x	x	-
30. Leptoceridae	x	x	x	x
31. Kokiriidae	-	x	-	-
32. Tasimiidae	-	x	-	-
33. Molannidae	x	-	-	x
34. Anomalopsychidae	-	-	-	-
35. Antipodoeciidae	-	x	-	-
36. Barbarochthonidae	-	-	x	-
37. Beraeidae	-	-	x	x
38. Calocidae	-	x	-	-
39. Coneosucidae	-	x	-	-
40. Cathamiidae	-	x	-	-
41. Helicophidae	-	x	-	-
42. Helicopsychidae	x	x	x	x
43. Hydropsalpingidae	-	-	x	-
44. Petrothrincidae	-	-	x	-
45. Sericostomatidae	x	-	x	x
	28	28	22	21

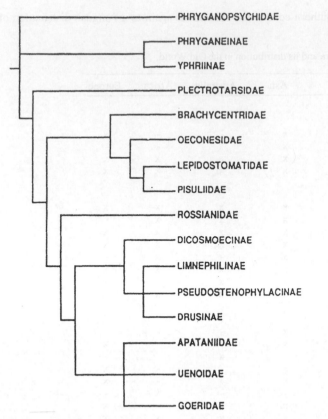

- PHRYGANOPSYCHIDAE
- PHRYGANEINAE
- YPHRIINAE
- PLECTROTARSIDAE
- BRACHYCENTRIDAE
- OECONESIDAE
- LEPIDOSTOMATIDAE
- PISULIIDAE
- ROSSIANIDAE
- DICOSMOECINAE
- LIMNEPHILINAE
- PSEUDOSTENOPHYLACINAE
- DRUSINAE
- APATANIIDAE
- UENOIDAE
- GOERIDAE

Fig. 6: Phylogeny of families of the infraorder Plenitentoria proposed by Wiggins & Gall (after Gall 1997).

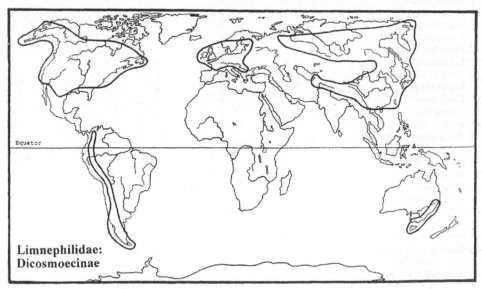

Limnephilidae:
Dicosmoecinae

Fig. 7: Distribution area of the subfamily Dicosmoecinae.

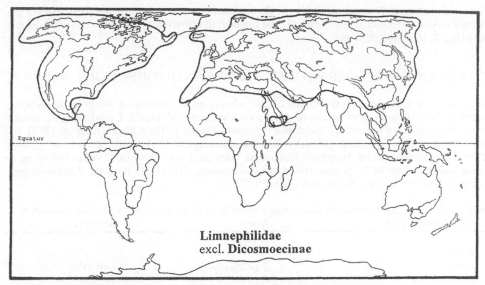

Fig. 8: Distribution area of the Limnephilidae without the range of the subfamily Dicosmoecinae.

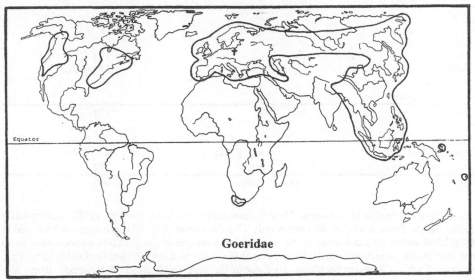

Fig. 9: Distribution area of Goeridae.

In conclusion we probably have to accept that nearly all families ramifying basally from Limnephilidae in the family cladogram are of Mesozoic age. The changing configuration of continents and terranes subsequently affected their distribution. On a global scale the present day distribution patterns of families contain the biogeographic signal which was imprinted with the splitting of the continents during the Mesozoic.

Apart from the 10 uninformative families the remaining 7 families with an Australasian distribution have continuous or disjunct range pattern (figs. 9-12). The families Goeridae, Lepidostomatidae and Dipseudopsidae are clearly of Asian orgin and reach only the northern parts of the Australian region. Reciprocally, only the family Hydrobiosidae has a substantial extension from Australia into the Oriental Region (fig. 3). The ranges of the Stenopsychidae and Limnephilidae: Dicosmoecinae represent old disjunctions which antedates the coming together of

Asia and Australia. They are examples of the stationary groups, which do not participate in the Austral-Asian interchange via dispersal. The bimodal distribution of Odontoceridae in Australasia (fig. 4) is of uncertain age.

5 GENERA AND SPECIES WITH AUSTRALASIAN DISTRIBUTIONS

Turning to the genus level, it is difficult to give a figure on the portion of Australasian patterns, because the total number of genera in Asia has not yet been calculated. I can only give an estimate which is based on country checklists for Japan (Tanida 1987), China (Dudgeon 1987), India (Higler 1992), Thailand (Malicky & Chantaramongkol 1999) and Sunda-Land (Ulmer 1951). The figures on the Australian fauna were calculated from Neboiss (1986). According to these sources a total of 35 genera have Australasian ranges. This is 23 % of the Australian genera and about 19 % of the Asian ones (Tab. 2).

Table 2: Percentage of Australasian distribution patterns in families, genera and species of Trichoptera.

Distributional Type		Asia		Australia
Families				
endemic		2 (= 7%)		6 (= 21%)
cosmopolitan		10		10
Australasian		7 (= 61 %)		7 (= 61 %)
non-Australasian		9		5
total		28		28
Genera				
endemic	ca.	80		113
Australasian		35 (= 19%)		35 (=23%)
Papua/NG		-		6
non-Australasian		65		-
Total	ca.	180		154
Species				
Australasian		8 (< 0,5%)		8 (= 1%)
non-Australasian		3992		792
Total	ca.	4000	ca.	800

In comparing faunas of Sulawesi, New Guinea and Cape York Neboiss (1987) calculated a similar figure. From a total of 80 genera only 17 were common to all three areas, which makes 21 %. Most of the genera belong to the ancient and widespread families and contain species on both continents, which obviously have been there prior to the collision in the Miocene (e.g. *Agapetus* Curtis, *Chimarra* Stephens, *Hydroptila* Dalman, *Diplectrona* Westwood). Every single genus has to be examined to assess whether it is of Asian or of Australian stock. As an example, the distribution of 3 genera of Hydropsychidae is depicted in fig. 2. *Macrostemum* Kolenati is an invasive element to Australia from Asia, whereas its Australian counterpart *Baliomorpha* Neboiss remains restricted to New Guinea and East Australia.The genera are related and demonstrate the bicentric distribution of the family in Australasia. Interestingly, the third genus *Leptopsyche* McLachlan is endemic to New Guinea. In addition to this genus, there are 5 further genera which are endemic on New Guinea and neighbouring islands: *Herbertorossia* Ulmer, *Abacaria* Mosely, *Tanorus* Neboiss, *Symphitoneurina* Schmid, *Oecetinella* Ulmer. Their phylogenetic relationships remain unstudied, but these groups are clear evidence for New Guinea as an secondary diversity centre in Trichoptera.

Table 3: Species of Trichoptera with Australasian distribution patterns (AUS = Australia, HAL = Halmahera, JA = Java, NG = New Guinea, SU = Sulawesi, SUM = Sumatra)

species	range
Apsilochorema gisbum Mosely, 1953	AUS, NG, SUL
Oxyethira incana Ulmer, 1906	SUM, JA, NG, AUS,
Aethaloptera sexpunctata (Kolenati, 1859)	Asia to AUS
Oestropsyche vitrina (Hagen, 1859)	Asia to NG
Dinarthropsis picea Ulmer, 1913	JA, NG
Triplectides gilolensis McLachlan, 1866	JA, HAL, NG
Adicella pulcherrima Ulmer, 1906	SU, JA, NG
Oecetis buitenzorgensis Ulmer, 1951	JA, NG

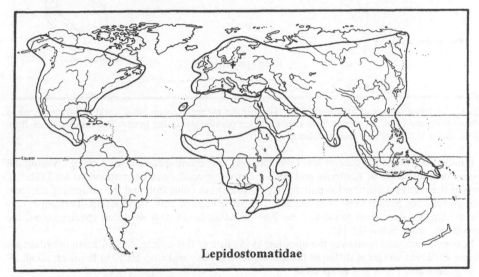

Fig. 10: Distribution area of Lepidostomatidae.

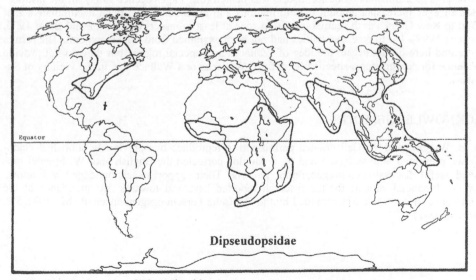

Fig. 11: Distribution area of Dipseudopsidae.

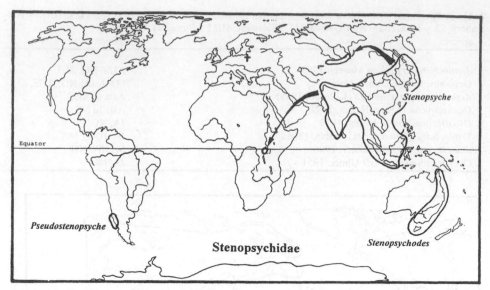

Fig. 12: Distribution area of Stenopsychidae. The disjunct ranges in South Siberia and in Africa belong to *Stenopsyche* and represent relicts from a much larger distribution of the genus. The fossil species from Baltic amber is *Stenopsyche imitata* Ulmer.

Finally, looking to the species level one can ask, how many species have a range crossing the continent boundaries of Australia and Asia ? Only 8 species have been found so far (Tab. 3). Two of them have a distribution pattern from continent to continent, and the remaining six have a range between Sumatra and New Guinea. The percentage of this Australasian distribution pattern within the Australian fauna is 1 %. For the Asian fauna with its richer species count, the percentage drops below 0,5 %.

In conclusion, and to answer the question in the title of this article, Australasian distributions carry a different weight at different taxonomic levels. They make up 60 % in families, about 20 % in genera and 1% or less in species.

Finally, what does this all mean concerning Wallacea? For about 100 years the significance of the area as a transition zone for insects was denied (e.g. Pagenstecher 1909). In the last decades an ever increasing amount of informations and arguments have been accumulated, that point to New Guinea as the actual transition zone at least in insects (e.g. Franz and Beier 1970, Parsons 1999). Up to now, caddisflies had not been examined from this viewpoint. The results presented here do not negate the idea of Wallacea as a special region, but they do not provide evidence for recognising a specifically defined Wallacea or a Wallace line in the vicinity of Sulawesi.

ACKNOWLEDGEMENTS

G. B. Wiggins and W. Gall granted permission to reproduce two previously published cladograms, presented here as figs. 5 and 6. J. Dunlop corrected the English text. W. Speidel provided useful discussion on biogeographic patterns. Their support is acknowledged with appreciation. Financial support for the research reported here and towards my attendance at the conference in Armidale was provided by the Deutsche Forschungsgemeinschaft (Me 1085/5-2, Me 1085/8-1).

REFERENCES

Burret, C., Duhig, N., Berry, R. & Varne, R. 1991: Asian and South-western Pacific continental terranes derived from Gondwana, and their biogeographic significance. *Aus. Syst. Bot.* 4: 13-24.

Dudgeon, D. 1987. Preliminary investigations on the faunistics and ecology of Hong Kong Trichoptera. In M. Bournaud & H. Tachet (eds), *Proceedings of the fifth International Symposium on Trichoptera, Lyon*: 111-117. The Hague: Dr. W. Junk Publishers.

Frania, H. E. & Wiggins, G. B. 1997. Analysis of morphological and behavioural evidence for the phylogeny and higher classification of Trichoptera (Insecta). *Royal Ontario Museum, Life Sci. Contr.* 160: 1-67.

Franz, H. & Beier, M. 1970. Die geographische Verbreitung der Insekten. In J. G. Helmcke, D. Starck & H. Wermuth (eds), *Handbuch der Zoologie, Arthropoda, Bd. IV, (2), nr. 6*, Berlin: Walter de Gruyter Verlag.

Gall, W. K. 1997. Biogeographic and ecologic relationships in the Plenitentoria (Trichoptera). In R. Holzenthal & O. S. Flint (eds), *Proceedings of the eighth International Symposium on Trichoptera*: 109-116. Minneapolis: Ohio Biological Survey.

Hall, R., 1996. Reconstructing Cenozoic SE Asia. In R. Hall & D. Blundell (eds), *Tectonic evolution of SE Asia*: 153-184. Geological Society London, Special Publication 106.

Heads, M. 1999. Vicariance biogeography and terrane tectonics in the South Pacific: analysis of the genus *Abrotanella* (Compositae). *Biol. J. Linn. Soc.* 67: 391-432.

Hennig, W. 1950. *Grundzüge einer Theorie der phylogenetischen Systematik.* Berlin: Deutscher Zentralverlag.

Higler, L. W. G. 1992. A checklist of the Trichoptera recorded from India and a larval key to the families. *Oriental Insects* 26: 67-106.

Malicky, H. & Chantaramongkol, P. 1999. A preliminary survey of caddisflies (Trichoptera) of Thailand., In H. Malicky, & P. Chantaramongkol (eds), *Proceedings of the ninth International Symposium on Trichoptera*: 205-216. Chiang Mai: Chiang Mai University.

Metcalfe, I. 1998. Palaeozoic and Mesozoic geological evolution of the SE Asian region: multidisciplinary constraints and implications for biogeography. In R. Hall, & J.D. Holloway (eds) *Biogeography and Geological Evolution of SE Asia*: 25-41. Amsterdam, The Netherlands: Backhuys Publishers.

Neboiss, A. 1986. *Atlas of Trichoptera of the SW Pacific-Australian Region.* Dordrecht: Dr. W. Junk Publishers.

Neboiss, A. 1987. Preliminary comparison of New Guinea Trichoptera with the faunas of Sulawesi and Cape York Peninsula. In M. Bournaud, & H. Tachet (eds*), Proceedings of the fifth International Symposium on Trichoptera, Lyon*: 103-108. The Hague: Dr. W. Junk Publishers.

Novokshonov, V. & Sukatcheva, I. 1993. Early evolution of caddisflies. In C. Otto (ed), *Proceedings of the seventh International Symposium on Trichoptera, Umea*: 95-99. Leiden: Backhuys Publishers.

Packham, G. 1996. Cenozoic SE Asia: reconstructing its aggregation and reorganisation. In R. Hall & D. Blundell (eds), *Tectonic evolution of SE Asia*: 123-152. Geological Society London, Special Publication 106.

Pagenstecher, A. 1909. *Die geographische Verbreitung der Schmetterlinge.* Jena: Gustav Fischer Verlag.

Parsons, M. 1999. *The butterflies of Papua New Guinea, their systematics and biology.* San Diego: Academic Press.

Polhemus, D. A. 1996. Island arcs, and their influence on Indo-Pacific biogeography. In A. Keast, & S. E. Miller (eds), *The origin and evolution of Pacific Island biotas, New Guinea to Eastern Polynesia: patterns and processes*: 51-66. Amsterdam: SPB Academic Publishing.

Schmid, F. 1980. *Genera des Trichoptères du Canada et des États adjacents.* Ottawa: Agriculture Canada.

Sukatcheva, I. D. 1999. The early Cretaceous caddisfly fauna of England. In H. Malicky, H. & P. Chantaramongkol (eds.): *Proceedings of the ninth International Symposium on Trichoptera*: Chiang Mai: 371-375. Chiang Mai University, Thailand.

Tanida, K. 1987. An introduction to the Trichoptera Fauna of Japan, with a tentative checklist of genera In M. Bournaud, M. & H. Tachet (eds), *Proceedings of the fifth International Symposium on Trichoptera, Lyon*: 119-124. The Hague: Dr. W. Junk Publishers.

Ulmer, G. 1930. Trichopteren von den Philippinen und von den Sunda-Inseln. *Treubia* 11: 373-498.

Ulmer, G. 1951. Köcherfliegen (Trichopteren) von den Sunda-Inseln. Teil 1. *Arch. Hydrobiol. Suppl.* 19: 1-528.

Weaver, J. 1984. The evolution and classification of Trichoptera. Part 1: The groundplan of Trichoptera. In J.C. Morse (ed), *Proceedings of the fourth International Symposium on Trichoptera, Clemson*: 413-419. The Hague: Dr. W. Junk Publishers.

Wiggins, G. B. 1984. Keynote address: Trichoptera, some concepts and questions. In J.C. Morse (ed), *Proceedings of the fourth International Symposium on Trichoptera, Clemson*: 1-12, The Hague Dr. W. Junk Publishers.

Wiggins, G. B. & Wichard, W. 1989. Phylogeny of pupation in Trichoptera, with proposals on the origin and higher classification of the order. *J. N. Amer. Benthol. Soc.* 8: 260-276.

Butterflies and Wallace's Line: faunistic patterns and explanatory hypotheses within the south-east Asian butterflies

R. L. Kitching, R. Eastwood & K. Hurley

Cooperative Research Centre for Tropical Rainforest Ecology and Management, Griffith University, Brisbane, Qld 4111, Australia

ABSTRACT: This paper reanalyses the distribution of the butterflies (Lepidoptera: Papilionoidea and Hesperioidea) in the south-east Asian region and adjacent mainland land masses using the most recent published information on distributions. The overall butterfly faunas of the Malay Peninsula, Borneo, Sulawesi, New Guinea and Australia are compared at the Family, Sub-family, Tribe, Genus and Species level. Distinct contrasts among these faunas arise only at the tribal level and below. Close linkages between the Malay Peninsula and Borneo, Australia and New Guinea are demonstrated at these taxonomic levels. The expected isolation of Sulawesi again emerges from the analyses. Detailed analyses of the Papilionidae, Nymphalidae-Danainae and Nymphalidae-Satyrinae are presented as both isopleth maps showing variation in species richness across the region, and in terms of Sørensen similarity values between adjacent islands (and other land masses). A set of discontinuities emerge in one, two or all three of these taxa which may be explained using an appropriate combination of tectonic, evolutionary and ecological mechanisms. Hypotheses are posed to explain each of the observed discontinuities and, where appropriate, further analyses are suggested.

1 INTRODUCTION

It was Henry Walter Bates, writing in *A Naturalist on the River Amazons,* who urged us to study butterflies and, in particular, the brilliantly coloured patterns on their wings. "On these expanded membranes nature writes as if on a tablet, the story of the species so truly that all changes of the organisation register thereon..."(Bates 1892). Alfred Russel Wallace, whose eponymous line is the focus for this volume, was equally aware of the scientific power to be derived from the study of butterflies but used them to gain insight into biogeography as well as the processes of evolution (Wallace 1876). The reasons for this choice of subject matter are as compelling today as they were in Wallace's day. The butterflies are taxonomically well known, their distributions even in the most obscure of locations are relatively well documented thanks to legions of collectors who seek them, and they are sufficiently species-rich that patterns can be sought, and found, at the level of whole faunas.

The butterflies, in fact, are but two superfamilies of the insect Order Lepidoptera – the Papilionoidea or 'true' butterflies, and the Hesperioidea or skippers. These two superfamilies are generally thought to be monophyletic and, together with the Neotropical Hedyloidea form a sister-clade to the Geometroidea (Minet 1991, Scoble 1982, Weller & Pashley 1995, Ackery *et al.* 1998). It is current practise to recognize five major families within the Papilionoidea with up to three minor groups assigned family or sub-family status reflecting the convictions or otherwise of different workers (Ehrlich 1958, Kristensen 1976, Ackery 1984, Scott 1985, Kitching 1999). The Hesperioidea are accepted to comprise but the single family Hesperiidae (Evans 1937 *et seq.*). In total about 20 000 species of butterfly have been described (Smart 1976) and, although new species are described annually, they represent by far the best known of all insect groups. Whereas we could argue that as few as a half (or less) of the world's beetles have been described, the currently named butterflies probably account for over 90% of extant species.

The downside to using the butterflies for biogeographic analysis is that there is a dearth of fossils in the group and those that exist (perhaps 80 or so specimens in all, R. de Jong, *pers. comm.*) are rather uninformative in revealing higher level relationships. The oldest known butterfly fossils are from about the Paleocene/Eocene transition and these can be placed comfortably within modern sub-families (Shields 1976). Virtually all butterfly fossils are Caenozoic with Colorado's Floris-sant beds providing the richest source (Emmel *et al.* 1992). The supposition is that the butterflies diverged from their sister-group sometime in the Cretaceous and co-radiated with the angiosperms (Shields 1976, Feltwell 1986). A few sub-families are currently associated with particular conti-nents but most are virtually cosmopolitan in occurrence. Interesting parallels at the generic level have led authors to suggest Gondwanic connections (see, for example, Kitching & Dunn 1999) but the requisite molecular analyses that might confirm such linkages have not been carried out.

In this paper we present an analysis of the distribution of the butterflies of Wallacea and adjacent regions as a whole (aggregated into families, sub-families, tribes, genera and species) then examine in more detail the patterns of species richness and endemism in the Papilionidae, Danainae and Satyrinae. We identify major biogeographical discontinuities within our results and present testable hypotheses that may account for the observed patterns. We present these analyses against newly available tectonic information (Metcalfe 1998, Hall 1998) which provides a much stronger underpinning to understanding both the history of the current land masses themselves and the butterfly species which occur upon them than has been available to most previous workers.

Of course, many key contributions to understanding the distribution of the butterflies of the region have been made since Wallace's time and we build upon these. Most notable of these predecessor works is that of Holloway and Jardine (1968) who used multi-variate methods based on taxonomic coincidence of species within the Indo-Australian region. They identified both zooge-ographic sub-regions in this fashion and sets of species having similar distributions, which they called faunal elements. Holloway (1998) substantially expanded these earlier analyses adding phylogenetic information on selected groups of butterflies (and cicadas) as well as reviewing the various philosophical approaches and explanatory tools available for explaining observed biogeo-graphic patterns. The approach taken in this paper falls within the paradigms defined by Holloway although we use both different methods and somewhat updated data sets. We conclude by posing a series of testable hypotheses to account for key discontinuities in the distributions of butterflies in the region.

2 METHODS, DATA BASES AND SUBJECT TAXA

We present here two analyses of butterfly distributions. In the first we compare and contrast the entire butterfly faunas of five major land-masses spanning the region – the "Major Land-Mass Analyses" and, in the second, we present an island by island analysis of three taxa for which detailed distribution data was readily available – "Selected Taxa Analyses".

2.1 *Major Land-Mass Analyses*

Regional species' lists (see below) are utilised to compare the species richness across five major land masses within the Indo-Australian region: the Malay Peninsula, Borneo, Sulawesi, New Guinea and continental Australia. We examine the similarities among these faunas at the family, sub-family, generic and species levels. For the Lycaenidae alone we add a tribal level analysis. The following key works are employed: Malay Peninsula – Corbet *et al.* (1992); Borneo – Otsuka (1988), Seki *et al.* (1991) and Maruyama & Otsuka (1991); Sulawesi – a working list kindly provided as a personal communication by R. de Jong & R. I. Vane-Wright; New Guinea – Parsons (1999); Australia – Nielsen *et al.* (1996).

The grouping of species into higher taxa largely follows the scheme used by Nielsen *et al.* (1996) with additional decisions for the Nymphalidae based on Harvey (1991) and for the Lycaenidae, Eliot (1973). Authorities for names and other nomenclatural details can be found in Ackery & Vane-Wright (1984, Danainae), Miller (1968, Satyrinae), Nielsen *et al.* (1996 Papilionidae) and Tsukada (1982, 1985a, 1985b, all groups). We have recognised the Libytheidae and Riodinidae as family-level designations while being fully aware that these are arbitrary decisions.

For similarity analyses at the family, sub-family and tribal levels we have calculated species counts for each higher category and then computed parametric correlation coefficients ('r') among all pairs of land-masses. At the generic and species level we calculated Sørensen's indices to estimate the level of similarity among the five land masses. We use the standard formula:

$$\text{Index Value} = \frac{2c}{a+b}$$

where c is the number of taxa shared across the two land masses being compared, and a and b are the total number of taxa in the first and second land masses respectively.

2.2 *Analyses of selected taxa*

We selected three higher taxa for which detailed island by island distribution maps are available (or can readily be constructed). We relied heavily on the published maps of Tsukada's compendium *The Butterflies of the South-east Asian Islands* (1982, 1985a, 1985b) augmented for extralimital regions by the handbooks already referenced plus D'Abrera (1982, 1985, 1986). The three taxa, Paplionidae, Nymphalidae-Danainae and Nymphalidae-Satyrinae, were chosen partly on the grounds of availability of data but also because they represent groups with different histories, life-cycles and general biologies.

The *Papilionidae* are swallowtail butterflies and, as a family, are distributed world-wide. They include the largest and most robust of butterflies and are potentially highly mobile. There are 700+ species worldwide (Smart 1976) but are not particularly associated with any one continent in terms of species richness. All Indo-Australian species belong to the nominate sub-family, the Papilioninae. Additional subfamilies occur in North (Parnassiinae) and South America (Baroniinae) and in Europe (Parnassiinae). Papilionidae are represented in the Wallacean region and adjacent areas by 207 species and while a few species (including, within the region *Papilio demoleus* and *P. fuscus*) are known to be migratory most are not. Papilionids utilise a wide range of foodplants ranging from weedy herbs to rain-forest trees. However, within the region, species are particularly associated with plant families such as the Rutaceae, Lauraceae and Monimiaceae with a significant (possibly monophyletic) subset associated with the highly toxic vine genus *Aristolochia* (Aristolochiaceae).

The *Danainae* are a sub-family of the Nymphalidae comprising about 160 species world-wide. Their biology and systematics have been described in detail by Ackery and Vane-Wright (1984) on whose work this brief account is based. Although cosmopolitan as a family there is a marked focus in terms of both generic and species richness in south-east Asia (see below). Like the papilionids these are, for the most part, large robust species with a propensity for long-distance movement. Species of *Danaus, Euploea, Tirumala, Ideopsis* and *Parantica* are known migrants with the prodigious north-south flights of *D. plexippus* in North America having been widely studied (see Urquhart 1960, Malcolm & Zalucki 1993). As their common name, 'milkweed butterflies', implies members of this family feed as larvae almost exclusively on plants of the family Asclepiadaceae and the allied families Apocynaceae and Moraceae. Plants belonging to these families frequently contain toxic latexes and the use of these toxins by the butterflies as anti-predator devices is well-known. Many species are boldly marked presumably as advertisement of their toxic status.

The *Satyrinae* are a large sub-family of the Nymphalidae, commonly referred to as the browns or satyrs, with between 1200 and 1500 species worldwide (Ehrlich & Raven 1965). There are 394 species in the region of concern to our analysis. In general the satyrines are small to medium sized butterflies as adults and are not particularly strong fliers. No species are known to be migratory, and highly speciose genera have evolved by adaptive radiation particularly in mountainous (eg. *Erebia* spp.) and forested (eg. *Mycalesis* spp., *Hypocysta* spp.) regions. Satyrines exploit monocotyledonous plants such as grasses, sedges, rushes, bamboos and palms as larval hosts.

For each of these three taxa contour maps were constructed for the regions with isopleths indicating species richness (following the techniques described previously in Kitching 1981, Kitching & Scheermeyer 1993 and Kitching & Dunn 1999). We have also constructed a global isopleth map for danaine genera and separate maps for the two major tribes of Satyrinae in the region to

clarify the basis of particular patterns seen in the local sub-family level analyses. We calculated
Sørensen similarity coefficients for the Papilionidae, Daninae and Satyrinae among all pairs of
islands and other land masses in the region. These were transcribed to a second map in each case
summarising the coefficient values. We used these maps to identify the major faunal discontinuities (ie those transitions with lower Sørensen Index values) for which we propose explanatory
hypotheses (see Discussion).

3 RESULTS

3.1 Major Land-Mass Analyses

The results of these analyses are presented in Tables 1 and 2. Table 1 shows similarities among
species distributions at the family, sub-family and tribal level based on the calculated correlation
coefficients. Table 2 presents similarity coefficients at the generic and species level across all
pairs of land-masses.

Table 1. Comparisons of the butterfly faunas of five major land masses spanning Wallace's line in
the Indo-Australian region. Correlation coefficients are presented based on the numbers of species
found in each land mass at (a) the family level (entire faunas), (b) the sub-family level (entire
faunas), and (c) the tribal level (Lycaenidae only).

(a) Species within Families (all butterflies)

	Malay Peninsula	Borneo	Sulawesi	New Guinea	Australia
Malay Peninsula	1.00	-	-	-	-
Borneo	1.00	1.00	-	-	-
Sulawesi	0.93	0.92	1.00	-	-
New Guinea	0.94	0.95	0.92	1.00	-
Australia	0.96	0.95	0.82	0.91	1.00

(b) Species within Sub-families (all butterflies)

	Malay Peninsula	Borneo	Sulawesi	New Guinea	Australia
Malay Peninsula	1.00	-	-	-	-
Borneo	1.00	1.00	-	-	-
Sulawesi	0.70	0.70	1.00	-	-
New Guinea	0.81	0.81	0.86	1.00	-
Australia	0.66	0.66	0.88	0.77	1.00

(c) Species within Tribes (Lycaenidae only)

	Malay Peninsula	Borneo	Sulawesi	New Guinea	Australia
Malay Peninsula	1.00	-	-	-	-
Borneo	0.99	1.00	-	-	-
Sulawesi	0.73	0.89	1.00	-	-
New Guinea	0.49	0.56	0.72	1.00	-
Australia	0.46	0.50	0.68	0.96	1.00

Table 2. Similarity values for the butterfly faunas of five major land masses spanning Wallace's Line based on co-occurrence of taxa across pairs of land masses at (a) the generic level and (b) the species level. The numbers of cosmopolitan taxa, taxa unique to a single land-mass and taxa unique to any two land masses are also presented.

(a) Genera

	Malay Peninsula	Borneo	Sulawesi	New Guinea	Australia
Malay Peninsula	1.00	-	-	-	-
Borneo	0.936	1.00	-	-	-
Sulawesi	0.685	0.695	1.00	-	-
New Guinea	0.445	0.447	0.557	1.00	-
Australia	0.347	0.338	0.448	0.656	1.00

Cosmopolitan Genera	64
Genera restricted to one land mass	69
Genera restricted to two land masses	92

(b) Species

	Malay Peninsula	Borneo	Sulawesi	New Guinea	Australia
Malay Peninsula	1.00	-	-	-	-
Borneo	0.744	1.00	-	-	-
Sulawesi	0.219	0.232	1.00	-	-
New Guinea	0.083	0.08	0.139	1.00	-
Australia	0.092	0.085	0.148	0.264	1.00

Cosmopolitan Species	41
Species restricted to one land mass	278
Species restricted to two land masses	465

One pattern is outstanding at all levels of this analysis. The extremely close affinity between the faunas of the Malay Peninsula and Borneo is apparent throughout with correlations of close to 1.0 at the family, subfamily and tribal levels. These two landmasses share 94% of genera and 74% of species of butterflies so the higher level correlations are perhaps not surprising. Apart from this, useful (ie. interpretable) pattern does not emerge until the lower taxonomic levels.

At the family level little pattern is evident. The distribution of species across families differs little among the major land masses with an average correlation of 0.93 ± 0.01. The same lack of pattern is evident at the subfamily level with correlations overall averaging 0.79 ± 0.03 in a range from 0.66 to 1.00 (Table 1b).

At the tribal, generic and species level however, patterns are evident and readily interpretable. At the tribal level, for the Lycaenidae (Table 1c), the faunas of the Malay Peninsula, Borneo and Sulawesi form a highly similar group (r values from 0.73 to 0.99). Similarly the faunas of Australia and New Guinea are highly correlated (r = 0.96). The least similar distribution of species across Tribes is evident when comparing the Malay Peninsula and Australia (r=0.46).

Similarity analyses at the generic and species level (Table 2), inasmuch as they deal taxon by taxon, are in general more revealing than the correlation analyses based on aggregated numbers of species. At these lower taxonomic levels an increasing range of similarity values is observed. At the generic level 55% ($\pm6\%$) of taxa are shared with a range of values from 34% to 94%. At the species level there is an average similarity of only 21% ($\pm6\%$) of species and a range from 8% to 74%.

As with the analysis of lycaenid tribes, the faunas of the Malay Peninsula, Borneo and Sulawesi share many genera (68-94%). Australia and New Guinea share 66% of genera and New Guinea and Sulawesi a high 56%. Again the lowest levels of generic similarity are between the Malay

Peninsula and Borneo on the one hand and Australia on the other (34-35%).
At the species level there is a marked contrast between the similarity index of the Malay Penin-
sula and Borneo (74% of species shared) and all other comparisons (8-26%). The next highest
values are between Australia and New Guinea (26%), Borneo and Sulawesi (23%) and the Malay
Peninsula and Sulawesi (22%). All other pairs show less than 16% similarity and all pair-wise
comparisons involving the Malay Peninsula and Borneo, on the one hand, and New Guinea and
Australia, on the other, indicate 9% or less shared species.
In interpreting the generic and species level similarities the relatively high number of cosmopolitan
taxa must be borne in mind (64 genera and 41 species).

3.2 *Analyses of selected taxa*

In this section we present pairs of maps, the isopleth diagrams for each taxon with the similarity
analyses carried out upon species numbers across adjacent islands and other land masses. Table 3
represents a summary of the discontinuities suggested by the similarity analyses and indicates the
relative importance of each particular geographical boundary within and across the selected taxa.

TABLE 3. Discontinuities in butterfly faunas in south-east Asia and adjacent regions. Scores in
the body of the table represent ranked differences between the faunas of adjacent regions following similarity analyses with 4 indicating lowest levels of similarity and 0 indicating very similar
faunas.

Comparison	Papilio- nidae	Danai- nae	Satyri nae	Sum of Scores	Rank[1]	Explanatory hypotheses
Borneo/Sulawesi	4	3	4	11	1	Edge of Sunda Shelf, tectonic isolation, endemism on Sulawesi
New Guinea/Australia	3	2	4	9	2	Contrasting climates, vegetation
Sulawesi/Lesser Sundas	2	2	4	8	3	Contrasting geological, histories, climates, vegetation
Moluccas/Lesser Sundas	4	1	2	7	4=	Contrasting geological, histories, climates, vegetation
Sulawesi/Moluccas	3	1	3	7	4=	Palaeogeography, Island effects, endemism, New Guinea elements
Sula/Moluccas	2	1	3	6	6	Isolation, island size, Satyrine low vagility
Borneo/Southern Philippines	4	0	3	7	7	Sunda Shelf, tectonic isolation, endemism in Philippines
Australia/Lesser Sundas	3	0	2	5	8	Geological history, species/area effects, Asian elements in Sundas
Sulawesi/Sula	0	2	1	3	9	Isolation, island size, Sulawesi endemism
New Guinea/Moluccas	0	0	4	4	10	Small vs continental islands, Endemism in new Guinea
Palawan/Northern Philippines	0	0	1	1	11=	Isolation, geological history, endemism in Philippines
Palawan/Borneo	0	0	1	1	11=	Attenuation, island size

[1] Based on both the frequency of the pattern across taxa and its strength within taxa.

3.2.1 *Papilionidae*
The distribution of species richness in the swallowtail butterflies is presented in Figure 1. Local
faunas in the region range from less than 10 species on much of the Australian continent to over
55 species in the Indo-Chinese interior (parts of Thailand, Myanmar and Laos see Pinratana 1979,
1992; D'Abrera 1982). This region represents a clear 'peak' of the richness contours with, gener-

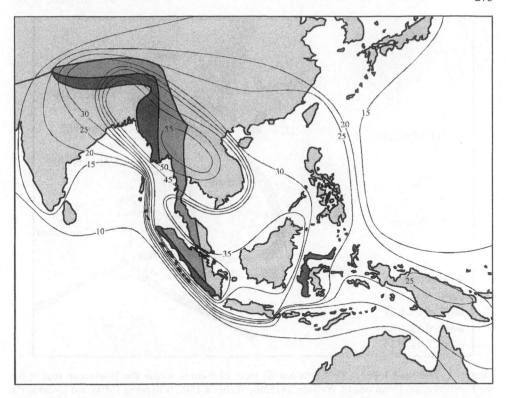

Figure 1. Species isopleths for the Papilionidae within the Indo-Australian region (isopleth increment = 5). Shading on the background map represents putative tectonic terranes derived from
Gondwanaland (after Metcalfe 1998).

ally speaking a falling off in numbers outwards from this region across the Wallacean region
to Australasia. Two anomalies in this otherwise relatively simple situation are apparent. First,
Sumatra is particularly rich in species with over 40 species (compared with, for example, 30-35 in
the Malay Peninsula and in Java, and 35-40 in Borneo). Second, there is a sub-peak in papilionid
numbers in northern New Guinea (25+ species), largely due to the New Guinea radiation of the
troidine bird-wing butterflies. In summary the isopleths for the Papilionidae show the group reaching peaks in the higher elevation, rain-forested mainland of Asia with secondary peaks in the large,
wet, topographically diverse islands of Sumatra, Borneo and New Guinea.

The similarity analyses for the Papilionidae are presented in Figure 2. The most notable discontinuities (Sørensen's Index values less than 24%) in the faunas are between Borneo (and the southern Philippines) and Sulawesi, Borneo and the southern Philippines, Sulawesi and the Lesser Sundas, and between Australia and New Guinea. Secondary discontinuities (SI values from 25-34%)
are present between Sulawesi and the Moluccas, and between Australia and the Lesser Sundas.
In traditional terms these results suggest a more complex pattern than is usefully summarised by
any of the named 'lines'. In addition to the Borneo/Sulawesi divide the most compelling feature
emerging from this analysis is the west to east divide between the Lesser Sundas and islands to
their north which extends eastwards as a break between New Guinea and continental Australia. For
the Papilionidae there are no major discontinuities between the Moluccas and New Guinea, and
the island of Sula is clearly linked to Sulawesi rather than to the Moluccan islands to the East.

3.2.2 Nymphalidae-Danainae
Numbers of species within the subfamily Danainae range from six on the mainland of Australia
up to 35 on the great Sundanese islands of Sumatra and Java (Fig. 3). Unlike the Papilionidae the

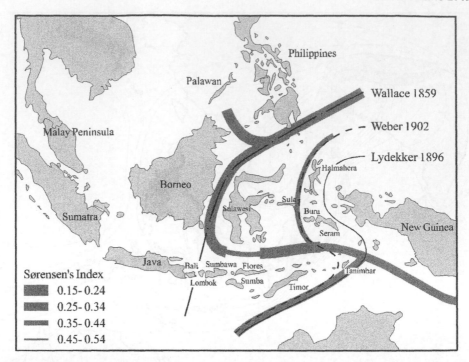

Figure. 2. Sørensen Index values between all pairs of islands within the Wallacean region for the Papili-onidae. Locations of Wallace's (1859), Weber's 1902 (Pilseneer 1904) and Lydekker's (1896) lines are also shown.

Danainae have a distinctly Wallacean focus with numbers falling off across the Asian mainland and to the north and west. As is the case in the Papilionidae, there is a distinct secondary peak of numbers in northern New Guinea with 25 species. Sulawesi is also particularly rich in danaines with 33 species whereas, surprisingly, Borneo has only 27 species recorded. Again Sumatra is richer than the adjacent Malay Peninsula. The 'Wallacean' focus in danaine richness is a global rather than simply a local – that is: Asian - phenomenon. Figure 5 shows the global distribution of the sub-family in terms of generic diversity highlighting the peak of six genera found in New Guinea and the Greater Sunda Islands in contrast to but three in Africa and Central America.

Discontinuities identified from the similarity analyses are presented as Figure 4. Again the greatest divergence seen in the fauna is that between Borneo (and the southern Philippines) and Sulawesi which, in this taxon, extends north to discriminate between the Philippine fauna and those to the east. Notable differences are also apparent between the Sulawesi fauna and those in Sula and the Moluccas to the East, between Sulawesi (and the Moluccas) and the lesser Sundas to the South, and between Australia and New Guinea. Lesser discontinuities are apparent between Sula and the Moluccas, between Borneo and the southern Philippines and between Timor and the rest of the Lesser Sundas.

3.2.3 Nymphalidae-Satyrinae
There are more species of Satyrinae in these analyses than of either of the other two higher taxa analysed here. Figure 6 shows that species richness is geographically bimodal with a peak of over 85 species in the Assam/Myanmar region and a second slightly smaller one of more than 55 species in central New Guinea. Geographically intermediate regions have intermediate numbers of species although, as in the previous two analyses the island of Sumatra again stands out as richer than the adjacent Malaysian mainland and, in this case, than any of the other Greater Sunda Islands. In an

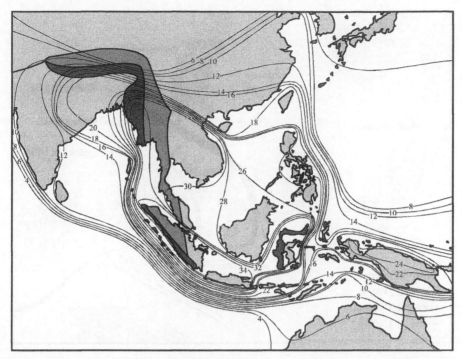

Figure. 3. Species isopleths for the Nymphalidae-Danainae within the Indo-Australian region (isopleth increment = 2). Shading on the background map represents putative tectonic terranes derived from Gondwanaland (after Metcalf 1998).

attempt to understand this pattern further we examined the distribution of species in the satyrine tribes represented in the region. The Indo-Australian Satyrinae divide into five tribes (following the scheme of Miller 1968): the Biini with 13 species, the Elymniini with 248 species, the Eritini with 8 species, the Ragadiini with 16 species and the Satyrini with 109 species. We have mapped the richness patterns for the Satyrini and Elymniini (Figure 8a and 8b). The bimodal distributions noted earlier persist in the maps for both of these tribes although the relative contribution of each to the two 'peaks' is different. The Elymniini, although diverse in New Guinea, are particularly so in the Assam/Myanmar region (60+ species compared with 30+ species) whereas the Satyrini show the reverse pattern (24+ species in New Guinea compared with a peak of 14+ species in the Assam/Myanmar region.

Returning to the analyses of the entire Satyrinae, Figure 7 summarises the Sørensen similarity indices. As with the other taxa one of the largest discontinuities observed is between the faunas of Borneo and Sulawesi. This is matched in magnitude by that between Sulawesi and the Lesser Sundas, between the Moluccas and New Guinea, and between New Guinea and Australia. Major discontinuities of somewhat lesser magnitude are present between Borneo and the Philippines, between Sulawesi (plus Sula) and the Moluccas, and between the Moluccas and the Lesser Sundas. Finally, minor discontinuities isolate both Sula and Palawan in the Satyrinae.

4 DISCUSSION

In this discussion we review briefly the three classes of mechanism within which explanations for biogeographic patterns may be found. We then bring together these possible explanations with the patterns we have observed (mostly summarised in Table 3). Of course many of the discontinuities observed in the results of our analyses can be aligned with one or other of the many lines, or parts

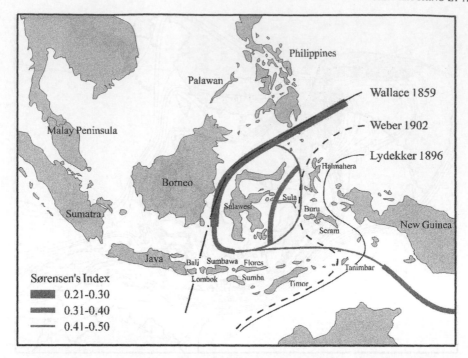

Figure. 4. Sørensen Index values between all pairs of islands within the Wallacean region for the Nym-phalidae-Danainae. Locations of Wallace's, Weber's and Lydekker's lines are also shown.

of them, that have blessed the classificatory approach to biogeography in this region (Simpson 1977). We cannot but agree with Simpson that Huxley's (1868) Line (that is: Wallace's line modified to align with the edge of the Sunda Shelf and extend west of the Philippines) and Lydekker's (1896) Line (aligned with the boundary of the Sahul Shelf) are real biogeographical boundaries, the region in between, called 'Wallacea' by Dickerson (1928), Holloway and Jardine (1968) and many other authors, divides up in an almost endless number of ways depending on the taxa being examined and the statistical approach taken. Simpson (1977) sensibly suggests that this intervening region be excluded from sub-regional classification and that it should be studied for what it is, one of the most biogeographically dynamic and varied regions on the face of the earth. Accordingly we treat each of the discontinuities we have identified on its own merits and propose mechanistic hypotheses to account for them. Where possible we suggest further analyses that might be used to test these hypotheses.

4.1 General Considerations

In trying to account for the natural patterns of distribution of taxa across a geographical region three types of process are involved: the tectonic (including eustatic), the evolutionary and the ecological (including climatic) (Holloway 1998, Kitching & Dunn 1999 etc).

4.1.1 Tectonic explanations

These assume that the movement of continental plates and associated terranes remained, at least in part, above sea level and carried the ancestors of the taxa we observe currently. We suppose that the descendent taxa have remained closely attached to the original land mass and that we can use this to account for their present distributions. When dealing with a group such as the butterflies we must be convinced that the taxa we are dealing with are of sufficient age as to have existed throughout relevant periods of tectonic movement. Secondly, highly mobile species may

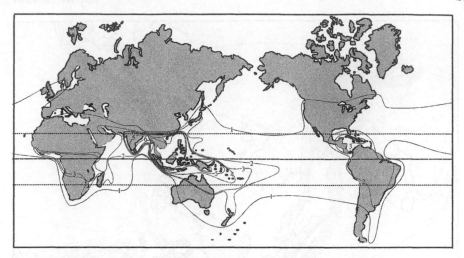

Figure. 5. Genera isopleths for the Nymphalidae-Danainae globally (isopleth increment = 1)

well confound any tectonically based pattern as a result of recent dispersal among land masses. We assume butterflies have radiated to their current levels of diversity since the Cretaceous and that the basic clades within the butterfly superfamilies are also relatively ancient. Accordingly we expect to see tectonically induced distribution patterns only in the lower levels of classification in which clades at say the tribal or generic level have originated following post-Pangaean break-up. It is also likely that many highly diverse species-groups that are currently encountered are relatively recent in origin. To put it another way we are unlikely to be able to detect tectonic impacts at either extreme of the taxonomic hierarchy on the grounds of extreme age of taxa on the one hand, and extreme youth on the other.

4.1.2 Evolutionary processes
In seeking explanations for patterns of butterfly diversity evolutionary processes will be important. An ancestor species, isolated on a large island such as Sulawesi or Sumatra, for example, may well give rise to a group of descendent species as the processes of adaptive radiation driven by climatic cycles, vegetation and topographic change lead to speciation events. Such 'clouds' of closely related species then act as potential emigrant species to smaller and more remote land-masses were further, perhaps unidirectional, adaptation may occur to produce highly local endemic species. In general the more diverse the land-mass in terms of vegetation or topography the more likely it is that such adaptive radiation will have occurred. More isolated land-masses (or their constituent terranes) are also more likely to have produced such species-groups since the occasional rare arrivals of ancestor species would remain isolated for long periods of time. Any consequent genetic change would not be swamped by frequent further arrivals of original stock from the source population. Such adaptive radiations are a likely explanation where we observe several closely related species, often with very limited distributions on a single land mass (eg., *Euploea* spp. in Sulawesi, *Oreixenica* spp. in Australia, *Arhopala* spp. in Borneo and so on).

4.1.3 Ecological processes
Lastly, distributions on islands may reflect simple, current time, ecological processes as the constituent dynamics of immigration and extinction interact in the fashion so clearly encom-passed by the Theory of Island Biogeography (MacArthur & Wilson 1967). In these situations co-occurring faunas interact across barriers until an equilibrium number of species is reached where the characteristic immigration and extinction curves intersect. In situations driven by these mechanisms we expect, in general, to see more species on larger, less remote islands, than on smaller, more distant ones. The actual equilibrium number of species, however, will also be affected by the climate, topography and vegetation of the land-mass concerned. To summarise, where ecological

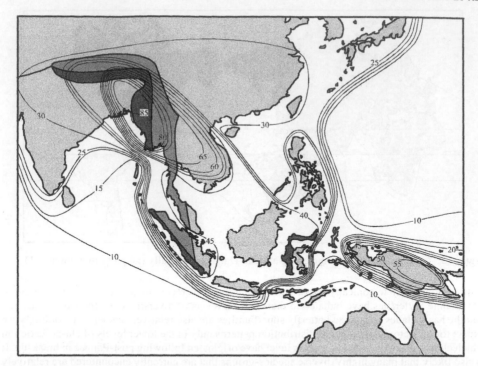

Figure. 6. Species isopleths for the Nymphalidae-Satyrinae within the Indo-Australian region (isopleth increment = 5). Shading on the background map represents putative tectonic terranes derived from Gondwanaland (after Metcalf 1998).

effects are pre-eminent we expect to see species shared across land masses (endemic radiations may subsequently occur but only over a much longer time scale) with a marked species-area effect, perhaps modified by local environmental conditions.

4.2 Wallaciam Discontinuities

Turning briefly to each of the patterns noted in our analyses we can identify the most likely explanations for the differences (or lack of differences) that emerge.

4.2.1 Malay Peninsula/Borneo
The extraordinarily high levels of similarity between these two land masses reflect a deep time common tectonic history and substantial, perhaps repeated, periods of land-bridge connection during glacial maxima. Moss and Wilson (1998) suggest that there was a land connection between Borneo and mainland south-east Asia throughout the Tertiary period. There are no grounds for regarding these faunas as separate at any fundamental level. Some levels of endemism at the species level have, of course, arisen but many of the major radiations in species groups (eg. Arhopala, Elymnias, Euthalia) are shared by the two land masses. We conclude that the best explanation for this level of similarity is one of shared tectonic history and long term geographical connectedness.

4.2.2 Borneo/Sulawesi, Borneo/southern Philippines
The contrast in richness and faunal composition between Borneo and Sulawesi and between Borneo and the southern Philippines is marked in all our analyses. Indeed this is the strongest repeated pattern across all three of the taxa analysed in detail (Table 3). This segment of Wallace's Line (as modified by Huxley) is alive and well as far as the butterflies are concerned! The reasons for

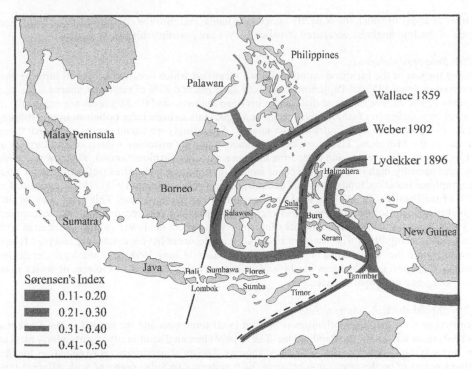

Figure. 7. Sørensen Index values between all pairs of islands within the Wallacean region for the Nym-phalidae-Satyrinae. Locations of Wallace's, Weber's and Lydekker's lines are also shown.

this division have been discussed extensively by others (see Simpson 1977 and related references). Basically this segment of Wallace's line represents the edge of the Sunda shelf projecting from the Asian mainland. Essentially there is a deep time isolation of the two land masses implying separate developments of fauna at least since the early Tertiary (Moss & Wilson 1998). In addition non-'Asian' terranes with closest affinities in the Moluccas, New Guinea and the Philippines gradually accreted onto the 'Asian' parts of Sulawesi to build-up the present land mass. Such diverse terranes will have carried with them species ancestral to those curently seen in Sulawesi and, further, permitted island-hopping for additional Philippine or Australasian faunal elements. These species, added to Asian elements (for which some Bornean affinity might be expected) produce an overall fauna quite distinct from the Bornean one. In addition the topography and shape of the Sulawesi complex in recent times will have enhanced processes of specialisation and development of very local endemisms – differentiating even further the butterfly fauna. In similar fashion the southern Philippines are geologically diverse in origin and represent a quite separate geological development from more western land masses.

4.2.3 *Sulawesi/Lesser Sundas, Moluccas/Lesser Sundas, Australia/New Guinea*

We group these three discontinuities together because climatic explanations can be adduced for each. The northern elements in each pair, which have considerably richer faunas in each case, are moister, more mesic islands with a basic rainforest vegetation (prior to human clearing). The southern elements are distinctly drier with rainforest elements scattered in a matrix of scleromorphic, savannah vegetation. These contrasts are probably sufficient to ensure the major differences observed between the faunas. This west/east climatic transition extends further to differentiate the Australian and New Guinea faunas in the east. In addition, the contrasting geological histories of the Lesser Sundas in comparison with both the Moluccas and Sulawesi further contributes to faunal differences. Hypothesising, however, that this discontinuity is climate-driven suggests further analysis in which the butterfly faunas of islands, matched for size and position, north and south of the line might be compared statistically. We are pursuing this line of inquiry.

4.2.4 Sulawesi/Moluccas
This is the last of the various discontinuities we identified which occurred across all three groups analysed in detail (35% of Papilionidae, 45% of danaines and 28% of satyrines, shared – see also Figures 2, 4 & 7). The observed differences between Sulawesi and the Moluccas are explicable in several complementary fashions. First, Sulawesi represents a single large (albeit strangely shaped) island compared with many much smaller islands. Accordingly we would expect a less rich fauna overall in the Moluccas. This is eminently testable again by matching islands of similar size in different parts of the region and comparing them in a pair-wise fashion. Second, we know that there are unexpectedly high levels of endemism among the Sulawesi butterflies (reflecting, *inter alia*, geographical isolation, topography and geomorphology) (Vane Wright 1991). This is expressed as 40% of the Papilionidae, 47% of the Danainae and 74% of the Satyrinae. This alone may be sufficient to account for the differences between the Sulawesi and Moluccan faunas. The exceptionally high levels of endemism in the Satyrinae may well reflect the lower vagility associated with this taxon in comparison with the other two. Last, the Moluccas have undoubtedly received faunal elements from the New Guinea area to the east (see also de Jong 1998) to a much greater degree than has Sulawesi (which shares but 14% of its butterfly fauna with New Guinea, of which most are widespread cosmopolitan species).

4.2.5 Sula/Moluccas, Sulawesi/Sula
The first of these two discontinuities is present in all three taxa and the second in the Danainae and Satyrinae but not for the Papilionidae. The Sula/Moluccan discontinuity is particularly marked for the Satyrinae (only 21-30% of species shared). The most parsimonious explanation for this pattern seems to be the interaction between the remoteness of Sula compared with adjacent land masses and the general low levels of mobility to be expected of the Satyrinae. For the Danainae only 31-40% of the fauna are shared between Sulawesi and Sula and for the Satyrinae this figure is 41-50%. These differences are likely to reflect island size and the high degree of endemism shown by the Sulawesi fauna discussed above.

4.2.6 Australia/Lesser Sundas
There are clear discontinuities between Australia and the Lesser Sundas (here taken to include Timor and Tanimbar) for both the Papilionidae (25-34% shared species) and the Satyrinae (31-40% shared). This is particularly so given that only the north western Australian regional fauna was used for comparison. The Lesser Sundas present both small island effects and contain an attenuated Asian element contributed by dispersal from the larger islands of the Indonesian arc. In addition, there is dispersal from New Guinea in the west along the Banda Arc island chain as suggested by Michaux (1991) since many of these islands rose from the ocean floor. These effects too will build to the observed differences.

4.2.7 Palawan/northern Philippines, Palawan/Borneo
These two differences are present only in our analyses of the Satyrinae and hence, a priori, are likely to reflect the lower vagility of this sub-family. They indicate a modest degree of isolation of the island of Palawan both from Borneo to the south and the northern Philippines to the north. In both cases 40-50% of species are shared. There are three levels of likely explanation for these differences. First, and probably least important in the case of the Satyrinae, is the geological contrasts. Palawan was in direct land connection with Borneo during recent glacial maxima. The islands of the Philippine archipelago have a complex geological history but have no long-standing affinity with the Asian mainland. Second, Palawan is an isolated island and hence would be expected to have both lower equilibrium species numbers and relatively high numbers of endemic forms. Third, both Borneo and the Philippines have faunas that are of diverse origins and which include many local endemics. The long island of Palawan has acted as a faunal bridge between Borneo and the northern Philippines containing significant elements from both faunas which are individually quite different.

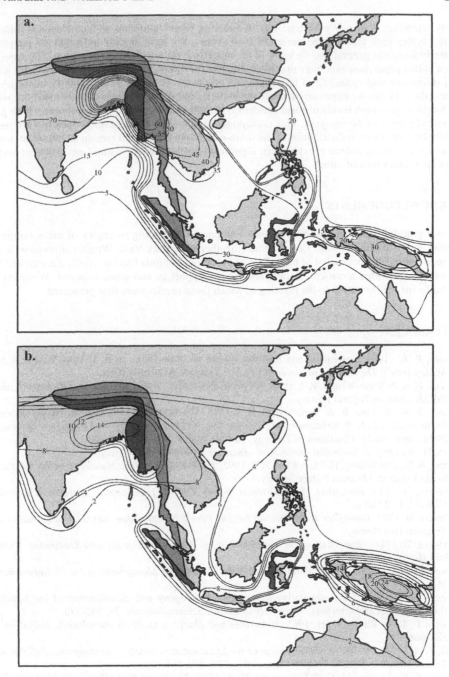

Figure. 8. Species isopleths for two satyrine tribes within the Indo-Australian region. (a) The Elymniini (isopleth increment = 5) (b) The Satyrini (isopleth increment = 2).

We suggest that our results confirm Bates' and Wallace's contentions that the butterflies remain vital instruments for biogeographical analyses. We suggest though that an all encom-passing, 'final' palaeo-biogeographic analysis of south-east Asian Lepidoptera will remain an elusive goal. Apart from the complexity of contributing historical factors, there remains a great deal of uncertainty regarding timing. We have no way of knowing what significant random events at different spatial scales took place or the timing of such events. We have already indicated the paucity of butterfly fossils in general and we know of no butterfly fossils at all from the region. The approach taken in this paper, however, may provide a template for future, more sophisticated, analyses which will refine our understanding of biogeographic processes. It would be particularly interesting to extend the analyses we have carried out to the butterfly families Lycaenidae and Hesperiidae where we might expect high levels of local endemism as well as the sheer species richness which gives statistical power to the analyses. This is currently not possible because of the lack of sufficiently detailed information on distributions. The south-east Asian region will remain an important testing ground for biogeographers provided that significant portions of the natural environment and the associated biota remain intact.

ACKNOWLEDGEMENTS

We are grateful to Henry Barlow of Kuala Lumpur for providing us copies of maps on the Satyrinae and to Rienk de Jong, Leiden, who (together with Dick Vane-Wright) allowed us access to the unpublished species list of butterflies of Sulawesi. Melinda Laidlaw, Sally Stewart and Guy Vickerman provided technical, critical and moral support as and when required. We thank Ian Metcalfe for the invitation to the meeting at which these results were first presented.

REFERENCES

Ackery, P. R. .1984. Systematic and faunistic studies on butterflies. In R. I. Vane-Wright & P. R. Ackery (eds), *The Biology of Butterflies*: 9-21. , London: Academic Press.

Ackery, P. R. & Vane-Wright, R. I. 1984. *Milkweed Butterflies: their Cladistics and Biology*. London: British Museum Natural History.

Ackery, P. R., de Jong, R. & Vane-Wright, R. I. 1998. The butterflies: Hedyloidea, Hesperioidea and Papilionoidea. In N. P. Kristensen (ed), *Lepidoptera, Moths and Butterflies. I Evolution, Systematics and Biogeography* Handbook of Zoology, Part 35. Berlin: Walter de Gruyter.

Bates, H. W. 1892. *A Naturalist on the River Amazons*. London: John Murray.

Corbet, A. S., Pendlebury, H. M. & Eliot, J. N. 1992. *The Butterflies of the Malay Peninsula*, 4th Edition Kuala Lumpur: Malayan Nature Society.

D'Abrera, B. 1982. *Butterflies of the Oriental Region. Part 1. Papilionidae, Pieridae & Danaidae*. Victoria: Hill House.

D'Abrera, B. 1985. *Butterflies of the Oriental Region. Part 2. Nymphalidae, Satyridae and Amathusidae*. Victoria: Hill House.

D'Abrera, B. 1986. *Butterflies of the Oriental Region. Part 3. Lycaenidae and Riodinidae*. Victoria: Hill House.

Dickerson, R. E. 1928. The distribution of life in the Philippines. *Monographs of the Philippine Bureau of Science* 21: 1-322.

Ehrlich, P. R. 1958. The comparative morphology, phylogeny and classification of the butterflies (Lepidoptera: Papilionoidea). *University of Kansas Science Bulletin* 39: 305-370.

Ehrlich, P. R. & Raven, P. H. 1965. Butterflies and plants: a study in coevolution. *Evolution* 18: 586-608.

Eliot, J. N. 1973. The higher classification of the Lycaenidae: a tentative arrangement. *Bulletin of the British Museum Natural History, Entomology* 28: 373-505.

Emmel, T. C., Monno, M. C. & Drummond, B. A. 1992. *Florissant Butterflies: a Guide to the Fossil and Present-day Species of Central Colorado*. Stanford: Stanford University Press.

Evans, W. H. 1937. *A Catalogue of the African Hesperiidae, indicating the classification and nomenclature adopted in the British Museum*, London: British Museum (Natural History).

Feltwell, J. 1986. *The Natural History of Butterflies*, Beckenham: Croom Helm.

Hall, R. 1998. The plate tectonics of Cenozoic SE Asia and the distribution of land and sea. In R. Hall & J.D. Holloway (eds), *Biogeography and Geological Evolution of SE Asia.*: 99-131. Leiden: Backhuys.

Harvey, D. J. 1991. Appendix B. Higher classification of the Nymphalidae. In H. F. Nijhout *The Development and Evolution of Butterfly Wing Patterns:* 255-273. Washington: Smithsonian Institution Press.

Holloway, J. D. 1998. Geological signal and dispersal noise in two contrasting insect groups: R-mode analysis of the Lepidoptera and cicadas In R. Hall & J.D. Holloway (eds), *Biogeography and Geological Evolution of SE Asia.*: 291-314. Leiden: Backhuys.

Holloway, J. D. & Jardine, N. 1968. Two approaches to zoogeography: a study based on the distributions of butterflies, birds and bats in the Indo-Australian area. *Proceedings of the Linnean Society of London* 179: 153-188.

Huxley, T. H. 1868. On the classification and distribution of the Alectromorphae and Heteromorphae. *Proceedings of the zoological Society of London* 1868: 294-319.

de Jong, R. 1998. Halmahera and Seram: different histories, but similar butterfly faunas In R. Hall & J.D. Holloway (eds), *Biogeography and Geological Evolution of SE Asia.*: 315-325. Leiden: Backhuys.

Kitching, R. L. 1981. The geography of the Australian Papilionoidea. In A. Keast (ed), *Ecological Biogeography of Australia:* 979-1005. The Hague: Junk.

Kitching, R. L. 1999. The higher classification of butterflies. In R. L. Kitching, E. Scheermeyer, R. E. Jones & N. E. Pierce (eds), *Biology of Australian Butterflies:* 25-32. Melbourne: CSIRO.

Kitching, R. L. & Dunn, K. L. 1999. The biogeography of Australian butterflies. In R. L. Kitching, E. Scheermeyer, R. E. Jones & N. E. Pierce (eds), *Biology of Australian Butterflies:* 53-74. Melbourne: CSIRO.

Kitching, R. L. & Scheermeyer, E. 1993. The comparative biology and ecology of the Australian danaines. In S. B. Malcolm & M. P. Zalucki (eds), *Biology and Conservation of the Monarch Butterfly:* 165-175. Los Angeles: Natural History Museum of Los Angeles County.

Kristensen, N. P. 1976. Remarks on the family-level phylogeny of butterflies (Insecta, Lepidoptera, Rhopalocera). *Zeitschrift für zoologische Systematik und Evolution-forschung* 14: 25-33.

Lydekker, R. 1896. *A Geographical History of Mammals.* Cambridge: Cambridge University Press.

MacArthur, R. & Wilson, E. O. 1967. *The Theory of Island Biogeography,* Princeton: Princeton University Press.

Malcolm, S. B. & Zalucki, M. P. (eds). 1993. *Biology and Conservation of the Monarch Butterfly.* Los Angeles: Natural History Museum of Los Angeles County.

Maruyama, K. & Otsuka, K. 1991. *Butterflies of Borneo Vol. 2, No. 2. Hesperiidae.* Tokyo: Tobishima Corporation.

Metcalfe, I. 1998. Palaeozoic and Mesozoic geological evolution of the SE Asian region: multidisciplinary constraints and implications for biogeography. In R. Hall & J.D. Holloway (eds), *Biogeography and Geological Evolution of SE Asia.*: 25-41. Leiden: Backhuys.

Michaux, B. 1991. Distributional patterns and tectonic development in Indonesia: Wallace reinterpreted. *Australian Systematic Botany* 4: 25-36.

Miller, L. G. 1968. The higher classification, phylogeny and zoogeography of the Satyridae. *Memoirs of the American entomological Society* 24: 1-174.

Minet, J. 1991. Tentative reconstruction of the ditrysian phylogeny (Lepidoptera: Glossata). *Entomologica Scandinavica* 22: 69-95.

Moss, S. & Wilson, M. 1998 Biogeographic implications from the Tertiary palaeographic evolution of Sulawesi and Borneo In R. Hall & J.D. Holloway (eds), *Biogeography and Geological Evolution of SE Asia.*: 133-164. Leiden: Backhuys.

Nielsen, E. S., Edwards, E. D. & Rangsi, T. V. (eds) . 1996. *Monographs on Australian Lepidoptera, 4: Checklist of the Lepidoptera of Australia.* Melbourne: CSIRO.

Otsuka, K. 1988. *Butterflies of Borneo.* Volume 1, Tokyo: Tobishima Corporation.

Parsons, M. 1999. *The Butterflies of Papua New Guinea: Their Systematics and Biology,* London: Academic Press.

Pilseneer, P. 1904. La 'ligne de Weber', limite zoologique de l'Asia et de l'Australie. *Bulletin. Classe de Sciences. Academie Royale de Belgique.* 1904: 1001-1022.

Pinratana, A. 1979. *Butterflies of Thailand* Volume 1, Bangkok: Viratham Press.

Pinratana, A. 1992. . *Butterflies of Thailand* Volume 3, Bangkok: Viratham Press.

Scoble, M. J. 1982. The structure and affinities of the Hedyloidea: a new concept of the butterflies. *Bulletin of the British Museum (Natural History), Entomology* 53: 251-286.

Scott, J. A. 1985. The phylogeny of butterflies (Papilionoidea and Hesperioidea). *Journal of Research in the Lepidoptera* 23: 241-281.

Seki, Y., Takanami, Y & Otsuka, K. 1991. *Butterflies of Borneo Vol. 2, No. 1. Lycaenidae.* Tokyo: Tobishima Corporation.

Shields, O. 1976. Fossil butterflies and the evolution of the Lepidoptera. *Journal of Research on the Lepidoptera* 15: 132-143.

Simpson, G. G. 1977. Too many lines; the limits of the Oriental and Australian zoogeographic regions. *Proceedings of the American philosophical Society* 121: 107-120.

Smart, P. 1976. *The Illustrated Encyclopaedia of the Butterfly World in Colour.* London: Hamlyn.

Tsukada, E. (ed). 1982. *Butterflies of the South-east Asian Islands. Vol. 1. Papilionidae,* Tokyo: Plapac.

Tsukada, E. (ed). 1985a. *Butterflies of the South-east Asian Islands. Vol. 2. Pieridae, Danaidae,* Tokyo: Plapac.

Tsukada, E. (ed). 1985b. *Butterflies of the South-east Asian Islands. Vol. 3. Satyridae . Satyrinae, Amathusiinae and Libytheidae.* Tokyo: Plapac..

Urquhart, F. A. 1960. *The Monarch Butterfly.* Toronto: University of Toronto Press.

Vane-Wright, R. I. 1991. Transcending the Wallace Line: do the western edges of the Australian region and the Australian plate coincide? *Australian Systematic Botany* 4: 183-197.

Wallace, A. W. 1859. Letter from Mr Wallace concerning the geographical distribution of birds. *Ibis* 1: 449-454.

Wallace, A. W. 1876. *The Geographical Distribution of Animals with a Study of the Relations of Living and Extinct Faunas, as Elucidating the Past Changes of the Earth's Surface.* London: Macmillan.

Weller, S. J. & Pashley, D. P. 1995. In search of butterfly origins. *Molecular Phylogenetics and Evolution* 4: 235-246.

The vertebrate fauna of the Wallacean Island Interchange Zone: the basis of inbalance and impoverishment

Allen Keast

Biology Department, Queen's University, Kingston, Ontario K7L 3N6, Canada

ABSTRACT: Freshwater fishes, Amphibia, mammals and birds are used to produce an updated biogeography of Southeast Asia, Wallacea, and Wallace's Line. The faunas are documented using newer data sets and interpreted relative to biological characteristics, and plate tectonics, geographical, and climatic, information. To give focus the material is arranged relative to two questions: (1) Why are the vertebrate faunas of Wallacea impoverished and unbalanced, that is, lacking in so many lineages dominant on the adjacent continents?; and (2) Why, given 10-15 million years of proximity has there not been a massive interchange of continental faunas through Wallacea. The survey confirms and amplifies earlier biogeographic knowledge. The extraordinarily rich the Oriental freshwater ichthyofauna is sharply delimited by Makassar Strait; it contrasts with the marine-derived ones of New Guinea - Australia, and Wallacea. In frogs contrasting Oriental and Australian lineages, only tenuously enter Wallacea from west and east. In mammals the exclusion of most lineages of the rich Borneoan assemblage from Sulawesi is explained on historic and ecological grounds. Birds, the best group for such analysis because of their diversity and abundance are represented by a series of faunal "strata" with different histories, distributions, and ecologies. The Wallacean avifauna, far from being just a limited attenuation of the continental ones proves to be an integrated, island adapted unit, with a generally common structure. The geological data provides a background to the evolution of the contemporary biota and in addition to *in toto* generalized effects the differently adapted vertebrate lineages have obviously responded to its different components. Relative to the two basic questions posed it is noted that the New Guinea flora and insect and, to some extent, reptilian faunas, are Oriental in origin. Historically, then there must have been a considerable movement of biotas through Wallacea. The exclusion from Wallacea indicated by birds and mammals would, hence, be less general than appears. Relative to the "impoverishment" question it is pointed out that if the Wallacean avifauna is rich and diversified relative to ecological opportunities available the term does not really apply. It is depauperate only when measured against the continents, rather different environments. The status of Wallace's Line is redefined relative to a cross-section of biological groups.

1 INTRODUCTION

The evolutionary interest in southeast Asia lies not so much in the rich Sundaland rainforest biota, nor that of New Guinea to the east, but in the area in between. Wallacea, the name given to that belt of tropical islands between Borneo (Kalimantan) and New Guinea, is of great interest because here the two great continental biotas, those of the Oriental and Australian. Regions approach each other and overlap. This is potentially one of the world's richest biological zones, with opportunity to gain elements from both west and east. Yet these expectations are not realized. The biotas of the Greater Sunda Islands (Borneo, Sumatra, Java), and those of New Guinea and Australia, remain largely independent.

Many of the major Oriental and Australian - New Guinea lineages are absent from Wallacea. The Region is seemingly faunally depauperate and unbalanced, with many major continental lineages absent. Just why this is so has not received the attention it merits. Most focus has been on Wallace's Line, the precipitous faunal and floral transition zone across the 40 km wide Makassar Strait between Borneo and Sulawesi, and its southern continuation, the Lombok Strait between Bali and Lombok. As fully documented, and argued, by Alfred Russel Wallace (1869, 1876), P.L. Sclater (1858), Bernard Rensch (1931), and others, the Line separates an Oriental vertebrate fauna of gibbons, orang-utans, bears, rhinoceroses, and cats, from an impoverished Australian one with marsupial phalangers and cockatoos prominent. On the basis of such highly visible groups Wallace's Line has achieved, in the eyes of the public, an almost mythical status. Indeed, the title of the conference "Where Worlds Collide", inferring a vigorous "confrontation" between the two continental biotas, supports this image. Reality is different. In mammals, birds, and some other groups, there is a major biotic turnover at Wallace's Line. The Line, however, is of very little importance in the distribution of plants and insects (Whitmore 1981a and b, 1987; Gressitt 1982). It represents the eastern edge of the Asian continental shelf with its rich biota. Makassar Strait is a striking physical barrier. The diversified Australian - New Guinean biota does not approach the Line. It remains far to the east within the continental limits of these landmasses. Only a few Australian forms reach Makassar Strait.

Basic biological questions surround Wallacea: (i) How does its biota compare with the continents on either side; (ii) What are the patterns of penetration of the Region from the two directions; what has dictated this; (iii) Why are its biotas impoverished and "unbalanced" (in a continental sense); (iv) Why, given that it has been fifteen million years since the northward moving Australian landmass reached its present proximity to Asia, has there not been a massive interchange of biotas between these two continents?

The brief review focusses on these questions. Analysis will be made using four vertebrate groups (freshwater fish, Amphibia, mammals, and birds) to represent a series of contrasting adaptive types. Newer faunal data will be used. Interpretations are made on the basis of the attributes of the groups, and newer geological and geographic data, including that contained in this volume.

2 THE VERTEBRATE FAUNAS OF SOUTHEAST ASIA, WALLACEA, AND NEW GUINEA - AUSTRALIA

From the times of Wallace (1869, 1876) and Sclater (1858), the vertebrate faunas of southeast Asia have been a target for research and documentation. There is now a huge literature on Southeast Asian faunas. Freshwater fish have been covered by Roberts (1978, 1989), McDowell (1981), Cranbrook (1981), Kottelat (1985), Allen (1991), Kottelat et al (1993), and Rainboth (1996). The amphibian faunas are reviewed by Inger (1966), Tyler et al (1981), Burton (1986), Menzies (1987), and Allison (1996); the land and freshwater tortoises by Iverson (1984) and Allison; and the snakes by Mertens (1930), Welch (1988), Allison (1996), and How and Kitchener (1997). Mammalian biogeography has been variously covered by Raven (1935), Groves (1976, 1980, 1985), Simpson (1977), Medway (1969), Cranbrook (1981), Payne et al (1985), Heaney (1985, 1986), Musser (1987, 1991), Schmidt et al (1995), and Kitchener and Suyanto (1996). Amongst the more detailed discussions of birds are those of Rensch (1931), Mayr (1944a), Lincoln (1975), White and Bruce (1986), Keast (1983, 1996), Michaux (1991, 1994), Coates and Bishop (1997), and Stattersfield et al (1998). A large number of regional bird lists from the area are now available. The biota of Sulawesi, the most interesting island, has been surveyed by Groves (1976), Holloway (1987, 1990), Musser (1987, 1991), Whitten et al (1987), Michaux (1996), and Evans et al (1999). Relevant here also are later reviews of southeast Asian biogeography in Whitmore (1975, 1981a and b, 1987).

2.1 Freshwater Fish

Freshwater fish can be grouped into three categories in terms of their physiological tolerance to salinity. These are "primary", "secondary", and "peripheral" division fish, representing, respectively: (i) forms belonging to exclusively freshwater lineages and that are unable to withstand salt; (ii) freshwater forms of marine derivation; and (iii) diadramous species that can use both fresh and saltwater. This classification, advanced by Myers (1949), though somewhat artificial, remains an important framework for ichthyofaunal analysis (Darlington 1957).

Southeast Asia has a very rich primary freshwater fish fauna. The Mekong, the largest river, has 1200 species (Zakaria-Ismail 1994), a richness exceeded only by the Amazon. Kampuchea has 215 species in 47 families; Peninsula Malaya, 384 in 50; Java 100 in 12; and Borneo, 394 in 12 (Mohsin and Ambak 1983; Kottelat 1985; Lowe-McConnell 1991). Dominant families in the Malay Peninsula are Cyprinidae (the most speciose of primary division lineages), Channidae, Clariidae, Siluridae, and Belontiidae. Secondary division cyprinodonts are diversified. Cyprinids constitute 38% of the fish fauna of Kampuchea, 39% in South Vietnam and Thailand, and 34-54% for individual Malaysian rivers. Endemism at the species level is 11% for Sumatra, 9% for Java, and 8% for Borneo (Kottelat et al 1993). The richness of the freshwater fish faunas and low levels of endemicity in the Greater Sunda Islands is explained by former habitat continuity. When Sundaland was emergent extensive freshwater river systems interconnected mainland Asia, Java and Borneo (see map in Kottelat et al 1993).

East of Makassar Strait the freshwater fish faunas are unbalanced and depauperate. Sulawesi has only two primary fish species, and these were probably introduced by man, and a few secondary groups (cyprinodonts, atherinids) of Asian origin, in its fauna of 68 species. Endemism at the species level is a high 78%, compared to 8% for Borneo. A few primary fishes (e.g. the cyprinid genus *Rasbora*) reach the Philippines. It has been suggested the colonization took place by two routes, along the Palowan-Mindoro, and Sula-Mindanao island chains when sea levels were lower and land more continuous (Cranbrook 1981). Eastwards from western Java and along the Lesser Sunda chain the dropoff in freshwater fish species richness is precipitous. (Bali actually has more species than dry eastern Java.) Lombok has only 5 primary freshwater fish species (in *Rasbora, Puntius,* Cyprinidae), three of which were probably introduced by humans. Included here is the part air-breathing catfish, *Clarias batrachus.* One primary species reaches Sumbawa.

Diadramous and vicarious fish species dominate the fresh waters of Sulawesi and the Lesser Sundas, entirely compose the freshwater faunas of the eastern Wallacean islands.

New Guinea and Australia are exclusively populated by secondary and diadramous fish species except for the ancient lungfish (*Dipnoi*) and *Scleropages* and, dubiously, one or two forms that may fit the primary definition (Munro 1967, McDowall 1981). The nature of any Early Tertiary Australian or ancient Gondwanan, freshwater fish fauna is unknown. Because of the absence of a major intra-continental river system like the Amazon or Mekong Australia probably has never had a very rich freshwater fish fauna. Certainly, the increasing aridity of the late Tertiary and Pleistocene would have spelled the death knell of an earlier diversity.

New Guinea has only some 316 freshwater species: 103 of these are basically estuarine forms (Allen 1991). These are believed to mostly have a marine larval stage. They are widespread outside of New Guinea, including northern Australia.

The huge Fly River system has a mere 77 species; the equally majestic Sepik to the north 35 species (Roberts 1978). All are marine derived (with the possible exception of *Scleropages.* The faunas are dominated by marine catfish (Ariidae, Plotosidae), Gobiidae, Eleotridae, Clupeidae, and Atherinidae). This compares with 290 species (in 120 genera and 40 families) for Borneo's comparable and most impressive river, the Kapaus (Roberts 1989). Occurring in the Kapaus but absent from New Guinea are minute-bodied forms, and morphologically specialized ant-eating, scale-eating, phytoplanktivorous, and frugivorous, species. There is a diversity of zooplanktivores and piscivores. The freshwater fish fauns of the Fly and Sepik are relatively morphologically and ecologically generalized.

The fish faunas of the Greater Sundas, and New Guinea - Australia thus have had different origins, and are quite differently structured.

Makassar and Lombok Straits are a near complete barrier to primary freshwater fish species. Wallace's Line separates the rich Asian fauna from the depauperate secondary Wallacean one. By contrast, 28 families of continental shelf marine fishes, not so limited, have a continuous distribution from the Mekong to New Guinea.

2.2 *Amphibia*

Borneo has five species of caecilians, the Philippines, three. Two genera are represented: *Caudacallia*, with 5 species extending from Malaya to Borneo and the Philippines, and *Ichthyophisa* with 30 species from India to the Philippines (Duellman and Trueb 1986). Sulawesi and the other Wallacean islands, and Australia - New Guinea, lack caecilians.

The Greater Sunda Islands have an exclusively Asian frog faunas (Inger 1966; Frost 1985; Duellman 1993). The New Guinea and Australian faunas are almost entirely endemic. Three families are restricted to the former, two are shared. One family and three subfamilies (four major groups) are centred on the latter (Tyler et al 1981; Frost 1985; Duellman and Trueb 1986; Burton 1986; Duellman 1993; Allison 1996).

The Australia - New Guinea frog fauna, besides the Asterophryinae and Genophryninae, consists of the Myobatrachidae (20 living genera, 99 species), the endemic sub-family Pelodryadinae (formerly placed in the Leptodactylidae), and Hylidae (Duellman and Trueb 1996). Australia's oldest frog is the "leptodactylid" *Lechriodotus casca* of Early Eocene age (Tyler and Godthelp 1993).

The major groups of frogs occurring are:

2.2.1 *Pelobatidae*
Europe and North America to Greater Sundas and Philippines with the subfamily Megophryinae being represented in Borneo by four genera (the genus *Leptobrachella* has at least six species);

2.2.2 *Rhacophoridae*
Widely distributed and speciose in the Old World tropics, with 7 species in the Asian tropics, extending east to Borneo and the Philippines;

2.2.3 *Bufonidae*
World-wide with the genera *Ansonia* and *Pelophryne* in Borneo and the Philippines, and *Leptophryne, Pedostibes*, and *Pseudobufo* limited to Malaya and the Greater Sundas;

2.2.4 *Ranidae*
Cosmopolitan, with some genera variously extending to New Guinea. There are secondary radiations of ranids in Sundaland, New Guinea, and islands to the east. Amongst the major ranid genera, as currently defined, *Amolops* is limited to the Greater Sundas; *Staurois* has three species in Borneo and the Philippines; *Platymantes* 38 species in New Guinea, the Philippines, Solomons and Fiji; whilst *Discodeles* is restricted to the Bismarks and Solomons, and *Palmotorappia* to the latter.

Within the New Guinea *Rana* Menzies (1987) distinguished twenty species in four groups. These, remarkably uniform in appearance, occur in all habitats up to 3000 metres. Newer molecular data is showing a much greater radiation of *Rana* in New Guinea than was suspected (Steve Donnellan, personal communication). The group, however, makes up only a small fraction of the 200-odd species of New Guinea frogs. Two species reach Timor, including the endemic *Rana timorensis*; two Seram, one known only from the unique hototype taken at 920 m on Mt. Marasela), and a race of the New Guinea *R. grisea*. *R. modesta* is Sulawesan and *R. grunniens* is on Ambon. Two other species reach Wietar. There is a limited westward penetration of presumed New Guinea lineages. Timor now reveals several new species (D.T. Iskandar, pers. com.)

2.2.5 *Microhylidae*
With 281 species in 61 genera (Frost 1985; Burton 1986) are nearly cosmopolitan.

Prominent sub-families are differentially developed in Southeast Asia and Australia, as follows:

Microhylinae (New World, Southeast Asia) include the following major genera *Chaperina*, one species (Malaya to Philippines); *Gastrophrynoides*, 1 (Borneo); *Glyphoglossus*, 1 (Southeast Asia); *Kalophrynus*, 9 (South China, Borneo); *Kaloula*, 9 (China, Lesser Sundas, Philippines); *Metaphrynella*, 2 (Malay Peninsula, Borneo), *Microhyla*, 21 (China to Bali), and *Phrynella*, 1 (Malay Peninsula, Sumatra).

Dyscophinae, southeast Asia to Borneo.

Asterophryninae, exclusive to the Australian - Papuan region; with 43 named taxa in 7 genera (Burton 1986; Burton and Zweiffel 1995) *Asterophrys*, 1; *Barygenys*, 7; *Hylophorbus*, 1; *Pherhaposis*, 1; *Phrynomantis*, 15; *Xanobatrachus*, 9; *Xenorhina*, 6.

Genyophryninae, most diverse in New Guinea, but reaching Sulawesi, Lesser Sundas, and southern Philippines, with 70 species in six genera, and 50 New Guinea species (Burton 1986). Major genera are *Choerophrae*, 1 (New Guinea); *Cophixalis*, 23 (New Guinea and Australia); *Copiula*, 3 (New Guinea); *Genophryne*, 1 (New Guinea); *Oreophryne*, 24 (South Philippines, Sulawesi, Lesser Sundas, New Guinea, New Britain); and *Sphenophryne*, 17 (New Guinea, New Britain, extreme northern Australia).

This data is from the standard list of Duellman and Trueb (1986). Newer taxonomic works have introduced some changes - see Burton and Zweiffel (1995). For example, three species have been removed from *Cophixalus* to form a new genus, *Albericus* (Burton and Zweiffel 1995). These changes do not alter the basic biogeographic features outlined above.

2.2.6 *Myobatrachidae*
20 living genera, 99 species, endemic to Australia - New Guinea, formerly placed in the Leptodactylidae, which name is now restricted to the South and Central American assemblages. Grouping is currently into two sub-families, each with either Australian or New Guinea genera and two are shared between Australia and New Guinea. The Indian fossil genus *Indobatrachus* has been placed here (Duellman and Trueb 1986).

2.2.7 *Hylidae*
World-wide, with sub-family Pelodryadinae being Australian - New Guinean. *Cyclorana* has 13 Australian species; *Litoria* 106 species shared between Australia, New Guinea, the Moluccas, Lesser Sundas, Timor and the Bismarcks; and *Nyctimystes*, has 26 species, inhabiting New Guinea, Australia and the Moluccas (Duellman and Trueb).

Borneo has about 100 frog species with various montane endemics. Sulawesi has about 40: these have overwhelmingly Asian affinities. About 57% of species endemic. Borneo and Sulawesi share ten species. To the south relatively wet western Java has 35 frog species, the dryer east 21; Bali, has 11. Seven Sundaic species reach Lombok and Flores (Inger).

The latest set of faunal figures for New Guinea (Allison 1996) lists 201 native frog species: 14 in the Ranidae (Menzies lists 20), 104 in the Microhylidae, 7 Myobatrachidae, and 76 Hylidae. A sizable number of species remain to be described, Allison stresses.

The concensus is that the Australian hylids and, probably, myobatrachids, are related to South American forms and hence are Gondwanan; (Tyler et al 1981; Allison 1996). The ranids and microhylids, by contrast, are of Southeast Asian derivation.

The New Guinea Genyophryninae reach the Lesser Sundas (Flores), Sulawesi, and the Philippines (genus *Oreophryne*). One genus (*Xenorhina*) of the Asterophryninae reaches the Moluccas and *Phrynomantis* reaches Seram (Frost 1985, Allison 1996).

Most Wallacean frogs are highland dwellers. They link to the moister conditions of the montane forests. Altitudinal occurrences of the genophrynines in major islands are given by Frost (1985). Many Sulawesan frogs are regional (Iskandar and Nio 1996).

For a group with moist skins and needing fresh water as a larval habitat frogs have shown modest ability to penetrate the Wallacean islands from both directions. This is most true of the ranids. The ranid radiation in New Guinea would suggest that these arrived in an earlier colonization.

2.3 Mammals

The Oriental mammal fauna is rich. Thailand has 117 genera and 244 species; Malaysia, 95 and 199; Borneo, 105 and 224 (Lekagul et al 1977; Medway 1969). With a typical subset of this fauna Borneo has moles, Dermoptera, lorises, gibbons, orang-utan, pangolins, porcupines, dog, bear, otters, weasels, cats, rhinoceroses, and mouse-deer, none of which cross Wallace's Line to Sulawesi. Bats and rodents are highly diversified.

Within Sundaland, despite former land continuity there is some regional endemism (Groves 1985). At the species level (%) this is as follows: Malaya, 4.5; Sumatra, 10.3; Java, 23; Borneo, 32.5; and Palawan, 63.3. Groves (1985) documents several distinct mammalian distribution patterns. The commonest are: (i) Malaya, Sumatra, and Borneo, this accounting for 18% of total species; and (ii) this distribution plus Java, 11%. Only four species of land mammals were found over the whole of Sundaland, including Palawan: a rat, a monkey, a cat, and an otter.

The mammal fauna of Sulawesi is highly depauperate and unbalanced. Its 127 indigenous species include: 9 shrews, 62 bats, 46 rats, 8 squirrels, 1 porcupine, 4 monkeys, 3 tarsiers, 1 or 2 civets, possibly introduced by man, 2 bovids and 2 suids, along with 3 phalanger species, marsupials of New Guinea origin (Groves 1976, Musser 1987, Whitten et al 1987). Endemism is exceptionally high, 98% if bats be excluded, and 62% with them. Many mammalian lineages (bovids, suids, phalangers) show primitive characters (Musser 1987), confirming long-standing isolation and the ancient nature of the island. The two bovids separate out into a lowland and a highland morphotypes, or species. The macaque monkeys and tarsier species are allopatric mainly replacing each other on the islands different "arms" (Groves 1976, 1980; Musser 1987).

The mammal faunas of the Lesser Sunda Islands and Malaku are very impoverished and are dominated by bats and rodents (Musser 1991). The best studied island fauna is that of Flores. It has a deer and a civet. Four of six genus level taxa are endemic, compared to 16 of 21 on Sulawesi, and 6 endemic species give an endemism level for species of 85.7% (Groves 1985).

To the south there is a marked dropoff in mammal species richness between Java and Bali, which does not share the former's fauna. Mammals that reach, or formerly reached, Bali but not Lombok, only 20 km, to the east, include the tree shrew (*Tupaia*), the scaley anteater, two species of squirrel, the leopard cat, *Felis bengalonsis*, and the tiger. The fauna of Lombok (Kitchener et al 1990) consists of 23 genera and 36 species of Chiroptera; 4 genera and 7 species of rodents (some Bali rodents and bats do not reach Lombok); and 3 genera and 3 species of carnivores (the civets *Viverricula* and *Paradoxurus* and felid *Felis bengalensis javanensis*); two primates (macaque and leaf monkey *Trachypithecus*); three artiodactils (pig, muntjak and deer *Cervus timorensis florescensis*); and an insectivore, a fauna of 53 species. On the basis of this revised faunal list Kitchener et al (1990) question that Lombok and Bali were separated by a deep strait throughout the Pleistocene and suggest that a dry land corridor may have extended through to them.

The primitive marsupial and monotreme endemic and "Gondwanan" faunas of Australia and New Guinea contrast strikingly with the placental faunas of Southeast Asia. Rodents and bats, originally of Asian origin, form a second faunal stratum. New Guinea has a rich mammal fauna of 86 genera and 227 living and extinct species that also includes monotremes, 2 species; marsupials, 68; bats, 73; and rodents, 69 species (Flannery 1996). This is a limited subset of the Australian one.

The New Guinea element penetrates westward to the Moluccas, which have an extraordinary radiation of phalangerids (Flannery 1996). There is an endemic genus and species of peroryctid bandicoot (*Rhynchomeles prattorumin*) in montane Seram, and fossils of a possibly related species on Halmahera. Two cuscus species on Seram were probably introduced by humans.

Fossil deposits on Halmahera, provionally dated at 3500 years, include a bandicoot, sugar glider (*Petaurus*) and a cuscus (Flannery 1995). Of the Sulawesan cuscuses one, *Ailurops ursinus,* is the most plesiomorphic of the group and Flannery suggests that it warrants its own subfamily. The second Sulawesan form, *Strigocuscus*, also retains many plesiomorphic features. Flannery also documents five indigenous rodents in Seram with their closest relatives in New Guinea. He found a sharp shift in a dominance of Asian to New Guinea elements between Halmahera and Sulawesi.

Mammals provide a classic support for Wallace's Line as an important biogeographic barrier - see, especially, the detailed field studies and faunal lists by Raven (1935). Sulawesi lacks the bulk of Sundaland lineages, its depauperate nature argues, by inference, that the fauna must be characterized by many unutilized adaptive zones. It is generally agreed that the more distinctive elements became isolated in the Pliocene or early Pleistocene. Whether the bovids and ancestral *Babirosa* from the south and phalanger from the east reached Sulawesi by continuous land, transport on "drifting terraines", or over water-gaps is unknown. Later arrived endemics, the tarsiers and macaques, presumably reached the island over-water.

Fossil mammal faunas from the Plio-Pleistocene are known from Java, Borneo, and Sulawesi (Cranbrook 1981; Groves 1985). Significantly Sulawesi never apparently acquired the diversified Javan fauna. The stegadont proboscideans that reached Sulawesi, Mindanao, Timor, and co-occurred with early man on Flores (Morwood 1998), were able to colonize across water gaps, and did not need land connections (Simpson 1977).

Mammalogists (Cranbrook 1981; Musser 1987), on the basis of this data argue that Sulawesi must have been an isolated island since the Miocene. Geological data (Hall, this volume) might, however, permit a direct land junction with the Lesser Sundas at some stage.

2.4 Birds

The avifaunas of Southeast Asia and Sundaland, and those of New Guinea - Australia contrast sharply. That Wallace's Line is a major distributional barrier to birds has been stressed by Sclater (1858), Rensch (1931), Stresemann (1939-1941), Mayr (1944a), Darlington (1957), Keast (1983), and others. The contrasting nature of a large segment of the New Guinea and Australian avifaunas, always obvious, has been confirmed by molecular studies (Christidis and Boles 1994). Many dominant groups are endemic. Some 15 families fall into this category.

Malaya has 248 genera and 442 species of breeding land and freshwater birds; Sumatra, 250 and 432; Java, 213 and 329; and Borneo 209 and 389 (Keast 1996). The smaller Javan avifauna can be linked to the smaller size of this island and its dry eastern half. The avifauna of Sumatra is similar to that of Malaysia. The Javan avifauna includes a few forms of Indo-Chinese origin, absent elsewhere in Malaysia (Delacour 1947). Their occurrence can be linked to the island's earlier separation and more open habitats. Within Java species richness drops from the wetter west (340 species) to the dryer eastern half (245 species); the eastern tip has only 170 species (Mayr 1944a).

Sulawesi has 159 genera (116 endemic), and 220 species; 84 of them endemic (Stresemann 1939-1941; Mayr 1944a; Coates and Bishop 1997; Stattersfield et al 1998). The avifauna is overwhelmingly Asian in origin, but distinct, from that of Borneo. There are a few Australian groups. Distributional analysis indicates that Sulawesi has contributed 38 bird species to the Moluccas, and received 25 back; 17 to the Philippines and received 58; 12 to Flores and received 18; 3 to Borneo and received 9; and obtained two from Java (Stresemann 1939-1944).

Bali (area of 5560 km^2), has about 99 genera and 127 bird species; Lombok (4729 km^2), 81 and 101; Sumbawa (15,448 km^2), 90 and 111; Flores (17,150 km^2), 96 and 119; Sumba (11,153 km^2), 78 and 89; and Timor (32,000 km^2), 95 and 113. Buru (9,064 km^2), has 94 genera and 115 species; Seram (18,623 km^2), 88 and 117; and Halmahera (17,779 km^2), 88 and 120 (Table 1) - see data in White and Bruce (1986), Coates and Bishop (1997), Stattersfield et al (1998).

The basic features of Wallacean ornithogeography have now been established (Mayr 1944a, Keast 1983), as follows:

Makassar and Lombak Straits are a complete barrier to the Asian Trogonidae, Capitonidae, Eurylaimidae, Chloropseidae, and Sittidae. Groups represented to the east of this (10-20% of species) include: Picidae, Phasianidae, Pittidae, Bucerotidae, Pycnonotidae, Timaliinae, Nectariniidae - see species numbers attenuation graphs in Keast (1983).

From west to east through Wallacea, the proportion of elements of western origin decreases sharply; the reverse applies outwards from New Guinea - Australia.

Table 1. Avifaunas of the major Wallacean islands relative to those of adjacent continental areas. Comparative representation of species belonging to four feeding guilds: (i) diurnal birds of prey (hawks); (ii) small insectivorous warblers and flycatchers; (iii) frugivores (pigeons, parrots, birds-of-paradise); and (iv) large water-birds.

Landmass	Area Km²	Elevation Max. m.	Genera No.	Species No.	No. species per 1000 km²	Endemic Exclusive to Island	Species shared 2-3 adjacent islands	No. endemics per 1000 km² Exclusive	Shared with adjacent islands	Diurnal birds of prey %	Small insectivores %	Frugivores %	Water-birds large %
Malay Peninsula	131,000	2180	216	429	3.3	3		0.02		6.4	21.7	16.5	5.8
Sumatra	473,570	3522	189	359	0.76	5		0.01		5.0	17.2	14.0	8.9
Java	125,615	3407	165	287	2.28	16		0.13		5.6	15.0	13.6	10.5
Borneo	293,000	3804	209	386	1.31	29		0.10		4.9	19.4	16.5	6.4
New Guinea, Vogelkop	300,000	4662	221	323	1.10	29				6.7	24.5	27.0	5.8
Sulawesi	181,286	2825	160	220	1.21	58	16	0.32		9.5	11.8	10.9	13.2
Halmahera	17,779	1412	88	120	6.67	4	22	0.23	1.10	11.7	9.2	20.8	14.2
Buru	9,064	2251	94	115	12.78	11	7	1.22	0.60	9.5	15.6	21.7	17.0
Seram	18,623	2798	88	117	6.5	14	5	0.78	0.66	8.5	11.1	21.4	16.1
Bali	5,560	3105	99	129	23.0	1		0.20		6.1	16.2	16.2	10.8
Lombok	4,729	3489	83	101	21.5	1	3		0.08	8.4	12.1	16.3	13.9
Sumbawa	15,448	2850	90	111	7.6	1	16	0.67	0.35	6.3	11.7	14.4	13.5
Flores	17,150	2203	96	119	7.0	4	15	0.24	0.51	8.4	16.0	16.0	14.2
Sumba	11,153	2100	78	89	8.1	6	3	0.55	0.20	7.6	18.2	16.9	15.7
Timor	32,000	1977	95	113	3.5	12	12	0.38	0.45	8.8	10.6	17.7	15.9

Most endemic Australian groups (Ratites, bower-birds) are confined to continental New Guinea-Australia. Birds-of-paradise reach Halmahera, cockatoos, and honeyeaters, Sulawesi and Lombok. Megapodes, the acanthizid *Gerygone*, and a *Pachycephala,* breach Wallace's Line.

The 50%-50% line, where western and eastern forms are in balance approximates to Weber's zoogeographic line, which passes east of Timor and west of Buru and Halmahera (Mayr 1944a). This is also true of snakes (How and Kitchener 1997).

Many wide-ranging avian families are well-represented both in Southeast Asia and Australia, and have undergone major radiations in each. Long occupancy of both regions is thus demonstrated. Included are hawks, herons, pigeons, parrots, cuckoos, kingfishers, and campephagids. Most of these also have species on the Wallacean islands. The pre-eminent such group are the pigeons. Sulawesi has 19 species; Lombok, 15; Flores, 13; Seram, 12; Halmahera, 17; Timor, 9; and the Vogelkop, New Guinea, 19.

Birds, in having large numbers of species on islands, are the most successful island inhabiting vertebrates. Explaining this are such avian attributes as: (a) mobility (many large-bodied species simultaneously occupy several islands); (b) capacity (of small forms) for species replication in different habitats and at different elevations; and (c) ability to divide food resources finely.

Species richness relative to island area correlations are not close for the larger Wallacean islands, despite the pervasiveness of this concept in island biogeography theory (MacArthur and Wilson 1963; Heaney 1986), and its obvious broad applicability when very large and small islands are compared (Table 2). Numbers of species per 1000 km^2 of area figures range from 0.76 to 1.31 apply for the large continental islands of Sumatra, Borneo, and New Guinea, and also Sulawesi (Table 1). For the "dryer" Lesser Sunda islands of Sumbawa, Flores, Sumba, and Timor the figures are 7.6, 7.0, 8.1, and 3.5. Halmahera (6.67) and Seram (6.50) fall into this range. Small Bali and Lombok, adjacent to the Greater Sundas have, however, high figures, 23.0 and 21.5 species per 1000 km^2 of area). Buru, half the size of 60 km distant Seram, has just as many species.

Proximity to a rich species source area ensures high species richness in these cases. In turn, habitat diversity, especially occurrence of highlands, and isolation greatly influence species richness. Lawlor (1986), in a comprehensive review of small mammals on islands, found island area influences to be minor compared to colonization extinction effects.

Clusters of endemic bird species are a feature of all the larger islands. Borneo has 5 endemic genera and 29 endemic species (Smythies 1960). Java has 16 endemic species. Mainland Malaya, lacking isolation, has only 3 endemic species and Sumatra, until recently connected to it, 5. The New Guinea avifauna is over 90% endemic at the species level (Keast 1981). Endemism in Wallacea relates both to individual islands and to clusters of islands: specifically to Buru plus Seram; Sumbawa-Lombok-Flores; Sulawesi-Banggai-Sula, and Timor-Wetar-Damar (Coates and Bishop 1997).

Sulawesi has 58 endemic species; it shares and additional 16 with adjacent islands. Buru has 11 endemic species and shares 7 more with Seram; Sumba has 6 and shares 3, Timor 12, and shares 12 with its contiguous islands (Coates and Bishop 1997, Table 1). Relative to island area these figures transform as follows. Buru has 1.22 endemic species per 1000 km2; when the endemics shared with Seram are added the figures for the combined area, drop to 0.60. For Sumbawa, plus Lombok and Flores the figure is 0.35. For Timor the figure is 0.38; when Wetar and Damar, which have small areas, are added the figure rises to 0.45 endemic species per 1000 km^2.

Table 2. Bird communities and faunas of some major Wallacean island study sites and island, families and sub-families, and numbers of genera and species.

Site	Sulawesi Lore Lindu	Sulawesi Dumoga Bone	Sulawesi Rawa Aopa Watumohai	Seram Manusela	Buru	Flores	Sumba	Tanimbar	Sangihi	Sula Taliabu	Obi	Kasiruta
Source	Watling (1983)	Rozendaal & Dekker (1989)	Wardell et al (1998)	Bowler & Taylor (1989)	Jepson (1993)	Verhoeye & Holmes (1998)	Linsley et al (1998)	Bishop & Brickle (1998)	Ribey (1997)	Jones et al (1991)	Linsley (1995)	Lambert (1994)
Latitude, longitude degs.	1.4 S 120.1 E	0.3 S 123-124 E	4.4 S 22-125 E	3.0 S 129 E	3.0-3.3 S 124-127 E	8.0-8.3 S 120-123 E	10.0 S 120.0 E	7.3 S 131.3 E	0.1 S 129 E	1.8 S 124.6 E	1.3 S 127.4 E	0.25 S 127.1 E
Area km²	2,290	3,000	1,052	1,860	8,000	17,150	10,900	5,058	1,700	3,000	2,200	500
Habitats main	Rain for., montane	Rain forest	Decid. For., Wooded savannah	Rain for., montane	Rain forest	Decid. For., Thorn scrub	Moist & dry evergreen, for., Woodland	Dry mons., Sem-everg. For.	Rain forest	Rain forest	Rain forest	Rain forest
Elevation max. m.	2,356	1,970	980	2,490	2,251	2,400	1,225	200	1,300	1,650	1,280	
Bird groups												
Casuariidae				1-1								
Podicipedidae	1-1				1-1	1-1	1-1					
Phalacrocoracidae	1-1		1-1	1-1	1-1	1-2	1-1			1-1	1-1	
Anhingidae	1-1	1-1	1-1	1-2				1-1		1-2		
Ardeidae	7-7	7-9	8-12	4-6	4-4	10-12	8-10	5-6	4-4	4-4	3-3	
Threskiornithidae	1-1	1-1	1-1	1-1			1-1					
Ciconiidae	1-1	1-1	1-1	1-1	1-1	1-1	1-1	1-1	1-1	1-1	1-1	
Pandionidae	1-1	1-1	1-1	1-1		1-1						
Accipitridae	12-14	12-15	8-12	5-7	6-6	9-11	6-7	5-5	3-3	6-6	3-3	3-3
Falconidae	1-1	1-2	2-2	1-2	1-1	1-2	2-2	1-2	1-1	1-1	1-1	1-1
Anatidae	2-3	1-1	2-3	2-2	2-2	3-5	2-3	3-3		2-2		
Megapodiidae	2-2	2-2	2-2	1-2	1-1	1-1	1-1	1-1	1-1	1-1	1-1	1-1
Phasianidae	1-1	1-1	1-1	1-1		1-2	1-2	1-1		1-2		
Turnicidae	1-1	1-1				1-2	1-2					
Rallidae	7-10	6-8	6-9	4-4	2-2	7-7	4-4		1-1	3-3		
Jacanidae	1-1		1-1	1-1	1-1	1-1	1-1					
Burhinidae				1-1		1-1		1-1				
Columbidae	10-17	9-17	8-12	8-12	9-12	8-13	3-3	5-8	7-11	8-11	5-8	4-5
Psittacidae	5-7	4-7	5-6	10-11	9-9	4-4	5-5	5-5	4-6	5-5	5-5	5-5
Apodidae	2-4	3-4	4-4	3-4	3-4	3-3	2-2	1-1	3-4	3-3	2-2	2-2
Caprimulgidae	2-2	2-2	2-2	1-1	1-1	1-2	3-3	1-1		2-2	1-1	1-1
Alcedinidae	5-7	5-7	4-4	4-5	2-2	5-6	2-3	1-2	3-4	4-5	2-3	1-1
Meropidae	2-2	1-1	1-1			1-1	1-1	1-1				
Coraciidae	1-1	1-1	1-1									
Bucerotidae	2-2	2-2	2-2				1-1				1-1	
Picidae	2-2	2-2	2-2	1-1		1-1	1-1					1-1

Table 2 (cont.)

Pittidae	1-1	1-2				1-1	1-1	1-1	1-2	1-2	1-1	1-2
Alaudidae						1-1						
Hirudinidae	1-1	1-1	1-1	1-1	1-1	1-2	1-1	1-1	1-1	1-1		
Motacillidae						1-1	1-1					
Campephagidae	2-7	2-6	2-3	1-3	1-2	3-4	3-3	2-3	1-2	2-3	2-3	2-3
Pycnonotidae	1-1	1-1		1-1	1-1	1-1			1-1	1-1	1-1	1-1
Sylviidae	6-6	4-4	3-4	4-4	4-4	6-7	4-4	3-3	3-3	3-3	3-3	
Muscicapinae	4-7	4-7	2-2	2-4	2-4	2-4	3-4	1-1	1-2	4-4	1-1	1-1
Timaliinae	1-1	1-1	1-1			1-1				1-1		
Turdinae	5-5	1-1	1-2		3-5	3-5	1-1	1-2	1-1	1-1		
Acanthizidae	1-1		1-1		1-1	1-1		1-1				
Monarchinae	1-1	1-1	1-1	2-3	2-3	3-5	2-2	3-5	1-1	2-2	3-3	1-1
Rhipidurinae	1-1	1-1		1-3	1-3	1-2	1-1	1-1		1-1	1-2	
Dicrurinae	1-2	1-2	1-1	1-1	1-1	1-1	1-1		1-1	1-1		1-1
Pachycephalinae	2-3	2-2		1-2	1-2	1-2	1-1				1-1	
Paridae						1-1						
Dicaeidae	2-3	1-3	1-1	1-1	1-1	1-2	1-1	1-2	1-1	1-1	1-1	1-1
Nectariniidae	4-4	2-3	1-2	1-2	1-2	2-3	1-1	1-1	3-3	2-3	1-2	1-2
Zosteropidae	2-4	2-2	3-4	3-4	1-1	3-6	1-2	1-1	1-2	1-2	1-1	2-2
Meliphagidae	2-3	2-2	3-5	3-5	3-3	2-3	3-3	3-3	1-1	1-1	1-1	
Estrildinae	5-7	1-2	1-4	1-1	1-1	5-8	3-5	2-4	1-1	2-2	1-1	1-1
Sturnidae	5-6	5-5	3-3	2-3	1-2	3-3	1-2	3-3	1-1	3-3	1-1	1-1
Oriolidae		1-1	1-1	1-1	1-1	1-1	1-1	1-1	1-1	1-1		
Artamidae	1-2	1-2	1-2	1-1	1-1	1-1	1-1	1-1	1-1	1-1		
Paradisaeidae										1-1	1-1	2-2
Corvidae	1-2	1-1	1-1	1-1	1-1	1-2	1-1	1-1	1-1	1-1	1-1	1-1
No. higher categories	47	43	39	43	36	47	44	36	27	36	29	23
No. genera and species	131-172	108-149	95-124	89-117	74-86	118-159	92-106	67-81	50-63	76-86	47-54	37-44
Species/1000 km²	86	49	124	58	48	9	10	16.2	48	28	25	88
Species/genus	1.38	1.38	1.31	1.31	1.16	1.35	1.15	1.21	1.26	1.13	1.08	1.19
% with one species	38	37	38	42	50	36	43	53	48	42	69	61

Distinct montane clusters of species characterize all the major islands. Many endemics are montane forms. In Borneo four (of the five) endemic genera are montane, along with 17 of the 29 endemic species (Smythies 1960). On Sulawesi 23 of the 58 endemic species are montane; Seram 9 of 14; Buru 9 of 10. In the Lesser Sunda island endemics chain (Lombok to Alor) of 31 endemics 9 are montane. Timor, with only low mountains, has only one montane endemic in its 23. For Halmahera, again with few high mountains, the figure is 2 out of 28. In New Guinea of 397 endemic species 132 are restricted to montane forest and highlands.

In his comprehensive analysis of montane floras van Steenis (1979) notes that montane habitats (land above 1000 m) represent only a minor fraction of total areas. They account, however, for a high proportion of endemic plant species. The biogeographic importance of montane systems in increasing biotic diversity on islands is, hence, great. There is no simple correlation between island area and number of endemic species. Physiographic diversity is important. Age of isolation is critical, as illustrated by Sulawesi.

2.5 *The Nature and Structure of the Wallacean island Bird Communities*

How are the land and freshwater avifaunas of the Wallacean islands structured in terms adaptive zones occupied (ecomorphological lineages), and relative representation of these? Do the avifaunas represent a unique faunal set, distinct from those of mainland set Asia and Australia? To consider these questions the avifaunas must be analyzed at the community level. I do this by drawing on twelve recently published regional faunal lists (Table 2). To achieve breadth they include three Sulawesan site lists (National Park sites), two of them vegetated by rainforest, and the third mainly by monsoon forest; a Seram National Park (rainforest) site; sites on the dryer islands of Flores and Sumba, with rainforest, thorn scrub and savannah; plus some on large (17,000 km^2 Flores) and small (500 km^2 Kasiruta) islands. The communities are considered in terms of numbers of higher categories (families and sub-families), and numbers of genera and species.

The three levels in the classificatory hierarchy have different evolutionary and ecological implications. Classification is based on morphology. Families, by definition, are much more morphologically distinct than genera. They identify major adaptive groups, in the case of birds, for example; long-billed aquatic waders; hook billed, clawed diurnal predators; generalist frugivores; and small-billed, small-bodied insectivores. The genus represents a lower stratum of structural and adaptive differentiation within the families; multiple genera might demonstrate that an area has the capacity to support "variation within the basic theme". The species has been regarded as the smallest genetically-distinct entity; species commonly separate out on habitat; their numbers could provide some indication of habitat or space adversity and/or a monitor of past opportunities for allopatric speciation.

2.5.1 *Sulawesi, Lore Lindu (Watling 1983): area, 2000 km^2*
Mountainous with only 10% of area below 1,000 m; lowlands covered with rainforest; highlands by montane rainforest, and with some elfin moss forest; climate wet, seldom with more than one or two completely dry months in a year

2.5.2 *Sulawesi, Dumoga - Bone (Rozendaal and Dekker 1989); area, 3,000 km^2*
Broadly comparable

2.5.3 *Sulawesi, Rawa Aopa Watumohai (Wardell et al 1998) area, 1,052 km^2;*
At the southern tip of the southeastern peninsula; dryer; maximum height only 980 m; deciduous forest at higher elevations and with mangrove forest, hill forest, swamp forest, and peat swamps; with 40,000 ha covered by wooded savannah, probably largely anthropogenic.

For Lore Lindu the numbers of breeding land and freshwater birds are: families and sub-families, 43: genera, 108; and species, 149. For Dumoga Bone the figures are 47, 131, and 172; for Rawa Aopa Watumohai 39, 95, and 124. The percentages of families represented by single species are 37, 38, and 38. Ratios of numbers of species per genus are 1.38, 1.38, and 1.31. The three Sulawesi sites are thus, faunally comparable.

Basic taxonomic categories of birds present in the Lore Lindu community are as follows: herons (7 genera and 9 species), hawks (12 and 15), rails (6 and 8), pigeons (9 and 17); parrots (4 and 7); owls (3 and 6); cuckoos (6 and 8); swifts (3 and 4); kingfishers (5 and 7); sylviid warblers (4 and 4); muscicapid flycatchers (4 and 7); and starlings (5 genera and 5 species). Families (37%) represented by single species are grebes, cormorants, anhingas, ibises, storks, osprey, phasianid button quails, turnicid button-quails, pittas, swallows, bulbuls, babblers, Australian acanthizid warblers, monarchs, and rhipidurine flycatchers. Such dominant Asian groups as woodpeckers (2 species), pittas, and timaliids, are poorly represented, as are Australian-derived meliphagids (3 species), and acanthizid warblers.

There is a limited development of highland species in pigeons (*Ptilinopus finscheri, Cryptophaps poecilorrhoa*); sylviids (*Orthotomus cuculatus, Phylloscopus sarasinorum*) and flycatchers *Ficedula westermanni, F. hyperythra*. Species pairs in *Myza* meliphagids, dicrurids (drongos), and *Zosterops*, replace each other in elevation-determined habitats.

The Dumoga-Bone bird community is comparable in structure, with again a high diversity of herons, accipitres, pigeons, parrots, cuckoos, kingfishers, and also starlings. However, some groups are represented by fewer species. There is elevational species replacement in pairs of kingfishers, drongos, flycatchers, *Myza* honeyeaters; and single/species of highland warbler, *Rhipidura* fantails, pachycephalines and *Myzomela* honeyeaters.

The Rawa Aopa Watumohai community is also rich in numbers and genera of species of herons, accipitres, rails, pigeons, parrots, cuckoos, and owls. There are two genera and 3 species of ducks, and four genera and species each of swifts and kingfishers. The Sulawesan pattern of two species of woodpecker, a single babbler, and two hornbills is maintained. There are few campephagids, dicaeids and zosteropids.

The lists thus reflect the basic features of the Sulawesan avifauna as a whole, including development of endemics, absence of many Greater Sunda island forms, and a few Australian forms. The highland endemics include forms of western, eastern, and local derivation. The island support a diversified avifauna. However, species replication within genera is not high.

2.5.4 Seram,Manusela National Park (Bowler and Taylor 1989); area, 1860 km²

Covering 11% of the island's surface area; retain much original primary lowland rain forest, with limited montane rainforest beginning at 1000m.

The bird community includes representatives of 43 higher categories, 38% with only single species. There are 89 genera and 117 species of breeding birds, giving a species to genus ratio of 1.31. Numerically dominant groups are, again, herons, accipitres, rails, pigeons, parrots (10 genera, 11 species, and hence diversified), cuckoos, swifts and, being closer to Australia, more meliphagids. Patterns of highland verses lowland species replacements are comparable to Sulawesi occurring in *Ptilinopus* pigeons and *Ficedula* flycatchers. There are highland species of warblers (*Bradypterus castaneus, Phylloscopus poliocephalus, Orthotomus cuculatus*); a meliphagid (*Lichmera monticolus*), and *Zosterops*.

2.5.5 Buru (Jepson 1993); area, 9,000 km².

This rugged island, rising to 2760 m, now much cleared, is vegetated by rain forest, riverine forest, and secondary grassland.

Half the size of Seram it never the less has an equivalent avifauna (Tables 1, 2), with 36 families and sub-families; half represented by single species; 74 genera, and 86 species; and a species/genus ratio of 1.16. The dominant lineages are the same but in the communities analyzed by Jepson (1993) there were fewer rail, kingfisher, thrush, and starling species.

2.5.6 Flores (Verhoeye and Holmes 1998), 17,150 km².

2.5.7 Sumba (Linsley et al. 1998), 10,900 km².

The eastern Lesser Sunda Islands, including Timor (34,000 km2 - see Stattisfield et al 1998) lie in the rain shadow of the Australian continent. They receive little rain between April and November. Some coastal regions of Flores have an 8-month dry season, an annual rainfall of under 1000mm and are vegetated by dry deciduous forest and thorn scrub. In the mountains (maximum height, 2400 m) the annual rainfall reaches 3.500 m, the dry season is limited to two months and the forests are as verdant as those to the west (Verhoeye and Holmes 1998). Here, at about 800 m on the south-facing sides of the mountains semi-evergreen rain forest gives rise to montane forest.

Sumba, 45 km distant at the nearest point; non-volcanic with a limestone, mudstone, and basalt substratum; is hilly with deeply dissected plateaux; the highest peak only 1,225 m; annual rainfall is 500-200 mm with a long dry and short wet season. The vegetation is deciduous monsoon forest with pockets of semi-evergreen forest, and extensive grassland at least partly of anthropogenic origin.

Flores has 47 higher categories of birds (36% of families being represented by single species), 118 genera and 159 species of breeding land and freshwater birds (Table 2). Verhoeye and Holmes list 10 genera and 12 species of herons; 9 and 11 accipitres; 3 and 5 ducks; 7 and 7 rallids; 8 and 13 pigeons; 4 and 4 parrots; 6 and 6 cuckoos; 3 and 5 owls; 3 and 3 swifts; 5 and 6 kingfishers; 3 and 4 campephagids; 3 and 5 thrushes; 6 and 7 sylviids; 2 and 4 muscicapids; 4 and 5 monarchs; 1 and 4 dicaeids; 3 and 6 zosteropids; 5 and 8 finches; 2 and 3 meliphagids. Most bird species have broad elevational ranges. Stattersfield et al (1998) list 17 "restricted range" species, 8 confined mainly to the lowlands below 1000 m, and whilst 9 are associated with the montane slope pockets of semi-evergreen rain forest. There are highland/lowland species replacements in the pigeon genera *Ptilinopus* and *Macropygia*; and exclusively highland dwelling warblers (*Orthotemus cuculatus, Phylloscopus presbytes, Seicercus montis*) and muscicapine flycatchers (*Ficedula hyperythra, F. newtermanni* plus *Pachycephala nudigula, Zosterops montanus*, and *Lophozosterops superciliaris*. Again, included are species of both presumed Asian and Australian derivation.

Sumba, 45 km from Flores, has 44 higher categories (43% represented by single species); 92 genera and 106 breeding species; 24% of which are water-birds, and a prominent grassland element with Australian relationships (Mayr 1944b, Stattersfield et al. 1998, Linsley et al. 1998). The grassland includes *Circaetus, Coturnix, Turnix, Caprimulgus, Centropus, Mirafra, Cisticola, Anthus, Taeniopygia*, and *Lonchura*. Dominating the island's avifauna again are herons, accipitres, rails, parrots, cuckoos but with fewer species; plus a few species of quails, buttonquails, caprimulgids, campephagids, and three genera and species of meliphagids, and 3 and 5 of finches. There are single species of hornbill, pitta, pipit, parid tit, *Rhipidura, Pachycephala*, and two *dicaeids, nectarinids*, and *Zosterops*.

2.5.8 Tanimbar (Bishop and Brickle 1998); area 5,058 km².

This uplifted coralline archipelago, midway between Timor and New Guinea, represents the extreme southeast of Wallacea. Mountains are lacking. The eastern section supports diversified semi-evergreen monsoon forest, the west drier and structurally more simple forest; swamp and mangrove communities occur.

There are 10 endemic species and 16 endemic subspecies in the 86 species of land and freshwater birds (Bishop and Brickle 1998). Some 15 Australian and 13 Palearctic migrants over winter in Tanimbar. Occurring are multiple species of the same range of forms but single species of quails, megapodes, caprimulgids, swifts, pittas, swallows, drongos, acanthizids, muscicapids, rhipidurids, artamide, nectarinids, dicaeids, and zosteropids.

Australian elements are dominant; Asian ones largely absent.

2.5.9 Sangihi (Ribey 1997); area, 1,700 km².

2.5.10 Taliabu, Sula Archipelago (Jones et al 1991); area, 3,000 km².

2.5.11 Obi (Linsley 1995); area 2,200 km².

2.5.12 Kasiruta (Lambert 1994); area 500 km².

These four small islands, partly rain-forested, have 23-36 families and sub-families, 42-69% represented by single species, 37-76 genera; and 44-86 species. Species to genus ratios range from 1.08 to 1.26. When numbers of (assumed breeding) species are calculated for the arbitrary 1.000 km² high figures of 27, 29, 32, and 88 result. These are slightly greater than the figures for Bali and Lombok and much higher than those for the larger islands (Fig. 1).

Groups with proportionately the highest number of species in the larger sites and islands tend also to be dominant (pigeons, parrots) but there is a habitat effect (absence of water-birds from the Kasiruta list). Sub-dominants (accipitres, anatids, owls, swifts, campephagids), represented by 2-5 species here, are comparable. In accord with the scaling down in island size there is, as noted, an increase in the number of families represented by single species. Obi and Kasiruta, adjacent to Halmahera, have birds-of paradise.

2.6 Summary

The following general conclusions can be extracted from the foregoing:

The Wallacean island avifaunas represent a common set with comparable structures. These obviously relate to the islands, large and small, tending to provide the same sets of ecological opportunities. Dominant groups those with proportionately the largest number of species, are the same throughout notwithstanding some differences in the habitats available (lowland verses montane; rain forest verses monsoon forest. Also comparable is a tier of sub-dominants, represented by 3-5 species, and a wide range of families and sub-families represented by only one or two species.

In broader perspective the Wallacean avifauna is a modified subset of that of the adjacent continents. It is defined by certain lineages being disproportionately important, or uncommon, and a wide range of absentee. The islands provide somewhat different sets of ecological opportunities and constraints. Megapodes, for example, are prominent on islands where there are no mammalian ground predators, absent (at least in the contemporary sense) from the Greater Sunda Islands.

Has Wallacea generated endemic lineages? Dr. Hugh Ford has pointed out to me that the honeyeater genus *Lichmera* and, possibly, *Philemon*, might be an example of such. If there have been others they have, presumably, long since been absorbed in the adjacent continental avifauna.

Consistent with more limited areas and, hence, carrying capacities, species replication within groups is less. Numbers of species per genus ranges, for the 12 communities, from about 1.08 to 1.38. By contrast for larger areas this is greater; for the Malay Peninsula avifauna this is 1.99; Sumatra, 1.90; Borneo, 1.32, eastern New Guinea, 1.5 (Table 1). For the Australasian Region as a whole (Australia plus New Guinea) it is 2.43.

5. The relatively high numbers of families and sub families/represented on the larger Wallacean islands demonstrates that the islands provide a high diversity of distinctive "ecological opportunities". These are common to all the islands; with some extending down to even the smallest islands.

What are the attributes of the dominant bird groups on the Wallacean islands? Herons, frequenting lakes, marshes, and estuaries with the former tending to fluctuate seasonally must, by definition, be mobile. This attribute, critical in seasonally dry Australia, adopts these large water-birds to utilize small areas of habitat on islands. Raptors, restricted by body size or diet to low densities, must have large ranges and be seasonally nomadic: these attributes favour the use of clusters of islands, not just single ones. This swells the species richness figures for the individual islands. On the Wallacean islands raptores make up 7-10% of total bird species.

Pigeons, usually with 8-11% of total bird species (only 4% on Sumba) are ecologically versatile being frugivores and granivores, and arboreal and gronad feeders. Multiple species co-occurrence within habitats is favoured by species specializing on fruit of different sizes (Diamond 1975). Their mobility and the capacity of some to move between islands according to fruiting seasons makes them successful colonizers of event the most remote eastern Pacific islands (Keast 1996).

Parrots, commonly with about 9% of species, share many of the attributes of pigeons. Cuckoos (2.5-8%) are diet generalists, taking both insects and fruit; many optimize on ecological opportunities by migratory and nomadic behaviour. Kingfishers, averaging 2.3-5.5% of total species, are large-billed predators of small reptiles, large insects, and crabs, versatile in the use of strandline, forest, and river systems; many species are also good inter-island dispersers. Niches for aerial feeding swifts, flycatchers, foliage-gleaning warblers, berry-feeding thrushes, are also omnipresent on islands.

Another way of considering resource use patterns in communities is in terms of feeding guilds, the sum total clustering of species relative to a resource type. If members of several lineages are attracted to a particular resource, lineages that maybe structurally different, it argues that such is particularly available and diversified. The resource may, of course, be food or living space but discussion of the guild concept commonly focusses on the former. To better assess Wallacea as a living place for birds I calculate the number species in four broad guild types (Table 1). Figures are developed for: (i) large water-birds (herons, spoonbills, ducks, etc); (ii) raptors, the diurnal birds of prey (hawks, falcons, eagles); (iii) frugivores (pigeons - some of which are granivores, parrots, barbets bulbuls, thrushes, birds-of-paradise); and (iv) small insect-eaters (warbles, flycatchers, drongos).

Large water-birds prove to be proportionately less important (5.8-10.4%) on the Wallacean islands than on the larger land-masses (10.8-17.0%). This can be linked to, proportionately, less aquatic habitat. Equivalent proportionate differences apply to raptors. The small insect-eaters, with replication of species in different habitats and with capacity to divide living space in forest vertically are proportionately more abundant on the continents, where there is greater habitat diversity. Frugivores make up equal proportions in the avifaunas of Wallacea and on the continents and Greater Sundas.

The limitations of the guild method of analysis must be stressed. Allowance has to be made for the fact that some species are mixed feeders, taking more than the one or "central" resource. It would be appropriate to deal with biomasses not numbers of species. There is obviously a big difference in the impact of a large-bodied pigeon and small bodied barbet on a fruit resource. Numbers of species might best give some measure of the number of structural entitles or specialists being brought to bear on a resource type. Comparisons should also be made against a proper measure of resource availability.

3 THE WALLACEAN ENVIRONMENT RELATIVE TO VERTEBRATES

Major influences on vertebrates evolution have been:

3.1 *Tectonic history*

Wallacea is the site of a triple plate junction. It has long been recognised that this is an area of great geological instability, with islands appearing and being lost, of emergences and submergences, oceanic deeps, changing island areas, and violent volcanic activity. In later years various geologists (Katili 1975, Hamilton 1979, Audley-Charles 1987, Burrett et al 1991, Veevers 1991, Moss and Wilson 1998, and Hall 1998, 1999) have attempted reconstruction of land area changes. The latest reviewers (Hall 1998; Hall and Holloway 1998; Hall, this volume) have confirmed the following features: (i) there has never been direct land contact between Asia and Australia; (ii) Makassar Strait has been deep water and a major distributional barrier since probably the Eocene; in the past it was slightly wider than at present; (iii) Borneo has always lain within the Asian orbit, as has a western segment of Sulawesi; (iv) Throughout, the Wallacean region has been dynamic; vast areas of ocean crust have been lost by subduction in the course of this and, hence, it is unlikely that we will ever to able to precisely disentangle the sequential history of land in the area; (v) Many islands like Buru, and most mountain ranges, are relatively recent; (vi) Halmahera formerly lay closer to New Guinea and has migrated subsequently to its present position; the Vogelkop area of New Guinea was formerly an independent island; and Sulawesi was, until recently, smaller.

Such a history has obviously greatly influenced vertebrate distribution and diversity, and the development of new forms. It would have led to phases of extinction.

3.2 Sea level changes

Sea level fluctuations from the late Tertiary through the Pleistocene repeatedly changed inter-island distances. During glacial times sea levels world-wide dropped due to the locking up of vast amounts of waters in the northern glaciers. A 120 m drop at 18,000 B.P., at the time of the last glacial maximum (Fairbanks 1989) defines the amplitude of the changes. See maps of sea level changes in many publications including the general book of van Oosterzee (1977). At times of lower sea level the Greater Sunda islands were broadly connected to Asia; Australia and New Guinea were joined: the distances between Lesser Sunda Islands were reduced. Throughout the archipelagos island shorelines were extended seawards, increasing the areas of most of the islands. This must have facilitated dispersal and interchange, including with the Philippines.

There is evidence that degree of sea level drop may have differed somewhat from one glacial maximum to another. Heaney (1985, 1986), explains Philippine mammal evolution in terms of a greater drop in the Middle Pleistocene (which he links to the 120 m bathmetric line), and a lesser late Pleistocene one. Sea level drops changed oceanic current patterns, and, hence, regional climates.

During interglacials sea levels were slightly higher than at present with a 2.5-5.0 m elevation being commonly quoted (e.g. Tiji et al 1984), although a maximum figure of 25 m has been suggested. One effect of this would have been to isolate blocks of land on either side of the Tempe depression in Sulawesi.

Sea level changes, per se, would have periodically increased, and decreased, colonizing apportunities.

3.3 Climates and climatic shifts

Wallacea has been subject to vast and continuing climatic shifts - see comprehensive reviews Walker (1972), Walker and Hope (1982), Verstappen (1975), Flenley (1979), Whitmore (1981a), Morley and Flenley (1987), Whitten et al (1987), Walker and Chen (1987), Hope (1996), and various writers in this book, especially Peter Kershaw. When the Intertropical Convergence Zone, characterized by frequent rain, was south of the Equator, parts of northern Wallacea experienced a drier, more seasonal climate with lower rainfall and humidity, and seasonal changes in daily temperature might have been marked (Verstappen 1975). See level changes had a direct and striking effect on regional climates.

A figure of 30% lower rainfall (than at present) for tropical regions 10-12,000 years ago has variously been suggested. New Guinea data developed by Alan Walker, and colleagues, suggests temperatures at 2500 m being 10°C lower, and at sea-level 2-3°C lower for 18,000 years B.P. The tree line might have been 1500 m lower in New Guinea and 350-500 m lower at times of maximum temperature reductions. Island habitat changes, resulting from rainfall and temperature effects between the glacial and interglacial times would have been drastic.

Better insights into how climatic and sea levels effects differed between the successive glacial maxima and, in turn, successive interglacials are needed. Pollen cores (Flenley and Morley 1978; Walker and Hope 1982; Wang et al 1999; Kershaw et al 2000; and Peter Kershaw this volume) can potentially supply such: so far, however, such data is only available for the last glacial - interglacial phase.

Climatic oscillations obviously periodically had a drastic effect on biotas, favouring phases of expansion and evolution, and alternatively, causing extinctions.

3.4 Habitat types and habitat diversities

Malaysia - Borneo, and New Guinea are covered by huge rainforest blocks, and have relatively uniform rainfall and seasonal conditions. Lesser Sunda zone climates are mostly seasonal, dryer, and vegetated by monsoon forest and savannah (the latter is possibly largely anthropo-genic) - see the climate and vegetation maps and discussions in Whitmore (1975), and van Steenis (1979). Rainforest on these islands is restricted to pockets. Whitmore (1981a) points out that in Borneo and New Guinea the rainforests of the former have functioned over a long pe-riod as refugia for rich assemblages of rainforest species. However, other evidence (this vol-ume) argues that the area of rainforest in Borneo has fluctuated. Rainforest, and non rainforest species would have had to respond to this. Extinctions might not have been great in Borneo (if sufficient areas of all habitats remained) but this would certainly have applied on the islands.

3.5 *Distance effects and isolation*

Distance - species richness effects form one of the features of the MacArthur and Wilson (1963, 1967) theory of island biogeography. The idea needs quantification relative to other variables in the case of Wallacea. Within Sundaland a west-east dropoff in mammal species, combined with habitat features (Brandon-Jones 1996), and bats in the Lesser Sundas (Kitchener 1996) are examples.

3.6 *The area - species richness effect*

A direct relationship between the size of an island and the number of animal species it supports has been demonstrated (MacArthur-Wilson 1963, 1967). Heaney (1986) has found an island area species richness effect for small mammals on the Philippine Islands. On Australian offshore islands, numbers of species occurring has been adequately predicted from area alone (Burbidge et al 1997). Lawlor (1986), as noted, in a more comprehensive review of small mammals on islands, found any area effect to be minor compared to colonization/extinction effects. As demonstrated here for birds several factors (proximity to rich faunal source areas, presence of mountains) can be more important controllers of species richness than island size.

   Area effects obviously also have a historic dimension: if an island was smaller at one time its small species number may reflect this. Lesser Sunda small islands are disproportionately rich in bats, snakes, frogs and birds (Kitcherner 1996).

3.7 *Biological factors influencing the use of islands by vertebrates*

Biological interactions, animal and plant, animal and animal, would have been a continuing moulding influence throughout the history of Wallacea. It is suggested that the existence of a discrete, well-adapted and discrete Wallacean avifauna might have denied the access of new-comers to the islands (or, at least, made their establishment much more difficult). This could be part of the explanation for the absense from the Wallacean islands of many continental rainfor-est adapted lineages. See also discussion in Kitchener (1996).

4 THE VERTEBRATE BIOGEOGRAPHY OF WALLACEA: REAPPRAISAL AND RESTATEMENT

The newer biological and geological data sets reviewed here reiterate and elaborate on the his-torical biogeography of Wallacea as previously documented (e.g. Whitmore 1981a, b; 1987). Australia reached its present proximity to Asia in the Miocene. It and Southeast Asia have never been connected. The Wallacean area has continued to be tectonically active with islands appearing and changing size. Inter-island distances have fluctuated. Climates have varied markedly in a long-term cyclical framework and as a result of local events like the closure of marine straits. The extent to which these long-term environmental shifts are known has been summarized in detail to bring their full extent to the fore. There is little need here to speculate as to their impact; individually and cumulatively it has obviously been vast, leading to surges of range spread, drastic extinctions, speciation, and a great diversity of local and differential ef-fects.

   An optimist may argue that we have now a fair outline of the biogeographic history of Wallacea. Alternatively one can note that we can only guess at the major biogeographic events of the last 15 million years, and that, again, pending more phylogenetic and molecular studies, knowledge of biological relationships remains inadequate. There is some reassurance in that our insights are logical and consistent with findings in other systems.

The rich Southeast Asian vertebrate fauna extends as far east as the continental edge a Makassar Strait. This Strait is an ancient barrier extending back to the time when Australia first approached Asia. The reason that broad land connection to Asia at times of lower sea level, this has occurred repeatedly and applied until recently. The Greater Sundas are large enough to maintain a rich biota and this has been repeatedly rejuvenated from the mainland. There is some loss of biogeographic diversity from mainland Asia to Borneo, and some regionalization of faunas occur within the Greater Sundas. Habitat influences on occurrence are marked, particularly in the occurrence of distinct montane biotas.

At the eastern end of Wallacea the Australian - New Guinean biota is distinct. It retains the hallmark of long-standing isolation, and retains some archaic, endemic, and Gondwanan, features. This biota is, in turn, delimited in the north by the edge of that continental shelf (Lyddekker's Line).

The rich Oriental freshwater fish fauna of "primary" forms is delimited by Makassar Strait. Australia - New Guinea have an unrelated fauna of marine derivation. Any initial Gondwanan element has been largely lost. Wallacea has only a minor and "peripheral" freshwater fish fauna.

The frog faunas of Southeast Asia and Australia are largely distinct. One Asian lineage, the Ranidae, has reached New Guinea to radiate there. This obviously represents an older colonization. There is a minor penetration of Wallacea both from west and east by the opposing continental amphibian lineages. The circumstances of these inter-island colonizations would be interesting to know.

Mammalian biogeography provides a different set of dimensions. The highly diversified Sundaland fauna stops short at Makassar Strait. The Sulawesan one, impoverished and depauperate, is of restricted diversity. It has suids, bovids and marsupial phalangers that are regarded as some of the more primitive in their respective lineages. Early arrival and persistence in isolation is, hence, indicated. Sulawesi was, until recently, smaller. The mammal faunas of Wallacea are dominated by bats and rodents that, it is generally acknowledged, can cross inter-island water-gaps. There is both a limited penetration of Wallacea both from west and east by the respective continental faunas. Marsupials reach Seram, Halmahera, and Sulawesi.

Birds add a whole series of different dimensions to biogeographic discussion. A series of attributes combine to make them highly efficient island, as well as continental, inhabitants. They are especially fitted to exploiting the opportunities that islands provide. The islands, hence, are characterized by large numbers of bird species. Analysis brings out two major features, the occurrence of a distinct series of faunal "strata" and that Wallacea has a distinct, and definable, avifauna distinct from that of the continents on either side (even though it is drawn from them). Occurring are: (i) Oriental lineages that do not occur east of Wallace's Line; (ii) Families represented east of the Line by small sets of ecologically adventurous or "plastic" species; (iii) and (iv) Australian counterparts to the east; (v) Lineages that must have interchanged early that have radiated both in Sundaland and Australia; (vi) Groups that have Wallacea-based radiations; (vii) Species (mainly large water-birds) that occur, or wander at will through the whole area; and (viii) Migrants that occur in Wallacea seasonally. Birds thus exhibit a range of "responses" ranging from absence to a range of methods of utilization of the area.

In summary, Wallacean biogeography must be considered in many dimensions, the contemporary and the historic, and independently for the different groups. "Deductions" based on the contemporary situation may be irrelevant to the past. Note here the fact that New Guinea has largely a Malesian flora and insect fauna, and that forms of Oriental derivation are prominent in its reptilian fauna. On this basis there must have been a considerable initial west-east (and possibly east-west) interchange, probably beginning soon after Australian approached Asia and that has continued since, where good dispersing groups are concerned. We know nothing about the circumstances of these earlier interchanges. But they must be acknowledged in any assessment of the primacy of Wallace's Line as a biogeographic barrier.

Two questions were asked at the beginning of this paper: (i) Why is the vertebrate fauna of Wallacea impoverished and depauperate; and (ii) Why has Wallacea not been a highway for the massive interchange of biotas between Asian and Australia, given that there has now been 15 million years of proximity. The two issues are closely intertwined. If, as its biota suggests, the Wallacean islands are "inadequate" as environments for supporting diversified continental biotas they could not have been an inter-continental "highway". The questions and issues need rethinking.

Two dimensions are involved. Pervasive factors (extensive and persistent inter-continental water gaps; inter-island distances; limitations of island environments; sweeping climatic oscillations) have throughout affected all biotas. Individually, however, they could have effected the lineages differentially. It would be worth while to try to sort this out (e.g. by advancing and testing hypotheses?). Some enlightenment might result. However, assessments would suffer from being based on the contemporary situation. The environmental features and variables listed have operated, probably often in ways unknown, for millions of years.

Are the Wallacean faunas (specifically their bird faunas) truly depauperate? The present review strongly infers that the islands are "saturated" with a full quota of potential adaptive types. On this basis they are not depauperate. They are only depauperate when compared with the continents and in many basic ways continental and island environments are quite different. Use of the term is, hence, potentially misleading.

Relative to this issue there are still many things that it would be nice to know about Wallacean systems. What occupies the frog and tadpole niches on islands from which these are absent? Is the foliage-eating monkey niche, unused in Wallacea? Does the paucity of mammalian frugivores (monkeys, gibbons, fewer Megachiroptera) in New Guinea explain the island's rich fauna of frugivorous birds?

REFERENCES

Allen, G.R. 1991. *Field guide to the freshwater fishes of New Guinea*. Publ. No. 9. Christensen Research Institute, Madang.

Allison, A. 1996. Zoogeography of amphibians and reptiles of New Guinea and the Pacific region. In A. Keast & S.F. Miller (eds.), *The origin and evolution of Pacific Island biotas, New Guinea to eastern Polynesia; patterns and processes*: 407-436. Amsterdam: SPB Academic Publishing.

Audley-Charles, M.G. 1987. Dispersal of Gondwanaland: relevance to evolution of the angiosperms. In T.C. Whitmore (ed*)*, *Biogeographical evolution of the Malay Archipelago*: 5-25. Oxford: Clarendon Press.

Bishop, D. & Brickle, N.W. 1998. An annotated checklist of the birds of the Tanimbar Islands. *Kukila* 10: 115-150.

Bowler, J & Taylor, J. 1989. An annotated list of the birds of Manusela National Park, Seram. Birds recorded on the operation Raleigh Expedition. *Kukila* 4: 3-29.

Brandon-Jones, D. 1996. The Asian Colobinae (Mammalia: Cercopithecidae) as indicators of Quaternary climatic change. *Biol. J. Linn. Soc.* 59: 327-350.

Burbidge, A.A., Williams, M.R. & Abbott, I. 1997. Mammals of Australian islands: factors influencing species richness. *J. Biogeogr.* 24: 703-715.

Burrett, C., Duhig, N., Berry, R. & Varne, R. 1991. Asian and South-west Pacific continental terranes derived from Gondwana, and their biogeographic significance In P.Y Ladiges, D.J. Humphreys & L.W. Martinelli (eds.), *Austral Biogeography*: 101-123. Melbourne: CSIRO, Australia.

Burton, J.C. 1986. A reassessment of the Papuan subfamily Asterophryinae (Anura: Microhylidae). *Rec. S. Aust. Mus.* 19: 405-450.

Burton, J.C. & Zweiffel, R.G. 1995. A new genus of Genophrynine Microhylid frogs from New Guinea. *Am. Mus. Novit.* 3129: 1-7.

Christidis, L. & Boles, W.E. 1994. *The taxonomy and species of birds of Australia and its territories.* Royal Australian Ornithologists' Union Monograph 2, RAOU, Hawthorn East, Victoria.

Coates, B.J. & Bishop, K.D. 1997. *A guide to the birds of Wallacea. Sulawesi, the Moluccas and Lesser Sunda Islands, Indonesia*. Alderley, Queensland: Dove Pubs.

Cranbrook, the Earl. 1981. The vertebrate faunas. In T.C. Whitmore (ed*.)*, *Wallace's Line and Plate Tectonics*: 57-69 Oxford: Clarendon Press.

Darlington, P.J. 1957. *Zoogeography: the geographical distribution of animals*. New York: Wiley.

Delacour, J. 1947. *Birds of Malaysia*. New York: MacMillan.

Diamond, J.M. 1975. Assembly of species communities. In M.L. Cody, M & J.M. Diamond (eds), *Ecology and evolution of communities*: 342-444. Cambridge, Mass: Harvard Univ. Press.

Duellman, W.E. 1993. *Amphibian species of the world: additions and corrections.* University of Kansas Museum of Natural History, Special Publication 21.

Duellman, W.E. & Trueb, L. 1986. *Biology of Amphibians*. New Jersey: McGraw Hill.

Evans, B.J., Morales, J.C., Supriatna, J. & Melnick, D.J. 1999. Origin of the Sulawesi macaques (Cercopithecidae: *Macaca*) as suggested by mitochrondrial DNA phylogeny. *Biol. J. Linn. Soc* 66: 539-560.

Fairbanks, R.G. 1989. A 17000-year glacio-eustatic sea level record: influence of glacial melting rates on the Younger Dryas event and deep ocean circulation. *Nature* 342: 637-642.

Flannery, T.F. 1995. *Mammals of the South-west Pacific and Moluccan Islands*. Sydney: Reed Books.

Flannery, T.F. 1996. Mammalian zoogeography of New Guinea and the surrounding island. In A. Keast & S.E. Miller (eds), *The origin and evolution of Pacific Island biots, New Guinea to Eastern Polynesia: patterms and processes*: 399-406. Amsterdam: SPB Academic Publishing.

Flenley, J.R. 1979. *The equatorial rain forest - a geological history*. London: Butterworth.

Flenley, J.R. & Morley, R.J. 1978. A minimum age for the deglaciation of Mt. Kinabalu, East Malaysia. *Modern Quat. Res. SE Asia* 4: 57-61.

Frost, D.R. (ed.) 1985. *Amphibian species of the world. A taxonomic and geographic reference*. Lawrence, Kansas: Assoc. Syst. Coll., Allen Press.

Gressitt, J.L. 1982. Zoogeographical summary. In J.L. Gressitt (ed.), *Biogeography and ecology of New Guinea*: 897-918. The Hague: Dr. W. Junk.

Groves, C.P. 1976. The origin of the mammalian fauna of Sulawesi (Celebes). *Zeitschr. F. Sauget.* 41: 201-216.

Groves, C.P. 1980. Speciation in *Macaca*: the view from Sulawesi. In Lindburg, D.G. (ed.), *The macaques: studies in ecology, behavior and evolution*: 84-124. New York: Van Nostrand Reinhold.

Groves, C.P. 1985. Plio-Pleistocene mammals in island southeast Asia. *Modern Quat. Res. SE Asia* 9: 43-54.

Hall, R. 1998. The plate tectonics of Cenozoic southeast Asia and the distribution of land and sea. In R. Hall & J.D. Holloway (eds), *Biogeography and geological evolution of southeast Asia*: 99-131. Leiden: Backhuys.

Hall, R. 1999. Cenozoic reconstructions of southeast Asia and the southwest Pacific: changing patterns of land and sea. In Metcalfe, I. (ed.) *Where worlds collide: faunal and faunal migrations and evolution in southeast Asia - Australasia. Abstracts*: 33-35. Armidale: University of New England Asia Centre.

Hall, R. & Holloway, J.D. (eds) 1998. *Biogeography and geological evolution of South-east Asia. Leiden*: Backhuys Publishers.

Hamilton, W. 1979. Tectonics of Indonesian region. *Prof. Pap. U.S. Geol. Surv.* 1087: 1-338.

Heaney, L.R. 1985. Zoogeographic evidence for middle and late Pleistocene land ridges to the Philippine Islands. *Modern Quat. Res. SE Asia* 9: 127-143.

Heaney, L.R. 1986. Biogeography of SE Asian mammals: estimates of rates of colonization, extinction, and speciation. *Biol. J. Linn. Soc.* 28: 127-165.

Holloway, J.D. 1987. Lepidopteran patterns involving Sulawesi: what do they indicate of past geography. Pp. 103-111 in Whitmore, T.C. (ed.) *Biogeographical evolution of the Malay Archipelago*. Clarendon Press, Oxford.

Holloway, J.D. 1990. Patterns of moth speciation in the Indo-Australian archipelago. In E.C. Dudley (ed.), *The unity of evolutionary biology*: 340-372. Portland, Oregon: Discoides Press.

Hope, G. 1996. Quaternary change and the historical biogeography of Pacific Islands. In A. Keast & S.E. Miller (eds), *The origin and evolution of Pacific Island biots, New Guinea to Eastern Polynesia: patterms and processes*: 165-190. Amsterdam: SPB Academic Publishing.

How, R.A. & Kitchener, D.J. 1997. Biogeography of Indonesian snakes. *J. Biogeogr.* 24: 725-735.

Inger, R.F. 1966. The systematics and zoogeography of the Amphibia of Borneo. *Fieldiana, Zoology* 52: 1-402.

Iskandar, D.T. & Nio, T.K. 1996. The amphibians and reptiles of Sulawesi, with notes on the distribution and chromosomal number of frogs. Proc. 1st Intern. Conference on Eastern Indonesian-Australian vertebrate fauna, Manado, November 22-26, 1994: 39-46.

Iverson, J.B. 1984. *A revised Checklist with distribution maps of the turtles of the world*. Privately printed by author, Richmond, Indiana.

Jepson, P. 1993. Recent ornithological observations from Buru. *Kukila* 6: 85-109.

Jones, A.J., Lucking, R.S., Davidson, P.J. & Raharjangtrah, W. 1991. Checklist of the birds of the Sula Islands, with particular reference to Talaiabu Island. *Kukila* 9: 37-53.

Katili, J.A. 1975. Volcanism and plate tectonics in the Indonesian island arcs. *Tectonophysics* 26: 165-188.

Keast, A. 1981. The evolutionary biogeography of Australian birds. In A. Keast (ed.), *Ecological bio-geography of Australia*: 1585-1636. The Hague: Dr. W. Junk.

Keast, A. 1983. In the steps of Alfred Russel Wallace: biogeography of the Asian-Australian interchange zone. In R.W. Sims, J.H. Price & P.E.S. Whalley (eds), *Evolution, time and space: the emergence of the biosphere*: 367-407. Special Volume 23, Systematics Association. London: Academic Press,.

Keast, A. 1996. Avian geography: New Guinea to the eastern Pacific. In A. Keast & S.E. Miller (eds), *The origin and evolution of Pacific Island biots, New Guinea to Eastern Polynesia: patterms and processes*: 373-398. Amsterdam: SPB Academic Publishing.

Kershaw, P., van der Kaars, S., Moss, P. & Wang, X. 2001. Quaternary records of vegetation, biomass, climate, and possible human impact in the Indonesian - Northern Australian region. In press. Paleo-geogr. Palaeoclimatol, Palaeoecol.

Kitchener, D.J. 1996. Biological diversity in Eastern Indonesia: an essentially mammalian perspective. Proc. 2nd International Conference on Eastern Indonesian-Australian Vertebrate Fauna, Lombok, December 10-13, 1996: 1-12.

Kitchener, D.J., Boeadi, Charlton, L. & Maharadatunkamsi. 1990. *Wild mammals of Lombok*. Rec. W.A. Mus. Suppl. No. 33.

Kitchener, D.J. & Suyanto, A. 1996. Intraspecific morphological variation among island populations of small mammals in southern Indonesia. In D.J. Kitchener, & A. Suyanto (eds), *Proc. 1st. Conf. on eastern Indonesia - Australian vertebrate fauna, Manado, 1994*: 7-13

Kottelat, M. 1985. Freshwater fishes of Kampuchea - a provisory annotated check-list. *Hydrobiologia* 121: 249-279.

Kottelat, M., Whitten, A.J., Kartikasari, S.N. & Wirjoatmodjo, S. 1993. *Freshwater fishes of Western Indonesia and Sulawesi*. Singapore: Periplus.

Lambert, F.R. 1994. Notes on the avifauna of Bacan, Kasiruta, and Obi, North Moluccas. *Kukila* 7: 1-9.

Lawlor, T.E. 1986. Comparative biogeography of mammals on islands. *Biol. J. Linn. Soc.* 28: 99-125.

Lekagul, B., McNelly, J.A., Marshall, J.T. & Askins, R. 1977. *Mammals of Thailand*. Bangkok: Associa-tion for Conservation of Wild Life.

Lincoln, G.A. 1975. Bird counts either side of Wallace's Line. *J. Zool., London* 177: 349-361.

Linsley, M.D. 1995. Some records from Obi, Maluku. *Kukila* 7: 142-151.

Linsley, M.D., Jones, M.J. & Marsden, S.J. 1998. A review of the Sumba avifauna. *Kukila* 10: 60-90.

Lowe-McConnell, R.H. 1991. *Ecological studies in tropical fish communities*. Cambridge: Cambridge University Press.

MacArthur, R.H. & Wilson, E.O. 1963. An equilibrium theory of insular zoogeography. *Evolution* 17: 373-387.

MacArthur, R.J. & Wilson, E.O. 1967. *The theory of island biogeography. Monographs in population biology*. Princeton, N.J.: Princeton University Press,

Mayr, E. 1944a. Wallace's Line in the light of recent zoogeographic studies. *Quart. Rev. Biol.* 19: 1-14.

Mayr, E. 1944b. The birds of Timor and Sumba. *Bull. Amer. Mus. Nat. Hist.* 83: 127-194.

McDowell, R.M. 1981. The relationships of Australian freshwater fishes. In A. Keast (ed.), *Ecological biogeography of Australia*: 1251-1273. The Hague: Dr. W. Junk.

Medway, Lord 1969. *The wild mammals of Malaya and offshore islands including Singapore*. Kuala Lumpur: Oxford Univ. Press.

Menzies, J.O. 1987. A taxonomic revision of the Papuan *Rana* (Amphibia: Ranidae). *Aust. J. Zool.* 35: 373-418.

Mertens, R. 1930. Die amphibien und reptilien der inseln Bali, Lombok, Sumbawa und Flores. *Abh. Senckenb. Naturforsch. Ges.* 42: 115-344.

Michaux, B. 1991. Distributional patterns and tectonic development in Indonesia: Wallace reinterpreted. *Aust. Syst. Bot.* 4: 25-36.

Michaux, B. 1994. Land movements and animal distributions in east Malesia (eastern Indonesia, Papua New Guines, and Melanesia). *Palaeogeogr. Palaeoclimatol. Palaeoecol.* 112: 323-343.

Michaux, B. 1996. The origin of southwest Sulawesi and other Indonesian terranes: a biological view. *Palaeogeogr. Palaeoclimatol. Palaeoecol.* 122: 167-183.

Mohsin, A.K.M. & Ambak, M.A. 1983. *Freshwater fishes of Peninsular Malaysia*. Kuala Lumpur: Pertanian, Malaysia: Penerbit Universiti.

Morley, R.J. & Flenley, J.R. 1987. Late Cainozoic vegetational and environmental changes in the Malay Archipelago. In T.C. Whitmore (ed), *Biogeographical evolution of the Malay Archipelago*: 50-59. Oxford: Clarendon Press.

Morwood, M. 1998. *Stone tools and fossil elephants: the archaeology of eastern Indonesia and its implications for Australia*. Armidale, N.S.W.: University of New England.

Moss, S.J. & Wilson, M..E.J. 1998. Biogeographic implications of the Tertiary paleogeographic evolution of Sulawesi and Borneo. In R. Hall & J.D. Holloway (eds), *Biogeography and geological evolution of southeast Asia*: 133-163. Leiden: Backhuys.

Munro, I.S.R. 1967. *The fishes of New Guinea*. Port Moresby: Department of Agriculture, Stock and Fisheries,.

Musser, G.G. 1987. The mammals of Sulawesi. In T.C. Whitmore (ed.), *Biogeographical evolution of the Malay Archipelago*: 73-93. Oxford: Clarendon Press.

Musser, G.G. 1991. Sulawesi rodents: descriptions of new species of *Bunomys* and *Maxomys* (Muridae, Murinae). *Am. Mus. Novit.* 3001: 1-41.

Myers, G.S. 1949. Salt-tolerance of freshwater fish groups in relation to zoogeographical problems. *Bijdragen tot de Dierkunde* 28: 315-322.

Oosterzee, P. van 1997. *Where worlds collide. The Wallace Line*. Sydney: Reed Books.

Payne, J., Francis, C.M. and Phillipps, K. 1985. *A field guide to the mammals of Borneo*. Kota Kinabalu: The Sabah Society.

Rainboth, W.J. 1996. *Fishes of the Cambodian Mekong*. Rome: Food and Agriculture Organization of the United Nations.

Raven, H.C. 1935. Wallace's Line and the distribution of Indo - Australian mammals. *Bull. Am. Mus. Nat. Hist.* 68: 179-293.

Rensch, B. 1931. Die vogelwelt von Lombok, Sumbawa, under Flores. *Mitt. Zool. Mus. Berl.* 17: 451-537.

Ribey, J. 1997. The birds of Sanghi and Talaua, North Sulawesi. *Kukila* 9: 3-36.

Roberts, T.R. 1978. An ichthyological survey of the Fly River in Papua New Guinea. *Smithsonian Contributions to Zoology* 281: 1-172.

Roberts, T.R. 1989. The freshwater fishes of western Borneo (Kalimantan Barat, Indonesia). *Mem. Calif. Acad. Sci.* 14: 1-210.

Rozendaal, F.G. & Dekker, R.W.R.J. 1989. Annotated checklist of the birds of the Dumogo-Bone National Park, North Sulawesi. *Kukila* 4: 85-109.

Schmitt, L.H., Kitchener, D.J. & How, R.A. 1995. A genetical perspective of mammalian variation and evolution in the Indonesian Archipelago: biogeographical correlates in the fruit bat genus *Cynopterus*. *Evolution* 49: 399-412.

Sclater, P.L. 1858. On the general geographic distribution of the members of the class Aves. *J. Linn. Soc. (Zool.) Lond.* 2: 130-145.

Simpson, G.G. 1977. Too many lines: the limits of the Oriental and Australian zoogeographic regions. *Proc. Am. Philos. Soc.* 121: 107-120.

Smythies, B.E. 1960. *The birds of Borneo*. Edinburgh: Oliver and Boyd.

Steenis, C.G.G.J. van 1979. Plant-geography of east Malesia. *Bot. J. Linn. Soc.* 79: 97-178.

Stattersfield, A.J., Crosby, M.J., Long, A.J. & Wege, D.C. 1998. *Priorities in Biological Conservation*. Cambridge: Birdlife International.

Stresemann, E. 1939-1941. Die vogel von Celebes. *J. Ornith.* 87: 299-425; 88: 1-135, 389-487; 89: 1-102.

Tiji, H.D., Susitno, S., Sukija, Y., Harscino, R.A.F., Rachmat, A., Hainim, J. & Djundaedi, H. 1984. Holocene shorelines in the Indonesian tin islands. *Modern Quat. Res. SE Asia* 8: 103-117.

Tyler, M.J. & Godthelp, H. 1993. A new species of *Lechriodotus* Boulenger (Anura: Leptodactylidae) from the Early Eocene of Queensland. *Trans. Roy. Soc. S. Aust.* 117: 187-189.

Tyler, M.J., Watson, G.F. & Martin, A.A. 1981. The Amphibia: diversity and evolution. In A. Keast (ed.), *Ecological biogeography of Australia*: 1275-1302. The Hague: Dr. W. Junk.

Veevers, J.J. 1991. Phanerozoic Australia in the changing configuration of Proto-Pangea through Gondwanaland and Pangaea to the present dispersed continents. In P.Y. Ladiges, C.J. Humphries & L.W. Matinelli (eds), *Austral Biogeography*: 17-29. Canberra: CSIRO.

Verhoeye, J. & Holmes, D.A. 1998. The birds of the islands of Flores - a review. *Kukila* 10: 3-59.

Verstappen, H. Th. 1975. On palaeoclimates and landform development in Malesia. In P. Bartstra, & W.A. Casparie (eds), *Modern Quaternary research in southeast Asia*: 101-112. Rotterdam: Balkema.

Walker, D. (ed.) 1972. *Bridge and barrier: the natural and cultural history of Torres Strait*. Canberra: Australian National University Research School of Pacific Studies. Department of Biogeography and Geomorphology.

Walker, D. & Chen, Y. 1987. Palynological light on tropical rainforest dynamics. *Quaternary Science Reviews* 6: 77-92.

Walker, D. & Hope, G.S. 1982. TITLE? In Gressitt, J.L. (ed.), *Biogeography and ecology of New Guinea*: 263-286. The Hague: Dr. W. Junk.

Wallace, A.R. 1969. *The Malay Archipelago* (2 vols). London: MacMillan.

Wallace, A.R. 1976. *The geographical distribution of animals* (2 vols). London: MacMillan.

Wang, X., van der Kaars, S., Kershaw, P., Bird, M. & Jansen, F. 1999. A record of fire, vegetation and climate through the last three glacial cycles from Lombak Ridge core G6-4, eastern Indian Ocean, Indonesia. *Palaeogeogr. Palaeoclimatol. Palaeoecol.* 147: 241-256.

Wardell, J.C., Fox, P.S., Hoare, D.J., Marthy, W. & Anggraini, K. 1998. Birds of the Rawa Aopa Watumohai National Park, Southeast Sulawesi. *Kukila* 10: 91-114.

Watling, D. 1983. Ornithological notes from Sulawesi. *Emu* 83: 247-261.

Welch, K.R.G. 1988. *Snakes of the Orient: a checklist*. Florida: R.E. Krieger Publ. Co.

White, C.M.N. & Bruce, M.D. 1986. *The birds of Wallacea (Sulawesi, The Moluccas and Lesser Sunda Islands, Indonesia)*. London: British Ornithologists' Union.

Whitmore, T.C. 1975. *Tropical rain forests of the Far East*. Oxford: Clarendon Press.

Whitmore, T.C. 1981a. Wallace's Line and some other plants. In T.C. Whitmore (ed.), *Wallace's Line and Plate Tectonics*: 70-80. Oxford: Clarendon Press.

Whitmore, T.C. 1981b. Paleoclimate and vegetation history. In T.C. Whitmore (ed.), *Wallace's Line and Plate Tectonics*: 36- Oxford: Clarendon Press.

Whitmore, T.C. (ed.) 1987. *Biogeographic evolution of the Malay Archipelago*. Oxford: Clarendon Press.

Whitten, A.J., Mustafa, M. & Henderson, G.S. 1987. *The ecology of Sulawesi*. Djakarta: Gadjak Mada University Press.

Zakaria -Ismail, M. 1994. Zoogeography and biodiversity of the freshwater fishes of southeast Asia. *Hydrobiologia* 285: 41-48.

Dispersal versus vicariance, artifice rather than contest

B. Michaux
Private Bag, Kaukapakapa, New Zealand

ABSTRACT: The opposition between dispersal and vicariance is shown to be an artifact of poorly defined concepts. In place of a simplified opposition, five processes are identified that affect plant and animal distributions. These processes –modification, movement, mixing, splitting and juxtaposition - are not logically equivalent as they operate over different time scales. The time scales discussed are ecological time, geomorphological time, and geological time. Processes operating in geological time provide the context in which shorter time scale processes operate, and as such take logical priority. Dispersal as a process operating in ecological time is discussed in relation to the modification of the New Zealand avifauna by introduced species. Dispersal as a process operating over geomorphological time is discussed by reference to transitional biotas in Wallacea. Finally, dispersal in geological time is illustrated by the historical biogeography of the gymnosperm genus *Agathis* (Araucariaceae).

1 INTRODUCTION

While I remain opposed to any form of biogeographic explanation based on extraordinary events, I agree with Holloway (1998) that it is time to move beyond the opposition of vicariance versus dispersal. Holloway's (1998) solution for overcoming this opposition was to develop the middle ground, that is to sort species living at any locality into those that dispersed there and those that evolved *in situ*. This solution reminds me of a previous attempt to resolve another infamous opposition, namely the role of nature versus nurture in the origin of complex traits like intelligence. Historically this opposition arose between genetic determinists, who contended that intelligence was all in the genes, and environmentalists who contended it was all in the upbringing. Early attempts at resolution tried to quantify genetic and environmental components. For example, intelligence could be 60% genetic and 40% environmental. However, it proved impossible to do any such thing because genes and environment are not independent entities. A resolution of the nature-nurture dichotomy came about when traits like intelligence were viewed as outcomes of a developing complex in which both genetic and environmental inputs interact. Models of neural networks, central to AI research, place the emphasis firmly on systems and their interactions with real time inputs. This example shows that the nature-nurture opposition resulted entirely from poorly developed concepts. I would now like to examine the concepts of vicariance and dispersal in the light of this example.

2 CLARIFYING CONCEPTS, TIME SCALES AND LOGICAL PRIORITY

Table 1 shows five processes operating on three different time scales. Modification of a biota is defined as a process operating on an ecological time scale and involves individual species crossing physical barriers to establish themselves in new environments. Two processes operating on the longer geomorphological time scale are identified. The first – movement – is defined as the response by many species to long-term environmental changes. Range expansion (or contraction) takes place on a time scale less than an average species' lifetime and where the relative positions of landmasses are fixed. The second process - mixing – is defined as the response by species of two or more biota to colonizing opportunities that arise as new land is created, for example in collision zones. Mixing may also be an outcome of movement. Finally, operating over the longest time scale, are the processes of splitting a biota, by rifting along continental margins for example, or by the juxtaposition of two or more biota as terranes, arcs or cratons amalgamate in convergence or strike-slip zones. Splitting and juxtaposition operate over a time scale in which landmass dynamics are changing and speciation occurs. Indeed, these dynamics have provided to be precisely the conditions that favour speciation.

Table 1: Clarification of concepts.

ECOLOGICAL TIME (Here and now)	MODIFICATION of a biota through self-introduction of species	Relative	
GEOMORPHOLOGICAL TIME (10^3 - 10^5 years)	MOVEMENT of biotas in response to changing climate MIXING of biotas in response to change in landscape during marine regressions	position of land-masses fixed	Within a single species' lifetime
GEOLOGICAL TIME (Millions of years)	SPLITTING of a biota JUXTAPOSITION of biota	Mobility	Speciation

Opposing dispersal and vicariance simplifies what in reality is a more complex situation. Four of the five processes discussed above could be defined as 'dispersal' – i. e. modification, movement, mixing and juxtaposition, while two could be termed 'vicariance' – i. e. movement and splitting. The opposition is a consequence of conceptualising 'vicariance' and 'dispersal' as independent outcomes of different processes, rather than as dependent outcomes of linked processes.

Finally, the identification of a time scale over which each process operates links them in such a way as to form a logical hierarchy. Splitting and juxtaposition have logical priority because they precede the other processes historically. Movement, mixing and modification take place in a context. For example, modification of a biota by individual species crossing physical barriers is as much to do with establishment within a preexisting biota as it has to do with ability to cross physical barriers. I would now like to examine the process of dispersal over different time scales in more detail.

3 DISPERSAL IN ECOLOGICAL TIME

Since European settlement of New Zealand last century, over one hundred exotic bird species have been liberated and of these 36 have become established (Soper, 1972). A smaller number of self-introduced species have also become established during this century. Since the mid-1800's New Zealand's indigenous avifauna has been extensively modified. Soper (1972) described this modification thus:

> "We have, so to speak, two ornithological worlds: (1) the man-made environment, largely the province of birds that man himself has introduced and, of recent years, of birds such as the Spur-winged Plover and Welcome Swallow that

have introduced themselves, (2) the native environment still largely the province of native birds."

The New Zealand experience is instructive to any discussion about avian dispersal over an ecological time scale. Firstly, it is clear that irrespective of whether these birds were self-introduced or deliberately released, they are dependent on man-made environments and have not independently established themselves in native environments.

Secondly, we can deduce that the self-introduced birds discussed by Soper (1972) would not have established themselves in New Zealand without the unwitting aid of humans. Unless we are prepared to accept that their arrival was an extraordinary event that just happened to coincide with the transformation of the New Zealand landscape, then we are forced to the conclusion that these birds have probably been flying to New Zealand since they first evolved. If this is the case, then one must conclude that they have not been able to establish themselves as part of the New Zealand avifauna prior to extensive habitat modification. The importance of habitat modification in the initial establishment (and probable long-term survival) of self-introduced birds is highlighted by details of the ecological requirements of the species concerned. The Welcome Swallow (*Hirundo neoxina*) nests on man-made structures, the Spur-winged Plover (*Lobibyx novaehollandiae*) feeds and nests in paddocks, introduced frogs (*Litoria aurea, L. raniformis* and *L. ewingii*) form a major dietary component of the White-faced Heron (*Ardea novaehollandiae*), and so on. A corollary is that birds that disperse and establish themselves do so as soon as opportunities for establishment arise. On a geological time scale the establishment of a species' initial range will appear instantaneous. Lastly, it seems that once an avifauna has become established it is, over ecological time, impervious to continuing colonization without a concomitant environmental catastrophe.

4 DISPERSAL IN GEOMORPHOLOGICAL TIME

A locality map of eastern Indonesia and Melanesia is shown in Figure 1. Borneo and Java in the west mark the eastern limit of Sundaland. Between Sundaland and the Australian craton lies Wallacea. Wallacea is regarded as a transitional region, and has been the subject of much biogeographical speculation since Wallace first described the biota in 1859 (van Oosterzee, 1997). The tracks in Figure 1 indicate similar plant and animal communities. For example, the Melanesian arc track (Figure 1: line), found in Melanesia through northern New Guinea to the southern Philippines, is distinguishable, to a greater or lesser degree, from other tracks identified in Figure 1.

Some Wallacean islands, such as the Philippines, Halmahera, New Guinea, Timor and Sulawesi have more than one track associated with them and can thus be said to be biologically composite. Sulawesi and Timor are highlighted in Figure 1 because the composite nature of their biota is particularly clear. In both islands Asian-derived placental mammals coexist with Austro-Papuan marsupials (Audley-Charles & Hooijer, 1973; Musser, 1987; Cranbrook, 1991), and Austro-Papuan warblers and honeyeaters coexist with bulbuls and Old World warblers and flycatchers. The composite character of the mammalian and avian faunas of Sulawesi and Timor is mirrored by these islands' composite geological makeup. Terranes derived from the Australian continental margin and transported westwards within shear zones, collided with southwest Sulawesi in the Miocene (Hall, 1998). Modern Timor was formed by a Pliocene collision between a southwards migrating arc of Asian derivation with an Australian derived rift fragment (McCabe & Cole, 1989).

Mixing seems an inevitable outcome in such situations, given the presence of Australian and Wallacean biota together with environments open to colonization by species from both. As new land emerged within a collision zone, dispersal over an ecological time scale would ensure these new environments were colonized. While I suppose it is theoretically possible that some long distance colonization could occur, probability arguments suggest that colonists will be derived from adjacent biotas. The degree of intermixing between two biota will lie somewhere between the extremes of total fusion and perfect terrane fidelity. In Sulawesi there is evidence that avian genera have retained terrane fidelity, with twelve of fourteen endemic genera restricted to the highlands of central and southern Sulawesi (White and Bruce, 1986). Irrespective of the degree of intermixing, phylogenetic connections are not lost and it is, therefore, possible to deduce that an island is composite without species necessarily remaining attached to particular geological terranes.

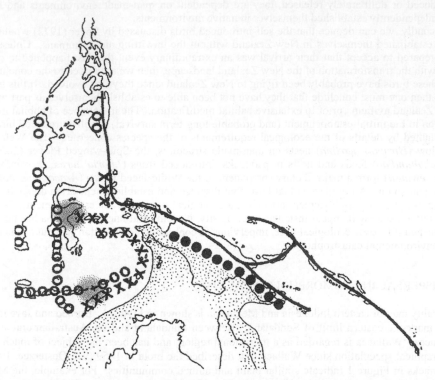

Figure 1: Map of eastern Indonesia and Melanesia. Symbols indicate biological (genealogical) patterns referred to as tracks in the text. Line = Melanesian arc track, closed circle = Melanesian rift track, cross = Banda rift-arc track, open circles = Sumba terrane track, stippled = Australian craton.

5 DISPERSAL IN GEOLOGICAL TIME

Geological models reconstructing the breakup of east Gondwana produce 'snapshots' of deep history, providing a pictorial representation of the Indo-Pacific region's geological development. Evolution has taken place in this developmental context. I would like to examine this interplay between geological development and biological evolution with an example that illustrates dispersal on a geological time scale.

Agathis species are emergent forest trees with massive columnar trunks and large spreading crowns. The spectacular sight of mature *Agathis* is made all the more impressive by the habit of continuously shedding lower branches and bark, leaving the trunk clear of epiphytes and vines. The bark of all *Agathis* species exudes characteristic terpentene resins which Whitmore (1980) suggested might yield useful taxonomic characters. Bark fragments and other organic detritus form a humus mound around the base of mature trees. This mound is penetrated by feeding roots and must remain intact for the tree to remain healthy.

Reproduction is by cone-born seed. Trees start to produce fertile seed at 30 years and mature trees are prolific seed producers. Nurserymen report that only 50% of the seed produced is fertile, and of these only 100 out of 25 000 will germinate (Adams, 1977). The reason for low germination is the narrow window of opportunity between the female cone opening and the seed losing viability. In the case of *A. australis* this window is as short as a few weeks. In New Zealand *A. australis* has expanded and contracted within its present range repeatedly over the past 30,000 years, but it is inconceivable - given the details of its ecology and short-lived viability of its seed - that it has done so over water. Interestingly, despite the tree growing successfully throughout New Zealand as cultivated specimens, it never appears to have naturally colonised south of its present range of approximately 38°S (Salmon, 1980).

Agathis, Araucaria and the recently discovered *Wollemia* (Jones et al., 1995) are extant members of the family Araucariaceae. Fossils of *Araucaria* are amongst the oldest of any extant coniferous genus, being known from the Jurassic (Setoguchi *et al.*, 1998). An extinct genus *Araucariacites* is also reported from the Triassic (Suggate, 1978). *Agathis* fossils are known from the Cretaceous. Setoguchi *et al.* (1998) confirmed monophyly for the Araucaraceae in a cladistic analysis of rbcL gene sequences. While their study was primarily concerned with *Araucaria*, it also showed that *Agathis* is monophyletic and provided some limited phylogenetic information (see below).

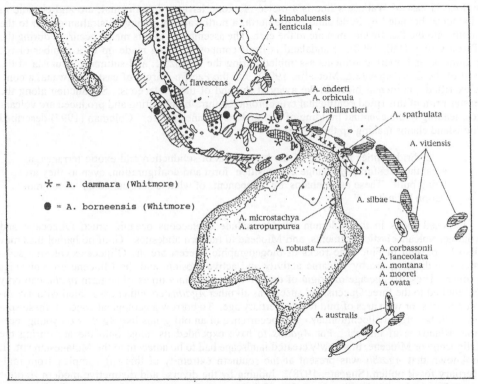

Figure 2: Distribution of extant Agathis. Horizontal hatching = Melanesian arc, cross = Melanesian rift, vertical hatching = Banda rift-arc complex, diagonal hatching = Sumba terrane, cross-hatching = Indochina terrane, dots = Sibumasu terrane, stippled = cratonic fragments.

The taxonomy of *Agathis* is far from clear and the literature can be confusing and contradictory. While the Melanesian and Pacific species appear well defined and the nomenclature mostly stable, the same cannot be said for Indonesian species. Setoguchi et al's. (1998) study adds to this confusion by using the name *palmerstoni* for a Malaysian species, which Whitmore (1980) had synonymised with *robusta* from Australia. Figure 2 shows the distribution of extant *Agathis* species. Following Whitmore (1980), *Agathis* found in central Indonesia are referred to *A. dammara*. In contrast, de Laubenfels (1988) recognised two species, *philippensis* and *celebica*, as well as a number of endemics on Borneo. These endemics are shown on Figure 2, but note that Whitmore (1980) synonymised *endertii* with *borneensis* and only recognised *flavescens* as a subspecies of *dammara*. Two species are found on islands of the Melanesian arc (*vitiensis* in Fiji, Vanuatu and the southern Solomon Islands, and *silbae* in Vanuatu). Eight species live on modern remnants of the Melanesian rift, five in New Caledonia, one in New Zealand and two species at mid-montane altitudes in the highlands of central New Guinea. There are three Australian species restricted to the rain forest of Queensland.

Setoguchi et al's (1998) strict consensus tree of ten species supports only two groupings –two Malaysian species form one grouping (but see above) and the two Melanesian arc species and *robusta* from northeast Australia the other. Setoguchi et al. (1998) also presented a more resolved cladogram based on additional morphological character analysis. This cladogram implies that the Melanesian arc is a sister area to Australia + Sunda Shelf. These areas in turn show a sister group relationship to the Melanesian rift + Sumba terrane. This latter relationship is potentially important in resolving the biogeography of key Indonesian areas like Sulawesi, but uncertainty in basic taxonomy makes caution necessary.

The key to understanding the present distribution pattern of *Agathis*, I would argue, is dispersal in geological time. While it is beyond the scope of this essay to adequately summarise the geological literature, the main outlines of geological development in the Indo-Pacific are clear. Indonesia east of Wallace's line is a meleé of exotic terranes (continental, arc and oceanic in character) bounded by Sundaland in the north, a northward migrating Australian craton to the south, and the Pacific Ocean plate to the east. The assemblage of this meleé occurred during the Tertiary (e.g. Hall, 1998). Sundaland is itself composite, being made up of a number of rift fragments and micro-continents assembled during the Mesozoic, and sutured to Eurasia at the end of the Cretaceous (e.g. Metcalfe, 1998). On the Pacific margin of east Gondwana, a complex rifted continental block broke away at the end of the Cretaceous. Subduction along the eastern rim of this rifted continental crust promoted intra-block rifting and produced arc volcanics, leading to the eventual building of the modern Melanesian arc. Coleman (1997) described the island chains making up this arc as:

> "hybrid entities made up of the byproducts of subduction and exotic terranes; as arcs, they continually change composition, form and configuration, even as they are being built. These outer chains are components of what is here called an 'accommodation boundary'."

Exposed rocks in the Solomon Islands include Cretaceous oceanic crust, Oligocene and younger volcanoclastic sediments, and Miocene to modern andesites. Granitic batholithes may also be present in Fiji. The rocks of biogeographic interest are the Oligocene volcanic sediments, indicating nearby volcanic activity and active erosion, and the Miocene arc volcanics proper. These fix the age of origin of the arc to mid-Tertiary times. The term *in situ* can only be applied to the three Queensland species as all other *Agathis* are either associated with assembled blocks, or with arcs of mid to late Tertiary age. To borrow a geological concept, these species can be termed allochthonous. The occurrence of an old genus like *Agathis* on young, oceanic islands is of interest. For *Agathis* to have expanded its range onto the arc during the Oligocene or Miocene, this newly created landscape had to be adjacent to the Melanesian rift. It is known that *Agathis* was present at the southern extremity of this rift complex from mid-Tertiary fossil pollen (Suggate, 1978). Judging by the diverse and distinctive modern *Agathis* assemblage associated with present day rift fragments, the genus was also present in the central and northern sections of the rift. It is also likely that *Agathis* was associated with *Araucaria*, as it still is today in New Guinea and New Caledonia, or as in the past in New Zealand. The absence of *Araucaria* on the modern arc can only be an accident of history. Note that the biogeographic evidence from *Agathis* falsifies a hypothesis of oceanic origin for the Melanesian arc (e. g. Yan and Kroenke, 1993).

Taxonomic confusion within Indonesian *Agathis* results in more questions than insight. The most pressing needs are to confirm the status of *borneensis* as the Sunda shelf taxon, and *dammara* as the Sumba terrane taxon. Setoguchi et al's (1998) hypothesis that *dammara* is most closely related to the New Caledonian species and *borneensis* with the Australian species *robusta* can then be tested with more certainty. Does the taxon *damarra* represent a species complex? What is the status of *A. flavescens*, and is its association with the east Malay platelet (Gatinsky & Hutchison, 1993) primary or the result of dispersal in geomorphological time? Are the endemics that de Laubenfels (1988) recognised valid? These important timber trees, with the potential for commercial planting, warrant further study of basic taxonomic and systematic relationships, which would have important implications for the biogeography of this biodiversity hotspot.

6 CONCLUSION

Many groups other than *Agathis* have distributions extending from SE Asia or Sundaland, through Indonesia to New Guinea and the Pacific. The Pacific range is routinely interpreted as a result of dispersal in ecological time from Asia via Indonesia and New Guinea/Australia eastwards. I have attempted to clarify the concept of dispersal, and to explain why dispersal events occurring in ecological time are improbable explanations to account for present day distributions of indigenous species. The modern distribution of *Agathis* in Asia, Indonesia, Australia, and the Pacific is, I would argue, a result of dispersal of ancestral *Agathis* in geological time. In this view modern *Agathis* are derived from ancestors that formed part of an east Gondwanic maritime forest during the Cretaceous, or possibly earlier. This forest did not extend westwards into the Indian segment of east Gondwana or eastwards into the Antarctic or Chilean sectors. As the continental margin of east Gondwana became disrupted and dispersed, so too did this ancestral biota.

New Zealand's kauri forest is an evolved remnant of a 100 million-year old ecosystem. Other remnants of this once widespread Gondwanic maritime forest are dispersed throughout Indonesia and the Pacific. This panbiota, which exhibits a stunning modern biodiversity, is under threat of destruction almost everywhere. Even in rich countries like Australia and New Zealand logging still goes on and pests run rampant. When species and habitats are destroyed it is not just a national tragedy, but also a loss of part of something bigger. Preservation of relationship elements between the individual constituents of this panbiota, which is an international problem, requires some adjustment to conservation thinking, which is primarily concerned with the protection and promotion of biodiversity within national boundaries. We have a model for such a conservation approach in the international efforts to protect migratory waders such as the bar-tailed godwit.

ACKNOWLEDGEMENTS

My thanks to Ian Metcalfe and the co-organisers of the conference at UNE for the work they put into organising such a successful event, for the editorial assistance, and for the financial help that made my attendance possible. Jeremy Holloway provided a thoughtful review of an earlier draft.

REFERENCES

Adams, J. G. E. 1977. *Kauri. A King Among Kings*. Auckland: Wilson & Horton,.
Audley-Charles, M. G. and Hooijer, D. A. 1973. Relation of Pleistocene migrations of pigmy stegodonts to island arc tectonics in eastern Indonesia. *Nature* 241: 197-198.
Cranbrook, Earl of. 1991. *Mammals of South-East Asia*. Oxford: Oxford University Press.
Coleman, P. J. 1997. Australia and the Melanesian arcs: a review of tectonic settings. *AGSO Journal of Geology and geophysics* 17: 113-125.
Gatinsky. Y. G. & Hutchison, C. S. 1986. Cathaysia, Gondwanaland, and the Paleotethys in the evolution of continental Southeast Asia. *Geological Society of Malaysia Bulletin* 20: 179-199.
Hall, R. 1998. The plate tectonics of Cenozoic SE Asia and the distribution of land and sea. In R. Hall & J.D. Holloway (eds), *Biogeography and Geological Evolution of SE Asia.*: 99-131. Leiden: Backhuys.
Holloway, J. D. 1998. Geological signal and dispersal noise in two contrasting insect groups: R-mode analysis of the Lepidoptera and cicadas In R. Hall & J.D. Holloway (eds), *Biogeography and Geological Evolution of SE Asia.*: 291-314. Leiden: Backhuys.

Jones, W. G., Hill, K. D., & Allen, J. M. 1995. *Wollemia nobilis*, a new living Australian genus and species in the Araucariaceae. *Telopea* 6: 173-176.

de Laubenfels, D. J. 1988. *Agathis*. Flora Malesiana, Series 1 – Spermatophyta, 10(3): 429-442.

McCabe, R. & Cole, J. 1989. Speculations on the late Mesozoic and Cenozoic evolution of the Southeast Asian margin. In Z. Ben-Avraham, (ed), *The evolution of Pacific Ocean margins*: 143-160. New York: Oxford Monographs in Geology and Geophysics 8. Oxford University Press.

Metcalfe, I. 1998. Palaeozoic and Mesozoic geological evolution of the SE Asian region: multidisciplinary constraints and implications for biogeography. In R. Hall & J.D. Holloway (eds), *Biogeography and Geological Evolution of SE Asia.*: 25-41. Leiden: Backhuys.

Musser, G. G. 1987. The mammals of Sulawesi. In T. C. Whitmore (ed), *Biogeographical evolution of the Malay Archipelago*: 73-94. Oxford: Clarendon Press.

van Oosterzee, P. 1997. *Where Worlds Collide*. Australia: Reed Books.

Rangin, C. & others. 1990. The quest for Tethys in the western Pacific. 8 paleogeodynamic maps for Cenozoic time. *Bullétin de la Société gólogique Française* 6: 907-913.

Salmon, J.T. 1980. *The Native Trees of New Zealand*. Auckland: Reed.

Setoguchi, H., Osawa, T. A., Pintaud, J-C, Jaffré, T. & Veillon, J-M. 1998. Phylogenetic relationships within Araucariaceae on RBCL gene sequences. *American Journal of Botany* 85: 1507-1516.

Soper, M. F. 1972. *New Zealand Birds*. Auckland: Whitcombe & Tombs.

Suggate, R. P. (ed). 1978. *The Geology of New Zealand, volumes 1 & 2*. Wellington: Government Printer.

Wallace, A. R. 1859. On the Zoological Geography of the Malay Archipelago. Read at the Linnean Society, London, November 3, 1859.

White, C. M. N., & Bruce, M. D. 1986. *The Birds of Wallacea*. B. O. U. Checklist number 7.

Whitmore, T. C. 1980. A monograph of *Agathis*. *Plant Systematics and Evolution* 135: 41-69.

Yan, C. Y. & Kroenke, L. W. 1993. A plate tectonic reconstruction of the southwest Pacific 0-100 Ma. *Proceedings of the Ocean Drilling programme, Scientific Results* 30: 697-707.

The Australian rodent fauna, flotilla's, flotsam or just fleet footed?

H. Godthelp

School of Biological Science. University of New South Wales. Sydney. 2052

ABSTRACT: Until recently the lack of fossil evidence has lead to a complete reliance on molecular data for hypotheses relating to the origin, evolution and paleobiogeography of the Australian rodents. This has inevitably lead to a simplistic view of these processes. Data from newly available fossil material allow us to review the existing hypotheses and propose alternative views. Fossils show us that the evolutionary process in Australia has been complex with at least three phases of immigration, two from SE Asia directly and one from New Guinea. The new information allows us to make comparisons to known geological and biological events in the SE Asian region and see the Australian rodent radiation in a wider perspective.

In Australia, the "land of the marsupial", about 25% of modern land mammal species are murid rodents. Rodents have spread into all the ecosystems of Australia and are frequently the most abundant and conspicuous mammal present. All this in less than 6 million years of occupancy. But then rodents are nature's over-achievers and we shouldn't be surprised that they have succeeded in Australia as they have everywhere else in the world.

And yet even with this success and the numeric dominance of rodents in the Pleistocene and Quaternary vertebrate accumulations in Australia and their abundance in the contemporary environment we have an appallingly sparse fossil record for them. Of the 70+ living species nearly half of those have no fossil record and so far there are only two extinct species (*Pseudomys vandycki* Godthelp. 1988 and *Zyzomys rackhami*, Godthelp1996.) described from Australia, both from Pliocene sites in Queensland.

A consequence of the paucity of rodent fossils has been a near total reliance on molecular techniques in attempts to understand the evolution of Australia's rodent fauna. This reliance has lead, in the past, to what might be considered a simplistic view of the nature of the earliest evolutionary events that have shaped our rodent fauna.

Other difficulties encountered in the palaeontological study of rodents are the fragmentary nature of the material available and the inevitable reliance on dental characters that are sometimes prone to convergences and parallelisms. Despite this, the single defining synapomorphy for the murinae is a dental feature (the presence of the anterio-lingual cusp T1 on the first upper molar) and dental characters are often the only ones available.

Watts and Baverstock (1992,1994) using microcompliment fixation of albumin techniques have determined the Australo-New Guinean radiation to consist of 3 clades. The identification of these clades, the *Hydromys, Lorentzimys* and *Anisomys* clades prompted these authors to propose a classification containing 2 tribes. These tribes are the Hydromiini, containing the Australian genera plus some of the New Guinean genera including *Melomys* and *Uromys*, and the Anisomyini for the genera *Anisomys, Chiruromys, Hyomys, Macruromys, Mallomys* and *Pogonomys*. Within the Australian clade the most provocative grouping is that of the *Mesembriomys/Zyzomys* group and the *Melomys/Uromys* group. Both groups have the most derived dentitions in the Australian radiation but in ways that are radically different from each other.

Recent work involving amplified repetitive DNA (Pascale et al. 1990) has identified an element of the genome called Lx, which appears to be restricted to members of the true Murinae. This synapomorphy may define a monophyletic group of the Indo-Australian and African Murinae. Evidence of an extraordinary evolutionary event involving the rapid rise of the most successful mammalian group and the subsequent invasion of three continents.

The rich and diverse fauna being recovered from the Early Pliocene Rackham's Roost Local Fauna at Riversleigh is providing us with our best view of the early evolution of rodents in

Australia. To date there are more than 12 taxa of rodents represented in the fauna. Among them are members of the genera *Leggadina, Zyzomys* and a number of species of *Pseudomys*. Some of the species of *Pseudomys* can be assigned to a range ofspecies groups present in the modern fauna which I propose to be monophyletic, on the basis of dental characteristics. These groupings are *Pseudomys gracilicaudatus-P. nanus, P. australis-P. desertor-P. shortridgei, P. delicatulus* and *P. albocinereus-P. hermannsburgensis-P. novaehollandiae-P. pumilis-P. pilligaensis*. There are also *Psuedomys* species present that I cannot assign to a particular species grouping.

Additionally there are several new genera that show no close affinities to modern forms. Perhaps the greatest surprise has been the discovery of a *Potwarmus*-like "Dendromurine" in the deposit. This is to date the latest surviving member known of this primitive group that does not seem to have survived competition with the more derived murines elsewhere in Asia. Another taxon in the fauna that may have its affinities nearer to Asian murines seems to have affinities with the widespread, primitive *Chiropodomys* group from South East Asia.

The diversity of endemic murine genera in the Rackham's Roost assemblage suggests one of two things. Murids invaded Australia much earlier than Rackham's Roost time to enable evolution of these distinctive groups on this continent. Alternatively, murids may have entered Australia already differentiated (somewhere in southeastern Asia) into at least this many groups. Considering the first hypothesis, the potential for a much earlier origin is limited by the total absence of rodents in the late Miocene Alcoota and Ongeva Local Faunas of the Northern Territory (8 and 7 my respectively; e.g. Megirian *et al.* 1994), despite years of collecting at those sites. This does not rule out a time of origin between latest late Miocene and early Pliocene (7 to 5 million years ago), a time of climatic crisis in Australia and elsewhere (Archer *et al.* 1995). At present, however, there is no way to test these alternative scenarios.

The questions of how and when rodents entered Australia have been asked many times and there is a long list of hypotheses as a consequence. Ellerman (1941) suggested that murid rodents actually evolved in Australia on the basis of the known diversity. Most others, Tate (1951), Simpson (1961), Wood (1983), Watts and Aslin (1981), to name just a few, have however proposed a South East Asian source. The hypotheses concerning the timing and number of incursions into the continent have varied much more widely, with entry ranging from about 15 million years ago (Watts and Aslin, 1981) too less than one million. Most modern researchers especially those utilizing molecular data support a single major dispersal into Australia with New Guinea as a staging point, all derived from a single ancestor. Few of these researchers have had the benefit of data from the fossil record in Australia.

The fossil record indicates three major periods of evolution in the rodent fauna of Australia. The first, an initial burst of speciation, at or about the beginning of the Pliocene. The next occurred near the end of the Pliocene when many of the modern species emerged and *Rattus* first entered the biota and finally during the later part of the Pleistocene when there was an interchange between Australia and New Guinea. These immigrations could only have occurred during the brief interludes when there were either land connections or when the distances between landmasses in the region were significantly shorter than they are today. Work by Hall (1998) presents a series of tectonic reconstructions of the region over a wide temporal range. These show that on the basis of current understanding the most plausible configuration of the landmasses for terrestrial mammal migrations occurred at about 7mya. Additionally it is from this time on that the ocean currents (Kennett et al. 1985) become favorable for short water dispersals by rafting or swimming. Prior to this period rafts would have been taken out into the Indian Ocean rather than towards land.

Species of *Rattus* in the Plio-Pliestocene deposits at Floraville, northern Queensland, are the earliest record of the genus in Australia. This record post dates an explosion of diversity in southeastern Asia (Chaimanee and Jaeger 1999). It seems probable that the Floraville *Rattus* lineage is part of this Southeast Asian adaptive radiation. Other endemic Australian species of *Rattus* appear, in contrast, to be descendants of a second wave of immigrants that arrived from New Guinea. No representatives of this group are known from pre-Pleistocene deposits anywhere in Australia.

The presence in the fossil record of taxa with strong Asian affinities persuade me that there is a more complex sequence of events than a single "waif" dispersal or even dispersal through a Torresean land bridge that has brought rodents into Australia. Data from Thailand regarding the timing and effect of the appearance of *Rattus* into that area closely mirror the emerging story of the history of that group in Australia. The similarity of the sequence of events and the nature of the taxa involved suggest that little filtration occurred and that movement through the region at that time was relatively unimpeded. One is tempted to ask why other groups of mammals especially other rodent families had not then also made their way to Australia? Perhaps they did and merely failed to survive, any other group would have found themselves competing directly with a well-established and vigorous marsupial fauna not yet suffering from the ravages of the Pleistocene aridity and the depravations of human intervention. Only the discovery of Tertiary aged rodent fossils from the Indonesian region will enable us to delve into the earliest origins of Australian rodents.

REFERENCES

Archer, M.,et al. 1995. Tertiary Environmental and Biotic Change in Australia. In E.S. Verba, G.H. Denton, T. C. Partridge & L. H. Buckle (eds), *Paleoclimate and Evolution, with an emphasis on human origins*: 77-90. New Haven: Yale University Press.

Chaimanee, Y. & Jaeger, J.J. 1999. The evolution of *Rattus* (Mammalia, Rodentia) during the Plio-Pleistocene in Thailand. *Historical Biology* (in press).

Ellerman, J. R. 1941. *The families and genera of living rodents. Vol. 2, Family Muridae*. London, British Museum (Nat. Hist.).

Hall, R. 1998. The plate tectonics of Cenozoic SE Asia and the distribution of land and sea. In R. Hall & J.D. Holloway (eds), *Biogeography and Geological Evolution of SE Asia*: 99-131. Leiden: Backhuys.

Godthelp, H.J. 1988. *Pseudomys vandycki*, a Tertiary Murid from Australia. *Mem. Qd. Mus.* 28: 171-173.

Godthelp, H.J. 1996. *Zyzomys rackhami* sp. nov. (Rodentia, Muridae) a Rockrat from Pliocene Rackham's Roost Site, Riversleigh, Northwestern Queensland. *Mem Qd Mus.* 41: 329-333.

Kennett, J.P., Keller, G., & Srinivasan, M.S. 1985. Miocene planktonic foraminiferal biogeography and paleogeographic development of the Indo-Pacific region. *In* The Miocene ocean: paleogeography and biogeography, Geological Society of America Memoir 163: 197-236

Megirian, D., Murray, P. E, and Wells, R. T. 1994. The Late Miocene Ongeva Local Fauna from the Waite Formation of Central Australia. *Rec. South Aus. Mus.* 27: 225.

Pascale, E., Valle, E., & Furano, A.V. 1990. Amplification of an ancestral mammalian L1 family of long interspersed repeated DNA occurred just before the murine radiation. *Proc. Natl. Acad. USA.* 87: 9481-9485.

Simpson, G.G., 1961. Historical zoogeography of Australian mammals. *Evolution* 15: 431-446.

Tate, A. 1951. The Rodents of Australia and New Guinea. *Bull. Amer. Mus. Nat. Hist.* 97: 187-430.

Wood, A.E. 1983. The radiation of the Order Rodentia in the Southern Continents; the dates, numbers and sources of the invasions. *Schriftenr. geol. Wiss. Berlin.* 19: 381-394.

Watts, C. H. S. & Aslin.H. J., 1981. *The rodents of Australia*. Australia: Angus & Robertson Publishers.

Watts, C.H.S. & Baverstock, P. 1992. Phylogeny of the Australian rodents (Muridae): a molecular approach using microcompliment fixation of albumin. *Aus. J. Zool.* 40: 81-90.

Watts, C.H.S. & Baverstock, P. 1994. Evolution in the New Guinean Muridae (Rodentia) assesed by microcompliment fixation of albumin. *Aus. J. Zool.* 42: 295-306.

Corroboration of the Garden of Eden Hypothesis

Thomas H. Rich
Museum Victoria, P.O. Box 666E, Melbourne, Victoria 3001, Australia

Timothy F. Flannery
South Australian Museum, North Terrace, Adelaide, South Australia 5000, Australia

Peter Trusler
9 Whitehorse Road, Blackburn, Victoria 3130, Australia

Patricia Vickers-Rich
Earth Sciences Department, P.O. Box 28E, Monash University, Clayton, Victoria 3800, Australia

ABSTRACT: The Garden of Eden hypothesis states that, "Modern eutherian lineages diversified in regions that have no known Late Cretaceous mammals (such as Africa, Australia, and Antarctica) and suddenly dispersed widely during the early Tertiary," (Foote, Hunter, Janis & Sepkoski 1999). The occurrence of *Ausktribosphenos nyktos,* a possible erinaceid within the Eutheria in the Early Cretaceous of Australia, is consistent with that hypothesis.

Foote, Hunter, Janis & Sepkoski (1999) coined the phrase, "'Garden of Eden' hy pothesis" to explain an anomaly that exists between the time of origin of the extant eutherian orders as estimated from the fossil record and the substantially older evidence indicated by molecular clock data. Vertebrate paleontological evidence puts the time of appearance of most of the earliest representatives of the extant eutherian orders as being after the Mesozoic-Cenozoic boundary, 65 million years ago, that is, after the non-avian dinosaurs had become extinct, e.g. Benton (1999a-b). Molecular clock estimates, on the other hand, place this fundamental division amongst the eutherians significantly earlier, around the mid-Cretaceous, e.g. Easteal (1999a-b), Hedges et al. (1996), Kumar & Hedges (1998), Springer (1997).

Foote, Hunter, Janis & Sepkoski (1999) give an elaborate statistical argument as to why the fossil record should be relied upon in this matter. The mathematical rationale for their argument is presented in footnote 13 on page 1313. Near the end of their paper, they suggest four hypotheses to reconcile the anomaly between the paleontological and molecular clock evidence. The third of these was that the eutherians had, in fact, originated at the time estimated by the molecular clock data and that this occurred in areas where there was no Late Cretaceous mammalian paleontological record. They specifically singled out Africa, Australia and Antarctica as three potential places where this event could have happened. This is their "'Garden of Eden' hypothesis".

In addition to providing estimates of when the extant eutherian orders may have arisen, molecular evidence has also been invoked to recognise groupings among the eutherian orders. One of the most prominent of these is the Afrotheria (Stanhope *et al.* 1998a-b) (fig. 1). This is in recognition of the close molecular affinity between the five extant orders: Tubulidentata, Afrosoricida (tenrecs and golden moles), Sirenia, Proboscidea, and Hyracoidea, and part of a sixth: Macroscelidea. All of them, as the name implies, have an African base. The Afrotheria are hypothesized to have differentiated into these six groups on the African continent at some time in the Cretaceous and spread from there. That the Afrotheria have not been found in the Cretaceous of Africa as yet to document their presence at the time when they are hypothesized to have arisen there can be attributed to the same factors which have limited the recovery to date of Cretaceous mammals of any kind on that continent.

A group which molecular evidence places well outside of the Afrotheria are the Erinaceidae or hedgehogs (Springer *et al.* 1997, Stanhope *et al.* 1998b). The Erinaceidae are first represented in the northern hemisphere in the Palaeocene of North America (McKenna & Bell 1997). In the Australian late Early Cretaceous (Aptian) there occurs *Ausktribosphenos nyktos* (fig. 2), a possible eutherian mammal with a dentition remarkably similar to primitive erinaceids emplaced in a much more primitive jaw (Rich *et al.* 1999) (fig. 3). Erinaceid features of the dentition of *Ausktribosphenos nyktos* are the progressive reduction in molar size from M_1 to M_3; molars semi-rectangular in outline with some degree of exaenodonty (i.e. bases of the trigonid and talonid cusps are significantly lower on the labial than lingual side of the posterior premolars and anterior molars); M_1 paraconid salient and anteriorly projecting, elongating the prevallid shearing wall; and hypoconulids markedly reduced on M_{1-2}. Primitive features of the jaw of *Ausktribosphenos nyktos* are the presence of a Meckelian groove, a facet for the coronoid bone, and a shallow, vestigial groove for a post-dentary bone (= surangular facet). The relative proportions of the lower molars (low trigonids, M_3 markedly smaller than M_1), whilst common amongst Cainozoic eutherians, are quite atypical of Cretaceous forms from the northern hemisphere (fig. 4).

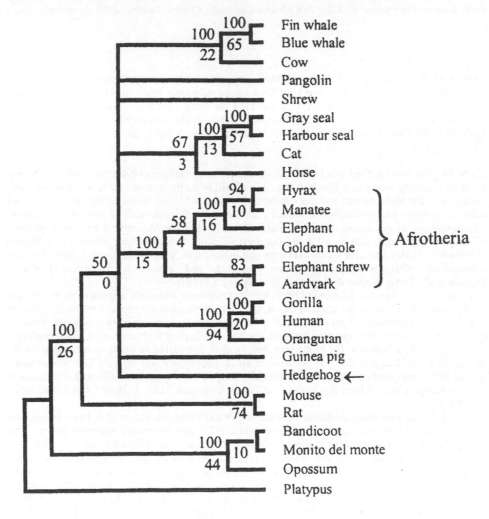

Figure 1. Cladogram of the eutherians, modified from Stanhope et al (1998a, fig. 1). Note the separation of the hedgehogs (Erinaceidae) from the Afrotheria.

During the Cretaceous, micro-continents or terranes were splitting off from the northern edge of Gondwana and drifting north from Africa (Polcyn, Tchernov & Jacobs 1999) on the west to

Australia (Metcalfe 1996, 1998) on the east. Many of these terranes have been identified as existing regions, now parts of southern Eurasia. Such terranes could easily have served as Noah's Arks (*sensu* McKenna 1973), transporting ancestral stocks of eutherians northwards during the Cretaceous.

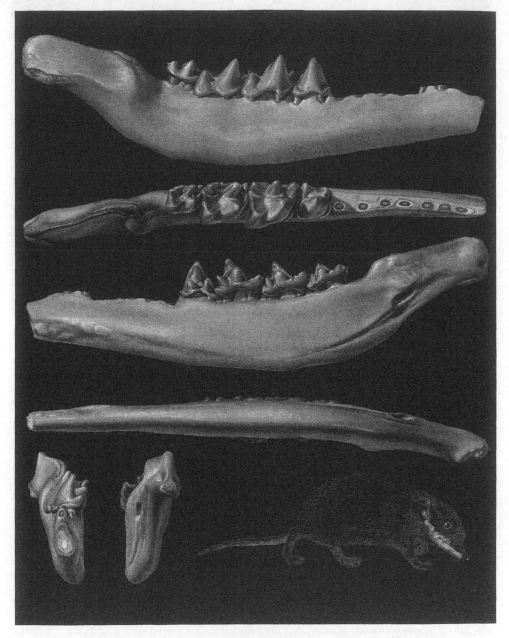

Figure 2. Holotype of *Ausktribosphenos nyktos*. Maximum length of specimen, 16 mm. Top four images from top to bottom: labial view, occlusal view, lingual view, and ventral view. Lower lefthand image, anterior view. Lowermost middle image: posterior view. In the lower right hand corner, the holotype is superimposed on a restoration of what the living animal (body length about 85 millimetres) might have looked like (modelled on the erinaceid *Neotetracus sinensis*). Technical art by P. Trusler, reconstruction by D. Gelt.

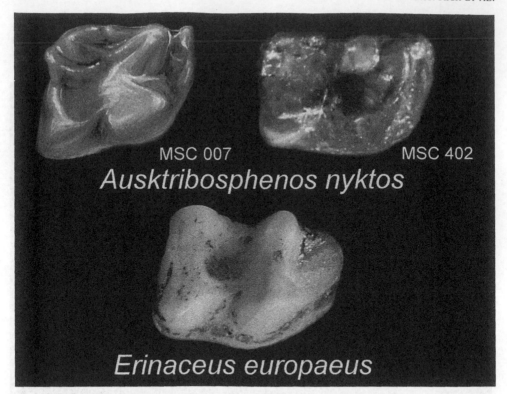

Figure 3. Two M₂s of *Ausktribosphenos nyktos* in occlusal view compared with the remarkably similar M₂ of the living European hedgehog, *Erinaceus europaeus*.

The transit time for terranes leaving the northern margin of eastern Gondwana to reach southeastern Asia would appear to be on the order of 80 million years (Metcalfe 1996, 1998). The West Burma terrane, for example, is shown as part of eastern Gondwana in the Late Jurassic (fig. 5a) (Metcalfe 1996, 1998) and part of southeast Asia by Late Cretaceous, (fig. 5c) (Metcalfe 1996, 1998). Although *A. nyktos* is only about 50 million years older than the oldest northern hemisphere erinaceids, it must be remembered that there is no earlier record of mammals of any kind in Australia. Therefore, the splitting of the stocks which give rise to the specimens we have in hand of *A. nyktos* and the earliest northern hemisphere erinaceids may have taken place 30 million years prior to the age of the oldest fossil. The presence in the Middle Jurassic of Madagascar of a tribosphenic mammal, *Ambondro mahabo* Fly nn *et al.* 1999, and *Tribotherium africanum* Sigogneau-Russell 1991 in the Early Cretaceous of Morocco, a tribosphenic mammal which has been tentatively interpreted as a eutherian (Kielan-Jaworowska 1992), makes it not unreasonable to hypothesise that such an earlier splitting might have taken place somewhere on the Gondwana continents.

Thus, for three reasons it would seem that the Garden of Eden hypothesis may have merit. First, Early   Cretaceous fossils exist in southeastern Australia which possess features suggestive that they either belong to or are closely related to an extant family of eutherians, the Erinaceidae. Second, a mechanism exists which can account for the timing and direction of movement of at least one part of one extant eutherian order during the Cretaceous that is in accord with the Garden of Eden hypothesis. Finally, that the Erinaceidae are placed well outside the Afrotheria is concordant with the physical separation of Australia and Africa by the Early Cretaceous.

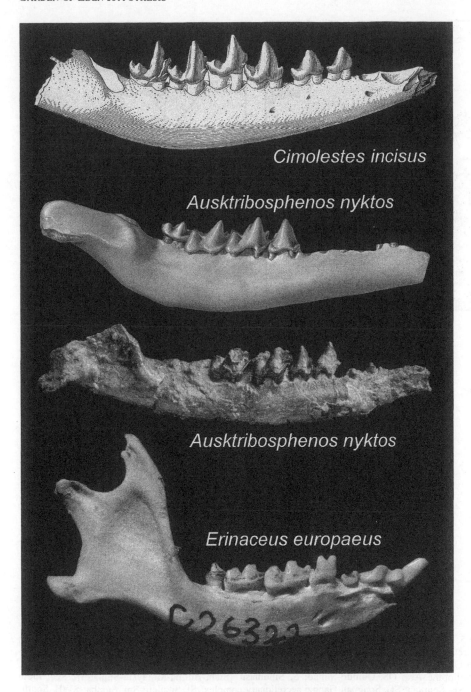

Figure 4. Two jaws of *Ausktribosphenos nyktos* in lateral view compared with a typical Cretaceous eutherian mammal *Cimolestes incisus* and the extant European hedgehog *Erinaceus europaeus*. The Early Cretaceous *A. nyktos* is more similar to the extant hedgehog than to the Late Cretaceous *C. incisus* in that its molars have notably lower crowns and the M_3 is markedly smaller than the M_1. In these features, *A. nyktos* has a tooth morphology common in the Cainozoic and all but unknown amongst the Cretaceous eutherian mammals of the northern hemisphere. (The jaw above of *A. nyktos* is 16 mm in length, the one below, 24 mm.)

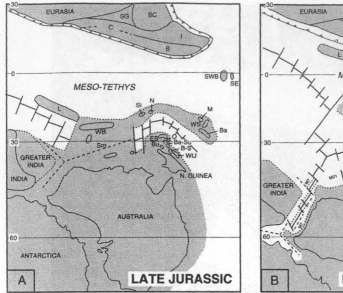

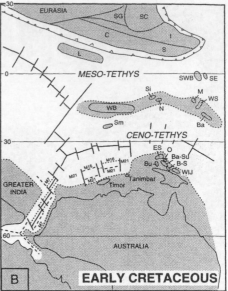

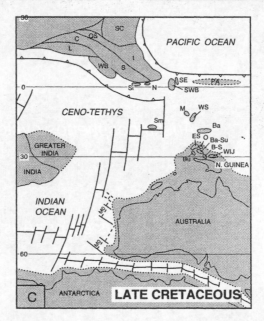

Figure 5. Palaeogeographic reconstructions for eastern Tethys: (a) Late Jurassic, (b) Early Cretaceous, (c) Late Cretaceous. B-S, Buru-Seram; Ba, Banda Allochthon; Ba-Su, Bangai-Sula; Bu, Buton; C, Changtang or Qiangtang; ES, East Sulawesi; I, Indochina; L, Lhasa; M, Mangkalihat; N, Natal; O, Obi-Bacan; PA, Philippine Arc; QS, Qamdo-Simao; S, Sibumasu {Siam-Burma-Malaysia-Sumatra}; SC, South China; SE, Semitau; SG, Songpan Gangzi accretionary complex; Si, Sikuleh; Sm, Sumba; SWB, Southwest Borneo; WB, West Burma; WIJ, West Irian Jaya; WS, West Sulawesi. M numbers refer to Indian Ocean magnetic anomalies. Modified from fig. 16 in Metcalfe (1996).

Contrasting opinions about Foote, Hunter, Janis & Sepkoski (1999) were discussed at length in Foote *et al.* (1999). Among other things, in their response, Foote, Hunter, Janis and Sepkoski in Foote *et al.* (1999) addressed the interpretation put forward above that the existence of *Ausktribosphenos nyktos*, a possible erinaceid from the Early Cretaceous of Australia, corroborates their Garden of Eden hypothesis. They rejected that interpetation, "...because (i) the fossil mammals [Rich, Vickers-Rich & Flannery in Foote *et al.* (1999)] discuss may not be relevant to the origin of modern eutherian orders; (ii) even if the fossils are modern eutherians, this would imply only a few missing lineages rather than than the large number of lineages required by the early origins scenario that we originally sought to to test."

We readily agree that eventually evidence may be forthcoming that will be sufficient to reject with confidence the hypotheses that *Ausktribosphenos nyktos* is closely allied to the Erinaceidae or even a eutherian mammal of any kind. However, we do not agree that the cases made in either of the two papers published to date where evidence was mustered to reject those hypotheses (Kielan-Jaworowska, Cifelli & Luo 1998, Archer *et al.* 1999), are strong enough to warrant that action at this time..

The points raised by Kielan-Jaworowska, Cifelli & Luo (1998) were addressed by Rich, Flannery & Vickers-Rich (1998).

Archer in Archer *et al.* (1999), "...suggests that *A. nyktos* may represent either an archaic monotreme, autapomorphic peramurid or a unique ordinal-level group that has converged on therian mammals." Regarding monotremes, Archer *et al.* (1999) listed three characters of *A. nyktos* that are found both in them and eutherians along with a fourth that might occur in monotremes as well. They did not point out a single feature of *A. nyktos* to be found in monotremes but not also in eutherians. Neither did they discuss the features of *A. nyktos* that distinguish it from the monotremes; *e.g.* (1) presence of a paraconid on M_1, (2) relatively expanded nature of all molar trigonids, (3) well-developed tribosphenic wear pattern on lower molars, particularly within the talonid basin, and (4) P_5 with three trigonid cusps and trigonid subequal in size to the M_1 trigonid (Rich *et al.* 1997, 1999). The presence of these features in *A. nyktos* is consistent with its being a eutherian mammal.

A single character was mentioned in Archer *et al.* (1999) as supporting a link between *A. nyktos* and the peramurids, "...the trigonid-like nature of the posterior premolar as in other peramurids including *Peramus* and the putative peramurid *Vincelestes*." The posterior lower premolars are, in fact, quite different. That of *Vincelestes neuquenianus* and the P_4 of *Peramus tenuirostris* are mediolaterally compressed blades, lacking any indication of a metaconid and with the paraconid represented by a diminutive cusp, far smaller than the protoconid and almost directly in front of it if present at all (fig. 6). In contrast, the most posterior premolar of *A. nyktos* is subequal in length and width, and the three trigonid cusps are all quite prominent with both the paraconid and metaconid much more lingually placed on the tooth than the protoconid (fig. 2). However, in the case of *P. tenuirostris,* Archer's intended meaning of, "...posterior premolar...," in Archer *et al.* (1999) is ambiguous. This is because as he pointed out fifteen years before (Archer 1984), the tooth designated as the M_1 rather than the P_4 by Clemens and Mills (1971) may be in fact the most posterior premolar; *i.e.* P_5. If, in his 1999 statement, Archer was referring to the P_5 [= M_1] rather than the P_4 of *P. tenuirostris,* the differences between the two species are the following: unlike the P_5 of *A. nyktos,* that of *P. tenuirostris* is a mediolaterally compressed blade with a much more anteroposteriorly expanded trigonid lacking a precingulum, and instead of merely a post cingulum, the presence of an elongated talonid on which is located a well developed cristid and at least one well developed cuspid (fig. 6a-c). A fundamental distinction not mentioned by Archer *et al.* (1999) is that *A. nyktos*, as is typical of eutherians, has a tribosphenic dentition, whereas the peramurids do not.

The characteristics of the mandible suggestive of *Ausktribosphenos nyktos* having been derived from an early symmetrodont stock mentioned by Archer *et al.* (1999) as having been pointed out by Kielan-Jaworowska, Cifelli & Luo (1998), were shown by Rich, Flannery & Vickers-Rich (1998) to be also present for the most part in *Prokennalestes*, an unquestioned eutherian mammal from the Early Cretaceous of Mongolia and in all instances, to be vestigial and hence, plesiomorphic features in any case.

Archer *et al.* (1999) do not specify a single feature of *A. nyktos* in support of the hypothesis that it may be, "...a unique ordinal-level group that has converged on therian mammals."

As to the absence of the large number of eutherian lineages in the Early Cretaceous of Australia expected by Foote, Hunter, Janis and Sepkoski *in* Foote et al. (1999) if the Garden of Eden hypothesis is valid, all we can say is that the total number of Mesozoic mammalian specimens now known from Australia is less than twenty-five, and almost half of those belong to the Ausktribosphenidae, *Ausktribosphenos nyktos* and a second genus and species as yet undescribed. All the rest are monotremes. All the monotremes but one were veritable giants of the Mesozoic mammalian fauna. Certainly, being five times or more the size of *A. nyktos*, they were larger than most known Cretaceous eutherians. To date all specimens of mammals from Australia of a size to be expected for eutherians during the Cretaceous come from a 20 metre^2 area at a single site, Flat Rocks. Thus, a bias towards two closely related species at the single site known to produce fossil mammals of the proper size would readily explain the lack of diversity in the known Cretaceous eutherian material from Australia, even if there were a myriad of different eutherians on the continent at that time.

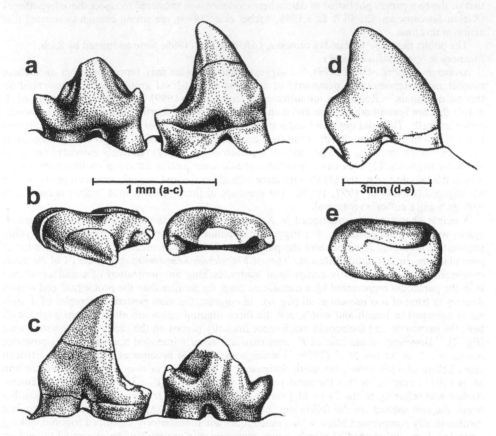

Figure 6. P₄-M₁ of *Peramus tenuirostris* in (a) labial, (b) occlusal, and (c) lingual views (left redrawn as right by Peter Trusler from Clemens & Mills 1971). P₄? of *Vincelestes neuquenianus* in (d) labial and (e) occlusal views (left redrawn as right by Peter Trusler after Bonaparte & Rougier 1987).

Given the small amount of Early Cretaceous mammalian material known from Australia, rather than being dismayed that it fails to establish the Garden of Eden beyond a doubt, we find it remarkable that what little fossil evidence has been found to date there lends any support to it at all. The fact that the anticipated diversity of Cretaceous Australian eutherians under the Garden Eden hypothesis cannot be documented at this time on the basis of those few fossils is to be expected. Likewise, given the current state of knowledge of the fossil record, that there should be uncertainty as to whether or not to the hypothesis of a close relationship between *Ausktribosphenos nyktos* and the Erinaceidae or the more general one that *A. nyktos* belongs within the eutherians at all will ultimately be accepted or rejected is also to be expected.

At least the affinities of *Ausktribosphenos nyktos* or the as yet undescribed ausktribosphenid should be resolved one way or other in the foreseeable future when an upper dentition of one of those species is eventually found. Unless new Cretaceous mammal-bearing sites are discovered on the continent or the pattern of discovery changes at Flat Rocks, even a single multifarious assemblage of Australian Cretaceous mammals in the size range to be expected for eutherians of that age may never be forthcoming to test whether or not a diversity of them was there at that time.

That a more diverse mammalian assemblage may eventually be forthcoming from the Flat Rocks assemblage is indicated by the presence there of a third mammal in addition to the two ausktribosphenids: *Teinolophos trusleri* Rich *et al.* 1999. When first described, *T. trusleri* was regarded as a eupantothere. It now appears to be an unusually small monotreme (Rich *et al.* 2001). This aloocation of *T. teinolophos* to the monotremes is based on the similarity of the form of the crown of the single molar preserved on the holotype of that species to that of the lower molars of the Early Cretaceous monotreme *Steropodon galmani* Archer *et al.* 1985 from Lightning Ridge, N.S.W.

The first word of the title of this paper was chosen with care. We do not intend to assert that the existence of possible eutherian mammals in the Early Cretaceous of Australia that may be erinaceids proves the Garden of Eden Hypothesis. Rather, we intended that "corroborate" be understood in the following sense, "To strengthen (an opinion, statement, argument, etc.) by concurrent or agreeing statements or evidence; to make more sure or certain; to support, confirm: said **a.** of a person; **b.** of the confirming statement" (Oxford Unabridged Dictionary).

ACKNOWLEDGEMENTS

We wish to thank Michael J. Benton, Richard Cifelli, Francisco Goin, Zofia Kielan-Jaworowska, Luo Zhexi, Denise Sigogneau-Russell, Richard H. Tedford, and Michael O. Woodburne for their helpful comments about a draft of this manuscript.

The Committee for Research and Exploration of the National Geographic Society has continuously funded *The Ghastly Blank* project for 18 tears. We gratefull acknowledge that long term, consistent financial, and equally important, moral support for without them, the fossils that are central to this report would not have been recovered. We also wish to thank the literally hundreds of volunteers who together have donated approximately thirty person-years of effort to the fieldwork necessary to find these fossils. Finally, we wish to thank the Australian Research Committee and Atlas Copco Australia for their long term assistance.

REFERENCES

Archer, M. 1984. Origins and early radiations of mammals. In *Vertebrate Zoogeography & Evolution in Australasia. Animals in Space & Time*: 477-515. Carlisle, Western Australia: Hesperian Press.

Archer, M., Arena, R., Bassarova, M., Black, K., Brammall, J., Cooke, B., Creaser, P., Crosby, K., Gillespie, A., Godthelp, H., Gott, M., Hand, S.J., Kear, B., Krikmann, A., Mackness, B., Muirhead, J., Musser, A., Myers, T., Pledge, N., Wang, Y., & Wroe, S. 1999. The evolutionary history and diversity of Australian mammals. *Australian Mammalogy* 21:1-45.

Archer, M., Flannery, T. F., Ritchie, A. & Molnar, R. E., 1985. First Mesozoic mammal from Australia-an Early Cretaceous monotreme. *Nature* 318: 363-366.

Benton, M.J. 1999a. Early origins of modern birds and mammals: molecules vs. morphology. *BioEssays* 21:1043-1051.

Benton, M.J. 1999b. Reply to Easteal. *BioEssays* 21: 1059.

Bonaparte, J.F. & Rougier, G. 1987. Mamiferos del Cretacico Inferior de Patagonia. *IV Congreso Latino-americano de Paleontologia, Bolivia* 1: 343-359.

Clemens, W.A. & Mills, J.R.E. 1971. Review of *Peramus tenuirostris* Owen (Eupantotheria, Mammalia). *Bulletin of the British Museum (Natural History) Geology* 20: 89-113.

Easteal, S. 1999a. Molecular evidence for the early divergence of placental mammals. *BioEssays* 21: 1052-1058.

Easteal, S. 1999b. Reply to Benton. *BioEssays* 21: 1060.

Flynn, J.J., Parrish, J.M., Rakotosamimanana, B., Simpson, W.F. & Wyss, A.R. 1999. A Middle Jurassic mammal from Madagascar. *Nature* 401: 57-60.

Foote, M., Hunter, J.P., Janis, C.M. & Sepkoski Jr., J.J. 1999. Evolutionary and preservational constraints on origins of biologic groups: divergence times of eutherian mammals. *Science* 283: 1310-1314.

Foote, M., Hunter, J.P., Janis, C.M., Sepkoski Jr., J.J., Archibald, J.D., Hedges, S.B., Kumar, S., Rich, T.H., Vickers-Rich, P. & Flannery, T.F. 1999. Divergence Times of Eutherian Mammals. *Science on Line* 285: 2031.

Hedges, S.B., Parker, P.H., Sibley, C.G. & Kumar, S. 1996. Continental breakup and the ordinal diversification of birds and mammals. *Nature* 381: 226-229.

Kielan-Jaworowska, Z. 1992. Interrelationships of Mesozoic mammals. *Historical Biology* 6: 185-202.

Kielan-Jaworowska, Z., Cifelli, R. L. & Luo, Z. 1998. Alleged Cretaceous placental from down under. *Lethaia* 31: 267-268.

Kumar, S. & Hedges, S.B. 1998. A molecular timescale for vertebrate evolution. *Nature* 392: 917-920.

McKenna, M.C. 1973. Sweepstakes, filters, corridors, Noah's Arks, and beached viking funeral ships in paleogeography. In D.H. Tarling, & S.K. Runcorn (eds.), *Implications of Continental Drift to the Earth Sciences* 1:295-308. London: Academic Press.

McKenna, M. C. & Bell, S. K. 1997. *Classification of Mammals above the Species Level.* New York: Columbia University Press.

Metcalfe, I. 1996. Gondwanaland dispersion, Asian accretion and evolution of eastern Tethys. *Australian Journal of Earth Science* 43: 605-623.

Metcalfe, I. 1998. Palaeozoic and Mesozoic geological evolution of the SE Asian region: multidisciplinary constraints and implications for biogeography. In R. Hall, & J.D. Holloway (eds) *Biogeography and Geological Evolution of SE Asia*: 25-41. Amsterdam, The Netherlands: Backhuys Publishers.

Polcyn, M.J., Tchernov, E. & Jacobs, L.L. 1999. The Cretaceous biogeography of the eastern Mediterrean with a description of a new basal mosasaurid from 'Ein Yabrud, Israel. In Y. Tomida, T.H. Rich, & P. Vickers-Rich, (eds.) Proceedings of the Second Gondwana Dinosaur Symposium. *National Science Museum Monographs 15*: 259-290. Toyko, Japan.

Rich, T. H., Flannery, T. F., & Vickers-Rich, P. 1998. Alleged Cretaceous placental from down under: Reply. *Lethaia* 31: 346-348.

Rich, T.H., Vickers-Rich, P., Constantine, A., Flannery, T.F., Kool, L. & van Klaveren, N.A. 1997. A tribosphenic mammal from the Mesozoic of Australia. *Science* 278: 1438-1442.

Rich, T.H., Vickers-Rich, P., Constantine, A., Flannery, T.F., Kool, L. & van Klaveren, N. 1999. Early Cretaceous mammals from Flat Rocks, Victoria, Australia. *Records of the Queen Victoria Museum* 106: 1-34.

Rich, T.H., Vickers-Rich, P., Trusler, P., Flannery, T.F., Cifelli, R., Constantine, A., Kool, L., & van Klaveren, N. 2001. Monotreme nature of the Australian Early Cretaceous mammal *Teinolophos. Acta Palaeontologica Polonica* 46: 113-118.

Sigogneau-Russell, D. 1991. Découverte du premier mammifère tribosphénique du Mésozoïque. *Comptes Rendus de l'Academie des Sciences de Paris* (2) 313: 1635-1640.

Springer, M. S. 1997. Molecular clocks and the timing of the placental and marsupial radiations in relation to the Cretaceous-Tertiary boundary. *Journal of Mammalian Evolution* 4: 285-302.

Springer, M.S., Cleven, G.C., Madsen, O., de Jong, W.W., Waddell, V.G., Amrine, H.M., & Stanhope, M.J. 1997. Endemic African mammals shake the phylogenetic tree. *Nature* 388: 61-64.

Stanhope, M.J., Madsen, O., Waddell, V.G., Cleven, G.C., de Jong, W., & Springer, M.S. 1998a. Highly congruent molecular support for a diverse superordinal clade of endemic African mammals. *Molecular Phylogenetics and Evolution.* 501-508.

Stanhope, M.J., Waddell, V.G., Madsen, O., de Jong, W., Hedges, S.B., Cleven, G.C., Kao, D. & Springer, M.S. 1998b. Molecular evidence for multiple origins of Insectivora and for a new order of endemic African insectivore mammals. *Proceedings of the National Academy of Science of the United States of America.* 95: 9967-9972.

Mammals in Sulawesi: where did they come from and when, and what happened to them when they got there?

Colin Groves

Department of Archaeology & Anthropology, Australian National University, Canberra, ACT 0200

ABSTRACT: It is evident that part, at least, of Sulawesi is of Sundaland origin. If there was a dry-land connection between Sulawesi and Sundaland in the Pliocene, this would be compatible with the observation that most of Sulawesi's mammals are endemic at the level of genera, sub-genera or well-differentiated species-groups. In some cases their closest relations seem to be with mammals from the Tatrot stage of the Siwaliks, presumably via the poorly-known "Siva-Malayan" fauna of Java. Within Sulawesi both altitudinal and geographic speciation are well-marked. The groups in which geographic variation is most easily appreciated are monkeys and squirrels. Combining the data from these groups indicates that species of the southern and southeastern peninsulas separated earliest; those of the central part and the northern peninsula form a single clade. What little is known of the fossil mammals of Sulawesi comes from the Pliocene Walanae Formation on the southern peninsula. Surprisingly, this is not known to contain anoa, babirusa or any of the other present-day endemics, but pygmy elephantids and the giant pig *Celebochoerus*. Until Neogene palaeontology gets started on other parts of Sulawesi we have no explanation for this apparent anomaly, but it has been proposed more than once that Sulawesi was a cluster of separate islands until quite late in geologic time.

1 INTRODUCTION

Sulawesi is the fourth largest island of the Indo-Australian archipelago. At 189,000 km^2, it is smaller than New Guinea, Borneo or Sumatra, and larger than Java, Luzon or Mindanao. Its shape is bizarre and spidery. From the central mass a long, thin (20 to 50 km wide) northern peninsula reaches 200 km north, then turns west for 500 km, and northeast again for its final 200 km. Shorter peninsulas project south, southeast and east from the central mass. In each of the four peninsulas mountains rear up to 2700 to 2900 m, and in the central mass of the island altitudes of over 3000 m occur (the highest peak is Nokilolaki, at 3311 m). Raised coral reefs, as much as 700 m. on Butung, and the presence of recent marine clays and sea shells in the Tempe Depression (which separates the southern peninsula from the central mass), indicate a history of local uplift until very recent times.

The Makassar Strait separates it from Borneo to the west; mostly 200 to 300 km wide, this narrows to just over 100 km in the northwest; elsewhere, offshore islands reduce the distances to other islands, though not by all that much. To the north, beginning 50 km offshore, the Sangihe and Talaud archipelagos are strung out for 300 km along an approximately south-north axis, and 250 km beyond them across open water lies Mindanao in the Philippines. To the east Peleng island lies some 20 km offshore, the Sula archipelago begin 50 km beyond that, and the islands of the Maluku group begin about 100 km east and southeast of the Sulas. Three fairly large islands – Butung, Muna and Kabaena – lie off the southeastern peninsula, and another, Salayar, off the southern peninsula, and beyond them the Nusatenggara group are 200 km and more across open sea.

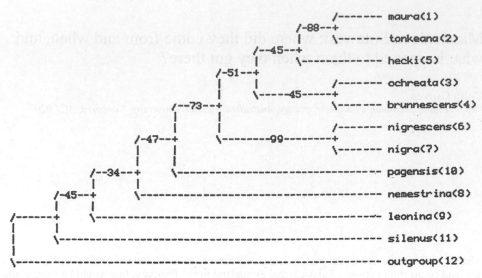

Figure1. Bootstrap tree for Sulawesi and related macaques. Bootstrap values are given along the branches. The figures in brackets after the names of the taxa indicate the order in which they were entered.

Deep water in all directions separates Sulawesi (including its offshore islands) from other parts of the Indo-Australian archipelago. The Makassar Strait is nowhere less than 1000 m deep (and in many parts reaches 2000 m), though at latitude 3°S such depths are found for only about 20 km offshore before the Sunda Shelf is reached. Similar sea depths occur all around Sulawesi, and around the Maluku to the east. These deep waters contrast with the the Sunda Shelf, on which Borneo, Java, Bali, Palawan and Sumatra stand, and which unites these western islands to the mainland of Southeast Asia; Palawan is separated from Borneo by seas 150 m deep, but at few other points does the depth of the Sunda Shelf reach even 50 m. Similar shallow seas characterize the Sahul Shelf, around Australia, New Guinea and the Aru Islands. The result is that when sea levels fell, at intervals during the Pleistocene coincident with ice ages at high latitudes, the islands from Borneo and Bali westward were connected to the Southeast Asian mainland (as Sundaland), and New Guinea and Australia were connected (as Sahul or Meganesia), while the lands in between remained isolated and separate. These isolated lands, long-isolated islands, included Sulawesi with its offshore islands.

It is not surprising, therefore, that the fauna (at least the mammals) of Sundaland, the Oriental fauna, have an overall similarity with each other and with those of mainland Southeast Asia: such groups as primates, dermopterans, insectivorans, ungulates, carnivores and rodents. The eastern edge of the Sunda Shelf corresponds more or less to the zoogeographic concept known as Wallace's Line, and its northern edge to Huxley's Line. Similarly, the Meganesian mammal fauna includes marsupials, monotremes and restricted group of murid rodents; the western margin of the Sahul Shelf corresponds to another zoogeographic boundary, Lydekker's Line. The history of these and other faunal lines in the region is recounted, together with a discussion of their significance, by Simpson (1977).

The mammals of Sulawesi are in the main a subset of the Oriental fauna, but very different from their nearest relatives west of the Wallace Line. Groves (1976) listed 100 mammal species, of which 71% are endemic at specific or even at generic level; if bats (Chiroptera) are excluded, fully 92% of of the remaining 59 species are endemic, an extraordinarily high total. Just 14% (or 9% if bats are excluded) have their closest relatives to the east (Australia and/or New Guinea), while all the rest are of Oriental affinities.

This list was updated by Musser (1987), who listed 123 native species, including 3 diprotodont marsupials, 8 shrews, 59 bats, 6 primates, 42 rodents (6 sciurids and 36 murids), 1 carnivore, and 4 artiodactyls (2 suids and 2 bovids). He too agreed that the overall affinities of the

Sulawesi mammals are heavily Oriental. He also pointed out that there is much diversity within Sulawesi: some mammals have highland and lowland vicariants, but most have distinct variants on the northern, southern and southeastern arms of this spidery-shaped island, and some have still further geographic divisions.

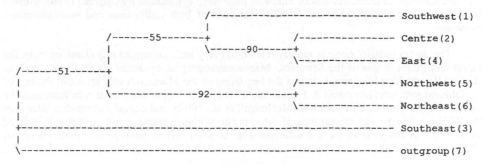

```
Bootstrap 50% majority-rule consensus tree
   (plus other groups compatible with this tree)
                                 /----------------------------- Southwest(1)
                                 |
              /------55-------+                /---------------- Centre(2)
              |               \------90------+
              |                              \---------------- East(4)
      /------51------+
      |             |                        /---------------- Northwest(5)
      |             \--------------92--------------+
      |                                      \---------------- Northeast(6)
      |
      +--------------------------------------------------------- Southeast(3)
      |
      \--------------------------------------------------------- outgroup(7)
```

(1) UNORDERED
Length 116
Shortest of 1000 random trees, 116
Consistency index* 0.756

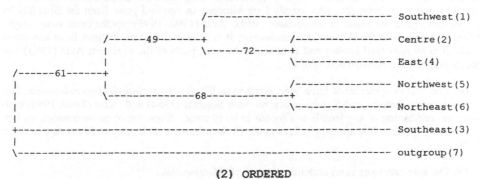

```
Bootstrap 50% majority-rule consensus tree
   (plus other groups compatible with this tree)
                                 /----------------------------- Southwest(1)
                                 |
              /------49-------+                /---------------- Centre(2)
              |               \------72------+
              |                              \---------------- East(4)
      /------61------+
      |             |                        /---------------- Northwest(5)
      |             \--------------68--------------+
      |                                      \---------------- Northeast(6)
      |
      +--------------------------------------------------------- Southeast(3)
      |
      \--------------------------------------------------------- outgroup(7)
```

(2) ORDERED
Length 150
Shortest of 1000 random trees, 149
Consistency index* 0.637

Figure 2. Bootstrap tree, values, and comparative lengths for random tree, based on Sulawesi macaques and squirrels.

There are also fossil mammals known from Sulawesi. The Walanae fauna of southwestern Sulawesi, the only known fossil site, consists of two species of *Stegodon* and (perhaps) one of *Elephas,* and the giant suid *Celebochoerus,* along with a giant tortoise, a crocodile, sharks and stingrays. Size changes, leading in this case to dwarf elephants and giant pigs, are typical of small island faunas, and it was mentioned above that there are indications of marine deposits in the Tempe Depression, suggesting that the southern peninsula was a separate island at one stage in the past. According to Sartono (1979), the Walanae fauna is coeval with Foraminiferal zones N19-20, and so has an age of between 4.6 and 3.0 Ma, i.e. Early Pliocene.

The questions, then, are: how did all these mammals get to Sulawesi, and when; how did present-day geographic patterns of composition and distribution arise; and what happened to those that are known only as fossils?

2 HOW DO THEY RELATE TO MAMMALS ELSEWHERE?

The relationships of the mammals of Sulawesi have been considered by Groves (1976), Musser (1987) and Aziz (1990). Musser (1987) gives lists of both indigenous and non-indigenous mammals.

(1) The dwarf buffalo *Anoa* is closely related to (very little advanced on) *Hemibos* from the Tatrot and Pinjor stages of the Siwaliks. Magnetostratigraphic evidence places these stages in the Middle and Late Pliocene, between the beginning of the Mammoth subchron and the end of the Olduvai subchron (Azzaroli & Napoleone, 1981); according to the recent recalibrations, the span would be from 3.20 to 1.78 Ma (McDougall et al., 1992), and actually somewhat later than the Walanae fauna. The occurrance of *Anoa* in the Walanae fauna is not confirmed: isolated teeth found in the vicinity have a different state of fossilization from *in situ* material (Aziz, 1990).

(2) The babirusa *Babyrousa* is difficult to classify with respect to other pigs, and is generally placed in its own subfamily Babyrousinae (Groves, 1980). It does not occur in the Walanae fauna; some molar teeth thought to be of babirusa are more likely those of *Celebochoerus* (Hooijer, 1954). There is also an "ordinary" pig, endemic at the species level (*Sus celebensis*), living in Sulawesi.

(3) The giant pig *Celebochoerus* of the Walanae fauna is not related to *Babyrousa*. It was regarded by Hooijer (1954) as related to *Propotamochoerus* from the Dhok Pathan zone of the Siwaliks and elsewhere; this was mainly Late Miocene in age and gone from the Siwaliks by Tatrot times (van der Made & Moyà-Solà, 1989). Aziz (1990, 1993), on the other hand, aligns *Celebochoerus* with the African *Phacochoerus*; it is at any rate quite distinct from any other suid. It is by now well known and represented by most parts of the skeleton; Aziz (1993) figures a magnificent reconstruction of it.

(4) The Walanae elephants have been ascribed to *Elephas (=Archidiskodon) celebensis*, *Stegodon trigonocephalus* and *S.sompoensis*, but both Sondaar (1984) and Aziz (1990, 1993) consider the attribution of any fossils to *Elephas* to be in error. *Stegodon trigonocephalus*, an Upper Siwalik species, first appears in Java in the Ci Saat fauna, which is latest Pliocene (Sondaar, 1984).

(5) The sole carnivore is an endemic giant civet, *Macrogalidia*.

(6) The macaques (*Macaca* spp.) were considered probably diphyletic by Evans et al. (1999): those of the southwestern and southeastern arms (*M.maura, ochreata, brunnescens*) versus those of the center, eastern and northern arms (*M.tonkeana, hecki, nigrescens, nigra*). Dr John Trueman (personal communication) has performed a t-PTP test on their dataset, and considers the diphyletic model not secure.

(7) The tarsiers of Sulawesi (*Tarsius spectrum* group) are the sister-group to those of Sundaland and the Philippines (Groves, 1998). They are not known as fossils on Sulawesi.

(8) The several species of shrews (*Crocidura*) in Sulawesi represent at least two dispersals: the "old endemics", and a much more recent dispersal resulting in the single species *Crocidura nigripes* (Ruedi et al., 1998). No dates could be placed on either of these two waves, but inferentially the second dispersal must have been an overwater one.

(9) Moore (1959) showed that there are three endemic genera of squirrels in Sulawesi: *Rubrisciurus*, *Prosciurillus* and *Hyosciurus*. Their interrelationships are unknown; they could represent a single, ancient stock, or they could have entered Sulawesi separately. Their species-level taxonomy is unclear.

(10) The Sulawesi murids (Musser, 1987) are very diverse. Some are belong to genera represented in Sundaland (*Rattus, Maxomys, Haeromys*) or even in the Philippines (*Crunomys*) or Flores (*Bunomys* – see Kitchener et al., 1991), but there are also several endemic genera, not closely related to any elsewhere (*Melasmothrix, Tateomys, Echiothrix, Margaretamys*).

(11) Finally the cuscuses are of two endemic genera (*Ailurops, Strigocuscus*), not closely related to any in Australia, New Guinea or Maluku.

There would appear, therefore, to be two faunal layers as far as mammals are concerned. First, there are the "old endemics", forming genera or species-groups: *Anoa, Babirusa, Celebochoerus, Macrogalidia*, macaques, tarsiers, "old endemic" shrews, the three squirrel genera, the endemic murid genera, and the cuscuses. Secondly, there are the "new endemics", species of more widespread genera, not too distantly related to species elsewhere: *Sus celebensis, Crocidura nigripes*, and some of the other murids. It is interesting to note that not one non-volant mammal species is shared with Sundaland or elsewhere, except for some that are human introductions (deer, palm-civet, commensal rats and mice).

#NEXUS

[Sulawesi biogeography: *Macaca, P.leucomus* gp., *P.murinus* gp., *Rubrisciurus, Hyosciurus*]

```
begin data;
  dimensions ntax=7 nchar=60;
  format missing=? symbols="0~4";
  matrix

Southwest  02220303000021100001330101003100 2120002011 01101000111 00010002
Centre     22220302100111120103302022000221200010201 01101003120 11111010
Southeast  10200302122101011011120000100210011002320 1201120011? ????????
East       22220302100111120103312001020202??????????0????????0????????
Northwest  00011413121211121013212112110120200001111 01220101100 ????????
Northeast  30002414100011132020002112100111120001201 01010011100 00001111
outgroup   000000??000?000000000 000?00200? 000000???0 0???00????? 000?????
;
end;
```

Figure 3. Character coding for Sulawesi mammals. Characters 1-21 are for *Macaca*, 22-31 for the *Prosciurillus leucomus* group, 32-41 for the *P.murinus* group, 42-52 for *Rubrisciurus*, and 53-60 for *Hyosciurus*.

3 HOW DID THEY GET TO SULAWESI?

Two models of arrival in Sulawesi are possible: by land or across the sea. Groves (1976) drew attention to the "Pulau Laut Centre of Diastrophism" reconstructed by Van Bemmelen (1949), the only geological candidate for a land bridge from Sundaland in the days before Plate Tectonics. The mammals would have had to cross the land-bridge in the late Pliocene, because according to Van Bemmelen it arose at about that time and submerged at the beginning of the Pleistocene. The only Pliocene fauna known from anywhere else in Indonesia was the "Siva-Malayan fauna" (Ci Saat fauna of Sondaar, 1984) from the lower levels of Java, which in turn was derived from the fauna of the Tatrot zone of the Siwaliks of northern India and Pakistan.

Cranbrook (1981) and Musser (1987), by contrast, could see no evidence that a land bridge had ever existed; for them, the Sulawesi mammal fauna is "depauperate and unbalanced", such as would be typical for one derived by a sweepstake route across a sea barrier, and bats, shrews and rodents are typical of such faunas. Musser (1987) likewise disparaged the idea that the sole Australasian elements, the cuscuses (Phalangeridae), might have clung to moving chunks of Gondwanaland that drifted westward to become part of Sulawesi; why are there no other marsupials on Sulawesi, he asked?

Yet primates and viverrids, at least, would hardly be predicted for a "depauperate, unbalanced" fauna (even if a rather restricted set of each), and particularly significant is that while the affinities of almost all the mammals are with Sundaland, the "old endemics" are relicts, sister-groups of the Sundaland genera, and only the "new endemics" might be derived from within that pool as might be expected if there were chance dispersals across the Macassar Strait. Moreover, while the "new endemics" are the sort of resilient animals that might be expected to cross wide sea barriers - a pig, a shrew, some rats - the "old" endemics are certainly not like this. Most of the broad phylogenetic groups to which the "old endemics" belong are rainforest, yet low sea-level periods, when rafting would have been easiest, are precisely the times when forest cover was most restricted (Heaney, 1991). Similarly, both cuscus genera are primitive, and while the absence of other marsupials is perhaps puzzling, differential (climate-driven?) extinctions are perfectly feasible. So, where does the "Pulau Laut Centre of Diastrophism" stand today?

It has been overtaken by plate tectonics. Geologically, Sulawesi is the result of several collisions of micro-continental fragments from the Australian plate. Hall (1996, Fig.2) depicts Southeast Asian palaeogeography at the end of the Miocene, 5 Ma. Southwestern/central Sulawesi is still part of Sundaland; the southeastern arm has accreted onto this, but the eastern/northern arm fragmented has not yet joined. Two models exist to explain the Makassar Strait. For Hall (1996) and others, the Makassar Strait is attenuated continental crust; the southwestern arm of Sulawesi was alongside eastern Borneo and drifted eastward, probably some time before the Early Miocene although the dating evidence is very poor. For Bergman et al. (1996), however, the Makassar Strait is a foreland basin, pushed down by the loading of the eastern/northern arm micro-continental fragments onto the southwestern/central arm, and this took place in the Late Neogene. But Professor Hall, in a comment during the conference, cautioned against relying on the Makassar Strait, suggesting instead that the land connections with Sundaland would better be sought to the south (Java).

Any of these scenarios could get mammals into Sulawesi dry-shod in time to appear in the Walanae fauna, but whether we can support such an early inferred separation of the Sulawesi endemics from their Sundaland sister-groups is unclear.

4 WHAT HAPPENED WHEN THEY GOT TO SULAWESI?

Most of the present-day mammalian fauna of Sulawesi is, at the generic level, more or less pan-Sulawesian, although some genera, such as the shrew-mice *Melasmothrix* and *Tateomys*, are restricted to high altitudes (Musser, 1987). What is true, however, is that there are marked geographic divisions, and the representatives of many genera and species-groups are taxonomically distinct on the different peninsulae, and often even within them. Full resolution of this geographic variation is a long way off, but some trends are already clear. Neighbouring vicariants can in most instances be shown to differ absolutely so that, even though they may form narrow hybrid zones where their ranges meet, they are most appropriately distinguished at species level.

Babyrousa has distinct taxa on the northern peninsula (with no differentiation along it), in the center, on the Togean Islands, and on the Sula Islands and Buru. Groves (1980) classified them all as subspecies of a single species, but the northern and Togean taxa at least are diagnosably distinct and should rank as distinct species, *Babyrousa celebensis* and *B.togeanensis* respectively. The species *B.babyrussa* occurs on the Sula Is. and Buru, to both of which it has been introduced, but as we know so little of babirusa from other parts of the Sulawesi mainland it is not possible to say from where it actually originated. Finally a single skull from central Sulawesi may or may not represent the species known otherwise only as a subfossil from the southern peninsula, *B.bolabatuensis*.

Characters for *Macaca*

1. Crown hair (not lengthened, short but erect, short crest, long crest), O
2. Cheek tufts (absent, short, bushy) O
3. Gluteal fields (absent, small, large) O
4. Rump patch (absent, present but not bushy, pale and bushy) O
5. Ischial callosities (flat, stand out) U
6. Tail length cf head + body (>50, 28-46, 24-34, 5-12, 3-6) O
7. Baculum length, mm (16-22, 20-25) O
8. Baculum dorsoventral basal diameter, mm (1.7-2.5, 2.2-3.0, 3.3, 3.3-4.8?, 4.5-4.8) O
9. Dorsal colour (brown, black)
10. Colour of shanks (as body, slightly paler, much paler) O
11. Colour of forearms (as body, slightly paler, much paler) O
12. Age-greying (none, slight, much) O
13. Colour of ventral skin (pale, black)
14. Tail hair (thick, sparse)
15. Relative muzzle length (61% or less, 62% or more)
16. Lateral ridges on muzzle (absent, weak, strong) O
17. Relative zygomatic breadth (>65, 60-65, 60) O
18. Malar surface (flat, convex)
19. Temporal margin of malar (broad, narrow, very narrow) O
20. Depth of supraorbital ridgesmm (5-6, 6-7, 7-8, 9 or more) O
21. Sagittal crest height in old males, mm (absent, 1-2, 3, 5-6) O

Table 1. List of characters used in cladistic analysis of Sulawesi macaques. The states (primitive to derived) are in parentheses. O = Ordered, U = Unordered (for multistate characters).

Groves (1969) divided anoas into a lowland and a mountain species, *Bubalus* [perhaps, preferably, *Anoa*] *depressicornis* and *quarlesi* respectively, but the picture is probably more complicated than this. Thus Sugiri and Hidayat (1996) recorded at least one specimen, from G.Nokilolaki in central Sulawesi, whose appearance and karyotype were unlike any previously known. The differentiation between the common lowland and mountain anoas is very deep; Kakoi et. al. (1994) found that two lowland and eight mountain anoas, from Surabaya and Ragunan zoos, differed in five amino acid residues in the ß haemoglobin molecule, and hypothesized that they may have already diverged before they entered Sulawesi.

But the best-marked geographic variation is found in macaques and squirrels. The macaques are divided into at least seven distinct species in different parts of Sulawesi (Fooden, 1969). I listed 21 morphological characters, including some that distinguish Sulawesi macaques from other members of the *Macaca nemestrina* group (Table 1), and ran a trial cladistic analysis using PAUP. The results (Fig.1) did not entirely fit with those of Evans et al. (1999), but some of the bootstrap values are low.

I made preliminary examinations of Sulawesi squirrel skins, in the Natural History Museum (London), Rijkmuseum van Natuurlijke Historie (Leiden), and American Museum of Natural History (New York). All of the genera show distinct geographic variation. Limited diversity has been described for *Hyosciurus* and none whatever for *Rubrisciurus*, but some variation none the less exists, although I cannot document it fully because material from some parts of Sulawesi is very limited. In the genus *Prosciurillus*, by contrast, a number of species and subspecies have been described, and these fall into three well-marked species-groups: the large, brightly-coloured squirrels of the *P.leucomus* group; the smaller, drab-agouti forms of the *P.murinus* group, among which (unlike the *P.leucomus* group) size differentials permit occasional sympatry; and the rare but widespread *P.weberi*. I analysed the first two groups separately, but I could not detect any noticeable geographic variation in the very sparse available material of *P.weberi*, so I excluded it from the analysis. The characters are listed in Table 2.

Characters for *Prosciurillus leucomus* group

1. Colour of flanks (fawn-agouti, more reddish, red-agouti) O
2. Colour of midback (as flanks, more ochery, strongly chestnut) O
3. Light postauricular spots (no, yes)
4. Colour of ear-backs (as body, black, white) U
5. Ear-tufts (absent, slight, long)
6. Colour of feet (as body, paler [more orange])
7. Colour of underside (reddish, red-yellow, grey, creamy) U POLARITY
   UNCERTAIN
8. Grey bases on underside (barely visible, markedly showing through)
9. Tail banding (absent, vague striations, marked bands)
10. Head + body length, mm (<165, 170-172, >176) O POLARITY UNCERTAIN

Characters for *Prosciurillus murinus* group

1. Colour of upperside (brown, yellower, redder) U
2. Colour of hair shafts (brown, yellow, red) U
3. Colour of underside (creamy, [red-]grey, red) U
4. Colour of ear-backs (as body, white)
5. Hair on ear-backs (sparse, dense)
6. Colour of feet (agouti as body, paler)
7. Head + body length, mm (ca 105, 120-130, 130-140, >160) O POLARITY
   UNCERTAIN
8. Tail length, mm (<85, 95-100, 105-110, >115) O
9. Hindfoot length, mm (<30, 30-31, 32-34) O
10. Tail banding (weak, striking)

Characters for *Rubrisciurus*

1. Banding of dorsum (agouti, more saturated)
2. Colour of dorsum (not red, red-brown, deep brown) O POLARITY
   UNCERTAIN
3. Colour flanks (maroon-red, brighter red, deep fiery rufous) O
   POLARITY UNCERTAIN
4. Colour of tail (maroon, reddens towards tip, blackish) O POLARITY
   UNCERTAIN
5. Posterior half of dorsum (as anterior, more blackish)
6. Colour of ears (dark maroon brown, reddish black, black) O
7. Colour of underside (gingery, maroon-red) POLARITY UNCERTAIN
8. Condylobasal length, mm (<63, 63-64, >64) O POLARITY UNCERTAIN
9. Tail length, mm (<180, >185) POLARITY UNCERTAIN
10. Hindfoot length, mm (<61, 62-67, >68) O POLARITY UNCERTAIN
11. Maxillary toothrow length, mm (<12.7, >12.9) POLARITY UNCERTAIN

Characters for *Hyosciurus*

1. Colour of upperside (fuscous, maroon red)
2. Tail (as body, brindled)
3. Colour of hands and feet (darkened, blackish)
4. Colour of underside (pale yellowish, white) POLARITY UNCERTAIN
5. Border of light zone of underside (iregular, regular) POLARITY
   UNCERTAIN
6. Tail length, mm (<120, >120) POLARITY UNCERTAIN
7. Hindfoot length, mm (48.5-51, 50-55) POLARITY UNCERTAIN
8. Greatest skull length, mm (50-57, 57-57.5, >60) O POLARITY
   UNCERTAIN

Table 2. Lists of characters used in cladistic analysis of Sulawesi squirrels. Notations as for Table 1.

Then, as an experiment (suggested by Dr John Trueman), I ran a "total evidence" cladistic analysis combining data for both macaques and squirrels, using a generalized outgroup. The resulting trees are given in Fig.2. The consistency index for the Unordered run is 0.81 (excluding uninformative characters, 0.756), and for the Ordered run is 0.673 (excluding uninformative characters, 0.637). The length of the tree is in both cases equal to the shortest of 1000 random trees. The basic distinction, with over 50% bootstrap value, is between the southeastern peninsula and the rest: does this suggest that the southeastern arm was, early on, isolated from the rest of Sulawesi?

There are many problems awaiting study. When did the "old endemics" arrive, and was it really early enough to get them across when (part of) Sulawesi was connected to Sundaland? And why have anoas and babirusas not turned up in the Walanae fauna? Were there, perhaps, two "old endemic" waves, and the second, the one which survives, replaced the first, the *Stegodon* plus *Celebochoerus* fauna? How did geographic differentiation take place – was there a phase in which, as Fooden (1969) surmised, Sulawesi was not one but a series of islands? When did the cuscuses arrive; was it really on the Australian plate fragments?

All these questions relate strongly to the basic question: what, actually, is the meaning of Wallace's Line?

REFERENCES

Aziz, F. 1990. Pleistocene Mammal Faunas of Sulawesi and their bearing to Paleozoogeography. PhD thesis, Kyoto University.

Aziz, F. 1993. Fosil fauna Sulawesi dan Batas Wallace. J.Geol.Sumberdaya Mineral 3: 2-9.

Azzaroli, A. & Napoleone, G. 1981. Magnetostratigraphic investigation of the Upper Siwaliks near Pinjor, India. Riv.Intal.Paleont. 87: 739-762.

van Bemmelen, R.W. 1949. The Geology of Indonesia. IA. General Geology of Indonesia. The Hague: Martinus Nijhoff.

Bergman, S.M., Coffield, D.Q., Talbot, J.P. & Garrard, R.A. 1996. Tertiary tectonic and magmatic evolution of western Sulawesi and the Makassar Strait, Indonesia: evidence for a Miocene continent-continent collision. In R.Hall & D.J. Blundell (eds), Tectonic Evolution of Southeas t Asia: 391-429. London: Geological Society Special Publication No.106.

Cranbrook, Earl of. 1981. The vertebrate faunas. In T.C.Whitmore (ed), Wallace's Line and Plate Tectonics. Oxford: Clarendon Press.

Evans, B.J., Morales, J.C., Supriatna, J. & Melnick, D.J. 1999. Origin of Sulawesi macaques (Cercopithecidae: Macaca) as suggested by mitochondrial DNA phylogeny. Biol.J.Linn.Soc. 66: 539-560.

Fooden, J. 1969. Taxonomy and Evolution of the Monkeys of Celebes (Primates: Cercopithecidae). Basel: S.Karger (Bibliotheca Primatologica, No.10).

Groves, C.P. 1969. Systematics of the anoa (Mammalia, Bovidae). Beaufortia, 17: 1-11.

Groves, C.P. 1976. The origin of the mammalian fauna of Sulawesi (Celebes). Z.Säugetierkunde, 41: 201-216.

Groves, C.P. 1980. Notes on the systematics of Babyrousa (Artiodactyla, Suidae). Zool.Meded., Leiden 55: 29-46.

Groves, C.P. 1998. Systematics of tarsiers and lorises. Primates 39: 13-27.

Hall, R. 1996. Reconstructing Cenozoic SE Asia. In R.Hall & D.J.Blundell (eds), Tectonic Evolution of Southeast Asia: 153-184. London: Geological Society Special Publication No.106.

Heaney, L.R. 1991. A synopsis of climatic and vegetational change in Southeast Asia. Climatic Change 19: 53-61.

Hooijer, D.A. 1954. Pleistocene vertebrates from Celebes. VIII. Dentition and skeleton of Celebochoerus heekereni Hooijer. Zool.Verh., Leiden, No.24: 46pp.

Kakoi, H., Namikawa, T., Takenaka, O., Takenaka, A., Amano, T. & Martojo, H. 1994. Divergence between the Anoas of Sulawesi and the Asiatic Water Buffaloes, inferred from their complete amino acid sequences of hemoglobin ß chains. Z.zool.Syst.Evolut.-forsch. 32: 1-10.

Kitchener, D.J., How, R.A. & Maharadatunkamṣi. 1991. Paulamys sp. cf. P.naso (Musser, 1981) (Rodentia: Muridae) from Flores Island, Nusa Tenggara, Indonesia – description from a modern specimen and a consideration of its phylogenetic affinities. Rec.West Aust.Mus. 15: 171-189.

McDougall, I., Brown, F.H., Cerling, T.E. & Hillhouse, J.W. 1992. A reappraisal of the geomagnetic polarity time scale to 4 Ma using data from the Turkana Basin, East Africa. Geophys.Res.Letters, 19: 2349-2352.

van der Made, J. & Moyà-Solà, S. 1989. European Suinae (Artiodactyla) from the Late Miocene onwards. Boll.Soc.Paleont.Ital. 28: 329-339.

Moore, J.C. 1959. Relationships among the living squirrels of the Sciurinae. Bull. Amer. Mus. nat. Hist. 118: 157-206.

Musser, G.G. 1987. The mammals of Sulawesi. In T.C.Whitmore (ed), Biogeographical Evolution of the Malay Archipelago: 73-93 . Oxford: Clarendon Press.

Ruedi, M., Auberson, M. & Savolainen, V. 1998. Biogeography of Sulawesian shrews: testing for their origin with a parametric bootstrap on molecular data. Mol.Phyl.Evol. 9: 567-571.

Sartono, S. 1979. The age of the vertebrate fossils and artifacts from Cabenge in South Sulawesi, Indonesia. Modern Quat. Research in S.E.Asia 5: 65-81.

Simpson, G.G. 1977. Too many lines: the limits of the Oriental and Australian zoogeographic regions. Proc.Amer.Philos.Soc. 121: 107-120.

Sondaar, P.Y. 1984. Faunal evolution and the mammalian biostratigraphy of Java. Cour.Forsch.Inst.Senckenberg 69: 219-235.

Sugiri, N. & Hidayat, N. 1996. The diversity and haematology of Anoa from Sulawesi. Anoa Species Population and Habitat Viability Assessment Workshop, 22-26 July 1996. PHPA, Taman Safari Indonesia, ICUN/SSC/AWCSG and IUCN/SSC/CBSG.

Section 6

Primates

Radiation and Evolution of Three Macaque Species, *Macaca fascicularis*, *M. radiata* and *M. sinica*, as Related to Geographic Changes in the Pleistocene of Southeast Asia

R-L. Pan
The University of Western Australia, Perth, WA, 6907, Australia & Kunming Institute of Zoology, Kunming, China

C. E. Oxnard
The University of Western Australia, Perth, WA, 6907, Australia

ABSTRACT: In a study of intrageneric variation in the genus *Macaca*, 72 dental and cranial measures of 11 species were examined morphometrically. Results indicate that 3 species, *M. fascicularis*, *M. sinica* and *M. radiata*, display close morphometric similarities. These species also share more similar patterns of sexual dimorphism compared with other macaques, regardless of which cranial part is considered. These findings differ from those of Fooden (1976) and 1980) and Delson (1980) concerning species-groups in *Macaca*, but are similar to findings from biochemistry, genetics and external features. These three species may thus have undergone common evolutionary processes and shared common ancestors. The crab-eating monkey, *M. fascicularis*, may have originated from southern India or the area that originally linked with Sri Lanka. They may have first dispersed eastwards, and then turned south, to penetrate south-eastern Asia as they approached the geographic barriers caused by the up-lift of the Qinghai-Tibetan Plateau in the Pleistocene.

1 INTRODUCTION

Crab-eating macaques, *M. fascicularis,* are widely distributed in south-eastern Asia, the most southern extreme of all macaque species in Java. According to Fooden (1976 and 1980) and Delson (1980), the species shows close relationship with rhesus monkey (*M. mulatta*), Japanese monkey (*M. fuscata*), and Taiwan monkey (*M. cyclopis*). All seem to form one species group, characterised by a bluntly bilobed and narrow glans penis, and a corresponding female genital tract (Fooden, 1980). Toque monkeys (*M. sinica*) are endemic to Sri Lanka and may have close relationships with the Bonnet monkey (*M. radiata*) in Southern India. The reproductive organs are quite different from those of the crab-eating monkeys, with an apically acute and broad glans penis and corresponding female genital tract (Fooden, 1980). However, the relationships between these three species as derived from other research fields are not in agreement with this: for instance, from body size and tail length as listed by Pan (1998). These three species have very similar body weights and tail lengths. They are the macaque species with the longest tails, but smallest body size. A close relationship between them is also implied by studies of gene frequencies (Weiss *et al,* 1973; Melnick and Kidd, 1985), molecular composition (Cronin *et al.,* 1980), allele frequencies (Fooden and Lanyon, 1989), and external morphology (Chan, 1996). They exhibit very similar locomotion patterns (Fooden, 1995).

The morphometric relationships between these species may help to clarify these controversies. The purpose of this study is, therefore, to re-evaluate the phenetic relationships of these three species as compared with other macaques. The data will then be used to assess hypotheses about the evolution and phylogeny of these species in relation to other species in genus *Macaca*. This study also shows how geographic changes have affected the evolution of the animals.

2 MATERIALS AND METHODS

Eleven species of the genus *Macaca* (Table 1) were used in this study. The materials are housed in a number of institutes, universities and museums in various parts of the world (see Acknowledgments). All specimens were from adult macaques, as judged by the full eruption of M3. They were of wild shot animals or of animals that had died of natural causes in the wild. The sexes of the skulls were taken from available records. Since macaques exhibit significant canine sexual dimorphism (Pan and Jablonski, 1995; Plavcan, 1990 and, 1993), the sexes of the specimens without field records of sex were determined by comparisons of their canine size and structure with the sexes of the known materials. When this procedure was indeterminate, these materials were excluded from analysis.

Table 1 Species and numbers (by sex) used in this study

Species	Males	Female	Unknown	Total
M. mulatta	20	20		40
M. fuscata	11	11		22
M. sinica	14	8		22
M. assamensis	20	20		40
M. radiata	12	12		24
M. arctoides	20	20		40
M. sylvanus	11	10		21
M. nemestrina	20	20		40
M. nigra	8	8		16
M. fascicularis	20	20		40
M. thibetana	11	8	1	20
Totals	167	157	1	325

Seventy-two linear dental, mandibular and cranial variables were measured on each specimen using digital calipers accurate to 0.01mm. Length measurements were made parallel to the median sagittal plane and breadth (or width) measurements were made in the coronal plane; height was measured at the right angles to length and breadth. Mesiodistal length of teeth was measured as the greatest length from the most mesial to the most distal points; buccolingual width was the greatest width, from the most buccal to most lingual points. Unilateral measurements were taken on the left side of the jaw or cranium; the right side was substituted if the left was missing or damaged.

Consideration of the number of variables is an important first step in any study. Small numbers of variables may give misleading results because they are too few to fully represent the structure or species being analysed. Large numbers of variables may result in other problems because they may give rise to additional noise in the data. However, the optimum number of variables also depends on the complexity of the structure or species being studied. Thus, the skull, unlike most postcranial bones, is involved in several different functions: mastication, facial appearance and brain protection. Skull development and growth are also complex: the mandible is affected by the movements of the neural crest, the face is influenced by the branchial arches, the calvaria is modified by the growth of the brain. Sexual dimorphism is also more complex than in other anatomical regions involving different patterns in different local anatomical regions for varying functions (Pan, 1998). Thus, one of the features of this study compared to many others, is the use of a large enough number of variables to cover such complexity.

Two multivariate statistical analyses, Principal Components Analysis (PCA) and Discriminant Function Analysis (DFA) were applied to the data. PCA examines the degree to which groups of specimens and clusters of variables are present in the data. Clusters of major contributing variables are identified using, as a significance criterion, a minimum loading of 0.300 (Hair, et al.,1992, pp 223-264). DFA examines the relationships of *a priori* known groups (sexes, species).

The variables were studied in three successive steps to focus on the aforementioned complexity of the skull. As the first step variables were grouped into 7 units on the basis of morphology and function. These units - upper teeth, lower teeth, mandible, maxilla, upper face, calvaria and cranium (Pan, 1998; Pan *et al.*, 1998) - were each analysed by PCA to derive patterns of group clustering. As a second step, units exhibiting very similar patterns of separation by sex and species were then grouped together to carry out composite studies of anatomical regions. These were, again, examined using PCA. As a final step, a number of the variables representing the skull as a whole were selected from the prior analyses. At this step, although PCA was still used to determine the groups, and reveal patterns of species and sex separations, DFA was also used to analyze the differences between *a priori* designated groups on the basis of information about sex and species external to the study.

The division of skulls into functional units, anatomical regions and the skull as a whole, was aimed at investigating the ways in which the different parts with different functions separate the animals. The analysis of the patterns in single-sex and combined-sex groupings allows determination of how sexual dimorphism varies in different analyses.

The criteria for selecting variables for the analyses of the anatomical regions and the whole skull were based on their influences in the first two axes of the respective previous PCAs. Initially, the variables that exhibited significant contributions, ie, those with loading of more than 0.300 were chosen. Though this reduced the number of variables somewhat, there were still an unbalanced number of variables (more variables describing teeth than other areas). Variables of lesser importance were therefore further culled until there were sub-equal numbers of variables for each unit and region.

These procedures thus resulted in a nested series of analyses. They provide a specific observational design attempting to reveal how information content changes in proceeding from simpler to ever more complex analyses. It might be expected to provide better interpretation of the information for the final conclusions.

3 RESULTS

The results in this study thus come from the design strategy just described: functional units, anatomical regions and the whole skull. Eigenvalues and eigenvectors for each analysis in which the sexes were studied separately were obtained. In studies where the sexes were analysed together, however, since the names and numbers of the variables making significant contributions were similar to those either in females or males separately, the eigenvalues and eigenvectors are not listed.

3.1 *Functional Units*

Seven units: the mandible, lower teeth, upper teeth, maxilla, cranium, calvaria and the upper face were analyzed separately by PCA.

a: Females: The mean dispersions of the species in females are illustrated in Fig.1. The first three units, the mandible, lower teeth and the upper teeth, show almost identical patterns. That is, the separation between species is mainly in the first axis (PC1). The three species, *M. sinica*, *M. fascicularis* and *M. radiata*, are significantly separated from the rest of the species and clustered at the left end of PC1. A similar clustering of these three species with significant isolation from the remaining species is also found in the studies of maxilla, the cranium and the calvaria.

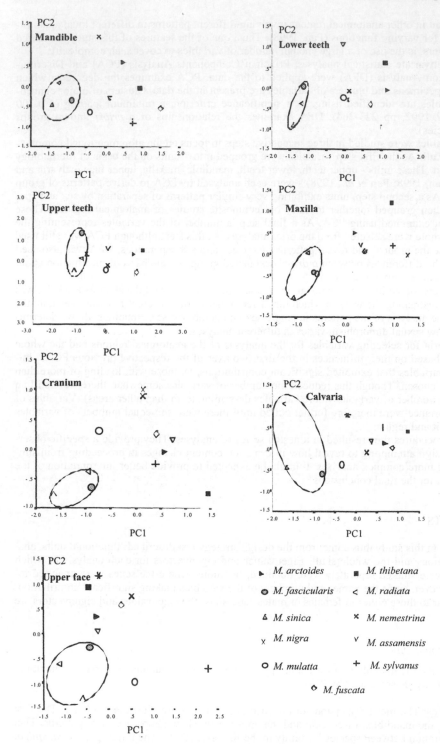

Figure 1 The plots of the first two principal components for functional units in the female macaques.

The analysis of the upper face is, however, quite different in that this same cluster of species is evident in PC2 (lower part of the axis) rather than PC1.

Table 2. Percentage of total variation, eigenvalues and eigenvectors for the first two PCA axes for variables of functional units in the female macaques. The variables reaching significant contribution level are bolded. See Pan (1998) and Pan *et al.*, (1998) for variable definitions.

Mandible

	1	2
Eigenvalue	10.94	1.46
Percentage	68.4	9.1
Cum. Per	68.4	77.5

Eigenvectors

	1	2
LBCB	**.682**	**.619**
LBMB	**.936**	.089
LIAW	**.430**	**.791**
CONDYLL	**.549**	-.281
CONDYLW	**.884**	-.201
CONM1	**.956**	-.143
CONM3	**.877**	-.221
MAM1	**.863**	-.296
MAT	**.725**	-.132
MANDLAL	**.896**	.011
MANDH	**.899**	-.070
MANDSYM	**.896**	.172
BICONDY	**.955**	-.144
MANCORW1	**.863**	-.068
SYMWID	**.884**	.055
LTRLALV	**.728**	.296

Lower teeth

	1	2
Eigenvalue	9.64	1.66
Percentage	60.3	10.4
Cum. Per	60.3	70.6

Eigenvectors

	1	2
LI1MDL	.249	**.921**
LI2MDL	**.562**	**.707**
LCMDL	**.509**	**-.310**
LP3MDL	**.701**	-.211
LP4MDL	**.888**	.099
LM1MDL	**.875**	.166
LM2MDL	**.929**	-.084
LM3MDL	**.909**	-.186
LI1BLL	**.623**	.002
LI2BLL	**.741**	-.071
LCBLL	**.544**	.160
LP3BLL	**.845**	.066
LP4BLL	**.914**	-.091
LM1BLLM	**.900**	-.062
LM2BLLM	**.935**	-.136
LM3BLLM	**.896**	-.148

Upper teeth

	1	2
Eigenvalue	10.68	1.33
Percentage	66.7	8.3
Cum. Per	66.7	75.0

Eigenvectors

	1	2
UI1MDL	**.458**	**.637**
UI2MDL	**.647**	**.423**
UCMDL	**.769**	.147
UP3MDL	**.851**	.040
UP4MDL	**.885**	-.156
UM1MDL	**.859**	-.214
UM2MDL	**.901**	-.194
UM3MDL	**.891**	-.196
UI1BLL	**.620**	**.569**
UI2BLL	**.764**	**.310**
UCBLL	**.814**	-.020
UP3BLL	**.867**	.043
UP4BLL	**.866**	-.056
UM1BLLM	**.895**	-.263
UM2BLLM	**.926**	-.214
UM3BLLM	**.904**	-.186

Maxilla

	1	2
Eigenvalue	4.64	.72
Percentage	77.4	11.9
Cum. Per	77.4	89.3

Eigenvectors

	1	2
MUZL	**.926**	-.107
PALLENG	**.959**	-.109
PALWID	**.905**	-.143
UBCB	**.914**	-.070
UBMB	**.933**	-.081
UIAW	**.582**	**.812**

Cranium

	1	2
Eigenvalue	6.10	.96
Percentage	76.3	12.1
Cum. Per	76.3	88.3

Eigenvectors

	1	2
BCRANL	**.951**	-.028
CRANL	**.980**	.015
CRANW	**.909**	.162
BIZYGW	**.934**	-.062
POSTORB	**.557**	**.689**
OCCH	**.567**	**-.674**
ANTBASI	**.976**	-.041

Calvaria

	1	2
Eigenvalue	4.30	.98
Percentage	61.4	14.0
Cum. per	61.4	75.5

Eigenvectors

	1	2
CALVL	**.926**	-.117
BPORW	**.911**	-.117
MIDPARW	**.895**	-.040
POSTORB	**.618**	.397
OCCH	**.594**	**-.663**
FORMAGL	**.550**	**.590**
FORMAGW	**.880**	.085

Upper face

	1	2
Eigenvalue	5.64	.71
Percentage	70.6	8.9
Cum. Per	70.6	79.5

Eigenvectors

	1	2
MUZL	**.863**	-.039
BIORBW	**.837**	.181
INTORBW	**.726**	**.633**
PIRH	**.861**	-.027
PIRW	**.755**	**-.476**
UFACEH	**.848**	-.197
INFRMAL	**.905**	.062
BIZYGW	**.908**	-.092

The percentage of the total variation, eigenvalues and eigenvectors are listed in Table 2. In PC1 only one variable, LI1MDL, does not show a significant positive contribution. In the second axis the variables showing significant positive contributions are LBCB, LIAW, LI1MDL,

LI2MDL, UI1MDL, UI2MDL, UI1BLL, UI2BLL, UIAW, POSTORB, FORMAGL and IN-
TORBW. Those with significant negative contributions are LCMDL, OCCH and PIRW.

Table 3. Percentage of total variation, eigenvalues and eigenvectors for the first two PCA axes for variables of functional units in the male macaques. The variables reaching significant contribution level are bolded.

Mandible

	1	2
Eigenvalue	10.79	1.27
Percentage	67.5	7.9
Cum. Per	67.5	75.4

Eigenvectors

	1	2
LBCB	.779	.505
LBMB	.914	.048
LIAW	.599	.700
CONDYLL	.618	-.040
CONDYLW	.825	-.147
CONM1	.940	-.163
CONM3	.865	-.252
MAM1	.745	-.335
MAT	.715	-.240
MANDLAL	.961	.002
MANDH	.897	-.071
MANDSYM	.865	.273
BICONDY	.922	-.182
MANCORW1	.800	-.163
SYMWID	.848	-.028
LTRLALV	.739	.311

Lower teeth

	1	2
Eigenvalue	9.31	1.94
Percentage	58.2	12.1
Cum. Per	58.2	70.3

Eigenvectors

	1	2
LI1MDL	.439	.660
LI2MDL	.624	.394
LCMDL	.601	.304
LP3MDL	.598	.531
LP4MDL	.886	.013
LM1MDL	.829	-.244
LM2MDL	.887	-.303
LM3MDL	.888	-.305
LI1BLL	.509	.473
LI2BLL	.772	.258
LCBLL	.562	.334
LP3BLL	.864	.049
LP4BLL	.868	-.179
LM1BLLM	.850	-.343
LM2BLLM	.890	-.313
LM3BLLM	.881	-.282

Upper teeth

	1	2
Eigenvalue	9.66	1.81
Percentage	60.4	11.3
Cum. Per	60.4	71.7

Eigenvectors

	1	2
UI1MDL	.484	.602
UI2MDL	.736	-.013
UCMDL	.590	.534
UP3MDL	.645	.046
UP4MDL	.874	-.105
UM1MDL	.867	-.184
UM2MDL	.905	-.207
UM3MDL	.899	-.180
UI1BLL	.484	.691
UI2BLL	.779	.241
UCBLL	.597	.458
UP3BLL	.854	.056
UP4BLL	.861	-.058
UM1BLLM	.876	-.310
UM2BLLM	.884	-.317
UM3BLLM	.874	-.304

Maxilla

	1	2
Eigenvalue	4.74	.52
Percentage	78.9	8.7
Cum. Per	78.9	87.6

Eigenvectors

	1	2
MUZL	.919	.081
PALLENG	.932	-.023
PALWID	.891	-.278
UBCB	.917	.205
UBMB	.852	-.429
UIAW	.812	.459

Cranium

	1	2
Eigenvalue	6.28	.73
Percentage	78.5	9.1
Cum. Per	78.5	87.6

Eigenvectors

	1	2
BCRANL	.953	-.097
CRANL	.930	-.104
CRANW	.949	-.039
BIZYGW	.926	.022
POSTORB	.714	-.407
OCCH	.617	.733
ANTBASI	.964	.023

Calvaria

	1	2
Eigenvalue	3.68	1.00
Percentage	52.5	14.2
Cum. per	52.5	66.7

Eigenvectors

	1	2
CALVL	.826	-.227
BPORW	.913	-.176
MIDPARW	.849	.094
POSTORB	.762	-.086
OCCH	.518	-.489
FORMAGL	.601	.419
FORMAGW	.482	.694

Upper face

	1	2
Eigenvalue	5.54	.92
Percentage	69.2	11.5
Cum. Per	69.2	80.8

Eigenvectors

	1	2
MUZL	.770	-.400
BIORBW	.826	.332
INTORBW	.701	.619
PIRH	.876	.016
PIRW	.761	-.313
UFACEH	.851	-.366
INFRMAL	.922	.186
BIZYGW	.923	-.038

b: <u>Males</u>: A similar set of findings exist in males (except in the analysis based on the lower teeth) as in females just described (Fig. 2).

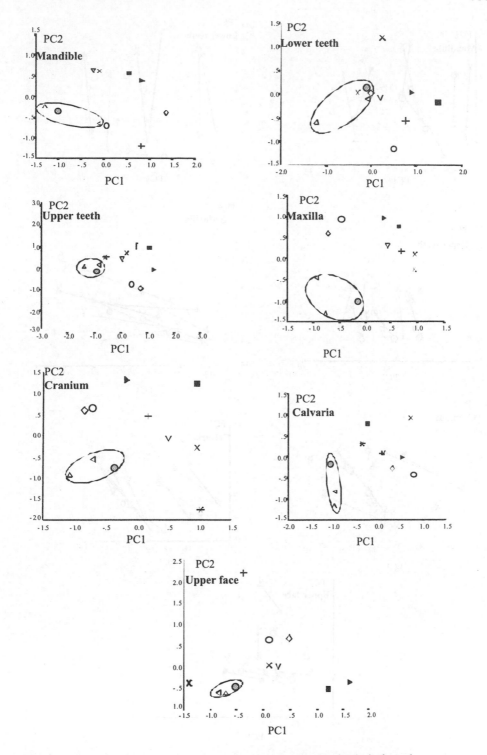

Figure 2 The plots of the first two principal components for functional units in the male macaques.

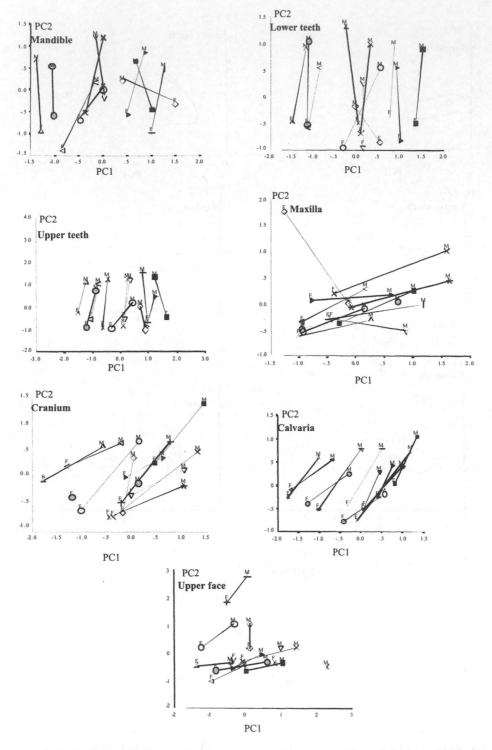

Figure 3 The differences by sex within, and by same sex between, species on the first two principal components based functional units.

All variables in PC1 show significant positive contributions (Table 3). The variables showing significant positive contributions to PC2 include LBCB, LIAW, LTRLALV, LI1MDL, LI2MDL, LCMDL, LP3MDL, LI1BLL, LCBLL, UI1MDL, UIAW, OCCH, FORMAGL, FORMAGW, UCMDL, UCBLL, BIORBW and INTORBW. Those with significant negative contributions are MAM1, LM2MDL, LM3MDL, LM1BLLM, LM2BLLM, UBMB, POSTORB, OCCH, MUZL, PIRW and UFACEH.

c: <u>Sexes taken together</u>: The variations between sexes in each species and among species are illustrated in Fig 3. Three patterns of variation exist. The first one is in the mandible, lower teeth and the upper teeth, in which the separations of species are mainly in PC1 but between sexes in PC2. The second pattern of variation, opposite to that just described, is in the maxilla and the upper face (sex separations in PC1, species separations in PC2. The third pattern, in the cranium and the calvaria, shows that the separations between both sexes and species are in both PC's simultaneously. That is: they are approximately diagonally aligned but at approximately 90 degrees to each other. Both PC's contribute to both sets of separations but in a reverse manner.

3.2 *Anatomical Regions*

The functional units showing the same pattern of differences in sex and species relationships were combined as described in the methods into anatomical regions. The first anatomical region, the masticatory apparatus included mandible, lower teeth and upper teeth. The second, the cranium as a whole consisted of cranium and calvaria. The third, the face as a whole, combined maxilla and upper face.

Regarding this last, the maxilla bearing, the upper teeth as it does, might have been expected to follow the pattern displayed by the mandible and both sets of teeth. But further inspection of the precise nature of the variables in the maxilla indicated that they are as much measures of the lower face as they are measures of the upper jaw. In addition, this area is the junction between the masticatory apparatus and the face as a whole. This perhaps explains why the maxilla is not similar to the rest of the masticatory apparatus.

a: <u>Females</u>: The three species are closely clustered and significantly separated from the rest of the species along PC1 in the analyses of the masticatory apparatus and the cranium as a whole (Fig. 4). The same grouping exists in the analysis of the face as a whole but in PC2.

Twenty-six variables from the three units represented the masticatory apparatus (Table 4) according to the criteria described in materials and methods. All have significant positive contributions to PC1. The variables showing significant positive contribution to PC2 include LIAW, UI1MDL, LI1MDL and LI2MDL. Those displaying significant negative contributions to this axis are CONM1, MAM1, MANDLALV, MANDH and BICONDY.

Thirteen variables were used to characterize the cranium as a whole. As with the masticatory apparatus, all variables on the first axis show significant positive contributions. Variables GLENOL and POSTORB show significant positive, but OCCH displays significant negative, major contributors to PC2.

The face as a whole is represented by 11 variables. Again, the contribution of the variables to PC1 are all positive, similar to both the masticatory apparatus and the cranium as a whole. For PC2 two variables, INTORBW and PIRH, show significant negative, and UIAW exhibits significant positive, contributions.

b: <u>Males</u>: The relationship between the three macaque species and the separation from the other species in the analysis of males are very similar to the aforementioned studies of females. That is, these three species are significantly separated from the other macaque species in the masticatory apparatus and the cranium as a whole on the first axis, but in the face as a whole on the second axis.

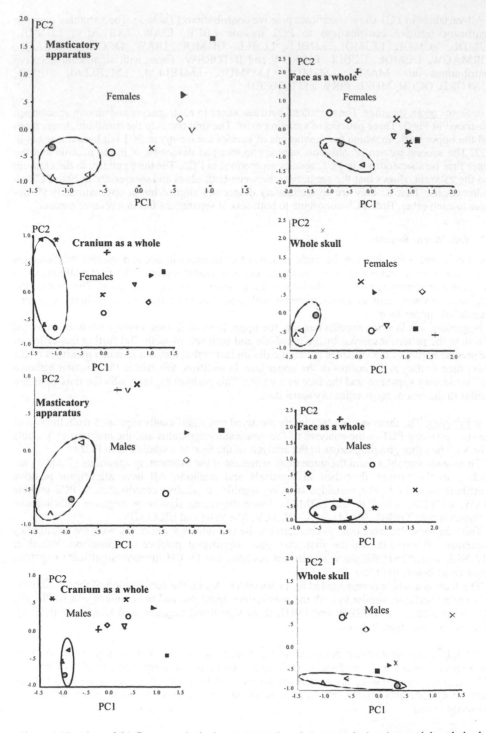

Figure 4 The plots of the first two principal components based on anatomical regions and the whole skull.

As also with the females, all variables of the three regions make significant positive contributions to PC1 (Table 4). In PC2, LIAW, LI1MDL, UCMDL, UI1BLL and LI1MDL make significant positive, but UM1BLLM and LM1BLLM show significant negative, contributions in the analysis of the masticatory apparatus; ZYGH, OCCH and GLENOL show significant positive, but MIDPARW and FORMAGW display significant negative, contributions in the analysis of the cranium as a whole. In the analysis of the face as a whole, the variables showing significant positive contribution to PC2 are BIORBW, INTORBW and PIRH. Those with significant negative contributions are UBCB and UIAW.

Table 4. Percentage of the total variation, eigenvalues and eigenvectors for the first two principal components in the three anatomic regions. The variables reaching significant contribution level are bolded.

Masticatory apparatus

	Female		Males	
	1	2	1	2
Eigenvalue	15.50	2.53	15.02	57.80
Percentage	59.6	9.7	2.71	10.4
Cum. Per	59.6	69.4	57.8	68.2
Eigenvectors				
LBCB	.736	.059	.792	.258
LBMB	.914	-.227	.888	-.219
LIAW	.615	.453	.666	.504
CONM1	.868	-.379	.853	-.164
MAM1	.742	-.468	.689	-.211
MANDLALV	.887	-.330	.883	-.021
MANDH	.819	-.312	.881	-.015
BICONDYL	.864	-.355	.883	-.211
UI1MDL	.531	.606	.606	.608
UI2MDL	.719	.278	.744	-.126
UCMDL	.793	.073	.729	.341
UP4MDL	.883	-.056	.859	-.259
UM1MDL	.831	-.028	.862	-.275
UI1BLL	.633	.258	.487	.630
UI2BLL	.773	.153	.760	.067
UCBLL	.767	-.103	.656	.248
UP4BLL	.872	-.012	.831	-.154
UM1BLLM	.898	-.094	.823	-.394
LI1MDL	.348	.803	.476	.734
LI2MDL	.630	.505	.626	.260
LCMDL	.525	-.226	.597	.168
LP4MDL	.922	.065	.861	-.090
LM1MDL	.856	.130	.812	-.234
LCBLL	.543	.204	.560	.263
LP4BLL	.886	.041	.861	-.202
LM1BLLM	.854	.009	.805	-.364

Cranium as a whole

	Females		Males	
	1	2	1	2
Eigenvalue	7.79	1.20	6.56	1.38
Percentage	59.9	9.2	50.5	10.60
Cum. Per	59.9	69.1	50.5	61.1
Eigenvectors				
CALVL	.953	-.053	.872	.175
BPORW	.925	-.104	.910	-.099
MIDPARW	.865	.006	.781	-.386
INTFRMAL	.887	-.243	.920	.097
POSTORB	.595	.598	.667	-.197
TEMFOSL	.864	-.058	.838	.006
TEMFOSW	.842	-.038	.762	-.002
ZYGH	.703	-.038	.650	.305
OCCH	.579	-.431	.530	.335
FORMAGL	.496	.049	.498	-.246
FORMAGW	.824	-.180	.367	-.623
GLENOL	.495	.722	.303	.685
GLNOW	.834	.152	.804	.059

Face as a whole

	Females		Males	
	1	2	1	2
Eigenvalue	7.30	1.00	7.07	1.33
Percege	66.4	9.1	64.3	12.1
Cum.	66.4	75.5	64.3	76.3
Eigenvectors				
MUZL	.925	.050	.893	-.237
BIORBW	.808	-.270	.720	.528
INTORBW	.676	-.340	.562	.639
PALLENG	.994	.106	.919	-.212
PALWID	.864	.166	.861	-.205
UBCB	.870	.204	.852	-.351
UBMB	.928	.091	.900	.093
UIAW	.493	.727	.699	-.478
PIRH	.832	-.303	.814	.345
PIRW	.731	-.253	.698	.128
UFACEH	.830	.030	.822	.028

c: The sexes taken together: In the analysis of both sexes for all species the patterns of separation are shown in Figure 5. In contrast to analyses where the sexes were examined in isolation, a different pattern was found in each anatomical region. That in the masticatory apparatus is identical to that found in each of individual anatomic unit of that region. The pattern in the cranium as a whole is somewhat similar to that of the face as a whole in terms of intraspecific variation, but quite different referring to the variation between species. Based on the masticatory apparatus and the cranium as a whole, the three species exhibit a very similar pattern that is quite different from those of the rest of the species.

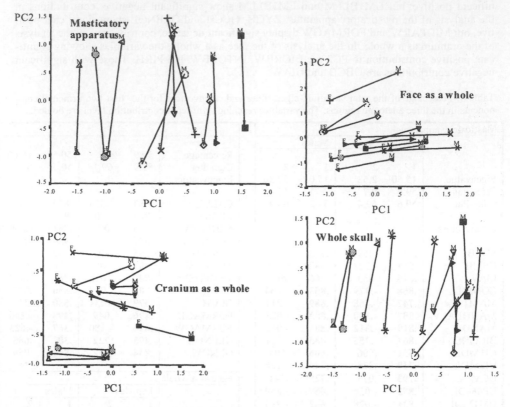

Figure 5 The difference by sex within, and by same sex between, macaque species revealed in the first two principal components in three anatomic regions and the whole skull.

3.3 The whole skull

Twenty-seven variables were selected for the analysis of the whole skull (Table 5).

Species dispersions in this analysis are displayed in Figs. 4 and 5. The three species are significantly separated from the remaining species along the first axis in the analysis of females. The separation in males is, however, quite different; occurring in PC2 and being less obvious than in females (Fig.4). When sexes are considered together, the differences between the sexes and the species are very similar to those shown in the masticatory apparatus (Fig. 5). However, in contrast to that analysis (and indeed to all other analyses in this study), additional statistically significant information about species and sex separations is included in axes PC3 and PC4 (not figured here). This indicates that the analysis for the whole skull, is not, in fact, identical to the masticatory apparatus. Rather, the information content of the masticatory apparatus must be larger and relatively independent of that in the other regional analyses, and is therefore placed into the first two axes in the analysis of the skull as a whole.

All variables show significant positive contributions to PC1 both in females and males. The variable showing significant contributions to PC2 in females are LBCB, LIAW, UI1MDL, UI1BLL, LI1MDL, LI2MDL and UIAW. In males they are LIAW, UI1MDL, UI1BLL, LI1MDL, LI2MDL, UBCB and UIAW. Variables PIRH in females, and UM1BLLM, LM1BLLM, BIORBW, PIRH, CALVL, MIDPARW, FORMAGL and FORMAGW in males, show significant negative contributions to PC2.

Table 5. Percentage of the total variation, eigenvalues and eigenvectors in the first two principal components based on the whole skull. The variables reaching significant contribution level are bolded.

Female			Male	
	1	2	1	2
Eigenvalue	14.72	3.46	12.84	3.58
Percentage	54.5	11.7	47.6	13.2
Cum. per	54.5	66.2	47.6	60.8
Eigenvectors				
LBCB	.765	.334	.834	.196
LIAW	.592	.677	.683	.480
CONM1	.919	-.197	.881	-.256
MAM1	.818	-.293	.675	-.229
UI1MDL	.362	.761	.623	.599
UCMDL	.835	.070	.743	.266
UI1BLL	.581	.414	.509	.603
UM1BLLM	.845	-.111	.701	-.328
LI1MDL	.245	.839	.492	.731
LI2MDL	.563	.575	.606	.310
LCBLL	.569	-.006	.636	.239
LM1BLLM	.807	.003	.707	-.340
MUZL	.908	-.077	.868	.162
BIORBW	.743	-.184	.709	-.382
INTORBW	.709	-.118	.559	-.299
PALWID	.846	-.133	.871	.009
UBCB	.910	.100	.827	.308
UIAW	.828	.428	.723	.474
PIRH	.793	-.335	.758	-.302
CALVL	.908	-.253	.852	-.355
MIDPARW	.848	-.247	.648	-.492
INFRMAL	.888	-.183	.878	-.251
POSTORB	.628	.012	.700	-.149
OCCH	.489	-.154	.404	-.239
FORMAGL	.469	-.162	.510	-.423
FORMAGW	.806	-.249	.260	-.336
GLENOL	.553	-.105	.492	.104

When each species is regarded as an independent group as in Discriminant Function analysis of females, the three species are similarly closely clustered as in the PCAs, and there is good separation from the other macaque species (Fig. 6). This result is even clearer in the analysis of males and in the analysis of the sexes taken together.

4 DISCUSSION

The results revealed in this study indicate that there is considerable variation in morphology between macaque species in different skull areas in each sex separately, and in both sexes taken together. These species and sex clusters may be related to the differences in evolution, ranging behavior, habitat selection and activity patterns, and influence, therefore, judgments about evolution and classification. They are more obvious than any view obtainable from examining the variables one by one or from viewing the anatomy without quantification. They provide a good example of the value of using morphometric methods in primate biology.

However, as indicated previously, the main goal of this study is to provide information about the relationship, among a number of species of the genus *Macaca,* of the three species, *M. fas-*

cicularis, M. sinica and *M. radiata*. Such information might be relevant to hypotheses about their origin and dispersion following tectonic movements in Asia during the Pleistocene.

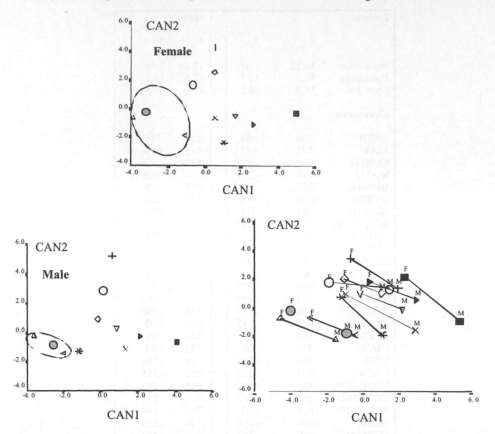

Figure 6 Biplots of the first two discriminant functions for the whole skull.

One unique character of this study is that a large number of variables (77) were filtered strategically at different stages in the study. Except LIMDL, all variables have significant contributions to the first axis of PCA. The following variables show significant contributions to the second axis regardless whether single sex, or both sexes were considered: incisor alveolar breadth (LIAW), the mesiodistal lengths of the first incisors (UI1MDL and LI1MDL), interorbital breadth (INTORBW), maxillary incisor alveolar breadth (UIAW) and glenoid length (GLENOL). In general, the anterior parts of the anatomy play an especially important role in the second axis.

The clusters of animals, both of sexes and species, occurring in different axes or combinations of axes in the hierarchical series of functional units, anatomical regions and the whole structure, provide a useful way to examine the complexity of the skull. Further details of this strategy are discussed in Pan and Oxnard (2000).

However, from the viewpoint of tectonic movements, climactic changes and related migrations, the most important findings here are the relationships among crab-eating, toque and bonnet monkeys, and their unique characteristics compared to other macaque species. A close relationship among three species has been found regardless of which functional unit, anatomical region or the whole skull is considered, and regardless of which sex is examined. They are significantly separated from the rest of the macaque species. When the sexes (separately) are included in the same analyses, they show a quite different pattern of variation and a clearer separation from the other species in the masticatory apparatus, the cranium as a whole and in the whole skull. This implies that the differences between the sexes (sexual dimorphism - though partly a matter of size) is relatively independent of differences between species (also partly a matter of size). And this implies, in turn, that even size is not a simple matter.

These morphometric results provide phenetic information and thus imply a combination of information due to evolution from common ancestry as well as parallels and divergences from ancestral states. Therefore conclusions from morphological analysis may reflect both existing and past evolutionary scenarios (Sneath and Sokal, 1973). It is possible that part of the close morphological relationship among the three macaque species stems from common retention of similar characteristics in which case their similarity would have little bearing on assessment of evolutionary and phylogenic relationship as proposed by (Fooden, 1976 and 1980, Delson, 1980).

On the other hand, the similarities may derive from parallel evolution of small body size: all the species were allocated along the first axis in relation to their body weight. This seems to be confirmed by the eigenvectors in the first axis: all variables, except for LI1MDL, showing a significant positive contribution to this axis. However, the first axis, by definition, contains mostly positive contributions of variables. Thus the fact that there are some situations where the first axis does not reflect body size differences between the species, but sex differences (also largely size but independent of body size) implies that the above conclusion is too simplistic. Nevertheless, it is possible that simple body size differences also explain the finding that *M. radiata* and *M. sinica*, of the *sinica* group, are widely separated from *M. arctoides*, *M. thibetana* and *M. assamensis*. The former species are of smaller body size than the latter. This may be due to divergent evolutionary pathways with adaptation to different geographical areas and environmental zones as a result of tectonic change (Delson, 1980; Fa, 1989). *Macaca radiata, M. sinica* may have undergone dwarfing and as a result share few of their size-based characteristics with their putative conspecific, *M. fascicularis* (Pan *et al.*, 1998). These three species, however, exhibit the longest tail length in macaques (Pan, 1998). <u>A convincing close relationship among the three species is also found in studies of biochemistry, genetics and external morphology.</u>

1) Thus Weiss *et al* (1973) analyzed the gene frequency data of macaques for a total of six erythrocyte enzymes and four serum proteins, and hemoglobin, each of which were screened for electrophoretic mobility. Close relationships between *M. fascicularis* and *M. radiata* were demonstrated graphically (their figures 1 and 2). With *M. sinica* they share the same trunk, and then separate into different branches based on 5 loci (6-PGD, Tf, Hb, TBPA, CA I) and loci 1-5, 6 (6-PGD, Tf, Hb, TBPA, CA I, CA II). In a similar manner, Melnick and Kidd (1985) indicate graphically that the two species groups, the group *fascicularis* and the group *sinica* including *M. sinica* and *M. radiata*, diverged from that same branch. They concluded that "The group of species genetically closest to the *fascicularis* species group belongs to Fooden's *sinica* group."

2) According to Cronin *et al*. (1980, their table 3-2) based on 25 proteins encoded for by 29 loci for 11 macaque species, the genetic distance between *M. fascicularis* and *M. radiata* is 0.41. This figure is less than those among the other three species: *M. radiata* and *M. nigra* (0.85), and *M. radiata* and *M. nemestrina* (0.54). The distance between *M. sinica* and *M. fascicularis* is 0.56, which is smaller than those between *M. sinica* and *M. nigra* (1.95), and between *M. sinica* and *M. nemestrina* (0.85). Another phenogram (Cronin *et al*, 1980, their figure. 3-3, p. 47) based on electrophoretic evidence, indicated that the divergence time between the branch with *M. radiata* and *M. sinica* and the one with *M. fascicularis* occurred about 2 MY ago.

3) Allele frequency data for 35 blood-proteins (Fooden and Lanyon, 1989) implied that the species in the *fascicularis* and *sinica* groups originated from the same ancestor(s). This result is also shown graphically in their figure 4 based on the jackknife strict-consensus tree derived from 16 pseudoreplicate Fitch-Margoliash analyses of macaque blood protein data using *Papio* as an outgroup.

4). These three species are predominantly arboreal (Fooden 1986 and 1995). They spend about 76% (*M. radiata* and *M. sinica*) and 98% (some populations of *M. fascicularis*) of their time in the trees.

5) Based on the combined data of external morphology, tail length, reproductive organs and molecular genetics from Delson (1980), Fooden (1980), Fooden and Lanyon, (1989) and Melnick and Kidd (1985), Chan (1996) computed a phylogenetic tree of macaque species. Chan

therefore also suggested that *M. fascicularis* and *M. sinica* were derived from the same ancestor(s).

Thus, the evidences about *M. fascicularis*, *M. sinica* and *M. radiata* found in this study, together with those reported by others, imply that these three species may have undergone similar evolutionary and phylogenetic experiences. This idea contradicts the results of the studies of Fooden (1976 and1980) and Delson (1980).

The evolution and phylogeny of macaques are mainly based on the fossil record, external morphology, the distribution of extant species and biochemical studies (Delson, 1980; Fa, 1989, Eudey, 1980; Fooden, 1991; Melnick and Kidd, 1985, Suzuki, *et al.*, 1996; Tanaka and Takenaka, 1996). The results of these studies show that there is a close relationship between *M. fascicularis*, *M. sinica* and *M. radiata* revealed by external morphology, craniodental morphometric, biochemical and genetic data. The findings suggest a reconsideration of some of the issues dealing with the evolution and phylogeny of macaques, especially the ancestor(s) and the patterns of radiation for the species in the groups *sinica* and *fascicularis* in which *M. sinica*, *M. radiata* and *M. fascicularis* are involved as defined by Delson (1980) and Fooden (1976 and 1980).

The fossil record implies that macaques dispersed from Africa to Europe and then radiated extensively in Asia during the Pliocene and Pleistocene and eventually occupying East Asia (Delson, 1980; Medway, 1970). It has been generally accepted that the *sylvanus* group first separated in the macaque radiation in Asia (Eudey, 1980; Fooden, 1991; Medway, 1970; Suzuki, *et al.*, 1996; Tanaka and Takenaka, 1996). The timing of the dispersion of the other species groups, however, is still controversial. Eudey (1980) assumed that the *fascicularis* group was the last one to disperse. Medway (1970), however, proposed that this species group was the first macaque to arrive in Asia. Fooden (1976) assumed that the *fascicularis* group dispersed earlier than the *sinica* group.

Another interesting issue concerning the evolution and phylogeny of macaques is the radiation of the *fascicularis* group and its ancestor(s). According to Eudey (1980), Delson (1980) and Fa (1989), the radiation and the origin of the *fascicularis* group in Asia were both quite different from those of the *silenus* and *sinica* groups. *Macaca fascicularis* was supposed to have originated in Southeast Asia, instead of south India, as was proposed for the other macaques. They dispersed northward to the other islands in Southeast Asia and other adjacent parts of Asia, and then gave rise to the species, *M. mulatta*, *M. cyclopis* and *M. fuscata*, of this group (Eudey, 1980, Fa, 1989). These hypotheses were based at least in part on the fact of the broad and continuous distribution of this group in Asia, especially in the southern-most areas (the Indonesian archipelago), and in eastern Java in the early deposits of the Holocene (fossil *M. fascicularis*: Fooden, 1975); Hooijer, 1962). This date is, however, still controversial: fossil *M. nemestrina*, was dated about at the Middle Pleistocene according to Aimi (1981) and Medway (1970). The wide distribution of *M. fascicularis* in Southeast Asia and its Middle Pleistocene fossil in Java suggested to Medway (1970) that this species or its ancestor(s) penetrated into Southeast Asia further than any other macaque species and adapted to different environmental conditions.

If *M. fascicularis* originated in Southeast Asia and then dispersed northward to the other parts in Asia, two questions need clarifying. First, how did *M. fascicularis* originate and specifically which ancestor(s) might it have come from? Second, if *M. nemestrina* in the *silenus* group was the first species to diverge and then penetrate into Southeast Asia, and the *fascicularis* group was the latest to diverge from Southeast Asia (Eudey, 1980; Fa, 1989), what are the relationships of the fossils of *M. nemestrina* and *M. fascicularis*, also found in these areas and dated at the same time? That time was believed to be the Middle Pleistocene (as implied by Aimi, 1981; Medway, 1970, or Holocene (as by Delson, 1980; Fooden, 1975). *Macaca nemestrina* has been postulated as derived from an ancestor similar to *M. silenus* (Delson, 1980, Eudey, 1980; Fa, 1989), and in turn, resulted in the origin of the Sulawesi macaques (Ablrecht, 1978; Fooden, 1969; Hamada *et al.*, 1988; Kawamoto, 1996; Takenaka *et al.*, 1987; Watanabe *et al.*, 1985).

The dispersion of the *fascicularis* group proposed by Medway (1970) may be plausible, that is, *M. fascicularis* had already arrived in Sundland by the Middle Pleistocene and adapted to different ecological conditions. The absence of *M. nemestrina* from some of the farthest islands of Southeast Asia, both recently and in the Middle Pleistocene, suggests that this species dispersed to this area later than *M. fascicularis*. According to this concept, it is possible that the ancestor(s) of *M. fascicularis* radiated from South India or the area between South India and Sri

Lanka where *M. radiata* and *M. sinica* are found. As for the radiation of the *silenus* group (Delson, 1980), they may have moved to the north India and turned to the east later. The direction of this dispersal route may have changed again to the south as they were approaching the gigantic geographic barriers in the south Hengdun Mountain in China, part of the Qinghai-Tibetan Plateau created by the collision between India and Asian continent about 4-5 my ago (Hall, 1999). The ancestors of *Macaca fascicularis* may have penetrated Sundland and finally reached Java. The results found here and from other studies (Cronin, *et al.*, 1980; Fooden and Lanyon, 1989; Melnick and Kidd, 1985; Weiss et al., 1973) seem to support this hypothesis. This process might have occurred earlier than that of *M. nemestrina* or at least at the same time. This species then spread to southwest Java and radiated among the archipelagos and parts of the continent except for Sulawesi in Southeastern Asia.

As a result of the above discussion and if the definitions of species-groups proposed by Fooden and Delson are accepted, new considerations about the phylogenetic relationship between four macaque species groups can be proposed:

Macaca sylvanus is a special species within the macaque genus. It separated from the Asian macaques about 3 MY ago remaining in Africa. It shows some unique external morphological and morphometric features, biochemical elements and sociosexual behavior and it has a greater genetic distance from the rest of the macaque species (Cronin, *et al.*, 1980; Delson,1980; Suzuki *et al.*, 1996; Tanaka and Takenaka, 1996; Taub, 1980).

Macaca fascicularis has a relatively primitive rounded glans penis which may therefore have characterised that of a pre-*silenus* ancestor of the group (Eudey, 1980). Fooden (1991) indicated that the female genital tract morphology is similarly relatively primitive among the species of the *silenus*, *sylvanus* and *fascicularis* groups. The reproductive tract is more derived in the species of the *sinica* group, and is most derived in *M. arctoides*. Relationships revealed by blood-protein allele frequencies are also generally concordant with those indicated by reproductive anatomy (Fooden and Lanyon, 1989). The similarities of external morphology, dental and cranial structures, and sexual dimorphism patterns found in this study, imply that *M. fascicularis*, *M. radiata* and *M. sinica* show very close relationships, especially between *M. fascicularis* and *M. sinica*. It is possible that the *sinica* group has a closer relationship with the *fascicularis* group than with the *silenus* and *sylvanus* groups. Many of the studies mentioned above support this assumption. The *silenus* group may have radiated first, as some authors assume (Delson, 1980; Eudey, 1980; Fa, 1989, Fooden, 1991), followed by the *fascicularis* group. The *sinica* group may have been the last group to disperse.

ACKNOWLEDGMENTS

Thanks are due to the Australian Research Council for its support through a Large Research Grant and Fellowship. Thanks are also owed to Professor Nina Jablonksi for part of the data analysed here and for helpful discussion of the problem. Dr Len Freedman is to be especially thanked for comments and suggestions on the work.

The following institutions, universities and museums are acknowledged for allowing access to specimens in their care: American Museum of Natural History, New York; Northwest Institute of Biology, Xining, Qinghai, China; Beijing Institute of Zoology, Beijing, China; Field Museum of Natural History, Chicago; Guangdong Institute of Entomology, Guangzhou, Guangdong, China; Guangxi Forest Department, Nanning, Guangxi, China; Guangxi Medical College, Nanning, Guangxi, China; Muséum National d'Histoire Naturelle, Paris; National Museum of Natural History, Washington, DC; Nanchong Teachers' College, Nanchong, Sichuan, China; Yunnan University, Kunming, Yunnan, China; Royal College of Surgeons, London; Department of Mammalogy, British Museum (Natural History), London; Zoological Reference Collection, National University of Singapore; Universität Zürich-Irchel.

REFERENCES

Aimi, M. 1981 Fossil *Macaca nemestrina* (Linnaeus, 1766) from Java, Indonesia. *Primate* 22: 409-413.

Albrecht, G. H. 1978 *The Craniofacial Morphology of the Sulawesi Macaque. Multivariate Approaches to Biological Problems*. Contributions to Primatology 13, Karger, Basel.

Chan, L. K. W. 1996. Phylogenetic interpretations of primate socioecology: with special reference to social land ecological diversity in *Macaca*. In Emilia Martins (ed*), Phylogenies and the Comparative Method in Animal* Behaviour: 327-363. Oxford, England, Oxford University Press.

Cronin, J. E., Cann, R. & Sarich, V. M. 1980. Molecular evolution and systematics of the genus *Macaca*. In D. G. Lindburg (ed), *The Macaques: Studies in Ecology, Behaviour and Evolution*: 31-51. New York: Van Nostrand Reinhold.

Delson, E. 1980. Fossil macaques, phylogenetic relationships and a scenario of development. In D. G. Lindburg (ed), *The Macaques: Studies in Ecology, Behaviour and Evolution*: 10-30. New York: Van Nostrand Reinhold.

Eudey, A. A. 1980. Pleistocene glacial phenomena and the evolution of Asian macaques In D. G. Lindburg (ed), *The Macaques: Studies in Ecology, Behaviour and Evolution*: 52-83. New York: Van Nostrand Reinhold.

Fa, J. E. 1989. The genus *Macaca*: a review of taxonomy and evolution. *Mammal Rev.* 19(2): 45-81.

Fooden, J. 1969. Taxonomy and evolution of monkeys of Celebes (Primates: Cercopithecidae) *Bibliotheca Primatologica* 10: 1-148.

Fooden, J. 1975. Taxonomy and evolution of liontail and pigtail macaques (Primates: Cercopithecidae). *Fieldiana Zoolog* 67: 1-167.

Fooden, J. 1976. Provisional classification and key to the living species of macaques (Primates: *Macaca*). *Folia Primatol.* 25: 225-236.

Fooden, J. 1980. Classification and distribution of living macaques (*Macaca* lecépède, 1799). In D. G. Lindburg (ed), *The Macaques: Studies in Ecology, Behaviour and Evolution*: 1-9. New York: Van Nostrand Reinhold.

Fooden, J. 1986. Taxonomy and Evolution of the Sinica Group of Macaques: 5. Overview of Natural History. *Fieldiana Zoology* 29: 1-22.

Fooden, J. 1991. Systematic review of Philippine Macaques (Primate, Cercopithecidae: *Macaca fascicularis* sub spp.). *Fieldiana Zoology* 64: 1-44.

Fooden, J. 1995. Systematic Review of Southeast Asian Longtail Macaques, *Macaca fascicularis* (Raffles , [1821]). *Fieldiana Zoology* 81.

Fooden, J. & Lanyon, M. M. 1989. Blood-protein allele frequencies and phylogenetic relationships in *Macaca*: a review. *Am. J. Primatol.* 17: 209-241.

Hair, J. F. Jr., Anderson, R. E. & Black, W. C. 1992. *Multivariate Data Analysis with Reading*: 193-264. Canada: Macmillan Publishing Company, Maxwell Macmillan.

Hamada, Y., Watanabe, T., Takenaka, O., Suryobroto, B. & Kawamoto, Y. 1988. Morphological studies on the Sulawesi macaques. I Phyletic analysis of body colour. *Primates* 29: 65-80.

Hooijer, D. A. 1962. Quaternary langurs and macaques from the Malay archipelago. *Zool. Verhanderlingen Mus. Leiden* 55: 3-64.

Kawamoto, Y. 1996. Population genetic study of Sulawesi macaques. In R. Shotake & K. Wada (eds), *Variation in the Asian Macaques*: 37-65. Tokyo: Tokai University Press.

Medway, L. 1970. The monkeys of Sundaland: ecology and systematics of the cercopithecids of humid environment. In J. Napier & P. H. Napier (eds), *Old World Monkey*: 513-553. London: Academic Press.

Melnick, D. J. & Kidd, K. K. 1985. Genetic and evolutionary relationships among Asian macaques. *Int. J. Primatol.* 6: 123-160.

Pan R. L. 1998. A Craniofacial Study of The Genus *Macaca*, with special Reference to the Stump-tailed Macaques, *M. arctoides* and *M. thibetana*: A Functional Approach. PhD Thesis, The University of Western Australia, Australia.

Pan, R. L. & Jablonski, N. G. 1985. Sexual dimorphism in canine size in *Macaca*. *Am. J. Phys. Anthropol.*, (Supplement 20): 166-167.

Pan, R. L., Jablonski, N. G., Oxnard, C. E. & Freedman, L. 1998. Morphometric analysis of *Macaca arctoides* and *M. thibetana* in relation to other macaque species. *Primates*. 39 (4): 519-537.

Pan, R. L & Oxnard, C. E. 2000. Craniodental Variation of Macaques (*Macaca*): Size, Function and Phylogeny. *Zool. Res.* 21 (4): 308-322.

Plavcan, J. M. 1990. *Sexual Dimorphism in the Dentition of Extant Anthropoid Primates*. Master Thesis. UMI Dissertation Information Service. University Microfilms International A Bell and Howell Information Company.

Plavcan, J. M. 1993. Canine size and shape in male anthropoid primates. *Am. J. Phys. Anthropol.*, 92: 201-216.

Sneath, P. H. A. & Sokal, R. R. 1973. *Numerical Taxonomy*. San Francisco: W. H. Freeman and Company.

Suzuki, H., Kawamoto, Y. & Takenaka, O. 1996. Phylogenetic relationships among the 19 species of genus *Macaca* based on restriction site variation in rDNA spacers. *International Symposium: Evolution of Asian Primates*: 18. Japan: Inuyama, Aichi.

Takenaka, O., Hotta, M., Takenaka, A., Kawamoto, Y., Suryobroto, B. & Brotoisworo, E. 1987. Origin and evolution of the Sulawesi macaques. 1. Electrophoretic analyses of haemoglobins. *Primates* 28: 87-98.

Tanaka, T. & Takenaka, O. 1996. Phylogenetic relationships of the genus *Macaca*, inferred from DNA sequence. *International Symposium: Evolution of Asian Primates*: 10. Japan: Inuyama, Aichi.

Taub, D. M. 1980. Female choice and mating strategies among wild Barbary macaques (*Macaca sylvanus* L.). In D. G. Lindburg (ed), *The Macaques: Studies in Ecology, Behaviour and Evolution*: 287-344. New York: Van Nostrand Reinhold.

Watanable, K., Hamada, Y. & Suryobroto, B. 1985. Morphological studies of the Sulawesi macaques: Allometric analysis. *Kyoto University Overseas Research Report of Studies on Asian Non-human Primates* 4: 79-85.

Weiss, M .L., Goodman, M., Prychodko, W., Moore, G. W. & Tanaka, T. 1973. An analysis of macaque systematics using gene frequency data. *J. Human Evol.* 2: 213-226.

Borneo as a biogeographic barrier to Asian-Australasian migration

Douglas Brandon-Jones

32A Back Lane, Ham, Surrey TW10 7LF, England

ABSTRACT: Widespread deforestations during cool dry glacial climates carved major discontinuities in the distribution of primates and other fauna and flora. Primates present and absent on Sumatra, Indonesia, indicate that two dry periods moulded their modern distribution. The earlier one eliminated all primates from Sumatra. The second allowed primates which had meanwhile returned, to survive in the wetter north of the island. These recolonizers first spread to Borneo during the interval between the deforestations, with subsequent postglacial diversification. Borneo's previous impoverished primate fauna suggests that faunal barriers between Borneo and western SE Asia had been as formidable as those which isolate Sulawesi from Borneo. Butterflies and Pleistocene stegodonts dispersing from Asia to Sulawesi, bypassed Borneo via the Philippines or Java. The failure of the leopard, tiger and wild dog to reach Borneo, discredits the notion that suitable habitats linked it and Sumatra during the glacial exposure of the Sunda Shelf. The belief that certain Malay mammals, which made it to the north Natuna islands, had been prevented by Sundaland drainage, from continuing their journey to Borneo, ignores indications that this did not deter the stink badger from reaching the Natuna islands from Borneo, and ignores the available dispersal route between the northern and eastern Sunda drainage. Even if this route was blocked, it remains inexplicable that a river system could obstruct a wild boar, but not a gibbon. Over twice as many Indian mammals in Malaya and Sumatra also range to Java rather than Borneo, although the strait between Sumatra and Java is deeper than that between Sumatra and Borneo. This pattern of distribution is incompatible with free overland dispersal. Colonization of the Malay archipelago was evidently primarily by rafting, but the shallow strait between Borneo and Sumatra was not the only barrier. Much more significant was the cool climate which, exacerbated by the desiccative effects of the glacial emergence of the Sunda Shelf, induced a severe and prolonged drought in south Borneo and south Sumatra. This caused the Pleistocene extirpation of south-west Bornean freshwater fish which were replaced by a younger fauna as the climate moderated. Bird distribution indicates grassland persisted in south Borneo after it had succumbed to forestation elsewhere in SE Asia. The principal Bornean glacial rainforest refugium extended from Mt Kinabalu to Mt Dulit. Two mammal genera, *Hylomys* and *Melogale* now restricted in Borneo to that area, were preglacially lowland in distribution. A similarly restricted Bornean monkey is seemingly montane only because its descendents have altered as they postglacially reoccupied adjacent areas. By the time rainforest had reattained the south coast, potential colonizers from Sumatra had first to negotiate the sea crossing. Moist climates in south Borneo were presumably infrequent before the last interglacial.

Widespread deforestations during cool dry glacial climates produced the pronounced and well-documented faunal and floral disjunction in the Indian subcontinent, and further disjunctions in SE Asia, now obscured by postglacial recolonization (Fig. 1; Brandon-Jones, 1996a). The survival in Sumatra, Indonesia, of primate genera restricted to moist rainforest, such as *Pongo* and *Presbytis*; but disappearance there of the ancestors of the relatively drought-tolerant primates, such as *Nasalis*, *Semnopithecus* and *Macaca*, which must have dispersed through Sumatra to the Mentawai islands and to Java; indicates that two dry periods moulded their modern distribution. The earlier, more severe one eliminated all primates from Sumatra. The later, lesser one allowed primates which had returned there interglacially, to survive in the wetter north of the island. The maritime climate of the Mentawai archipelago, which shielded it from glacial climatic extremes, enabled it to act as one reservoir for this recolonization (Brandon-Jones, 1993, 1996a, 1998). It is unlikely that Bornean rainforest completely disappeared, but some survivors of the earlier deforestation, *Hylobates klossii*, *Presbytis potenziani* and *Macaca pagensis* occur only on the Mentawai islands, and not on Borneo. It is therefore inferred that, rather than disappearing from Borneo, the genera *Hylobates*, *Presbytis* and *Macaca* were absent there until descendants of these Mentawai survivors later colonized it (Brandon-Jones, 1996a, 1998).

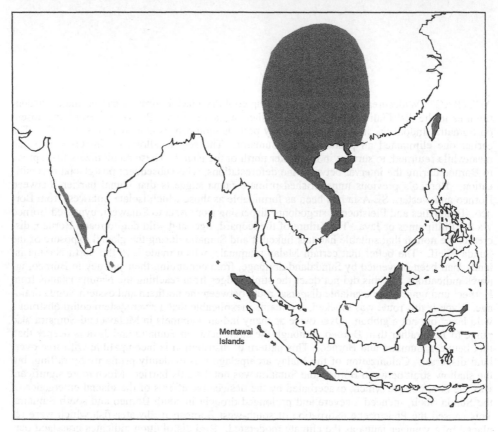

Figure 1. Estimated Asian forest distribution c. 80,000 years ago, based on primate distributions.

Macaca is known at Niah Cave, Sarawak, Borneo from levels predating the most recent glacial maximum (Brandon-Jones, 1996a). This glacial maximum is an unconvincing potential cause of the deforestation because there is no sign that it was milder than its predecessor. The aspect of the most recent glaciation which does seem less abrupt than that of its predecessor is its onset (Martinson *et al.*, 1987). The deforestations therefore appear correlated with the start of the two most recent glaciations at about 190,000 and 80,000 years ago. Near Kuala Lumpur, Malaysia, pine-grassland savannah replaced lowland rainforest, as grassland replaced woodland/fern in east Indonesia, 190,000-130,000 years ago. An exceptionally dry culmination to this period gave way to warm humid interglacial conditions from 126,000-81,000 years ago. Then in the Bandung plain, west Java, open grass-and-sedge dominated swamp vegetation replaced freshwater swamp forest, while *Asplenium* ferns declined in the nearby mountains. Fern resurgence 74,000-47,000 years ago, and scarcity from 47,000-20,000 years ago, signals a return to somewhat warmer conditions, followed by a distinctly cooler climate whose effects are evident in caves at Niah, Vietnam and the Philippines, when west Javan temperatures fell by 4-7°C, and grassland reappeared in east Indonesia. This glacial maximum may have finally annihilated the giant pangolin (*Manis palaeojavanica*) in north Borneo, but its dietary specialization on termites, presumably those building colossal mud edifices, would make it equally vulnerable to dense forestation. Its Middle Pleistocene Javan contemporaries corroborate this indication of extensive open (wooded) country in lowland north Borneo, 40,000 years ago. Such is clearly the natural habitat of the proboscis monkey (*Nasalis larvatus*) which, unlike the giant pangolin, has managed to cope with the vegetational transformation (Brandon-Jones, 1996a, 1998).

The north Sumatran recolonizing *Presbytis* and (more equivocally) *Hylobates* species morphologically more closely resemble their congeners in north Borneo than they do those geographically intervening. Parallel regional differences do not affect *Macaca* or *Semnopithecus* species. This indicates that *Presbytis* and *Hylobates* colonized Borneo during the interglacial, while *Macaca* and *Semnopithecus* first arrived when they recolonized Sumatra after the second deforestation (Brandon-Jones, 1996a, 1998). Bornean primates further diversified as rainforest regeneration subsequently enabled other species to evolve *in situ* or to invade (Brandon-Jones, 1996b, 1998). This implies that gibbons (*Hylobates*) and the monkey genus *Presbytis* were absent from Borneo until about 100,000 years ago. Macaques, the spangled leaf monkey (*Semnopithecus auratus*) and probably lorises and tarsiers, arrived less than 80,000 years ago (Brandon-Jones, 1998).

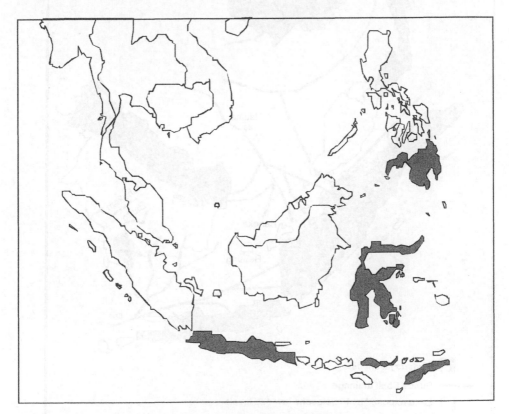

Figure 2. The known Pleistocene distribution of stegodonts in SE Asia.

Before this influx, Borneo evidently had an impoverished primate fauna, possibly comprising only the proboscis monkey, with probably the orang-utan (*Pongo pygmaeus*). This belies the impression from its equatorial situation and relatively large landmass (it is the third largest island in the world), that Borneo has been the permanent heartland of SE Asian everwet rainforest. Instead it suggests that, until then, faunal barriers between Borneo and western SE Asia were as formidable as those which isolate Sulawesi from Borneo. Access to Borneo was difficult not only for primates. Butterflies (Vane-Wright, 1991) and Pleistocene stegodonts (Fig. 2; Hooijer, 1975; Simpson, 1977) seem obliged to have found another route from Asia via the Philippines or Java to Sulawesi. Previously accepting that the glacial exposure of the Sunda Shelf enabled Malayan mammals to colonize Borneo (Banks, 1949), Banks (1976) was disconcerted by the Bornean absence of the tiger (*Panthera tigris*), leopard (*P. pardus*), dhole (or wild dog, *Cuon alpinus*) and Javan rhinoceros (*Rhinoceros sondaicus*). Cranbrook (1988) recorded the fossil presence there of three of them: the tiger, dhole and Javan rhino, and another currently absent large mammal, the Malay tapir (*Tapirus indicus*). The tiger, however, represented only by a deciduous canine, may have been imported as a human ornament, as could the conceivably domestic or feral dog, known there only from a (rather phallic) calcaneum and incomplete canine. The failure of the leopard, tiger and wild dog to reach Borneo, discredits the notion that suitable habitats glacially linked it and Sumatra, as does the absence on Sumatra of such Bornean endemics as the proboscis monkey and the monotypic ground squirrel genus *Rheithrosciurus*.

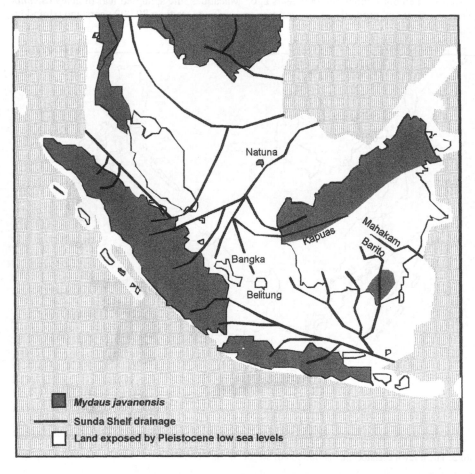

Figure 3. Mammal distributions relative to the glacial Sunda drainage. (River systems and exposed land after Verstappen, 1975.)

Banks (1949, 1976) conjectured that the black giant squirrel (*Ratufa bicolor*) and the wild boar (*Sus scrofa*) of the Malay peninsula, which accomplished the 560 km journey to the north Natuna islands, had been prevented from making the further 160 km to Borneo, by the north Sunda drainage which developed during the exposure of Sundaland (Fig. 3). The colobine monkey, *Presbytis siamensis* similarly reached Great Natuna island, but not Borneo (Fig. 3; Brandon-Jones, 1996a, 1996b). Banks' (1949, 1976) surmise ignores indications that the stink badger (*Mydaus javanensis*) appears unhindered in reaching the Natuna islands from the opposite direction. The stink badger occurs in Borneo, Java and Sumatra, but not in the Malay peninsula (Fig. 4; Corbet and Hill, 1992). Even if it is contended that the potential alternative dispersal route between the northern and eastern Sunda drainage was somehow blocked, it remains inexplicable that a river system could obstruct a wild boar, but not for example, a gibbon. Banks (1949) reported that over twice as many (twelve against five) Indian mammals in Malaya and Sumatra also range to Java rather than Borneo, although the strait between Sumatra and Java is deeper than that between Sumatra and Borneo.

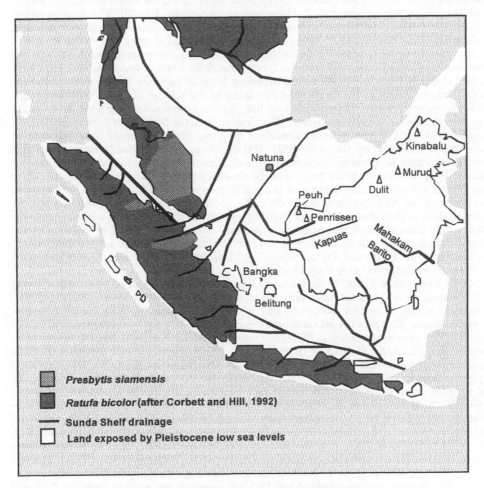

Figure 4. The distribution of the stink badger (after Corbett and Hill, 1992) relative to the glacial Sunda drainage. (River systems and exposed land after Verstappen, 1975.)

This pattern of distribution is incompatible with free overland dispersal. The clear influence of a sweepstake element, also contributing for example, to the haphazard island distribution of mammals and birds in the South China Sea (Banks, 1962), is compelling evidence that most colonization of the Malay archipelago occurred by rafting. The predominance of eastward over westward primate dispersal (Brandon-Jones, 1998) even intimates that it was assisted by a prevailing westerly current or wind direction, although other factors, such as access to a river bank from which floating islands were being detached, probably contributed to this directional bias. Much of this primate rafting probably emanated from rivers along the east Sumatran coast where it was still possible for Chasen (1940, p. viii) to see vegetation-covered floating islands, broken from river banks by floods in the 1920's or 1930's. The feasibility of rafting as a significant means of dispersal is discussed by Brandon-Jones (1998) and Smith (this volume). Brandon-Jones (1998) noted that animals such as proboscis monkeys, tigers and wild boars, are ironically disadvantaged in marine dispersal by their swimming ability which would encourage them to abandon their rafts as these are swept out to sea.

The greater distance from Sumatra to Borneo, may therefore have favoured dispersal to Java, but cannot be the sole cause of the difficulties facing potential Bornean colonizers. Much more problematic was the cool dry glacial climate which, exacerbated by the desiccative effects of the emergence of the Sunda Shelf (Steenis, 1935), probably prevailed for longer in southern than in north Borneo, presenting potential colonizers with an inhospitable landfall. This more plausibly explains the similarity between the southwest Bornean and east Sumatran freshwater fish than Molengraaff and Weber's (1921) belief that it reflects a shared river system during the glaciations. Beaufort (1951) noted the inconsistent lack of an equivalent relationship between the river faunas of south Borneo and north Java, even though the exposed Javan sea floor must have drained similarly (Fig. 4). The Barito and Mahakam rivers of southeast Borneo share an Eocene faunal element absent in the Pleistocene fish fauna of the Kapuas and east Sumatran rivers, while the more northerly situated Mahakam has more in common with the north Javan fauna than does the Barito. More Pleistocene fish are known from the Barito than from the Mahakam, suggesting that it bordered a central Sundaland dry zone where a younger fauna replaced an older one exterminated by glacial drought. This dry zone, which at glacial extremes perhaps even deteriorated to grassland-encompassed desert, can explain the faunal impoverishment (Banks, 1962) of such relatively large, centrally-situated islands as Bangka and Belitung. Bird distribution indicates grassland persisted in south Borneo after it had succumbed to forestation elsewhere in SE Asia (Stresemann, 1939). The north Javan fish fauna which contains no younger element (Beaufort, 1951), was probably replenished from west Java which escaped the worst of the drought. Such replenishment from east to west Borneo would be hindered by the central mountains. It is likely that gibbon taxa, which replace one another approximately along the courses of the Barito and Kapuas rivers (Fig. 5), are similarly revealing the area they postglacially reinvaded, after being glacially excluded.

Mount Kinabalu in Sabah, Malaysia, north Borneo, is a major focus of biodiversity (Wong and Phillipps, 1996). It is far richer than Timor in Australian temperate plants (Steenis, 1935). Prominent among numerous indications that it was the principal Bornean glacial refugium (Brandon-Jones, 1996a) is the presence on Christmas Island of a lowland subspecies of one of only two bird species restricted to 2000-3000 metres on Mount Kinabalu; and the nearest relatives of the other in the Philippines (Banks, 1937). The refugium probably extended, perhaps discontinuously, to Mount Dulit (Sarawak, Malaysia). Two mammal genera, the lesser gymnure *Hylomys* and the ferret-badger *Melogale*, now restricted in Borneo to this area (Fig. 5), are known from fossil evidence to have preglacially occurred in the Bornean lowlands (Medway, 1964, 1972). If their present distribution was the only part of their former range to remain habitable during the most recent glacial maximum, other animals and plants have probably similarly become "montane". The colobine monkey, *Presbytis comata everetti* with much the same Bornean distribution as *Hylomys*, is a seemingly montane subspecies only because its descendants have altered as they postglacially reoccupied adjacent areas (Brandon-Jones, 1997). *Hylomys* and *Melogale* in contrast, have evidently ceased Bornean dispersal since the glaciation.

The few distinct montane mammal and bird subspecies on the isolated peaks of Peuh and Penrissen on the western Kalimantan-Sarawak frontier (Chasen, 1935, 1940) indicate the existence there of subsidiary refugia. In five selected mountains, highland mammal and bird taxa decrease from Kinabalu through Dulit and Murud, to Peuh and Penrissen (see Fig. 3; Banks, 1933). A possible explanation is that several refugia survived the onset of the most recent glaciation, while only one Bornean refugium survived the more abrupt onset of the penultimate glaciation. The relative impoverishment of the Peuh and Penrissen refugia might reflect the failure of some species to interglacially migrate from the principal refugium. An alternative explanation is that this relative impoverishment reflects the relatively drier climate in this part of Borneo (close to the Kapuas river), which allowed only a relatively degraded class of forest to survive there, and therefore a less ecologically diverse fauna. An investigation into the ecology of these subspecies could probably discriminate between these hypotheses. Banks (1933) and Chasen (1935, 1940) did not supply such an analysis. Only if it was discovered that there is no close systematic relationship between the subspecies in different refugia, would one be forced to conclude that more than one refugium survived throughout.

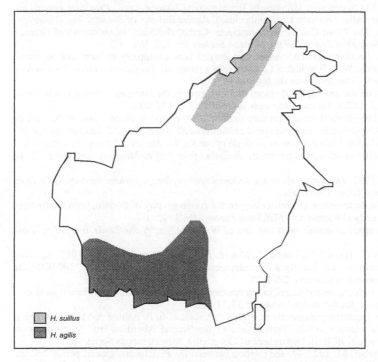

Figure 5. The distributions of *Hylomys suillus* (after Corbet and Hill, 1992) and *Hylobates agilis* in Borneo.

The inhibiting effect on Bornean colonization of a central Sundaland glacial climate generally perhaps resembling that of the present-day dry zone in north Burma, would have abated only gradually as the climate moderated. By the time precipitation had risen sufficiently to enable rainforest to reattain the south Bornean coast, potential colonizers had first to negotiate the sea crossing. Faunal and floral distributions indicate that before the last interglacial, such moist climates may have been infrequent, although even a coastal scrubland belt would impede rainforest flora and fauna. Fauna and flora rafting from east Sumatra may have floated past such inhospitable terrains in southern Sumatra, only to be foiled by similar ones in south Borneo.

ACKNOWLEDGEMENTS

I wish to thank an anonymous referee for constructive comments on the manuscript, and my wife, Chris Brandon-Jones, for preparing the figures.

REFERENCES

Banks, E. 1933. The distribution of mammals and birds in the Sarawak and adjacent parts of Borneo. *Proceedings of the Zoological Society of the London* 1933: 273-282.
Banks, E. 1937. The distribution of Bornean birds. *Sarawak Museum Journal* 4(15): 453-496.
Banks, E. 1949. *Bornean Mammals*. Kuching: Kuching Press.
Banks, E. 1962. The distribution of mammals and birds in the South China Sea and west Sumatran islands. *Bulletin of the National Museum, State of Singapore* 30: 92-96.
Banks, E. 1976. More mammals from Borneo. *Brunei Museum Journal* 3: 262-273.
Beaufort, L.F. de. 1951. *Zoogeography of the Land and Inland Waters*. London: Sidgwick and Jackson.
Brandon-Jones, D. 1993. The taxonomic affinities of the Mentawai Islands sureli, *Presbytis potenziani* (Bonaparte, 1856) (Mammalia: Primates: Cercopithecidae). *Raffles Bulletin of Zoology* 41: 331-357.
Brandon-Jones, D. 1996a. The Asian Colobinae (Mammalia: Cercopithecidae) as indicators of Quaternary climatic change. *Biological Journal of the Linnean Society* 59: 327-350.
Brandon-Jones, D. 1996b. *Presbytis* species sympatry in Borneo versus allopatry in Sumatera: an interpretation. In D.S. Edwards, W.E. Booth & S.C. Choy (eds.), *Tropical Rainforest Research - Current Issues*. Dordrecht: Kluwer. *Monographiae Biologicae* 74: 71-76.
Brandon-Jones, D. 1997. The zoogeography of sexual dichromatism in the Bornean grizzled sureli, *Presbytis comata* (Desmarest, 1822). *Sarawak Museum Journal* 50(71): 177-200.
Brandon-Jones, D. 1998. Pre-glacial Bornean primate impoverishment and Wallace's line. In R. Hall & J.D. Holloway (eds.), *Biogeography and geological evolution of SE Asia*: 393-403. Leiden: Backhuys.
Chasen, F.N. 1935. A handlist of Malaysian birds. *Bulletin of the Raffles Museum, Singapore* 11: 1-389.
Chasen, F.N. 1940. A handlist of Malaysian mammals. *Bulletin of the Raffles Museum, Singapore*, 15: 1-209.
Corbet, G.B. & Hill, J.E. 1992. *The mammals of the Indomalayan region: a systematic review*. Oxford: Natural History Museum/Oxford University.
Cranbrook, Earl. 1988. The contribution of archaeology to the zoogeography of Borneo, with the first record of a wild canid of early Holocene age. *Fieldiana Zoology* (N.S.)42: 1-7.
Hooijer, D.A. 1975. Quaternary mammals west and east of Wallace's line. *Netherlands Journal of Zoology* 25: 46-56.
Martinson, D.G., Pisias, N.G., Hays, J.D., Imbrie, J., Moore, T.C., Jr. & Shackleton, N.J. 1987. Age dating and the orbital theory of the Ice Ages: Development of a high-resolution 0 to 300,000-year chronostratigraphy. *Quaternary Research* 27: 1-29.
Medway, Lord. 1964. Post-Pleistocene changes in the mammalian fauna of Borneo: Archaeological evidence from the Niah caves. *Studies in Speleology* 1: 33-37.
Medway, Lord. 1972. The Quaternary mammals of Malesia: A review. In P. Ashton & M. Ashton (eds), *The Quaternary Era in Malesia*: 63-83. Transactions of the Second Aberdeen-Hull Symposium on Malesian Ecology. University of Hull, Department of Geography, Miscellaneous Series 13.
Molengraaff, G.A.F. & Weber, M. 1921. On the relation between the Pleistocene glacial period and the origin of the Sunda Sea (Java- and South China-Sea), and its influence on the distribution of coral reefs and on the land- and freshwater fauna. *Proceedings of the Section of Sciences. Koninklijke (Nederlandse) Akademie van Wetenschappen te Amsterdam* 23: 395-439.
Simpson, G.G. 1977. Too many lines: the limits of the Oriental and Australian zoogeographic regions. *Proceedings of the American Philosophical Society* 121: 107-120.
Smith, J.M.B., this volume.
Steenis, C.G.G.J. van. 1935. On the origin of the Malaysian mountain flora. 2. Altitudinal zones, general considerations and renewed statement of the problem. *Bulletin du Jardin Botanique de Buitenzorg* (3)13: 289-417.
Stresemann, E. 1939. Die Vögel von Celebes. *Journal für Ornithologie, Leipzig* 87: 299-425.
Vane-Wright, R.I. 1991. Transcending the Wallace line: do the western edges of the Australian Region and the Australian Plate coincide? *Australian Systematic Botany* 4: 183-197.
Verstappen, H. Th. 1975. On palaeo climates and landform development in Malesia. In G.-J. Bartstra & W.A. Casparie (eds), *Modern Quaternary Research in Southeast Asia*: 3-35. Rotterdam: Balkema.
Wong, K.M. and Phillipps, A. (eds). 1996. *Kinabalu: summit of Borneo, a revised and expanded edition*. Kota Kinabalu: the Sabah Society.

Modelling Divergence, Inter-breeding and Migration: Species Evolution in a Changing World

Charles Oxnard & Ken Wessen

Department of Anatomy and Human Biology, University of Western Australia.

ABSTRACT: This paper presents the results of a series of mathematical models of species and subspecies divergences over time. These models have been carried out so that factors like variations in extinction and fossilisation rates, lineage splitting and interbreeding rates, and migration and back-migration rates can be studied. The results imply that: fossils probably lie more frequently on extinct lineages than usually assumed, common ancestors are probably much further back in time than generally accepted, and migrations, as components of evolution, have occurred far more frequently and much earlier in time than is generally thought. To the degree that such models have verisimilitude, the findings have implications for fossil studies attempting to provide data relating to time and place of common ancestry, and for molecular evolutionary studies relying on assessments from fossil investigations in order to calibrate time and place of origin of groups. The models here used are more complex and include a greater number of factors than those used in the past. In particular they suggest that the commonly accepted assessments of the ancestor of humans and chimpanzees at 5 million years ago (Ma) and modern human origins at 150,000 years ago (Ka) need to be viewed with caution. They therefore confirm suspicions about origins as voiced by a number of molecular evolutionary biologists in recent years. The findings may have implications far more widely in biology for assessing common ancestry of species than just in human evolution. They may further provide better information about the complexity of "characters" and therefore improve their uses in cladistic analyses.

1 FOSSILS AND MOLECULES AND TIME

Calibrations of time in many molecular evolutionary studies depend, at some point or other, upon times obtained from studies of fossils. For instance, molecular investigations of the human/chimpanzee divergence are often calibrated from the oldest ape fossils that are assumed to be a little more recent than the time of the common ancestor of all great apes and humans. Likewise molecular assessments of the origins of modern humans are often calibrated from the oldest australopithecine fossils that are assumed to be a little more recent than the time of the common ancestor of humans and chimpanzees. Though such assessments depend upon estimations based upon fossil data, the actual fossil common ancestors themselves are rarely known.

For example, the common ancestor of all hominoids is usually set at circa 25 million years ago (Ma), that is, slightly earlier than the earliest known hominoid fossils (eg *Motopithecus* at 20 Ma, Gebo, 1997, though this is controversial). Likewise, the common ancestor of humans and chimpanzees is usually set at 5 Ma, that is slightly earlier than the earliest known prehuman ancestors, particular australopithecines (eg *Ardipithecus ramidus* at 4.4Ma, White, Suwa and Ausfaw, 1994, 1995); pre-chimpanzee fossils at about these times have never been recognised! Both of these assessments depend upon phylogenetic (including phylogenetic cladistic) evaluations of fossil lineages, and both involve recognition of migrations relating to environmental (geological, climatological, ecological, etc) changes.

2 AN INITIAL ATTEMPT AT MODELLING SPECIES EVOLUTION

In order to study some of the assumptions implied in assessing times of common ancestry from fossils, Oxnard (1995) used a simple model of evolutionary species divergence. This model examined the difference between phylogenies determined only from model "fossil" and "living species", and phylogenies based upon the totality of modelled species (fossil, unfossilised and living). It took into account, therefore, the implications of the effects of data that are missing because of the species that were never fossilised. The results implied that the patterns of evolution, based on all species, are very different from those based on fossil and extant species alone. The fossil-based reconstructions tended to greatly underestimate the distance in time back to the most recent common ancestor of living taxa.

In an elaboration of that study, Oxnard (1997) attempted to model the divergences of subspecies, ie a level of grouping that permits interbreeding (union) of groups as well as divergence, and where occasional "migrations" across "barriers" were permitted. In this case, when the living forms only were compared (usually the case in molecular studies) it was found that most migrations could not be identified.

A further model (also in Oxnard 1997) was undertaken to evolve populations (within a subspecies) with migrations between "continents". Identification of "mothers of mothers" (compare with mtDNA studies) and of "fathers of fathers" (compare with Y chromosome investigations) in this model produced results that placed the "mother of us all" on one continent, and the "father of us all" on another, and at a different time horizons. This is a result that, with tongue in cheek, requires not only a space machine but also a time machine. In parenthesis, this was exactly a study of mothers of mothers, and fathers of fathers; in this sense it does not replicate the molecular studies. The molecular studies, in spite of information in the popular press using such terms as "African Eve", do not provide information about the evolution of individuals, but about the evolution of molecules. As Ayala (1995) has so succinctly pointed out, these are not the same.

Such results imply need for caution in assessing time and place of origins based on fossils. Further, to the degree that the molecular studies of the living are calibrated using such fossil information, caution is also needed in evaluating the molecular studies.

Of course, these initial models were performed manually and thus carried out so few times that it was entirely possible that the results, coherent though they seemed to be, might be spurious. In addition, the simplicity of the models meant that they were, in reality, rather far removed from what might actually happen during real evolutionary events. However, because the results were so contrary to the conventional wisdom, it seemed that it might be worthwhile modifying the models, using more naturalistic assumptions, and carrying out large numbers of runs in order to obtain statistical descriptions of the phenomena.

3 THE CURRENT STUDY

The present study thus expands upon those models using computer implementations that allow thousands of simulations. The models are currently available in two distinct though related contexts: a non-interbreeding (species) evolutionary tree, and an interbreeding (subspecies) evolutionary tree. Both can integrate degrees of migration across barriers.

The computer model contains sub-routines that enable generation, extinction and fossilisation rates to be varied in different ways. The implications of changing these rates can thus be studied. For example, changing these rates in certain ways during the "evolving process" produces different shapes of the distribution of species over time. Some models have been run with generation, extinction and fossilisation rates that produce species distributions that are "bowl-shaped", that is, starting from a single species, the numbers of species are allowed to gradually increase over time. This simulates what might happen early in a radiation. Other models have used generation, extinction and fossilisation rates that produce "vase-shaped" distributions, that is, where the numbers of species show a gradual increase at earlier times and then a gradual decrease later. This simulates what might happen when a radiation has stabilised and then started to reduce through increased competition from species in other parallel radiations. These are both species shape distributions that are well known in the fossil record.

But a set of models of particular importance in relation to the evolution of many mammals, especially many primates of the present day, is where the generation, extinction and fossilisation rates are modified to produce "amphora-shaped" distributions. These are distributions with an initial increase in species numbers, as at the beginning of a radiation, followed by a gradual decrease in species numbers, as the radiation stabilises and niches fill, followed yet later by a long period of marked decrease in species numbers (the "neck of the amphora"). This results in a marked reduction of species at the most recent times. It models the situation in many primates and is especially the case for the hominoids of today. At the present time there are only a handful of hominoid species; in times past the fossils indicate that there have been much larger numbers.

The simulated species "evolve" through changes in "characters" (together with decisions to split, continue unchanged or go extinct). The characters for each species currently consist of two sets. One set includes the characters that decide the make-up of the next species-generation (therefore "hereditary" characters). The other set comprises characters that can also change, but whose changes do not contribute to the determination of the species in the next generation ("non-hereditary" characters). In the determination of the "real" phylogenies, only the "hereditary" characters in all species (living, fossil, and non-fossilised) are used. But the determinations of "reconstructed" phylogenies are based upon both types of characters but only from the fossil and living species.

In the models where interbreeding is allowed, the permitted outcomes are splitting, continuation unchanged, extinction and union (interbreeding). As with the other parameters, interbreeding rates between groups can be varied.

In both studies migrations can be allowed between up to four "continents". If migrations can occur then it is also possible for back-migrations to occur; and these can also be permitted or not. Migrations can be limited to specific periods thus mimicking the opening and closing of migration routes. Migration rates and times can be varied.

These computer models have been designed by Wessen and will be reported in detail in forthcoming papers and a thesis.

Many earlier studies (eg, Raup *et al.* 1973, and later, Sepkoski & Kendrick, 1993) also attempted to simulate the evolution of species but used an alternative computer technique, modelling phylogenetic branching as a Markov process with a preset equilibrium diversity, and then studying the taxonomy of the simulated clades. Their approaches differ from the models used in this study because they did not involve fossils, factors and characters, and because they used stochastic processes. Their studies were mainly aimed at understanding the taxonomies of clades, the evolution of higher taxonomic groups.

A very recent extended introduction and review of phylogeography (Avise, 2000) describes, through examination of empirical data, largely but not exclusively from molecular evolutionary studies, some of the principles and processes governing the geographic distributions of genealogical lineages. The work in this paper and that leading up to it (Oxnard 1995, 1997) are attempts to model phenomena like some of those in the schematic representations of Avise (ibid).

4 RUNNING THE MODELS

4.1 *The Species Model*

There were 1,000 simulations in first series of runs of the species evolution model. Figure 1 shows the average shape of the species distribution. The ancestral species is at the base and the extant species at the top, species being represented by open circles. In this series, the values of generation and extinction rates were arranged so as to produce an amphora-shaped distribution, thus modelling what may have happened in hominoid evolution from the time of the ancestral hominoids. Of course, being an average of a large number of runs, no fossils can be shown in the diagram though the average number of fossils over the full series of runs can be calculated and is given.

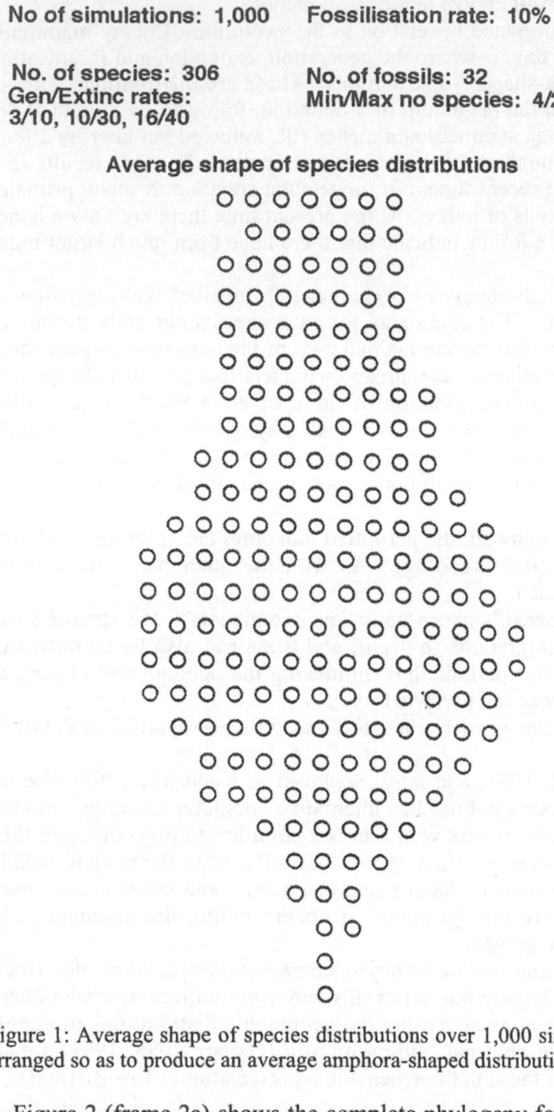

No of simulations: 1,000   Fossilisation rate: 10%

No. of species: 306   No. of fossils: 32
Gen/Extinc rates: Min/Max no species: 4/20
3/10, 10/30, 16/40

Average shape of species distributions

Figure 1: Average shape of species distributions over 1,000 simulations, with the parameters as indicated arranged so as to produce an average amphora-shaped distribution of species over time.

Figure 2 (frame 2a) shows the complete phylogeny for one of the models in this run. Phylogenetic relationships are indicated by the connections between one species and a later species in time. Divergence is indicated where a species gives rise to two species in the next species-generation. Continuation is indicated where a species continues into the next species-generation without a sibling species. Extinctions are indicated where the species has no onward connection in the subsequent species-generation. In these runs divergences were limited to pair-wise splits. (More complex divergences are permitted in the model. They would allow increased variation in the time depth of the model, so that where two pair-wise splits had occurred over the same period as a single different pair-wise split, they would show as a quadruple split assuming that both species split. A triple split would model the situation where one species split and the other did not, and so on). Species that are not chosen for fossilisation are shown as open circles, fossils and extant forms are shown as closed circles.

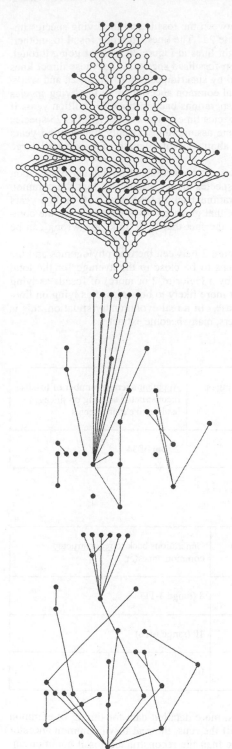

Figure 2: First Frame: One exemplar run from the thousand averaged in figure 1. Evolution of species over 25 species-generations is shown (25 million years if each species is given a time of 1 million years: see text. Species are represented by circles. Closed circles are fossil and extant species. Open circles are species for which fossils remains have not yet been discovered. The model parameters have been constrained so as to produce amphora-shaped distributions.

Second frame: The true links between the living and fossil species obtained by following the radiation of all species from the first frame (this can be checked by superimposition of this frame upon the first. Most of the fossils become extinct. The common ancestor of the living species is near the base of the model.

Third frame: The links visualised by analysing only the living and fossil species. The common ancestor of the living species appears to be very recent. Most of the fossils have links with other fossils or living species. Many fewer fossils are extinct.

Frame 2b shows the actual phylogenetic links between the fossils and the living species implied by the totality of phylogenetic branches in frame 2a. The lines from one fossil to another, and eventually to living species, are shown by straight lines in Figure 2b without going through the many stepped lines in Fig 2a that include the non-fossilised species. That these direct lines summarise the connections correctly can be checked by superimposition of 2b on 2a and working through the complexity of links in 2a. The actual common ancestor of all the living species in this particular run is a fossil that is 16 species-generations back in time (16 million years if we assume that 1 million years is an approximate species time). The entire model is 25 species generation in extent - 25 million years with the same assumption. Twenty five million years was chosen to simulate what is generally posited about the total length of time of all large hominoids.

In contrast is frame 2c of Figure 2. This provides the reconstructed phylogeny based upon analysis of the characters for the living and fossil species alone. It indicates that the common ancestor of all living species is only 6 species generations back in time (that is, 6 million years with the above assumption, a figure not at all unlike that given by fossil estimates for the common ancestor of humans and chimpanzees). Again, the fact that these links are wrong can be checked by superimposition on frames 2a and 2b.

Table 1 indicates that the difference shown in Figure 2 between the true phylogenies and the reconstructed phylogenies for that single run happens to be close to the average for the total 1,000 runs. There is a consistent over-assessment (by a factor of 2 or more) of fossils as lying on lineages leading to extant species. Fossils are far more likely to be assessed as lying on lineages leading to living species than is actually the case. In a real evolutionary situation, this is likely to be true - though some species lead on to others, many become extinct.

Table 1: Number of fossils lying on extant lineages

No of runs	True average number of fossils lying on lineages leading to extant species	Apparent average number of fossils reconstructed as lying on lineages leading to extant species
1,000	11 out of 34	25 out of 34

Table 2: Common ancestor of living species.

No of runs	Generations back to real common ancestor	Generations back to reconstructed common ancestor
52	18 (range 15-23)	8 (range 3-11)
28	0	10 (range 6-16)
20	0	0

Table 2 provides, from a smaller number of runs, more detailed data for timing of common ancestry. This shows that, in slightly more than half the runs, the real fossil common ancestor of all living species was much further back in time than the reconstructed fossil common ancestor. In about another quarter of the runs, the difference between the "real" and reconstructed situations was even greater: there was a reconstructed fossil common ancestor at a quite recent time, where, in reality, there was no common ancestor that was a fossil at all. (This excludes, of course, the occasional example where the starting species happened to be chosen as a fossil; the starting species, by definition, is always an ancestor of all living forms). It was only in a further quarter of the runs that the "real" and reconstructed situations gave the same answer - no common ancestor. But this occurred in those situations where there was no fossil common ancestor. We have not, so far, found an example in which there was a real fossil common ancestor but not a reconstructed one. To this degree, the reconstructions are correct.

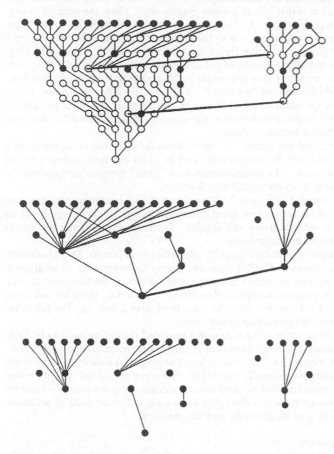

Figure 3: First frame: One exemplar of a model permitting both divergence and inter-breeding of subspecies, and allowing migrations (and back-migrations - although there is not a back-migration in this run) between two "continents". The numbers of groups and group-generations, and continents and migrations, have been kept artificially small so that the links can be visualised. Open and closed circles are defined as in Figure 2. Links between subspecies within continents are show by light lines, links involving migrations between continents by heavy lines. In this example the variables have been constrained so as to produce bowl-shaped distributions.

Second frame: The links between the living and fossil groups from the first frame as calculated from living and fossils groups alone. There is a very recent fossil common ancestor for all subspecies on each continent and a fossil common ancestor on the first continent for all living forms. Only one migration is found; it crosses several subspecies-generations; it is not one of the true migrations shown in the first frame.

Third frame: The actual links between living and fossils groups extracted from the first frame. There are no fossil common ancestors for the living forms on either continent or for the entire set of living forms. The migrations are not visualised because, of course, in this particular example, the migrations did not happen to involve any of the small number of groups that were fossilised. Eight of the 13 fossils become extinct.

At this stage we have not yet carried out runs of species evolution where there were migrations between continents.

4.2 The Sub-species Model

For the second series of runs, involving sub-species evolution, the total number of simulations was again 1,000. Here, there were four "continents" and migrations and back-migrations were permitted among them. However, these are such complex models with so many migrations and back migrations that interpretation is difficult to present graphically. Thus, the concept is presented first with a simpler example in Figure 3. Here there are only 11 subspecies-generations, only two "continents" and only two migrations. With the same conventions as before, Figure 3 (frame 3a) shows the complete phylogenetic tree for all forms, non-fossilised, fossils and living. In this case the pattern of links between sub-species indicates that union (interbreeding) as well as divergence occurred. There were only two migrations on this particular run, and these happened to involve sub-species which were not "fossilised". The number of fossils (shown, as before, by the closed circles) was set again at 10%. However, in this case, the other parameters were set so as to produce a bowl-shaped distribution of sub-species - this, presumably, mirroring more what may have occurred during human evolution.

Figure 3b shows the reconstructed phylogeny. It "finds" three fossil common ancestors, one each for the living groups on each continent separately, and an older common ancestor for all living sub-species on both continents. The reconstruction also "finds" a migration between the continents, although that migration is across several time levels!

This can be compared with frame c of Figure 3. This shows the true relationships amongst the living and fossil forms. There is actually no fossil common ancestor of all living groups on each continent, or of the entire suite of living sub-species. Furthermore, studying the fossils alone excludes information about the real migrations.

Figure 4 shows a full subspecies model involving 25 subspecies-generations, four continents, A, B, C and D, and several migrations and back-migrations among the continents. The diagram shows the total number of subspecies on each continent (as the width of the continent at each time slice). It also shows the migrations between the continents. However, there are too many sub-species to show each ind ivid ual one in a plot like those in Figures 2 and 3). The full information is, however, available from the printed computer output.

This example is one from a large series of such runs that provided the definitive data in Table 3. Thus, when a series of 50 such runs were examined in detail, the reconstructed phylogenies were unable to find the correct migrations. The reconstructed phylogenies usually found a much smaller number of incorrect migrations, usually migrations that were considerably more recent than in actuality, and always migrations that skipped several subspecies-generations. When the modelling has proceeded far enough for some characters to be assigned status such as primitive, shared derived, etc, the study may give insights into cladistic analyses.

Table 3: Number and times of migrations.

No of runs	Average no of real migrations	Median generation of real migrations	Average no of reconstructed migrations	Median generation of reconstructed migrations
50	23	16	12	9

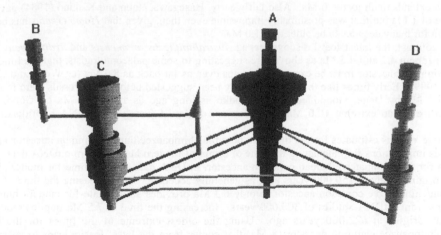

Figure 4: A summary picture of a model of a much larger number of subspecies-generations, with a larger number of migrations involving four continents (A to D). Individual subspecies are not shown, but the relative numbers of subspecies at each time level on each continent are represented by the width of the continental distribution at any particular subspecies-generation level. Links between species within continents are not figured but the links between continents are shown. In this example the subspecies distributions have not been constrained to be bowl-shaped, though this can be achieved. It is from 50 runs of models as complex as this that the data in Table 3 are derived.

5 CONCLUSIONS

The overall conclusion is that, whether employing information based on living species alone, or when information from fossils is included, reconstructed phylogenies consistently show a series of problems.

Reconstructions assess many more fossils as being on lineages leading to living forms than is actually the case. The corollary to this is that reconstructions recognise far fewer fossils as lying on lineages that go extinct.

Reconstructions place common ancestors of all living forms much more recently than is actually the case. In addition, in reconstructions, fossils themselves are commonly mis-identified as common ancestors.

When reconstructions find migrations, the numbers are seriously under-estimated. Those migrations that are found are quite wrongly timed, being postulated as being far more recent than is actually the case.

5.1 *Implications for human evolution*

To the degree that these models are reasonable, they imply the human/chimpanzee common ancestor is likely to have existed considerably earlier than the 5 Ma that is generally used for calibrating molecular studies.

It is true that others have obtained much more recent times for the common ancestor. Thus Hasegawa and Horai, (1991) obtained a figure of 4 Ma, but this was before the discoveries of *Australopithecus anamensis* at about 4.2 Ma and *Ardipithecus ramidus* at about 4.5 Ma. Very early investigators (eg Sarich and Wilson, 1976) obtained 3.0 Ma for the time of the chimpanzee/human common ancestor but that was before the discovery of *Australopithecus afarensis* which extends to almost 4 .0 Ma. Also fairly early, Hasegawa, Yano and Kishino (1984) gave a figure of 1 Ma but that was presumably impossible even then, given that *Homo erectus* has been known for many decades to be older than 1.0 Ma.

In contrast, the latest fossil discoveries, eg *Australopithecus anamensis* and *Ardipithecus ramidus* (about 4.2 and 4.5 Ma as above) are suggesting to some palaeontologists, that the time of the common ancestor must be extended, perhaps even as far back as 8.0 Ma (eg Wood and Collard, 1999). Early times like this had previously been suggested before these fossils were found, eg, 6.3 Ma by Dene, Goodman and Prychodko as long ago as 1976, 7.5 Ma by Goodman (1983), even an extreme 10.0 Ma by Sibley and Ahlquist (using DNA-DNA hybridisation, 1983).

These various estimates for the time of a possible chimpanzee/human common ancestor have serious implications for calculating the time of origin of modern humans from mtDNA data. At 5 Ma for the chimpanzee/human common ancestor, mtDNA data imply a time for modern human origins of about 150,000 years ago, as generally accepted. But increasing the time of the human/chimpanzee common ancestor to only 6.3 Ma provides an estimate for modern human origins from mtDNA studies of 300,000 years. Increasing the time to 7.5 Ma implies modern human origins at 622,000 years ago. Using the oldest extreme of the times for the human/chimpanzee common ancestor (8 Ma) that comes from the latest fossils leads to an even older assessment for modern human origins from the same mtDNA data.

These figures can also be much modified (and to similar degrees) by using different estimates of average population size over time or variable substitution rates for the different sites of the mtDNA genome (Ayala, 1995).

If these re-appraisals of modern human origins from mtDNA seem unlikely, they can be examined alongside various estimates for modern human origins based upon Y chromosome analyses (eg Fu and Li 1996). Though at first sight these estimates seem to confirm the initial figures obtained from mtDNA (eg for modern humans at 31,000-219,000 years = approximately the 150,000 years as obtained from mtDNA but with enormous limits) modifications making allowance for the average number of breeding males changes the figures enormously. Thus, with 10,000 breeding males the mean time of origin is raised to 313,000 years (114,000-721,000). Tripling the number of breeding males to just 30,000 (not really a very large number) raises the average time to 703,000 (284,000-1,507,000 years). The very large limits on these average times are also noteworthy.

Such reassessments would have serious implications for human evolution. For example, they would allow the molecular dates for Neandertalers to be well within the range of new times for modern human origins. They also imply that major reviews may be needed for assessments of more recent migration dates for present day human groups, eg, African populations migrating into Europe and Asia, Asian populations spreading to Australia, and Asian populating of the Americas. It is rather likely that migrations commenced much earlier than usually postulated, were frequent rather than single or rare events, and involved at least intermittent, and possibly, even, fairly continuous, additions to the gene pools of such peoples. The present day gene pools may be far more recent than the ages of the first migrating peoples. All of these possibilities are likely to have occurred during the long course of hominoid and human evolution, together with the geological, climactic, ecological and other changes that went with them.

5.2 *Development of the models*

There is also a different set of conclusions that relates to the modelling methodology. The findings are not due to the models being overly simplistic. In fact, these models include a far greater complexity of factors than it is currently possible to allow for in most fossil or molecular studies. Thus, they permit study of the effects of different extinction and fossilisation rates, of different group splitting and inter-breeding rates, of variations in migration and back-migration rates, and perhaps most exciting of all, of different blends of the hereditary and non-hereditary characters used in producing the relationships.

In relation to this last, work is in progress to model increasing complexity of the "characters". Thus, not only will future models include "hereditary characters" that determine phylogenetic relationships, and "non-hereditary characters" that add noise to the character complexes as at present, but so also will there be characters that display a degree of uni-directional change, that is, change that can continue for a period in a given direction. Such uni-directional changes will also be of two kinds. One kind will model adaptations to (say) continuous uni-directional changes in environment factors for period of time such as an ice age. Such modelling, for instance of characters related to gradually increasing cold, or gradually increasing dryness, might be included as hereditary due to selection. These would thus contribute to the determination of the lineages in the model. A second type will model those kinds of adaptations that are actually ontogenetic responses to environmental factors and which, therefore, do not immediately contribute to the lineages. These latter will provide naturalistic information clouding the reconstructed phylogenies.

The existence of these various types of characters in the models will allow us to study the degree to which cladistic methods of handling living and fossil data can be modified to disentangle the various components of characters. For though cladistics requires that characters be primitive or shared derived, in reality characters are usually observed features containing underlying complexes of characters each of which may be either primitive or shared-derived. As a result, the "character" used in cladistics is presumably only very rarely completely the true primitive or shared-derived character reflecting the gene pool. Much more commonly, it will be a complex mix of primitivenesses and shared-derivednesses. Recognition, through these models, that this may be the more usual situation, may help provide methods, for cladistic analysis, that disentangle or partition the information in such data.

Finally, the models are being further modified to simulate lineages of very large numbers of individuals (thousands of individuals) with the aim of examining implications for factors such as mtDNA and Y chromosome evolution. In addition to the effects of population size and bottlenecks (already known to influence determinations from molecular data) it is also hoped that there will result a better understanding of many other modulating factors. Thus, differences in reproductive sex ratios, mating patterns, degrees of consanguinity, types of migrations (whether largely of males or females or mixed), may all affect estimations based upon the different molecular studies. Their effects can all be studied using these models.

5.3 Broader implications for other studies

In addition, it has not escaped our attention that these models can be used to comment on, even illuminate, many of the matters raised at the conference and in this volume.

For instance, Long and Buffetaut's (2000) biogeographical comparison of dinosaurs and associated vertebrate faunas from the early Cretaceous, implies that dinosaur ancestors may have been present very much earlier in the south than in the rest of the world, a conclusion entirely consonant with the concepts presented in this paper. The models could easily be extended to cover such very much longer periods of time and different situations.

Evans, Morales and Melnick (1999) present molecular evidence for multiple dispersal, migration and hybridisation events in the macaques of Sulawesi and the Sunda Shelf. These are considerations that might be illuminated using these models.

Groves (2000) in studies of subspecific differences in many mammals, especially many primate species, determines that a far larger number of sub-species exist at the present time than has been generally accepted. Models like those described here might give insights into how stable, or how ephemeral, such sub-species may be, and if ephemeral, how difficult to detect them through studies of fossils.

The question of the survival or extinction of the orang-utan as discussed by Kaplan (1999) could be studied using these modelling processes. Up to now we have taken little notice of models in which entire groups became extinct. Perhaps we should look at the statistics of model extinctions; perhaps this might give further insight into extinction situations.

The question of multiple sequential tectonic events, as discussed by Metcalfe (2000) could be studied by appropriate modifications of the "continents" in the models. For example, the effects, on the evolution of species, of a series of tectonic events, producing a series of Tethys seas (early, middle and late), could be examined by allowing migrations, then preventing them, then allowing them again, and so on, in sequence.

The effects, on evolutionary processes, of differences in a variety of reproductive factors could be assessed. These include: (a) differences in sex ratio - most apes have adult sex ratios of 2 or 3 females or more to each male, sex ratios in humans have differed in different groups in the past. They include (b) differences in mating patterns. Contrast average mating patterns in orang-utans, bonobos and humans, together with the variations within each of these. They include (c) differences in types of "migration". In some species males are forced to "migrate" from their natal groups, in others, females. Many other such factors could be investigated.

Finally, as discussed above, these models may allow study of the methods of cladistics. Thus the models may allow assessments of the implications of "getting it wrong" when characters are assessed for cladistic analysis. For instance, the implications of treating a character as primitive or shared-derived, when that "character" is actually an observation that is a compound of several different underlying characters, each in a different state, could be assessed. The effects of the use of different out-groups in cladistic analyses could be determined. It is possible that factors like these could be the reason for some of the controversies that exist in cladistic studies.

ACKNOWLEDGEMENTS

We are most grateful to Dr Len Freedman and Professor F. Peter Lisowski for commenting on the work and manuscripts. The help of two anonymous referees has been most useful. The research is supported by Australian Research Council Large Grants to Professor Oxnard, and by the Centre for Human Biology, Director, Associate Professor James Chisholm, University of Western Australia.

REFERENCES

Avise, J.C. 2000. *Phylogeography: the History and Formation of Species.* Cambridge, Mass: Harvard University Press.
Ayala, F.J. 1995. The Myth of Eve: Molecular Biology and Human Origins. *Science,* 270: 1930-1936.
Dene, H.M., Goodman, M. & Prychodko, W. 1976. Immunodiffusion systematics of the primates. *Folia Primatologica* 25: 35-61.
Evans, B.J., Morales, J.C. & Melnick, D.J. 1999. Macaques in Sulawesi and the Sunda Shelf: a molecular analysis of dispersal, speciation and gene flow. In I. Metcalfe (ed), *Where Worlds Collide: Faunal and floral migrations and evolution in SE Asia-Australasia, Abstracts*: 25-26. Armidale: Asia Centre, University of New England.
Fu, Y-X., & Li, W.H. 1996. Estimating the age of the common ancestor of men from ZFY intron. *Science* 272: 1356-1357.
Gebo, D.L., MacLatchy, L., Kityo, R., Deino, A., Kingston, J. & Pilbeam, D. 1997. A hominoid genus from the early Miocene of Uganda. *Science* 276: 401-404.
Goodman, M. 1983. Macromolecular sequences in systematic and evolutionary biology. New York, Plenum.
Groves C. 2001. Mammals in Sulawesi: where did they come from and when, and what happened to them when they got there. In I. Metcalfe, J.M.B. Smith, M. Morwood & I. Davidson (eds), *Faunal and floral migrations and evolution in SE Asia-Australasia*. Rotterdam: Balkema. (This volume)
Hasegawa, M. & Horai, S. 1991. Time of the deepest root for polymorphism in human mitochondrial DNA. *Journal of Molecular Evolution* 32: 37-42.
Hasegawa, M. Yano, T. & Kishino, H. 1984. A new molecular clock of mitochondrial DNA and the evolution of the hominoids. *Proceedings of the Japanese Academy* 60: 95-105.
Kaplan, G. 1999. Survival, Adaptability and Evolution: the case of the orang-utan. In I. Metcalfe (ed), *Where Worlds Collide: Faunal and floral migrations and evolution in SE Asia-Australasia, Abstracts*: 41-42. Armidale: Asia Centre, University of New England.
Long, J.A. & Buffetaut, E. 2001. A biogeographic comparison of the dinosaurs and associated vertebrate faunas from the Mesozoic of Australia and Southeast Asia. In I. Metcalfe, J.M.B. Smith, M. Morwood & I. Davidson (eds), *Faunal and floral migrations and evolution in SE Asia-Australasia*. Rotterdam: Balkema. (This volume)
Metcalfe, I. 2000. Palaeozoic and Mesozoic tectonic evolution and biogeography of SE Asia-Australasia. In I. Metcalfe, J.M.B. Smith, M. Morwood & I. Davidson (eds), *Faunal and floral migrations and evolution in SE Asia-Australasia*. Rotterdam: Balkema. (This volume)
Oxnard, C.E. 1995. The challenge of human origins: molecules, morphology, morphometrics and modelling. In S. Brenner and K. Hanihara, (eds), series editor, C.E. Oxnard, *The Origin and Past of Modern Humans as Viewed from DNA*. In *Recent Advances in Human* Biology: 11-30. Singapore, New Jersey, London, Hong Kong: World Scientific.

Oxnard, C.E. 1997. The time and place of human origins: Implications from modelling. In G.A. Clark & C.M. Willermet (eds), *Conceptual Issues in Modern Human Origins Research:* 369-391. New York: Aldine de Gruyter.

Raup, D.M., Gould, S.J., Schopf, T.J.M. & Simberloff, D.S. 1973. Stochastic models of phylogeny and the evolution of diversity. *Journal of Geology* 81: 525-542.

Sarich, J.M. & Wilson, A.C. 1976. Rates of albumin evolution in primates. *Proceedings of the National Academy of Science, USA* 58: 142-162.

Sepkoski, J.J. Jr., & Kendrick, D.C. 1993. Numerical experiments with model monophyletic and para-phyletic taxa. *Paleobiology* 19(2): 168-184.

Sibley, A.G. & Ahlquist, J.E. 1984. The phylogenetics of the hominoids as indicated by DNA-DNA hy-bridisation. *Journal of Molecular Evolution* 20: 2-22.

White, T.D., Suwa, G. & Asfaw, B. 1994. *Australopithecus ramidus*, a new species of early hominid from Aramis, Ethiopia. *Nature* 371: 306-312.

White, T.D., Suwa, G. & Asfaw, B. 1995. *Australopithecus ramidus*, a new species of early hominid from Aramis, Ethiopia. *Nature* 375: 88. (letter of correction).

Wood, B. & Collard, M. 1999. The Human Genus. *Science* 284: 65-71.

Early hominid occupation of Flores, East Indonesia, and its wider significance

M.J. Morwood

Archaeology and Palaeoanthropology, School of Human and Environmental Studies, University of New England, NSW, Australia 2351

ABSTRACT: Dates for the arrival of hominids in Southeast Asia and their subsequent dispersal into Wallacea have implications for general models of our evolution, as well as the development of distinctly human traits – intelligence, forethought, co-operation, planning, technology and language. However, crucial parts of the archaeological sequence in the region are not clear. Recent archaeological discoveries on the island of Flores in east Indonesia rectify some of the problems.

1 EARLY HOMINID RESEARCH IN INDONESIA

Research on hominids in Southeast Asia began a little over 100 years ago with the work of Eugene Dubois, a Dutch anatomist, who came to Indonesia in 1887 looking for the missing link in Darwin's theory of human evolution. From 1890 to 1896 Dubois carried out a major excavation at Trinil, a site on the bank of the Solo River in central Java. He discovered a hominid skullcap and thighbone, and named the species *Pithecanthropus erectus* (or Upright ape-man) - since renamed *H. erectus*. Dubois' discovery is the type specimen for the species (Bellwood, 1997).

Over the past 100 years, many other hominid fossil localities have been discovered in Java - e.g. Ngandong, Mojokerto, Sambungmacan, Ngawi, Sangiran (e.g. Jacob 1975; Weidenreich 1951). Of these new localities, Sangiran is the best known and has produced many hominid fossils, including Sangiran 17, the most complete early hominid skull ever found in Java. Work at Sangiran continues to this day.

The Indonesian hominid sequence also shows evidence for *in situ* evolution in the 'modern' direction. Compared to specimens from Trinil and Sangiran, hominid skulls from Ngandong, Ngawi and Sambungmacan are less robust and have larger brains.

Problems with previous research in Java include –

1 The geological complexity of Lower and Middle Pleistocene deposits in central and eastern Java (Ithara *et al.* 1994).

2 The fact that the provenance of many fossils is uncertain.

3 Some of the claimed dates are on materials collected without due regard for site formation processes (Morwood *et al.* 2000).

4 There is minimal associated cultural evidence. In fact, some archaeologists have suggested that *H erectus* in Indonesia did not make stone artefacts (Bowdler 1993).

5   There is a virtual absence of archaeological evidence between 300,000 (?) and the first evidence of modern humans (i.e. *Homo sapiens*) in the region around 40,000 years ago. Song Terus, a limestone cave in central Java which has stone artefacts dated by uranium-thorium series to between 200,000 and 300,000, may herald a change in this situation (Francois Semah: pers. comm). However, the site has not yet yielded any early hominid bone.

As a result of these problems, dates for initial arrival of hominids in the region remain uncertain, as does the relationship between *H. erectus* and modern humans. There may be a genetic link between Javanese *H. erectus* and modern Australoid peoples, as the regional continuity model holds, or *H. erectus* in this part of the world may have been an evolutionary dead end that was replaced by a fresh wave of modern peoples dispersing from Africa about 130,000 years ago - the 'Out of Africa' model (Lahr and Foley 1998; Rightmire 1994). Unexpectedly, evidence from the east Indonesian island of Flores might provide answers to both these issues.

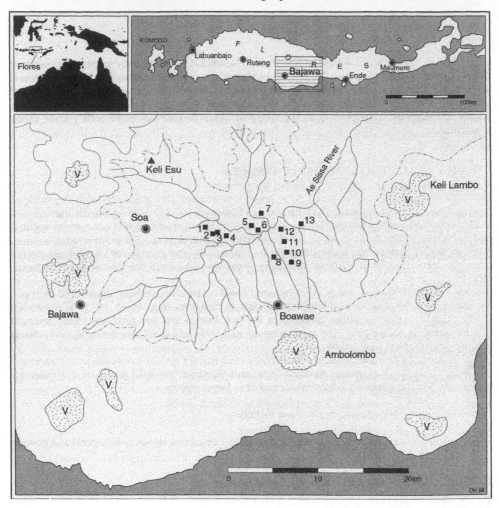

Figure 1. General location of Flores and fossils sites in the Soa Basin, central Flores. 1 Kobatuwa; 2 Mata Menge; 3 Lembahmenge; 4 Boa Leza; 5 Ola Bula; 6 Tangi Talo; 7 Wolo Milo; 8 Wolokeo; 9 Sagala; 10 Dozo Dhalu; 11 Kopowatu; 12 Ngamapa; 13 Pauphadhi. 14 Deko Weko, 15 Malahuma. The dashed "V" areas are volcanoes. The dotted line is the edge of the small basin drained by the Ae Sissa River.

2 GEOGRAPHICAL BACKGROUND

Flores lies midway between the continental areas of Australia (Sahul) and Southeast Asia (Sunda), on the most likely route for colonisation of Australia (Birdsell 1977). Even at times of low sea level at least two sea crossings were required to reach the island from the Southeast Asian mainland. The first and greatest of these deep-water sea barriers, between the islands of Bali and Lombok, was about 20km wide at times of low sea level. The second, between Sumbawa and Flores, was about 9km (Figure 1).

In contrast, the major islands of west Indonesia, such as Sumatra, Java, Bali and Kalimantan, were periodically connected by landbridges to the mainland. But it was never possible to walk further east to the islands of Lombok, Sumbawa, Flores, Timor, Sumba or Sulawesi. We know that there have never been landbridges connecting the islands of east Indonesia to either the Asian or Australian continental areas because, prior to recent human intervention, these islands had very depauperate land faunas. In fact, the sudden drop-off in animal species on islands east of Bali corresponds to a major biogeographical boundary known as Wallace's Line.

Only animals capable of strong swimming, rafting or flying were able to establish themselves east of Wallace's Lines. The Southeast Asian continental mammalian fauna is rich and varied – e.g. pigs, cattle, elephants, tigers, and rhinos - but the only land mammals to colonise islands east of Bali, along the Lesser Sunda Island chain, were elephants, such as *Stegodon*, and rodents. Elephants are large, buoyant, strong-swimming herd animals and are therefore particularly good island colonisers (Johnson 1980), whereas rodents must have crossed on natural rafts of flotsam.

Similarly, the unique nature of Australia's marsupial and monotreme mammals indicates that it has never been connected by landbridges to mainland or island Southeast Asia. Although there are a few marsupials found on east Indonesian islands and some are endemic (e.g. species of cuscus in Maluku), most were almost certainly transported there by humans.

It is generally thought that *H. erectus* populations lacked the capacity to make sea crossings, and that the islands of eastern Indonesia were occupied relatively recently, between 40,000 and 60,000 years ago, by fully modern humans (Davidson and Noble 1992). This general perception is now under serious challenge.

3 HISTORICAL BACKGROUND

Claims for *Homo erectus* on Flores were first made in the 1960s by Fr. Verhoeven, a Dutch priest and amateur archaeologist, who excavated fossil sites in the region. He reported stone artefacts at Mata Menge, Boa Lesa, Lembahmenge and Wolo Milo. He concluded that these sites were about 750,000 years old on the basis of their association in the fossil deposits with *Stegodon*, an ancestral elephant species region (Verhoeven 1968; Maringer and Verhoeven 1970a, b).

The range of evidence presented for his claims was generally judged inconclusive because of doubts about the identification of the stones nominated as artefacts, the uncertain stratigraphic association between these and *Stegodon* fossils, and the age of the strata (e.g. Allen 1991; Bellwood 1985: 66).

Palaeontological research undertaken by an Indonesian-Dutch team replicated Verhoeven's findings. They reported stone artefacts with the remains of *Stegodon* at Mata Menge and, on the basis of preliminary work, reported a similar association at Dozu Dhalu (van den Bergh 1997: 249; van den Bergh *et al* 1996).

Furthermore, their excavation at the stratigraphically older site of Tangi Talo, which lacked associated stone artefacts, revealed that there had been a major turn-over in fauna: Tangi Talo contained the remains of pygmy *Stegodon (S. sondaari),* giant tortoise *(Geochelone* sp.*)* and Komodo dragon *(Varanus komodoensis).* In contrast, Mata Menge, one of the sites containing stone artefacts, had large *Stegodon (S. trigoncephalus florensis),* crocodile and giant rat *(Hooijeromis nusatenggara).* Sondaar (1987) argued that this faunal turnover resulted from the arrival of a new predator, *H. erectus.* Palaeo-magnetic determinations from Tangi Talo and Mata Menge suggested that the former was 900,000 years old and the latter 'slightly less than' 730,000 (Sondaar *et al.* 1994:1260).

Reaction to these claims was muted, and where published, generally cautious (e.g. Bellwood 1997: 67-8). The major impediments to general acceptance of the conclusions were three-fold: the identification of stone artefacts, the lack of taphonomic detail, and the chronological ambivalence of a palaeo-magnetic transition three metres below the Mata Menge fossil/artefact deposit.

More recent work has shown that stone artefacts definitely occur *in situ* at a number of sites in the region, including Mata Menge and Boa Lesa, while a series of fission track dates show that the sites are of Early and Middle Pleistocene age, with the earliest stone artefacts dating to 840,000 BP (Morwood *et al.* 1997, 1998, 1999; O'Sullivan *et al.* 2000).

4 GEOLOGICAL CONTEXT

Evidence for early hominid presence on Flores comes from the Soa Basin on the upper Ae Sissa River (Figure 1). The basin is 40 by 26 km in size and is almost entirely surrounded by mountains and active volcanoes (Figure 2). At present it lies between 220 and 350 metres above sea level. There is one deeply incised river outlet.

Figure 2. General view of the Soa Basin, central Flores, with Minggus Siga, a Nagakeo assistant. (Photo: M.J. Morwood).

The work of Hartono (1961) provided a useful framework for establishing the general and specific geological context of the sites. He identified and mapped three basic stratigraphic units in the study area – the Ola Kile, Ola Bula and Gero Formations. (Localised deposits of recent tuffs and flows derived from the surrounding volcanoes also occur within the basin – but these will not be dealt with here).

The basal material of the Soa Basin, the Ola Kile Formation, is a volcanic laharic breccia and lava andesite with tuffaceous silt and sand intercalations. A date of 1.86 ma from the top of this formation, which is at least 75 metres thick, indicates that it is Pliocene in age. It dips at 7° to the south, while topographic changes from east to west mean that it forms the base of at least two sub-basins.

Unconformably overlying the Ola Kile Formation is the Ola Bula Formation, a series of tuffaceous sandstones and siltstones up to 120 m thick. It is characterised by distal volcanic deposits in the lower part and fluvial deposits in the upper, in which parallel laminations and cross bedding are sometimes evident. Strata of the Ola Bula Formation have basically remained horizontal and they lens out in the northern section of the Soa Basin. Elsewhere at the margins of the basin, it is unconformably overlain by recent volcanics. Dates for this formation range from 0.96 to 0.70 ma.

The lateral extent of some siltstone layers indicates that they may have been deposited in an extensive body of water. For much of its history the Soa Basin may have comprised a large lake or series of lakes, presumably when tectonic activity blocked the outlet. Periodically, however, a new river outlet appears to have been cut and the area drained to become a grassland savanna. Fossil and archaeological sites within the basin only formed during such 'dry-land periods'. At such times processes of erosion would have predominated, but localised tuffaceous sediments accumulated in rivers, creeks and waterholes. Fossils of land animals and stone artefacts occur only in these localised deposits. Sedimentary analysis of the strata at Boa Lesa confirms this general scenario: the fossils and stone artefacts at the site occurred in fluvial deposits in a riverbed, incised into a previously silicified, lacustrine mudstone.

The Ola Bula Formation is capped by the Gero Limestone. This comprises a series of alternating limestones, tuff, calcareous tuff and diatomaceous deposits. The limestone formed in freshwater, although there may also be a marine component. It is up to 40 metres thick, but is now patchy in distribution because of erosion. It appears to have been originally more extensive than the Ola Bula Formation because in the northern part of the basin, beyond the margins of the Ola Bula Formation, Gero limestone directly overlies the Ola Kile Formation. A tuff in the lower levels of the Gero Limestone dates to 0.68 ma, indicating that it is of Middle Pleistocene age.

The former extent and horizontal distribution of Gero Limestone across the Soa Basin suggest that it formed in a large lake. Some tuffaceous siltstone deposits occur within and overlying the Gero Limestone, indicating that small-scale volcanic eruptions continued, but these no longer provided a major source of sediments, as occurred during deposition of the Ola Bula Formation. Fossils of algae, fish and freshwater gastropods occur in the Gero Limestone but not those of land animals, which would have been confined to the lake margins. Land animals would have reoccupied the basin after the Ae Sissa River outlet formed and the lake drained.

An important point is that, except for the northern part of the Soa Basin, where a thrust fault has resulted in a dip to the south, the Ola Bula Formation and the Gero Limestone have remained horizontal. The relative heights of fossil sites within the Ola Bula Formation are, therefore, generally indicative of their relative ages, as confirmed by the fission track dates for the sites, as well as with previous age estimates obtained on the basis of fossil fauna, presence of tektites and the results of paleomagnetic determinations (Koenigswald 1957; Maringer and Verhoeven 1970a; Sondaar et al., 1994).

5 THE PROJECT

Joint Indonesian-Australian research on the Soa Basin fossil sites began in 1997. Participating institutions comprise the Indonesian Geological Research and Development Centre (GRDC), the Indonesian National Centre for Archaeological Research, Gadjah Madah University, the Indonesian Marine Geological Institute, the School of Earth Sciences at the University of Melbourne and the University of New England

The project, 'An archaeological and palaeontological investigation of sites in the Ola Bula Formation, central Flores', has now been funded for three years by the Australian Research Council. It is explicitly multi-disciplinary in emphasis involving archaeologists, geologists, palaeontologists, palynologists and local people.

We have now mapped the entire sedimentary basin at 1:50,000 scale, recorded a large number of column sections across the basin, collected samples for dating and analysis, and obtained 16 fission track dates for various stratigraphic units (O'Sullivan et al. 2000). We have also re-

corded 15 fossil sites within the Ola Bula Formation and undertaken excavations at 5 of these –
Tangi Talo, Dozu Dhalu, Kopowatu, Boa Lesa and Mata Menge (Figure 3; Morwood *et al*
1999).

Figure 3. Excavating remains of large *Stegodon* at Dozu Dhalu. The deposits are 850,000 years old. No *in
situ* stone artefacts were found (Photo: M.J. Morwood).

6 HISTORY OF THE SOA BASIN

So far four major dryland periods have been identified in the Soa Basin, three of which are associated with Pleistocene fossil sites. The dryland periods comprise -

Period 1: Associated with deposition of fossils at Tangi Talo 900,000 years ago. No *in situ* stone artefacts were recovered in excavations at the site. Fauna present included the pygmy *Stegodon sondaari*, giant tortoise *(Geochelone* sp.*)* and Komodo dragon *(Varanus komodoensis)* (Figure 4).

Figure 4. The endemic fauna found on Flores about 900,000 years ago included Pygmy *Stegodon*, Giant tortoise (*Geochelone* sp.) and Komodo dragon (*Varanus komodoensis*). The Komodo dragon depicted is about 3 metres in length.

Figure 5. By 850,000 BP many of the endemic species on Flores had become extinct and the Pygmy *Stegodon*, had been replaced by the full-sized *Stegodon trigonocephalus florensis,* which stood about 3 metres high at the shoulder. No stone artefacts have been recovered from sites of this age.

Period 2: Associated with deposition of fossils at Sagala, Dozu Dhalu and Ola Bula 850,000 years ago. Fauna present at the time included the large *Stegodon trigoncephalus florensis*, Komodo dragon, crocodile and giant rat (*Hooijeromis nusatenggara*) (Figure 5). No *in situ* stone artefacts were recovered from excavations at Ola Bula, Dozu Dhalu or Kopowatu.

Period 3: Associated with deposition of fossils at Boa Lesa, Mata Menge, Lembahmenge, Kobatuwa, Ngamapa, and Pauphadhi between 840,000 and 700,000 years ago. Fauna present at the time included large *Stegodon* and rodents. *In situ* stone artefacts are found at all these sites (Figure 6).

Figure 6. A stone artefact recovered *in situ* during excavations at Mata Menge. This basalt flake has a striking platform, bulb of percussion, two flake scars on the dorsal surface and retouch on both lateral margins.

Period 4: The present dryland period. Dates for this are currently being processed. This is associated with cultural deposits at Keli Esu Cave, Lowo Labo Shelter, Lio Husa Shelter, Ola Bula Village and other very recent sites. The presence of pottery in the basal levels of these sites indicates that they date to the last 4000 years at most. Local Nagakeo and Ngadha people still use some of the sites as temporary bases when hunting. An ethnoarchaeological study of the Nagakeo has been completed and excavations are now planned for associated sites (Figure 7; Sudarmadi 1999).

Figure 7. A Ngadha ceremony in progress at Bena village, central Flores. (Photo: Tular Sudarmadi).

In summary, our work has yielded unambiguous and relatively precise dates for the arrival of *H. erectus* on Flores by 840,000 BP. Artefacts are not present in deposits slightly older. It is also evident that specific geological circumstances have resulted in a depositional and cultural hiatus from about 600,000 BP to the Holocene. Adjacent geomorphic regions with the potential to document different parts of the sequence are currently being investigated.

7 IMPLICATIONS

Our results now provide the most reliable date for the initial appearance of hominids in an area of Southeast Asia. In contrast, dating the earliest of the well known fossil hominids in Java is more difficult because the region is much more geologically complex, and many of the deposits and fossils have clearly been redeposited (c.f. Hyodo *et al* 1993; Suzuki *et al* 1985). The Flores date is the same as that for the earliest evidence of hominids in Europe (Lahr and Foley 1998).

It is significant that evidence for arrival of hominids on the island post-dates a major change in the island's fauna, involving the extinction of two endemic species, the pygmy *Stegodon* and the giant tortoise sometime between 900,000 and 850,000 years ago. The extinction most likely relates to natural events such as volcanic eruption or drought rather than the impacts of early hominids. The chance of an animal population going extinct on an island like Flores must have been high.

There is also the question as to how hominids managed to reach Flores by 840,000 years ago. There are two possible explanations –

1. Temporary landbridges connected Wallacean islands to Sunda – or at the very least the crossing from Bali to Lombok and Sumbawa to Flores were much easier than today (Groves 1996).
2. Early hominids used some type of watercraft.

Concerning the first possibility, the relative coarseness of specific geological data means that the fossil biogeographical data is most relevant. In the Lower and Middle Pleistocene Java was virtually a Garden of Eden with great biodiversity and an abundance of terrestrial animals (e.g. pigs, deer, bovids, rhinos, hippos, elephants, feline carnivores, rodents, tortoises, canids, monkeys, apes, and hominids). At the same time Flores had a highly endemic fauna - Pygmy *Stegodon*, Komodo dragon and giant tortoise, and later Large *Stegodon*, Komodo dragon, rodents and hominids.

Basically, if an animal was not good at swimming, rafting, flying or being transported by wind, it did not reach Flores in the Lower or Middle Pleistocene, and this was beyond the capabilities of most Asian terrestrial species. Even animals which are relatively good swimmers, such as pig and deer, did not make it until much later with human intervention; nor did many animals which are much more likely than hominids to have travelled on natural rafts, such as monkeys (c.f. Smith 2001). Environmental differences between the Lesser Sunda islands and Sunda may be a factor here – but today pigs, deer and macaque monkeys all flourish on Flores.

8 CONCLUSIONS

I conclude on the basis of fossil biogeography that the water gaps in the Lesser Sunda Island chain represented major obstacles to terrestrial animal dispersal in the Lower and Middle Pleistocene. The appearance of hominids on Flores at this time therefore represents a remarkable achievement and one that must have required use of watercraft.

The biogeographical barrier represented by the Wallace Line and the dispersal of hominids across this barrier in the Lower Pleistocene indicate that *H. erectus* had the intelligence to make sea journeys. More specifically, the colonisation of Flores in the Early Pleistocene suggests that the technological capacity of early hominid populations may have been seriously underestimated. The recent discovery of well-designed, well -crafted and possibly composite spears at the 400,000 year old Schoningen kill site in Germany confirms this view (Thieme 1997).

Furthermore, the complex logistic organisation needed for people to build water-craft capable of transporting a biologically and socially viable group across significant water barriers, implies that by 840,000 years ago people had language, as persuasively argued by Davidson and Noble (1992). The degree of brain frontal lobe development and bilateral asymmetry evident in a late *H. erectus* cranium recently recovered from the Sambungmacan region of Java has similar implications – its morphology, specifically bilateral asymmetry of the brain and enlarged frontal lobes, indicates that the individual may have had the capability for language (Boedhihartono: pers. comm). Previously the organisational and linguistic capacity required for sea voyaging was thought to be the prerogative of modern humans and to have only appeared in the late Pleistocene.

In conclusion, evidence far from the African epicentre of hominid evolution, on the margins of continental Southeast Asia and beyond, has major implications for general models of hominid evolution and dispersal. An 840,000 year cultural sequence has now been established for Flores. Almost certainly hominid occupation of other east Indonesian islands, such as Timor, where *Stegodon* fossil sites with stone artefacts are reported, will be of similar antiquity (Verhoeven 1968). The islands of Wallacea, with their relatively isolated, bounded, simplified ecosystems and time depth of hominid occupation, thus seem ideally placed to investigate a whole range of fundamental questions concerning turning points in human evolution, dispersal, economy and technology.

ACKNOWLEDGMENTS

Grants from Australian Research Council and the UNE Vice Chancellor's Fund made the Flores fieldwork possible. Dr Paul O'Sullivan and Dr A. Raza at the School of Earth Sciences, Melbourne University, undertook the fission track dating. Fieldwork was conducted in collaborations with Dr Fachroel Aziz, Doug Hobbs, Dr Netty Polhaupessy, Suminto, Tular Sudarmadi,

Jackie Collins, Robert Bednarik and Iwan Kurniarsu. Doug Hobbs and Kathy Morwood drew the figures. Comments by Dr Boedhihartono (University of Indonesia) and Professor Russell L. Ciochon (University of Iowa) on aspects of early hominid research in Indonesia were very useful.

REFERENCES

Allen, H. 1991. Stegodonts and the dating of stone tool assemblages in island S.E. Asia. *Asian Perspectives* 30: 243-266.

Bellwood, P. 1985. *Prehistory of the Indo-Malaysian Archipelago*. Sydney: Academic Press.

Bellwood, P. 1997. *Prehistory of the Indo-Malaysian Archipelago*. Honolulu: University of Hawaii Press.

Birdsell, J.H. 1977. The recalibration of a paradigm for the first peopling of Greater Australia. In J. Allen, J. Golson and R. Jones (eds), Sunda and Sahul: prehistoric studeis in Southeast Asia, Melanesia and Australia: 113-168. London: Academic Press.

Bowdler, S. 1993. Sunda and Sahul: a 30 kyr BP culture area. In M.A. Smith *et al.* (eds), *Sahul in review*: 60-70. Canberra: Dept. of Prehistory, Research .School of .Pacific .Studies, Australian National University.

Davidson, I. & Noble, W. 1992. Why the first colonisation of the Australian region is the earliest evidence of modern human behaviour. *Archaeology in Oceania* 27(3): 135-142.

Groves, C.P. 1996. Hovering on the brink: nearly but not quite getting to Australia. In E. Rousham & L. Freedman (eds), *Perspectives in human biology, Vol 2:* 83-87. Perth: Centre for Human.Biology, University of Western Australia.

Hartono, H.M.S. 1961. Geological investigation at Olabula, Flores. Bandung: Unpublished report. Djawatau Geologi.

Hyodo, M., Watanbe, N., Sunata, W., Susanto, E.E. & Wahyono, H. 1993. Magnetostratigraphy of hominid fossil bearing formations in Sangiran and Mojokerto, Java. *Anthropol. Sci* 10(2): 157-186.

Ithara, M., Watanabe, N., Kadar, D. & Kumai, H. 1994. Quaternary stratigraphy of the hominid fossil bearing formations in the Sangiran area, central Java. In J.L. Frunzen (ed.), *100 years of Pithecanthropus: the Homo erectus problem:* 123-128. Frankfurt : Courier Forschungsinstitut Senckenberg.

Jacob, T. 1975 The Pithecanthropines of Indonesia. *Bull. Mem. Soc. Anthrop*, Paris 1.2 Ser 13: 243-256.

Johnson. D.L. 1980. Problems in the land vertebrate zoogeography of certain islands and the swimming powers of elephants. *J. Biogeography* 7: 383-98.

Koenigswald, G.H.R. von 1957 A tektite from the island of Flores (Indonesia). *Proc. Koninklijke Nederlandse Akademie van Wetenschappen*, Series B, 61(1), 44-46.

Lahr, M.M. & Foley, R.A. 1998 Towards a theory of modern human origins: geography, demography, and diversity in recent human evolution. *Yearbook of Physical Anthropology* 41: 137-176.

Maringer, J & Verhoeven, Th. 1970a. Die Steinartefake aus der *Stegodon*-fossilschicht von Mengeruda auf Flores, Indonesien. *Anthropos* 65:229-247.

Maringer, J & Verhoeven, Th. 1970b. Note on some stone artifacts in the National Archaeological Institute of Indonesia at Djakarta, collected from the *Stegodon*-fossil bed at Boaleza in Flores. *Anthropos* 65: 638-9.

Morwood, M.J., Aziz, F., Van Den Bergh, G.D., Sondaar, P.Y. & De Vos, J. 1997. Stone artefacts from the 1994 excavations at Mata Menge, west central Flores, Indonesia. *Australian Archaeology* 44: 26-34.

Morwood, M.J., O'Sullivan, P.B., Aziz, F. & Raza, A. 1998. Fission track age of stone tools and fossils on the east Indonesian island of Flores. *Nature* 392 (12 March):173-6.

Morwood, M.J., Aziz, F., O'Sullivan, P., Nasruddin, Hobbs, D.R. & Raza, A. 1999 Archaeological and palaeontological research in central Flores, east Indonesia: results of fieldwork, 1997-98. *Antiquity* 73:(280):273-86.

Morwood, M.J., O'Sullivan, P.B., Swisher, C., Susanto, E.E, Aziz, F. & Raza, A. 2000. Chronological hygiene: redating Mojokerto 1, East Java, Indonesia. (In prep.).

O'Sullivan, P.B, Morwood, M.J., Aziz, F., Suminto, Situmorang, M., Raza, A. & Hobbs, D.R..2000. Geology and age of the Soa Basin on the island of Flores, eastern Indonesia: archaeological implications. (Submitted to *Science*).

Rightmire, G.P. 1994 The relationship of *Homo erectus* to later Middle Pleistocene hominids. In J.L. Franzen (ed.), *100 years of Pithecanthropus: the Homo erectus problem*: 319-326. Frankfurt : Courier Forschungsinstitut Senckenberg.

Smith, J.M.B. 2001. Did primates, including early hominids, cross sea gaps in the Wallacean region on natural rafts? (This volume).

Sondaar, P.Y. 1987. Pleistocene Man and extinctions of island endemics. *Mem. du Societe Geologique de France. Nouveau Series* 150:159-63.

Sondaar, P.Y., Van Den Bergh, G.D., Mubroto, B., Aziz, F., De Vos, J., & Batu, U.L. 1994. Middle Pleistocene faunal turn-over and colonisation of Flores (Indonesia) by *Homo erectus*. *Comptes Residues de la Academie des Sciences*. Paris 319: 1255-1262.

Sudarmadi, T. 1999. An ethnoarchaeological study of the Ngadha: a megalithic culture in central Flores, Indonesia. Armidale: M.A. thesis, University of New England.

Suzuki, M., Wikarno, Budisantoso, Saefudin, L.& Itihara. M. 1985 Fission track ages of pumice tuff, tuff layers and Javites of hominid bearing formations in Sangiran, Central Java. In N. Watanbe & D. Kadar (eds), *Quaternary geology of the hominid bearing formations in Java*: 309-357. Bandung: Geology Research and Development Centre Special Publication 4.

Thieme, H. 1997. Lower Paleolithic hunting spears from Germany. *Nature* 385, 807-810.

Van Den Bergh, G.D. 1997. The late Neogene elephantoid-bearing fauna of Indonesia and their palaeozoogeographic implications: a study of the terrestrial fauna succession of Sulawesi, Flores and Java, including evidence for early hominid dispersal east of Wallace's Line. Utrecht: Ph.D. thesis, University of Utrecht.

Van Den Bergh, G.D., Mubroto, B., Aziz, F., Sondaar, P.Y. & J. De Vos. 1996. Did *Homo erectus* reach the island of Flores? In P. Bellwood (ed.), *Indo-Pacific prehistory: the Chiang Mai papers (Proceedings of the 15th IPPA Congress, Chiang Mai, Thailand 1994, Vol. 1*: 27-36. Canberra: Australian National University.

Verhoeven, Th. 1968. Pleistozane Funde auf Flores, Timor and Sumba. In *Anthropica Gedenkschrift zum 100. Geburtstag von P.W. Schmidt*: 393-403. St Augustin: Studia Instituti Anthropos No. 21.

Weidenreich, F. 1951. *Morphology of Solo Man*. New York : Anthropological Papers of the American Museum of Natural History 43(3).

The requirements for human colonisation of Australia

Iain Davidson
Archaeology and Palaeoanthropology, Human and Environmental Studies, University of New England, Armidale, Australia

ABSTRACT: When Alfred Wallace embarked on his voyage from Bali to Lombok he turned his back on the limitations of our hominin ancestry and did something distinctively human: he drew a line (Davidson 1999b) (Figure 1). Morwood's confirmation of dates for stone artefacts earlier than 800 thousand years ago (Morwood *et al.* 1999; Morwood *et al.* 1998) has shown how our hominin ancestors may have crossed similarly along the chain of islands of the Lesser Sundas. That may or may not have been a step of similarly human dimensions. In this paper, I argue that it is implausible to interpret that small step as an indicator of the emergence of human abilities and behaviour. The really "giant leap for mankind"—for our human ancestors—was the journey to Australia (Davidson & Noble 1992).

1 INTRODUCTION

The journey of our ancestors from the time of our common ancestor with other apes can be represented as a more prosaic one from Canberra (where we left the apes behind) to Armidale. In Sydney those ancestors became bipedal, they first left stone tools in Newcastle, made handaxes in Muswellbrook and continued to make them almost until they reached Armidale. For those unfamiliar with the geography of eastern Australia, a similar line can be drawn in the United States for a journey from San Francisco to New York (Figure 2). By this comparison, bipedalism emerged in San Francisco, stone tools in Salt Lake City. Acheulean handaxes first appeared around Kansas City and continued to be made until beyond Philadelphia. On this time scale, hominins crossed into Flores around Indianapolis.

Somewhere close to Armidale, or between Philadelphia and New York, our ancestors began that most distinctive of human behaviours: they began to talk to each other in languages only they understand. Language, as a means of communication using symbols, involves, permits and perhaps requires the generation of meanings shared among a community brought up to share the conventions of those symbolic meanings (Noble & Davidson 1996). This is true of those brought up speaking Australian rather than American English, or those who learned geology rather than archaeology, or who are university-educated rather than not. Such distinctions draw lines between communities as strong as or stronger than the line that Wallace crossed.

Once language emerged, all hell broke loose – all of it within Armidale, or between Philadelphia and New York. Language-using people colonised Australia; they painted caves, invented agriculture. They started to write down details of the wealth they had accumulated (on clay tablets in cuneiform of Linear B) and then they made up stories about it. They invented gods, colonialism and the Y2K bug.

My argument in this paper is that the early stone tools from Flores, while remarkable and important, indicate a sea-crossing by some means, but they do not constitute evidence, as the later sea-crossing into Australia did, of the beginning of that process of emergence of the full range of human abilities.

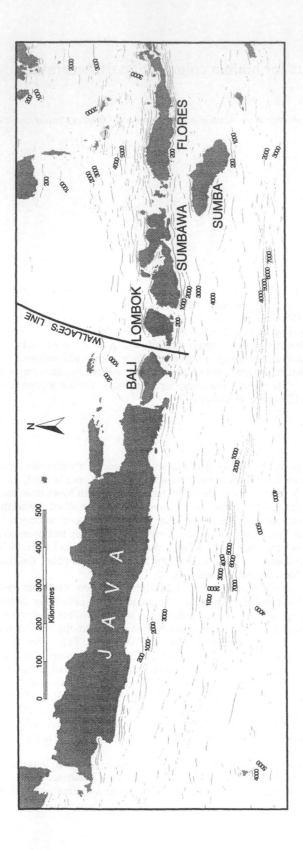

Figure 1. The Lesser Sunda Islands showing the bathymetric contours and the position of Wallace's Line. At periods of low sea-level during the Pleistocene, the available land area would be closer to the -200m contour than to the modern shore-line. (Drawing by Mike Roach)

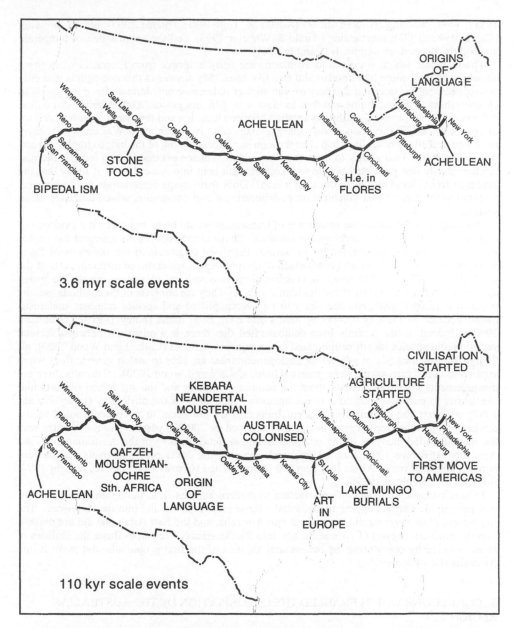

Figure 2. Time line for significant events in homnin and human evolution. The top map shows significant events as if they occurred on a journey from San Francisco to New York. Since so many significant events occurred between Philadelphi and New York (on this scale), the lower map shows that time expanded into another journey from San Francisco to New York. (Drawing by Mike Roach)

2 BEHAVIOUR OF EARLY *HOMO*

Our bipedal early ancestors were quite distinctive from the other apes of Africa. There are various ways of defining these distinctive differences (eg Isaac 1978). Noble and I (Noble & Davidson 1996) have pointed out several features: bipedalism (Hunt 1994); stone tool carrying, making and use (Schick & Toth 1993; Wynn & McGrew 1989); meat cutting (Keeley & Toth

1981); wood working (pointed sticks) (Keeley & Toth 1981; Thieme 1997); aimed throwing (Calvin 1982; 1993); meat eating (Aiello & Wheeler 1995); and the colonisation of temperate and lightly-wooded environments (Gamble 1986).

The question, which is crucial to evaluating the early evidence from Flores, is when these ancestors became more like humans and less like apes. My answer to this question is that only humans can ask the question and only we can answer it, because only humans have the ability to choose where to draw the line and thus to draw it in different places. Only humans can define different conventions for the arbitrary relation between meaning and the symbols with which we discuss it (Noble & Davidson 1996). Only humans can give meaning to Wallace's line, Huxley's line and all the other lines that slice through our understanding of the biogeographic diversity of Wallacea. That ability to draw the line, which Wallace exemplified so well, is the distinctive ability that permitted people to take that giant leap into Australia. Is it likely that the evidence from Flores also indicates such a leap? Does that voyage demonstrate the emergence, 800 000 years ago, of that combination of arbitrariness and convention which language alone permits?

One way to demonstrate the emergence of humanness would be to argue that the evidence of hominin evolution leaves unambiguous answers. Physical anthropologists interpret the phylogenetic relationships represented by the various fossilised fragments of the skeletons of those ancestors on the basis of careful comparison of morphological variations of different parts of the skull (Groves 1989). The resulting classifications into species, of course, cannot be tested against any other real record of species identification. They are artefacts of the methods used by physical anthropologists, and are likely to be oversimplified and species numbers underestimated because of the tiny number of fossils that have been found (Foley 1991; cf. Tattersall 1992). Indeed, it has recently been demonstrated that there is a mismatch between skeletal variation and species identification based on genetic comparisons (Collard and Wood 2000), although similar methods of generating phylogenetic trees are able to match genetic trees when applied to soft tissues of the same genera (Gibbs, Collard and Wood 2000). Naturally there are disagreements among specialists about the naming of species and the attribution of particular specimens to particular positions in our ancestry. One view of the phylogeny (Foley & Lahr 1997) suggests the hominins of modern Indonesia do not seem to have contributed to the modern populations in any region (though they survived for 700 000 years). This is consistent with the more detailed analyses of the relations among skeletal remains in the Australian and East Asia region (Brown 1992; Storm 1995), and also with the developing consensus that modern humans emerged out of Africa less than 800 000 years ago (Groves 1992; Groves & Lahr 1994; Rogers 1995; Rouhani 1989).

In understanding this lack of contribution to modern humans from the Asian *Homo erectus*, two geographic barriers appear fundamental—those into Australia and into the Americas. The Indonesian *Homo erectus* did not make it into Australia, and the East Asian one did not make it into the northern regions of Asia or further, into the Americas. We can evaluate the abilities of *Homo erectus* by considering the behavioural abilities of the first people who did make it into Australia (Davidson 1999a).

3 CONDITIONS WHICH ENABLED THE COLONISATION OF THE AUSTRALIAN REGION

The evidence suggests that both the Australian and the northern regions (including the Americas) were only colonised less than 60 thousand years ago, after the emergence of communication using symbols (Davidson & Noble 1998). Symbolic representation is necessary for the planning and anticipation involved in assembling the materials to build a watercraft and to have a purpose in building it (Davidson & Noble 1992). It is also necessary for ensuring the provisions necessary for the voyage. Similar planning and anticipation were necessary on a much more continuing basis for the colonisation of cold environments, to make suitable shelter and provision for the winter months.

The evidence around the world suggests remarkably similar dates (on a geological time-scale) for the earliest appearance of direct evidence for symbols: 90 thousand in South Africa (Henshilwood & Sealy 1997; Singer & Wymer 1982); 35 thousand in Europe (White 1989); 32 thousand in Australia (Morse 1993).

The Australian evidence, at the moment, is represented by the pierced sea-shells from Mandu Mandu, hundreds of kilometres from the sea at the time of use (Morse 1993). As in Europe at the same time (White 1989), this suggests that among the earliest uses of symbolic representation was the use of beads for personal decoration—arguably a means of drawing a line between bead wearers and others.

The use of symbols in communication, necessary for the colonisation of Australia, seems to be widespread around the world at about the time of that first colonisation (Davidson & Noble 1998), and was probably necessary for some of the other novel features of human behaviour that emerged in that time period. The issue for evaluating the evidence from Flores is whether there is evidence around the world for a similar context of emergence of language-based abilities from that time period.

4 WAS THERE EVIDENCEFOR SYMBOLIC BEHAVIOUR EARLIER THAN THE TIME OF COLONISATION OF AUSTRALIA?

Beads and painted caves are merely obvious signs of symbols. The watercraft that brought people to Australia was much earlier, and Noble and I (Davidson & Noble 1992) have argued that its construction involved symbolic representation, without direct evidence of such obvious signs. Was there evidence of such symbolic representation in the earlier archaeological record without the appearance of objects that were themselves symbolic?

The first, most obvious claim is from stone artefacts, of which the most remarkable are the bifacially flaked objects archaeologists call Acheulean handaxes.

4.1 Imposed form of artefacts

The refutation of this claim is still difficult, and very technical. I cannot go into details here (but see Davidson & Noble 1993; and Davidson 2000). The main claim to "symbolic" nature of Acheulean handaxes is their astonishing uniformity around the world for over a million years (Wynn 1979; 1995; Wynn & Tierson 1990). But the uniformity may mask a more complex reality which includes the selection by archaeologists of things which are uniform because they were selected (Dibble 1989), and may involve a convergence on a form without any intentional selection by the makers.

The clinching evidence is that Australia was colonised from a region said never to have had handaxes (Schick 1994) and after the final appearance of stone industries based on handaxes (Cabrera Valdes 1989), yet there are handaxes from Australia (Noble & Davidson 1996; Rainey 1991). The production of bifacially flaked objects occurred many times quite independently, as has been known and understood for many years by their appearance towards the end of the European Middle Palaeolithic (Mellars 1973). In this context, they were quite separated from Acheulean industries in time and thus in every other way. Such separation implies that they do not belong to a long tradition of culturally-designed objects.

4.2 Imposed form of Australian artefacts

The first Australians clearly had the ability to impose form on the objects they made. These were the ancestors of Australian peoples such as the Kalkadoons of Mt Isa (Davidson 1993) or the Anaiwan of the Armidale region (Godwin 1997). This was true both for the watercraft they used to get here, and the others they made to voyage off the northeast edge of the continent to the Bismarck Archipelago (Allen & Gosden 1991).

But early tools also indicate this ability. In early sites, there are bone tools shaped by grinding unrestrained by the limitations of the raw material, as were the ground-edged hatchet heads in the Huon Peninsula and in Arnhem Land (Davidson & Noble 1992). These hatchet heads were made at least ten thousand of years before the appearance of similar hatchets in western Asia and Europe marked the beginning of the destructive effects of agriculture.

4.3 *The question of burial*

The other great claim for early symbol use is the persisting belief that the Neandertals of Europe and western Asia buried their dead. I believe this claim cannot survive the recent reanalysis by Rob Gargett (1989; 1999). He has shown that there are only two circumstances in which the skeletal remains of these hominins (and the early modern humans of Qafzeh) survived with almost complete skeletons. Gargett has shown that if they were complete, then they were killed by rockfall when the sky fell in—"the Chicken Licken Effect". Sometimes the rock crushed them like a foot stamping onto an empty beer can. Gargett also showed that if they were not crushed by rockfall, then they seem to have decomposed in natural hollows in the sediment before those holes were naturally filled with more dirt. Whole disarticulated segments are always missing, indicating that partial decomposition of the body had taken place before burial was complete. There was something rotten in the state of Neandertals.

In Australia, the story is quite different. The earliest burials in the world are found here. The Mungo I cremation, burned, smashed, collected and buried (Bowler *et al.* 1970), indicating by the repeated attention to the body an order of magnitude different from anything even claimed for Neandertals, is dated to 26 thousand years ago (Bowler 1998). The Mungo III burial (Bowler & Thorne 1976), quite distinctively complete in a way that Neandertals are not, is probably as old as 40 000 years (by TL) (Bowler 1998) but there is uncertainty about the dating (Brown & Gillespie 2000).

5 WHAT HAPPENED ONCE AUSTRALIA WAS COLONISED?

Sometime after 100 thousand years ago something major changed. There were no people in the cold climates of the north at 50 thousand. After 40 thousand the north filled up, despite the worsening climate, and so did Australia. Are these two events related?

If the Americas were first colonised at 14 thousand years ago (Beaton 1991) (and there is now some doubt about this [Meltzer *et al.* 1997]), then it is worth pointing out that the previous time that such conditions existed was back at 80 thousand. I believe that this evidence will demonstrate that it was after 80 thousand that there was an emergence of planning abilities. These planning abilities enabled the colonisation of cold environments, sea-crossings such as were needed to get to Australia, and whatever combination of water-crossing and cold climate adaptation was needed to get to the Americas.

We should not exaggerate the ease with which the new achievements were won from the emergence of planning abilities. The dates for first colonisation of different regions in Australia show a rapid movement from first landfall to Tasmania, but a relatively slow occupation of "difficult" environments. The short time between the earliest dates in northern Australia and those for Tasmania fits with Birdsell's (1957) rapid colonisation model proposed long before archaeological evidence for the timing of colonisation on the basis of purely theoretical simulations of the demographic conditions for colonising all of Australia. Rindos & Webb (1992) suggested an "optimal maladaptation" model in which people *not* well adjusted to the poorly known and unpredictably varying Australian environments died out or moved on rather than staying to find ways of coping with resource variation locally.

6 THE SIGNIFICANCE OF EARLY STONE ARTEFACTS IN FLORES

This rapid assessment of hominin abilities around the world at the time of the stone artefacts of Flores suggests that there is little evidence around the world 800 000 years ago, or indeed until much later, for a hominin ability to impose form on artefacts. The evidence of the abilities of the first people in Australia suggests that they were engaged in planned activities, and had sufficient behavioural flexibility to cope rapidly with a wide range of environments—but not all of them. So what can we say about the early stone artefacts in Flores as a result of this rapid assessment of the comparisons with events contemporary with the artefacts, and with the major sea crossing much later in the story?

The crossing of Wallace's line by hominins was significant, but it seems unlikely that it involved the suite of abilities, based on planning, that are indicated by the first colonisation of Australia. Such planning does not seem to be characteristic of any hominin behaviour elsewhere at the time of the crossing to Flores. On the basis of the evidence from their other abilities reflected in the archaeological record, I doubt hominins constructed a watercraft.

It may be that the artefacts in Flores indicate a previously undocumented ability of early *Homo*, but if the sea crossing was short, we may need to make comparison with similar crossings elsewhere. It will be necessary to consider how early *Homo* reacted to the large rivers it undoubtedly encountered elsewhere in the world. If rafting on vegetation mats or logs is a serious possibility (see Smith this volume), would this have been an option in crossing the Nile, the Tigris or the Ganges?

What Morwood has demonstrated by showing an early date for stone artefacts east of the Wallace Line is that previous finds of such significance may have been ignored unjustifiably. Many other regions may not have been adequately explored as a result of the easy acceptance of models that appear to account for the known evidence. Where these models are aimed at understanding the evolutionary emergence of abilities we take for granted, they are often based on inadequate conceptualisation of those abilities. We need new conceptual models to provoke appropriate research.

It is claimed (Morwood et al 1998; Morwood this volume) that the Flores finds represent the achievement of human abilities at 800 000 years ago, which previously were only documented with the colonisation of Australia after 80 thousand years ago. In order to accept this, it is essential to explain why, as I have sought to show, the claim appears so surprising in the context of what else is known about the behaviour of early hominins. The ability to construct watercraft at this early date in the Early Pleistocene demands an explanation for the lack of colonisation of other islands around the world until the end of the Pleistocene or later. The Mediterranean islands, many of which are visible from the mainland, or become so at glacial period low sea-levels, were not colonised until after 11 thousand years ago (Cherry 1990). Any development of theory that arises from the Flores finds must also consider why there is such a difference in dates between the first sea-crossings in these two regions. Human ancestors were present in Europe at about the same time as they crossed into Flores (Pares & Pérez-Gonzalez 1995) but there were no crossings to the islands visible off the coast. Did Middle Pleistocene Mediterranean environments not produce rafts of vegetation in the way the islands of Wallacea do? Humans were present in Europe by 40 000 years ago (Bocquet-Appel & Demars 2000), but their behaviour does not appear to have included messing about in boats. It is suggested that fishing was an important behaviour before the crossing into Australia (Balme 1995; Davidson 1999a). Was there any evidence for fishing in the Mediterranean between 40 000 and 11 000 years ago? Humanness seems to be a necessary condition for regular sea crossing, but not a sufficient one.

7 THE SIGNIFICANCE OF AUSTRALIA

Noble and I have repeatedly pointed out that the evidence from the first human colonisation of Australia is crucial to understanding the evolutionary emergence of language. I persist in this claim in the face of the finds from Flores. Something different happened shortly before first colonisation of Australia, which has no parallel in the world of hominins at the time of the first stone tools in Flores. If we are right, then the nature of the first colonisation must be understood in terms of the symbolic construction of the physical and social environments by the first people to make the corssing. I have begun to develop that argument elsewhere (Davidson 1999a).

There are now sufficient numbers of sites excavated and dated in most regions of Australia that we can begin consider details of the pattern of colonisation. We need to consider the changing patterns of adaptation which made it possible for people to cope with difficult environments (Davidson 1990; O'Connor *et al*. 1993). But, along with the story of how the descendants of a small number of initial colonists (speaking a small number of languages) succeeded in occupying the whole of the continent, there is another story that has not been told. That is a story of the emergence of the many different peoples who lived in Australia at the time of the invasion from Europe. How did the Australian continent (Sahul) come to have so many different people, so many languages? In the period of historic documentation, there were more than

250 languages on the mainland subdivided (by some) into 27 families (Wurm 1972). New Guinea (connected to the mainland by Pleistocene low sea-level until 8 thousand years ago) had 200 languages in the Austronesian family, and about 700 Papuan languages in about 60 families (Dixon 1997). In seeking to explain the generation of such diversity, the theoretical arguments must be consistent with those that account for the long delay between the sea-crossing to Flores and that to Australia.

I believe the argument Noble and I have produced—that only the second crossing reflects the recent emergence of language—meets the requirement of accounting for different classes of evidence before and after colonisation (Noble & Davidson 1996). The diversity that emerged in Aboriginal Australia probably results from the human propensity to draw and define lines between categories, the common result of the use of language. Wallace was doing something distinctively human when he drew his line.

REFERENCES

Aiello, L.C. & Wheeler, P. 1995. The expensive-tissue hypothesis: the brain and the digestive system in human and primate evolution. *Current Anthropology* 36: 199-221.

Allen, J. & Gosden, C. (eds) 1991. *Report of the Lapita Homeland Project*. Canberra: Australian National University.

Balme, J. 1995. 30, 000 years of fishery in western New South Wales. *Archaeology in Oceania* 30: 1-21.

Beaton, J.M. 1991. Colonising continents: Some problems from Australia and the Americas. In T.D. Dillehay & D.J. Meltzer (eds), *The first Americans: search and research*: 209-229. Boca Raton: CRC Press.

Birdsell, J.B. 1957. Some population problems involving Pleistocene man. *Cold Spring Harbor Symposia on Quantitative Biology* 22: 47-69.

Bocquet-Appel, J.-P.& Demars, P.Y. 2000. Neanderthal contraction and modern human colonization of Europe. *Antiquity* 74: 544-52

Bowler, J.M. 1998. Willandra Lakes revisited: environmental framework for human occupation. *Archaeology in Oceania* 33: 120-155.

Bowler, J.M., Jones, R., Allen, H. & Thorne, A.G. 1970. Pleistocene human remains from Australia: A living site and human cremation from Lake Mungo, western New South Wales. *World Archaeology* 2: 39-60.

Bowler, J.M. & Thorne, A. 1976. Human remains from Lake Mungo: discovery and excavation of Lake Mungo III. In R.L.Kirk & A.G. Thorne (eds), *The origin of the Australians*: 127-138. Canberra.: Australian Institute of Aboriginal Studies.

Brown, P. 1992. Recent human evolution in East Asia and Australia. *Philosophical Transactions of the Royal Society of London B* 337: 235-242.

Brown, P. & Gillespie, R. 2000. Mungo claims unsubstantiated. *Australasian Science* 21 (5): 26-27.

Cabrera Valdes, V. 1989. Accelerator ^{14}C dates for early Upper Paleolithic (basal Aurignacian) at El Castillo Cave (Spain). *Journal of Archaeological Science* 16: 577-584.

Calvin, W.H. 1982. Did throwing stones shape hominid brain evolution? *Ethology and Sociobiology* 3: 115-124.

Calvin, W.H. 1993. The unitary hypothesis: A common neural circuitry for novel manipulations, language, plan-ahead and insight? In K.R. Gibson & T. Ingold (eds), *Tools, language and cognition in human evolution*: 230-250. Cambridge: Cambridge University Press.

Cherry, J.F. 1990. The first colonization of the Mediterranean islands: a review of recent research. *Journal of Mediterranean Archaeology* 3: 145-221.

Collard, M. & Wood, B. 2000. How reliable are human phylogenetic hypotheses? *Proceedings of the National Academy of the Sciences* 97: 5003-5006

Davidson, I. 1990. Prehistoric Australian demography. In B. Meehan & N. White (eds), *Hunter-gatherer demography: past and present*: 41-58. Sydney: Oceania Monographs.

Davidson, I. 1993. Archaeology of the Selwyn Ranges, Northwest Central Queensland. In G. Burenhult (ed), *The illustrated encyclopaedia of humankind. People of the stone age: hunter-gatherers and early farmers*: 210-211. Sydney: Weldon Owen.

Davidson, I. 1999a. First people becoming Australian. *Anthropologie (Brno)* 37(1): 125-141.

Davidson, I. 1999b. The game of the name: continuity and discontinuity in language origins. In B.J. King (ed), *The origins of language: what nonhuman primates can tell us*: 229-268. Santa Fe, NM: School of American Research.

Davidson, I. 2000 Tools, language and the origins of culture. http://www.infres.enst.fr/confs/evolang/actes/_actes17.html

Davidson, I. & Noble, W. 1992. Why the first colonisation of the Australian region is the earliest evidence of modern human behaviour. *Archaeology in Oceania* 27: 135-142.

Davidson, I. & Noble, W. 1993. Tools and language in human evolution. In K. Gibson & T. Ingold (eds), *Tools, language and cognition in human evolution*: 363-388. Cambridge: Cambridge University Press.

Davidson, I. & Noble, W. 1998. Two views on language origins. *Cambridge Archaeological Journal* 8: 82-88.

Dibble, H.L. 1989. The implications of stone tool types for the presence of language during the Lower and Middle Paleolithic. In P.A. Mellars & C.B. Stringer (eds), *The human revolution: behavioural and biological perspectives on the origins of modern humans*: 415-431. Princeton, NJ: Princeton University Press.

Dixon, R.M.W. 1997. *The rise and fall of languages*. Cambridge: Cambridge University Press.

Foley, R.A. 1991. How many species of hominid should there be? *Journal of Human Evolution* 20: 413-427.

Foley, R.A. & Lahr, M.M. 1997. Mode 3 technologies and the evolution of modern humans. *Cambridge Archaeological Journal* 7: 3-36.

Gamble, C. 1986. *The palaeolithic settlement of Europe*. Cambridge: Cambridge University Press.

Gargett, R.H. 1989. Grave shortcomings: The evidence for Neanderthal burial. *Current Anthropology* 30: 157-190.

Gargett, R.H. 1999. Middle Palaeolithic burial is not a dead issue: the view from Qafzeh, Saint-Césaire, Kebara, Amud, and Dederiyeh. *Journal of Human Evolution* 37: 27-90.

Gibbs, S., Collard, M. & Wood, B. 2000. Soft-tissue characters in higher primate phylogenetics. *Proceedings of the National Academy of the Sciences* 97: 11130-11132.

Godwin, L. 1997. Little Big Men: alliance and schism in northeastern New South Wales. In P. McConvell (ed), *Archaeology and linguistics: Aboriginal Australia in global perspective*: 297-309. Melbourne: Oxford University Press.

Groves, C.P. 1989. *A theory of human and primate evolution*. Oxford: Clarendon Press.

Groves, C.P. 1992. How old are subspecies? A tiger's eye-view of human evolution. *Archaeology in Oceania* 27: 153-160.

Groves, C.P. & Lahr, M.M. 1994. A bush not a ladder: speciation and replacement in human evolution. *Perspectives in Human Biology* 4: 1-11.

Henshilwood, C. & Healy, J. 1997. Bone artefacts from the Middle Stone Age at Blombos Cave, Southern Cape, South Africa. *Current Anthropology* 38: 890-895.

Hunt, K.D. 1994. The evolution of human bipedality: ecology and functional morphology. *Journal of Human Evolution* 26: 183-202.

Isaac, G.L. 1978. The food-sharing behavior of protohuman hominids. *Scientific American* 238: 90-108.

Keeley, L.H. & Toth, N. 1981. Microwear polishes on early stone tools from Koobi Fora, Kenya. *Nature* 293: 464-465.

Mellars, P.A. 1973. The character of the Middle-Upper Palaeolithic transition in south-west France. In C. Renfrew (ed), *The explanation of culture change: models in prehistory*: 255-276. London: Duckworth.

Meltzer, D.J., Grayson, D.K., Ardila, G., Barker, A.W., Dincauze, D.F., Haynes, C.V., Mena, F., Nunez, L. & Stanford, D.J. 1997. On the Pleistocene antiquity of Monte Verde, southern Chile. *American Antiquity* 64: 659-663.

Morse, K. 1993. Shell beads from Mandu Mandu rockshelter, Cape Range Peninsula, Western Australia, before 30,000 bp. *Antiquity* 67: 877-883.

Morwood, M.J., Aziz, F., O'Sullivan, P., Nasruddin, Hobbs, D.R. & Raza, A. 1999. Archaeological and palaeontological research in central Flores, east Indonesia: results of fieldwork 1997-98. *Antiquity* 73: 273-286.

Morwood, M.J., O'Sullivan, P., Aziz, F. & Raza, A. 1998. Fission track age of stone tools and fossils on the east Indonesian island of Flores. *Nature* 392: 172-176.

Noble, W. & Davidson, I. 1996. *Human evolution, language and mind*. Cambridge: Cambridge University Press.

O'Connor, S., Veth, P. & Hubbard, N. 1993. Changing interpretations of postglacial human subsistence and demography in Sahul. In M.A. Smith, B. Fankhauser & M. Spriggs (eds), *Sahul in Review: Pleistocene archaeology in Australia, New Guinea and Island Melanesia*: 95-105. Canberra: Australian National University.

Pares, J.M. & Pérez-Gonzalez, A. 1995. Paleomagnetic age for hominid fossils at Atapuerca archeological site, Spain. *Science* 269: 830-832.

Rainey, A. 1991. Some Australian bifaces. *Lithics* 12: 33-36.

Rindos, D. & Webb, E. 1992. Modelling the initial human colonisation of Australia: perfect adaptation, cultural variability, and cultural change. *Proceedings of the Australasian Society for Human Biology* 5: 441-454.

Rogers, A.R. 1995. How much can fossils tell us about regional continuity? *Current Anthropology* 36: 674-676.

Rouhani, S. 1989. Molecular genetics and the pattern of human evolution. In P.A. Mellars & C.B. Stringer (eds), *The human revolution: behavioural and biological perspectives on the origin of modern humans*: 47-61. Princeton, NJ: Princeton University Press.

Schick, K.D. 1994. The Movius Line reconsidered. Perspectives on the earlier paleolithic of Eastern Asia. In R.S. Coruccini & R.L. Ciochon (eds), *Integrative paths to the past: paleoanthropological advances in honor of F. Clark Howell*: 569-596. Englewood Cliffs, NJ: Prentice Hall.

Schick, K.D. & Toth, N. 1993. *Making silent stones speak*. New York: Simon & Schuster.

Singer, R. & Wymer, J. 1982. *The Middle Stone Age at Klasies River Mouth in South Africa*. Chicago: The University of Chicago Press.

Storm, P. 1995. The evolutionary significance of the Wajak skulls. *Scripta Geologica* 110: 1-247.

Tattersall, I. 1992. Species concepts and species identification in human evolution. *Journal of Human Evolution* 22: 341-349.

Thieme, H. 1997. Lower Palaeolithic hunting spears from Germany. *Nature* 385: 807-810.

White, R. 1989. Production complexity and standardization in Early Aurignacian bead and pendant manufacture: evolutionary implications. In P.A. Mellars & C.B. Stringer (eds), *The human revolution: behavioural and biological perspectives on the origin of modern humans*: 366-390. Princeton, NJ: Princeton University Press.

Wurm, S.A. 1972. *Languages of Australia and Tasmania*. The Hague: Mouton.

Wynn, T. 1979. The intelligence of later Acheulean hominids. *Man: Journal of the Royal Anthropological Institute* 14: 371-391.

Wynn, T. 1995. Handaxe enigmas. *World Archaeology* 27: 10-24.

Wynn, T. & McGrew, W.C. 1989. An ape's eye view of the Oldowan. *Man: Journal of the Royal Anthropological Institute* 24: 383-398.

Wynn, T. & Tierson, F. 1990. Regional comparison of the shapes of later Acheulean handaxes. *American Anthropologist* 92: 73-84.

Did early hominids cross sea gaps on natural rafts?

J.M.B. Smith
School of Human and Environmental Studies, University of New England, Armidale, N.S.W. 2351, Australia

ABSTRACT: Distributions of non-human primates suggest that some species have colonised across wide sea gaps. Many reports exist of natural rafts capable of transporting large terrestrial vertebrates across such gaps, including in Wallacea, and other data show that people can survive adrift for at least several days. Climatic and oceanographic features of this region are also consistent with the idea that natural rafts may have permitted *Homo erectus* to colonise Flores, and *H. sapiens* to colonise Australia. Speculations about linguistic or other skills based on the assumption that these hominids built watercraft should therefore be viewed cautiously.

1 INTRODUCTION

Humans probably reached Australia from southeast Asia between 60 000 and 40 000 years ago, during the last glacial period. The arrival involved the crossing of one or more relatively wide (*c*. 70-100 km) sea gaps, even via island stepping-stones at a time of low sea level (Birdsell 1977, Irwin 1992). It has generally been thought that the crossings must have been by some form of constructed watercraft, representing 'the first voyaging of its kind in the world' (Irwin 1992). 'The arrival in the Australian region may have been an accident, but it is not one that could have happened in the absence of the use of sea-going vessels constructed according to plan' (Noble & Davidson 1996). Davidson and Noble (1992; Davidson, this volume) further argued that as it would have required detailed social collaboration, this first demonstration of construction of watercraft also marked the first evidence for human language.

Recently Morwood *et al.* (1998) have confirmed that more primitive hominids *Homo erectus* reached the island of Flores by 840 000 years ago. This step similarly is likely to have involved crossing a sea gap (between Bali and Lombok). This raises the idea that '*Homo erectus* in this region was capable of repeated water crossings using watercraft' . If true, it in turn suggests 'that the cognitive capabilities of this species may be due for reappraisal' (Morwood *et al.* 1998), and that *Homo erectus* also used language (Morwood *et al.* 1999; Morwood, this volume). This conclusion if accepted would have fundamental implications for interpretation of human evolution and history.

The gap crossed by *Homo erectus* to reach Flores was substantially smaller than that surmounted by *H. sapiens* in reaching Australia. Inspection of bathymetric charts in the light of demonstrated Quaternary sea level fluctuations shows no barriers from continental Asia to Bali, and none between Lombok and Flores. The only sea gap, between Bali and Lombok, is some 30 km wide today, but would have been reduced to 'perhaps less than 400 m' (Kitchener *et al.* 1990) at times of glacially lowered sea level. These authors even suggest that this narrow, deep channel may be a result of Holocene erosion, and 'There is in fact every possibility that a southern corridor existed between S.E. Bali and S.W. Lombok, joined through Nusa Perida I., during the maximum lowering of Pleistocene seas' (Kitchener *et al.* 1990). A period of particularly low sea level occurred about 800 000 years ago, close to the time of hominid colonization of Flores

(Bergh *et al.* 1996, quoted by Brandon-Jones 1998). However, it seems more likely that the deep channel in Lombok Strait is old, and always carried a fierce current as it does today.

It is therefore probable that both *Homo erectus* and *H. sapiens* crossed sea gaps in Wallacea (the region lying between the Sunda and Sahul Shelves, including islands such as Flores and Sulawesi) in Pleistocene times. However, in both cases it might be unnecessary to invoke constructed watercraft. The most credible alternative mechanism is passive drift on rafts of vegetation carried to the sea by rivers in flood. Floating islands of this kind, to which people might have clung for survival without any intention to voyage, might even have been larger than simple watercraft, and able to carry more people. Nevertheless 'many people remain sceptical of over-water dispersal, believing that the use of (natural) rafts is improbable, unobservable and consequently untenable' (Censky *et al.* 1998).

To evaluate the likelihood that early hominids might have been dispersed in such a way, I attempt here briefly to:

a) review the natural distributions of some other, non-hominid primates (concerning whose lack of language and boat-building skills there is no argument) which might in the past have rafted across sea barriers;

b) assess whether natural rafts of sufficient size and durability to transport living hominids have actually existed; and

c) in the light of these findings, consider the likelihood of hominid crossings by natural rafts in Wallacea.

2 RAFTING AND THE DISTRIBUTIONS OF NON-HOMINID PRIMATES

Primates underwent their early evolution in Africa. Subsequent colonisations of South America by monkeys in the mid-Cainozoic, and of Asia by *Proconsul* at a similar time, appear to have required the crossing of significant sea gaps. However, these colonisations are remote in geological time and therefore not easy to reconstruct in any detail. Of greater relevance to the present case are Quaternary primate distributions and migrations in southeast Asia, including in and near Wallacea, recently reviewed and discussed by Brandon-Jones (1996a; 1996b; 1998; this volume) and Harcourt (1999).

Few mammal groups are found abundantly in both the northwest and the southeast parts of this region, therefore straddling the zoogeographical divide of Wallace's Line and its variants. Notable are murid rodents which have repeatedly crossed sea gaps to establish populations in Australia/New Guinea from Asia, probably by rafting (Godthelp, this volume; Hand 1984). No primates (except hominids) live on the Sahul Shelf (New Guinea and Australia), and their species richness on Wallacean islands is less than in Sunda lands (Borneo, Java etc). This pattern at first sight might seem to indicate that rafting by primates has been a very rare or even nonexistent event. However, as Brandon-Jones (1996a; 1996b; 1998) has elaborated, primate distributions in the region are of relatively recent origin, the factors constraining them appear to have had little to do with sea barriers, and there is good reason to suppose that rafting has contributed greatly to present distribution patterns.

During most of the Quaternary, rainforest had a substantially reduced distribution in southeast Asia due to drought, being replaced by 'degraded' forest or non-forest vegetation over large areas. This was particularly the case during glacial epochs when sea levels were lower and sea gaps narrower, and especially during two 'deforestation' periods beginning 190 000 and 80 000 years ago (Brandon-Jones 1996a; 1998; and included references; Brandon-Jones, this volume). Rainforests and their fauna may have been entirely eliminated from islands such as Sulawesi, and almost so from Borneo and Java. Most primates, being obligate rainforest animals, were confined to rainforest refugia, particularly in the Mentawai archipelago west of Sumatra but also in west Java, southern India, northern Vietnam, and perhaps elsewhere. Some but not all surviving rainforest primates have spread widely from these refugia during the postglacial epoch, the last 15-10 000 years.

During the past *c.* 800 000 years sea level fluctuations were of greater amplitude than previously, falling as low as 230 m (Batchelor 1979, quoted by Kitchener *et al.* 1990) and averaging 90 m below present level (Bergh *et al.* 1996, quoted by Brandon-Jones 1998). There has been during the Quaternary a general correlation between humid periods, with more extensive rainforest, and periods of high sea level with wider sea gaps. Those times when sea gaps would have been at their narrowest (and land bridges at their most extensive) were simultaneously the times when rainforest, and the distributions of rainforest-dependant organisms including most primates, would have been ecologically most restricted. Such a limitation would not have applied to savanna fauna, including hominids. For the same reason most mammal exchange between Australia and New Guinea during the Quaternary was by savanna (not rainforest) forms (Flannery 1995).

Brandon-Jones (1998 and this volume, but see also Kershaw *et al.*, this volume) has concluded from a review of the palynological evidence that, during glacial epochs when sea level was lower and there was much terrestrial continuity, climate was too dry to support rainforest except in limited refugia. As most primates cannot have migrated across Sunda when suitable habitats did not exist, he concludes that they must have done so during wetter, interglacial/postglacial epochs. For example, most primate species now found in Borneo reached there either between 130 000 and 80 000 years ago, or more likely since 20 000 years ago or less. At these times, when rainforest was more extensive but sea level was high, the animals must have crossed to the island by rafting (Brandon-Jones 1996b; 1998).

Muir *et al.* (1998) also recognise Sumatra-Borneo primate migration in the Quaternary, in this case by orang-utans, more than once, to account for the genetic linkages between their present populations. However, in contrast to Brandon-Jones they suggest that even during glacial epochs, there may at times have been a tropical rainforest or woodland cover extending between the two places and permitting overland migration. These authors therefore do not propose rafting, but they still acknowledge the difficulty for overland migration by this arboreal species posed by wide expanses of savanna, in particular during early interglacial/postglacial times when sea gaps had still not become fully re-established.

Macaques and tarsiers have crossed the further sea gap beyond the Sunda Shelf to colonise Sulawesi (*c.* 110 km from Borneo), where there are at least four, closely related species of *Macaca* and one of *Tarsius* (Harcourt 1999). In this case no glacial lowering of sea level could have permitted overland migration, and the sea gap appears to have been present for most or all of the Cainozoic (Hall, this volume). Both genera have survived there apparently because they do not require closed rainforest, which was not continuously present after their arrival. The largely ground-dwelling macaques of Sulawesi are taxonomically close to species which survived in refugia in the Mentawai Islands west of Sumatra and in India (Brandon-Jones 1998). However, the only non-human primate on islands of the Lesser Sunda chain east of Lombok, *Macaca fascicularis*, was probably introduced from western Indonesia in the mid-Holocene or later (Heinsohn, this volume).

Harcourt (1999) analysed distributions of primate genera and species on islands of the Sunda Shelf, and showed a close relationship between diversity and island area. Harcourt suggested there is no evidence for immigration to islands across water in postglacial time. However, when Sulawesi and Mentawai were included in the analysis the overall pattern was unchanged. This implies that these islands, despite being separated from Sunda by deep water, have acquired and retained as many primate taxa as can survive there for areal/ecological reasons. As persistent deepwater gaps have not affected overall primate richness, they cannot have been insuperable barriers to primate movement.

Dispersal patterns of primates other than hominids therefore appear to require at least some and possibly much colonisation across sea gaps. Rafting therefore emerges as a possible explanation also for hominid oversea migration. However, hominids are bigger than most other primates and it might be thought that bigger rafts would be needed to carry them. I now turn to the question of whether natural rafts capable of carrying hominids have existed.

3 NATURAL RAFTS

Natural rafts of tangled vegetation are inevitably ephemeral, and most are probably unobserved or unreported. Nevertheless, rafts demonstrably or apparently capable of carrying hominids or other large terrestrial vertebrates across sea gaps have been recorded many times, in various seas particularly within the tropics. Wallace (1876) noted that 'uprooted trees and rafts of driftwood often floated down great rivers and carried out to sea. Such rafts or islands are sometimes seen drifting a hundred miles from the mouth of the Ganges with living trees erect upon them; and the Amazon, the Orinoco, Mississippi, Congo and most great rivers produce similar rafts.'

Referring more specifically to the southeast Asian region Wallace (1892) later wrote, after discussing the swimming abilities of deer and pigs, 'there is a much more effectual way of passing over the sea, by means of floating trees, or those floating islands which are often formed at the mouths of great rivers. Sir Charles Lyell describes such floating islands which were encountered among the Moluccas, on which trees and shrubs were growing on a stratum of soil which even formed a white beach around the margin of each raft. Among the Philippine Islands similar rafts with trees growing on them have been seen after hurricanes'. Dammerman (1948) in similar vein wrote that 'Sometimes whole islands with trees still erect are met with and other growths such as nipa palms and clumps of bamboo, while whole chunks washed off from the coast, river-banks or mangroves may be found floating out to sea'. Brandon-Jones (1998) quoted two examples of floating nipa palm islands, originally noted by St. John (1862). One, with erect fronds, was initially mistaken for a three-masted vessel. Another was seen as it floated towards the sea by a pirate marooned on a riverbank in hostile territory; he swam to it and remained adrift for several days, subsisting on palm fruits. Hickson (1889, quoted by Dammerman 1948) recorded a macaque found on floating timber in the Sunda Strait in 1883 some days after the eruption of Krakatau, which was 'terribly scorched, but completely recovered'.

In South America, Dammerman (1948) quoting H. H. Kew related 'On the Amazon River islands of this kind have often been recorded; whole trees with the soil about their roots, and upon them all sorts of animals - birds, monkeys, squirrels and tiger-cats. Once no less than four pumas landed in this way at Montevideo one night to the great alarm of the inhabitants, who found them prowling about the streets in the morning'. Murphy (1926) recorded off the Peruvian coast during an El Niño year 'belts of murky, yellow-green water in which floated quantities of logs', with the harbour at Talara 'choked with ... flotsam in which the bodies of birds, fishes and other creatures were mingled with vegetation'. 'Live reptiles had also floated ashore on rafts' including a lizard *Dicrodon lentiginosus* and a fer-de-lance snake *Bothrops asper* which must have drifted at least 250 km from the north (the former specimen provided the first Peruvian record of the species). Murphy (1926) noted that these species also occur on islands of the Lesser Antilles to which they were supposed to have been transported by natural rafts.

Adding support to this supposition are further records from the Caribbean. In 1969, the Windward Passage Floating Island (Smithsonian Institute 1970, see also MacArthur 1972 and Peterscn 1991) was noted as being some 15 m in diameter and having 10-15 trees 10-12 m tall. It drifted for ten days and over more than 200 km after it was first noted between Haiti and Cuba about 100 km from land. In 1995, as a result of hurricane damage a month earlier, an extensive 'mat of logs and uprooted trees some of which were more than 30 feet long and had large root masses' washed up on the island of Anguilla, taking two days to pile up on shore (Censky *et al.* 1998). It was thought, on the basis of the distributions of lizards which travelled on it, that the mat had drifted from Guadaloupe, a distance of about 270 km. A large boa constrictor 'once floated to the island of St Vincent, twisted round the trunk of a cedar tree, and was so little injured on its voyage that it captured some sheep before it was killed', having crossed more than 320 km from Trinidad or the mainland coast of South America (Wallace 1892).

A raft was recorded near Aldabra (Indian Ocean) by R. J. Hnatiuk in 1973 and reported by Peterson (1991) as 'a half-submerged bamboo thicket carrying live fauna (e.g. crabs) accompanied by a large area of floating and submerged flotsam'. This floating bamboo island almost certainly came from either Madagascar or the East African coast and must have drifted at least 400 km before being observed. It was too large to be towed ashore by motorboat despite attempts to do so (R. J. Hnatiuk, pers. comm).

Because of the transient and unpredictable nature of rafting events, it is of course difficult to prove that terrestrial vertebrates can be transported alive in numbers sufficient to found a viable new colony. Nevertheless, a recent demonstration of this occurred when at least 15 large lizards (*Iguana iguana*) including several individuals of each sex colonised the Caribbean island Anguilla after about one month's drift on such a raft across 270 km of ocean. A female with enlarged ovarian follicles was seen 29 months after the colonisation, demonstrating post-dispersal survival and possible reproduction (Censky *et al.* 1998). An even more spectacular successful rafting can be deduced from the occurrence on Vanua Levu, Fiji, and on the Tonga

Islands, of the endemic, arboreal lizard *Brachylophus fasciatus*; its relatives in the iguanine group of iguanids are found on Galapagos and in the Americas (Etheridge & De Queiros 1988). Its ancestral founding population apparently crossed more than 10 000 km of ocean.

While reptiles such as iguanas might be able to survive on a raft without sustenance for longer than mammals, these examples nevertheless provide further support for the potential of rafting for dispersal of terrestrial vertebrates of all kinds. Perhaps especially for intelligent hominids, it can also be argued that some rafts themselves might provide sustenance, for example the nipa fruits which fed St. John's pirate mentioned above, and the freshwater lens which is likely to remain for several days in soil carried in larger rafts as suggested by Peterson (1991).

Sustenance can also be derived from the sea by people adrift, as shown by the records of many accidental drift voyages in the Pacific (Dening 1962, Levison *et al.* 1973). The longest such voyages span several months, and 'many of the Polynesian drift voyagers are quite successful in obtaining some food by catching fish and birds' (Levison *et al.* 1973). For example, five men whose canoe drifted more than 2400 km over about 70 days in 1947-48, from the Sangir group to New Guinea, subsisted on floating coconuts and flying fish and drank both rainwater and seawater. In several cases women were among the final survivors, in some cases all accompanying men having died (Dening 1962). But even single logs can transport living people without sustenance over significant periods and distances. For example, a woman was rescued 'clinging to a piece of driftwood' after she had been swept from her home in Honduras by floodwaters in the wake of Hurricane Mitch in 1998. She was found 80 km out to sea 'surrounded by wooden debris and dead farm animals' and had survived six days without food and water (Collcutt 1998).

Darlington (1957, quoting Powers 1911) mentioned a raft '100 feet square with trees 30 feet high, evidently tied together by the roots of living plants ... seen in the Atlantic off the coast of North America in 1892, and ... known to have drifted at least 1000 miles'. However, most accounts of natural rafts come from tropical latitudes. This is probably because it is mainly there that vegetation develops a very large biomass in humid environments where violent flooding can occasionally do substantial damage, tearing free large fragments of vegetation which might be floated to the sea. Furthermore, tropical vegetation includes some particularly buoyant, large plants such as nipa palms in the Sunda-Sahul region and bamboos more generally in warm regions, both being specifically mentioned in some accounts of large natural rafts. *Nypa fruticans* is a palm with tall, erect leaves growing from a subterranean trunk, which grows abundantly in estuarine situations throughout southeast Asia and New Guinea. Its edible fruits attract various primates including people, and its habitat renders it particularly liable to being torn free and launched into the ocean. Bamboos commonly grow in large thickets on riverbanks where both moisture and light are available, and where the plants' rhizomatous habit provides some ability to recover from physical disturbance as well as tending to interlock the plants. Bank erosion may easily result in whole bamboo clumps falling into the river. Their culms consist of robust, hollow, air-filled, watertight segments (today used to make watercraft, fluid containers and many other artefacts) allowing the plants to become durable rafts if launched into the ocean by floods.

With progressive destruction of riverine and estuarine vegetation in modern times, and with engineering works along rivers, rafts might reasonably be expected to be far less common today than they were in the past. The present cursory search of the literature nevertheless has yielded many examples of large rafts noted in the last two centuries, when they were probably less common than earlier. Large rafts have probably been a relatively more frequent phenomenon through past millennia, sufficient to account for much oversea transport of animals including hominids.

4 HOMINIDS AND RAFTING IN WALLACEA

Examples of natural rafts, frequently carrying terrestrial vertebrates, are mirrored by many other documented examples of dispersal of living animals across wide barriers, by swimming, wind, on or in flying animals, or by some other means. Long distance dispersal, viewed in the context of geological time, must be a rather frequent event. This appears to contradict the common observation of distribution patterns reflecting past and present barriers to such dispersal. The answer to this paradox may lie in the difficulties of post-dispersal establishment (Godthelp, this volume). It is well known that deliberate attempts to introduce species to places where they are not native have usually (but not always) failed when only a few individuals have been released. One factor causing this is likely to be competition by established animals whose populations are already at or near the carrying capacity of the receiving environment. Such competition, under these circumstances, might need to be only partial in order to exert a crucial effect on the colonists. It might explain, for example, why 'old endemic' mammal taxa have persisted on Sulawesi (Groves, this volume), having been the first to arrive and establish there, and having subsequently out-competed small numbers of newly arriving mammals before the latter managed to achieve long-term viable populations. Hominids might not have experienced such competition after crossing barriers in the Wallacean region, being generalist and adaptable mammals with no ecological equivalents in the previously present faunas.

Furthermore, physical conditions in the region appear conducive to rafting in the direction from southeast Asia towards Australia. Winds are generally more important in determining drift directions than the underlying current (Smith 1999). Winds in the Wallacean region blow persistently from the northwest for part of the year. Currents include the southward Indonesian Throughflow in the vicinity of the Moluccas and Timor, supplementing wind-driven drift. Furthermore, several records of drifted vertebrates refer specifically to events in the wake of climatic extremes including hurricanes (cyclones) and El Niño-related rains, and such events regularly occur in Wallacea. Therefore winds, currents and climatic extremes are all conducive to drift of natural rafts towards Australia from the Sunda shelf, a conclusion drawing support from the regular arrival of abundant buoyant fruits and seeds of Indonesian origin on northern Australian beaches (Smith 1999).

Rafting has been proposed previously by Peterson (1991) as a means whereby *Homo sapiens* might have colonised Australia. He quoted calculations by Burns *et al.* (1985) suggesting that floating islands with vegetation standing at least 2 m tall can be moved even by relatively low wind velocities. For example, such an island 30 m wide and 3-4 m tall if subjected to a persistent wind of 18 km/h will move 17 km in a day. This would carry it across a 30 km gap in just 42 hours, and a 100 km gap in less than six days, passages within the demonstrated endurance of hominids even if they lacked access to sustenance during the voyage.

The present brief review of primate distributions in Southeast Asia, of large natural rafts in tropical seas, of survival by people adrift, and of wind and current patterns, shows that drift colonisation of Australia was possible. It may of course be debated whether this was the more probable means than watercraft. After all, we know that humans did invent boats eventually, only the date and circumstances of that invention are under discussion. The rapid spread of *Homo sapiens* across the world 100-50 000 years ago itself implies some significant advance at that time, possibly involving language (Davidson, this volume). On the other hand, the only evidence that *H. erectus* might have built boats is the crossing of the narrow sea gap between Bali and Lombok, which might easily have been accomplished by natural raft. In both cases but particularly the latter, speculations concerning linguistic and social development based upon assumptions involving acquisition of watercraft construction skills should be viewed cautiously.

ACKNOWLEDGEMENTS

I thank Iain Davidson for initially sparking my interest in this topic, and Tim Bayliss-Smith, Douglas Brandon-Jones and Lesley Rogers for commenting on an earlier manuscript and pointing me towards further literature sources.

REFERENCES

Batchelor, B.C. 1979. Discontinuously rising late Cainozoic eustatic sea-levels, with special reference to Sundaland, Southeast Asia. *Geol. Mijn* 58: 1-20.
Bergh, G.D.van den, Vos, J.de, Sondaar, P.Y. & Aziz, F. 1996. Pleistocene zoogeographic evolution of Java (Indonesia) and glacio-eustatic sea level fluctuations: a background for the presence of *Homo*. *Indo-Pacific Pre-history Association Bulletin* 14 (Chiang Mai Papers 1): 7-21.

Birdsell, J.B. 1977. The recalibration of a paradigm for the first peopling of Greater Australia. In J.Allen, J.Golson & R.Jones (eds), *Sunda and Sahul: Prehistoric Studies in Southeast Asia, Melanesia and Australia*: 113-167. Canberra: Australian National University Press.

Brandon-Jones, D. 1996a. The Asian Colobinae (Mammalia: Cercopithecidae) as indicators of Quaternary climatic change. *Biological Journal of the Linnaean Society* 59: 327-350.

Brandon-Jones, D. 1996b. *Presbytis* species sympatry in Borneo versus allopatry in Sumatra: an interpretation. In D.S. Edwards, W.E. Booth & S.C. Choy (eds), *Tropical Rainforest Research: Current Issues*; *Monographiae Biologicae* 74: 71-76. Dordrecht: Kluwer.

Brandon-Jones, D. 1998. Pre-Glacial Bornean primate impoverishment and Wallace's Line. In R. Hall & J.D. Holloway (eds), *Biogeography and Geological Evolution of SE Asia*: 393-403. Leiden: Backhuys.

Brandon-Jones, D., this volume.

Burns, F.L., Moresby, J.F. & Peterson, J.A. 1985. The floating islands of Pirron Yallock, Victoria. *Australian Society of Limnology Bulletin* 10: 15-32.

Censky, E.J., Hodge, K. & Dudley, J. 1998. Over-water dispersal of lizards due to hurricanes. *Nature* 395: 556.

Collcutt, D. 1998. Mother safe after six days adrift. *The Times (London),* 5 Nov.: 1.

Dammerman, K.W. 1948. The Fauna of Krakatau 1883-1933. *Kon. Ned. Akad. Wet., Verhandlingen (Tweede Sectie), DI.* 44: 1-594.

Darlington, C.D. 1957. *Zoogeography: the Geographical Distribution of Animals*. New York: J.Wiley.

Davidson, I., this volume.

Davidson, I. & Noble, W. 1992. Why the first colonisation of the Australian region is the earliest evidence of modern human behaviour. *Archaeology in Oceania* 27: 135-142.

Dening, G.M. 1962. The geographical knowledge of the Polynesians and the nature of inter-island contact. In J.Golson (ed.), *Polynesian Navigation, Journal of the Polynesian Society (Suppl.)*, (reprinted Wellington: Reed, 1972): 102-154.

Etheridge, R. & De Queiros, K. 1988. A phylogeny of Iguanidae. In R. Estes & G. Pregill, (eds), *Phylogenetic Relationships of the Lizard Families*: 283-368. Stanford: Stanford University Press.

Flannery, T.F. 1995. *Mammals of New Guinea*, 2nd edition. Sydney: Reed.

Godthelp, H., this volume.

Groves, C., this volume.

Hall, R., this volume.

Hand, S. 1984. Australia's oldest rodents: master mariners from Malaysia. In M. Archer & G. Clayton (eds), *Vertebrate Zoogeography and Evolution in Australasia*: 905-912. Perth: Hesperian Press.

Harcourt, A.H. 1999. Biogeographic relationships of primates on South-East Asian islands. *Global Ecology and Biogeography* 8: 55-61.

Hickson, J.S. 1889. *A Naturalist in North Celebes*. London.

Irwin, G. 1992. *The Prehistoric Exploration and Colonisation of the Pacific*. Cambridge: Cambridge University Press.

Kershaw, A.P. *et al.*, this volume.

Kitchener, D.J., Boeadi, C. L., & Maharadatunkamsi 1990. *Wild Mammals of Lombok Island*. Records of the Western Australian Museum, Supplementary Series 33.

Levison, M., Ward, R.G. & Webb, J.W. 1973. *The Settlement of Polynesia. A Computer Simulation*. Canberra: ANU Press.

MacArthur, R.H. 1972. *Geographical Ecology. Patterns in the Distribution of Species*. New York: Harper & Row.

Morwood, M.J., this volume.

Morwood, M.J., O'Sullivan, P.B., Aziz, F. & Raza, A. 1998. Fission-track ages of stone tools and fossils on the east Indonesian island of Flores. *Nature* 392: 173-176.

Morwood, M.J., Aziz, F., O'Sullivan, P.B., Nasruddin, Hobbs, D.R.& Raza, A. 1999. Archaeological and palaeontological research in central flores, east Indonesia: results of fieldwork 1997-98. *Antiquity* 73: 273-286.

Moss, S.J. & Wilson, M.E.J. 1998. Tertiary palaeogeography of Sulawesi and Borneo. In R.Hall & J.D.Holloway (eds), *Biogeography and Geological Evolution of SE Asia*: 133-163. Leiden: Backhuys.

Muir, C.C., Watkins, R.F. & Ha, T.T. 1998. Map simulations, based on bathometric data and information about glacially induced sea-level changes, are used to test a palaeo-migratory hypothesis inferred from orang-utan population genetics data. *Assemblage 4* (University of Sheffield graduate student journal of archaeology), http://www.shef.ac.uk/~assem/4.html.

Murphy, R.C. 1926. Oceanic and climatic phenomena along the west coast of South America during 1925. *Geographical Review* 16: 26-54.

Noble, W. & Davidson, I. 1996. *Human Evolution, Language and Mind*. Cambridge: Cambridge University Press.

Peterson, J.A. 1991. Human dispersal from Wallacea to Sahul: a re-appraisal. In M.A.J.Williams, P.De Deckker & A.P.Kershaw (eds), *The Cainozoic in Australia: a Re-Appraisal of the Evidence. Special Publications of the Geological Society of Australia* 18: 339-346.

Powers, S. 1911. Floating islands. *Popular Science Monthly* 79: 303-307.

Smith, J. 1999. *Australian Driftseeds. A Compendium of Seeds and Fruits Commonly Found on Australian Beaches.* Armidale: School of Human & Environmental Studies, University of New England.

St.John, S. 1862. *Life in the Forests of the Far East. Vol. 1.* London: Smith, Elder & Co.

Wallace, A.R. 1892. *Island Life* (2nd Ed.). London: Macmillan.

Wallace, A.R. 1876. *The geographical distribution of animals, with a study of living and extinct faunas as elucidating the past changes of the earth's surface*, vols 1 & 2. London: Macmillan.

T - #0318 - 101024 - C0 - 254/178/22 [24] - CB - 9789058093493 - Gloss Lamination